Algebraic Curves
and Riemann Surfaces

Algebraic Curves and Riemann Surfaces

Rick Miranda

Graduate Studies
in Mathematics

Volume 5

American Mathematical Society

Editorial Board

James E. Humphreys
Robion C. Kirby
Lance W. Small

2000 *Mathematics Subject Classification*. Primary 14–01, 14Hxx; Secondary 14H55.

ABSTRACT. This text is an introduction to the theory of algebraic curves defined over the complex numbers. It begins with the definitions and first properties of Riemann surfaces, with special attention paid to the Riemann sphere, complex tori, hyperelliptic curves, smooth plane curves, and projective curves. The heart of the book is the treatment of divisors and rational functions, culminating in the theorems of Riemann-Roch and Abel and the analysis of the canonical map. Sheaves, cohomology, the Zariski topology, line bundles, and the Picard group are developed after these main theorems are proved and applied, as a bridge from the classical material to the modern language of algebraic geometry.

Library of Congress Cataloging-in-Publication Data
Miranda, Rick, 1953–
  Algebraic curves and Riemann surfaces / Rick Miranda.
    p. cm. — (Graduate studies in mathematics, ISSN 1065-7339; v. 5)
  Includes bibliographical references (p.   –   ) and index.
  ISBN 0-8218-0268-2 (acid-free)
  1. Curves, Algebraic. 2. Riemann surfaces. I. Title. II. Series.
QA565.M687   1995
516.3′52–dc20   95-1947
                                                                                      CIP

Copying and reprinting. Individual readers of this publication, and nonprofit libraries acting for them, are permitted to make fair use of the material, such as to copy a chapter for use in teaching or research. Permission is granted to quote brief passages from this publication in reviews, provided the customary acknowledgment of the source is given.

Republication, systematic copying, or multiple reproduction of any material in this publication is permitted only under license from the American Mathematical Society. Requests for such permission should be addressed to the Acquisitions Department, American Mathematical Society, 201 Charles Street, Providence, Rhode Island 02904-2294 USA. Requests can also be made by e-mail to reprint-permission@ams.org.

© 1995 by the American Mathematical Society. All rights reserved.
Reprinted with corrections 1997.
The American Mathematical Society retains all rights
except those granted to the United States Government.
Printed in the United States of America.
∞ The paper used in this book is acid-free and falls within the guidelines
established to ensure permanence and durability.
This publication was typeset by the author, with editorial assistance
from the American Mathematical Society, using $\mathcal{A}_{\mathcal{M}}\mathcal{S}$-LATEX,
the American Mathematical Society's TEX macro system.
Visit the AMS home page at http://www.ams.org/

10 9 8 7 6 5 4 3   17 16 15 14 13 12

To Jeanne

Contents

Preface	xix
Chapter I. Riemann Surfaces: Basic Definitions	1
1. Complex Charts and Complex Structures	1
Complex Charts	1
Complex Atlases	3
The Definition of a Riemann Surface	4
Real 2-Manifolds	5
The Genus of a Compact Riemann Surface	6
Complex Manifolds	6
Problems I.1	7
2. First Examples of Riemann Surfaces	7
A Remark on Defining Riemann Surfaces	7
The Projective Line	8
Complex Tori	9
Graphs of Holomorphic Functions	10
Smooth Affine Plane Curves	10
Problems I.2	12
3. Projective Curves	13
The Projective Plane $\mathbb{P}^2$	13
Smooth Projective Plane Curves	14
Higher-Dimensional Projective Spaces	16
Complete Intersections	17
Local Complete Intersections	17
Problems I.3	18
Further Reading	19
Chapter II. Functions and Maps	21
1. Functions on Riemann Surfaces	21
Holomorphic Functions	21
Singularities of Functions; Meromorphic Functions	23
Laurent Series	25
The Order of a Meromorphic Function at a Point	26

C^∞ Functions	27
Harmonic Functions	27
Theorems Inherited from One Complex Variable	28
Problems II.1	30
2. Examples of Meromorphic Functions	30
Meromorphic Functions on the Riemann Sphere	30
Meromorphic Functions on the Projective Line	31
Meromorphic Functions on a Complex Torus	33
Meromorphic Functions on Smooth Plane Curves	35
Smooth Projective Curves	36
Problems II.2	38
3. Holomorphic Maps Between Riemann Surfaces	38
The Definition of a Holomorphic Map	38
Isomorphisms and Automorphisms	40
Easy Theorems about Holomorphic Maps	40
Meromorphic Functions and Holomorphic Maps to the Riemann Sphere	41
Meromorphic Functions on a Complex Torus, Again	42
Problems II.3	43
4. Global Properties of Holomorphic Maps	44
Local Normal Form and Multiplicity	44
The Degree of a Holomorphic Map between Compact Riemann Surfaces	47
The Sum of the Orders of a Meromorphic Function	49
Meromorphic Functions on a Complex Torus, Yet Again	50
The Euler Number of a Compact Surface	50
Hurwitz's Formula	52
Problems II.4	53
Further Reading	54
Chapter III. More Examples of Riemann Surfaces	57
1. More Elementary Examples of Riemann Surfaces	57
Lines and Conics	57
Glueing Together Riemann Surfaces	59
Hyperelliptic Riemann Surfaces	60
Meromorphic Functions on Hyperelliptic Riemann Surfaces	61
Maps Between Complex Tori	62
Problems III.1	65
2. Less Elementary Examples of Riemann Surfaces	66
Plugging Holes in Riemann Surfaces	66
Nodes of a Plane Curve	67
Resolving a Node of a Plane Curve	69
The Genus of a Projective Plane Curve with Nodes	69

Resolving Monomial Singularities	71
Cyclic Coverings of the Line	73
Problems III.2	74
3. Group Actions on Riemann Surfaces	75
Finite Group Actions	75
Stabilizer Subgroups	76
The Quotient Riemann Surface	77
Ramification of the Quotient Map	79
Hurwitz's Theorem on Automorphisms	82
Infinite Groups	82
Problems III.3	83
4. Monodromy	84
Covering Spaces and the Fundamental Group	84
The Monodromy of a Finite Covering	86
The Monodromy of a Holomorphic Map	87
Coverings via Monodromy Representations	88
Holomorphic Maps via Monodromy Representations	90
Holomorphic Maps to $\mathbb{P}^1$	91
Hyperelliptic Surfaces	92
Problems III.4	93
5. Basic Projective Geometry	94
Homogeneous Coordinates and Polynomials	94
Projective Algebraic Sets	95
Linear Subspaces	95
The Ideal of a Projective Algebraic Set	96
Linear Automorphisms and Changing Coordinates	97
Projections	98
Secant and Tangent Lines	99
Projecting Projective Curves	101
Problems III.5	102
Further Reading	103
Chapter IV. Integration on Riemann Surfaces	105
1. Differential Forms	105
Holomorphic 1-Forms	105
Meromorphic 1-Forms	106
Defining Meromorphic Functions and Forms with a Formula	107
Using dz and $d\bar{z}$	108
$\mathcal{C}^\infty$ 1-Forms	109
1-Forms of Type $(1,0)$ and $(0,1)$	110
$\mathcal{C}^\infty$ 2-Forms	110
Problems IV.1	111
2. Operations on Differential Forms	112

Multiplication of 1-Forms by Functions	112
Differentials of Functions	113
The Wedge Product of Two 1-Forms	113
Differentiating 1-Forms	114
Pulling Back Differential Forms	114
Some Notation	115
The Poincaré and Dolbeault Lemmas	117
Problems IV.2	117
3. Integration on a Riemann Surface	118
Paths	118
Integration of 1-Forms Along Paths	119
Chains and Integration Along Chains	120
The Residue of a Meromorphic 1-Form	121
Integration of 2-Forms	122
Stoke's Theorem	123
The Residue Theorem	123
Homotopy	124
Homology	126
Problems IV.3	126
Further Reading	127

Chapter V. Divisors and Meromorphic Functions — 129

1. Divisors	129
The Definition of a Divisor	129
The Degree of a Divisor on a Compact Riemann Surface	129
The Divisor of a Meromorphic Function: Principal Divisors	130
The Divisor of a Meromorphic 1-Form: Canonical Divisors	131
The Degree of a Canonical Divisor on a Compact Riemann Surface	132
The Boundary Divisor of a Chain	133
The Inverse Image Divisor of a Holomorphic Map	133
The Ramification and Branch Divisor of a Holomorphic Map	134
Intersection Divisors on a Smooth Projective Curve	135
The Partial Ordering on Divisors	136
Problems V.1	137
2. Linear Equivalence of Divisors	138
The Definition of Linear Equivalence	138
Linear Equivalence for Divisors on the Riemann Sphere	140
Principal Divisors on a Complex Torus	140
The Degree of a Smooth Projective Curve	142
Bezout's Theorem for Smooth Projective Plane Curves	143
Plücker's Formula	143
Problems V.2	145
3. Spaces of Functions and Forms Associated to a Divisor	145

The Definition of the Space $L(D)$	145
Complete Linear Systems of Divisors	147
Isomorphisms between $L(D)$'s under Linear Equivalence	148
The Definition of the Space $L^{(1)}(D)$	148
The Isomorphism between $L^{(1)}(D)$ and $L(D+K)$	149
Computation of $L(D)$ for the Riemann Sphere	149
Computation of $L(D)$ for a Complex Torus	150
A Bound on the Dimension of $L(D)$	151
Problems V.3	152
4. Divisors and Maps to Projective Space	153
Holomorphic Maps to Projective Space	153
Maps to Projective Space Given By Meromorphic Functions	154
The Linear System of a Holomorphic Map	155
Base Points of Linear Systems	157
The Hyperplane Divisor of a Holomorphic Map to $\mathbb{P}^n$	158
Defining a Holomorphic Map via a Linear System	160
Removing the Base Points	160
Criteria for ϕ_D to be an Embedding	161
The Degree of the Image and of the Map	164
Rational and Elliptic Normal Curves	165
Working Without Coordinates	166
Problems V.4	166
Further Reading	167
Chapter VI. Algebraic Curves and the Riemann-Roch Theorem	169
1. Algebraic Curves	169
Separating Points and Tangents	169
Constructing Functions with Specified Laurent Tails	171
The Transcendence Degree of the Function Field $\mathcal{M}(X)$	174
Computing the Function Field $\mathcal{M}(X)$	177
Problems VI.1	178
2. Laurent Tail Divisors	178
Definition of Laurent Tail Divisors	178
Mittag-Leffler Problems and $H^1(D)$	180
Comparing H^1 Spaces	181
The Finite-Dimensionality of $H^1(D)$	182
Problems VI.2	184
3. The Riemann-Roch Theorem and Serre Duality	185
The Riemann-Roch Theorem I	185
The Residue Map	186
Serre Duality	188
The Equality of the Three Genera	191
The Riemann-Roch Theorem II	192

Problems VI.3	193
Further Reading	193

Chapter VII. Applications of Riemann-Roch	195
1. First Applications of Riemann-Roch	195
How Riemann-Roch implies Algebraicity	195
Criterion for a Divisor to be Very Ample	195
Every Algebraic Curve is Projective	196
Curves of Genus Zero are Isomorphic to the Riemann Sphere	196
Curves of Genus One are Cubic Plane Curves	197
Curves of Genus One are Complex Tori	197
Curves of Genus Two are Hyperelliptic	198
Clifford's Theorem	198
The Canonical System is Base-Point-Free	200
The Existence of Meromorphic 1-Forms	200
Problems VII.1	202
2. The Canonical Map	203
The Canonical Map for a Curve of Genus at Least Three	203
The Canonical Map for a Hyperelliptic Curve	203
Finding Equations for Smooth Projective Curves	204
Classification of Curves of Genus Three	205
Classification of Curves of Genus Four	206
The Geometric Form of Riemann-Roch	207
Classification of Curves of Genus Five	209
The Space $L(D)$ for a General Divisor	210
A Few Words on Counting Parameters	211
Riemann's Count of $3g - 3$ Parameters for Curves of Genus g	212
Problems VII.2	215
3. The Degree of Projective Curves	216
The Minimal Degree	216
Rational Normal Curves	216
Tangent Hyperplanes	217
Flexes and Bitangents	219
Monodromy of the Hyperplane Divisors	221
The Surjectivity of the Monodromy	222
The General Position Lemma	224
Points Imposing Conditions on Hypersurfaces	225
Castelnuovo's Bound	228
Curves of Maximal Genus	230
Problems VII.3	232
4. Inflection Points and Weierstrass Points	233
Gap Numbers and Inflection Points of a Linear System	233
The Wronskian Criterion	234

Higher-order Differentials	236
The Number of Inflection Points	238
Flex Points of Smooth Plane Curves	241
Weierstrass Points	241
Problems VII.4	243
Further Reading	245

Chapter VIII. Abel's Theorem — 247

1. Homology, Periods, and the Jacobian — 247

The First Homology Group	247
The Standard Identified Polygon	247
Periods of 1-Forms	247
The Jacobian of a Compact Riemann Surface	248
Problems VIII.1	249

2. The Abel-Jacobi Map — 249

The Abel-Jacobi Map A on X	249
The Extension of A to Divisors	250
Independence of the Base Point	250
Statement of Abel's Theorem	250
Problems VIII.2	250

3. Trace Operations — 251

The Trace of a Function	251
The Trace of a 1-Form	252
The Residue of a Trace	253
An Algebraic Proof of the Residue Theorem	253
Integration of a Trace	254
Proof of Necessity in Abel's Theorem	255
Problems VIII.3	256

4. Proof of Sufficiency in Abel's Theorem — 257

Lemmas Concerning Periods	257
The Proof of Sufficiency	260
Riemann's Bilinear Relations	262
The Jacobian and the Picard Group	263
Problems VIII.4	264

5. Abel's Theorem for Curves of Genus One — 265

The Abel-Jacobi Map is an Embedding	265
Every Curve of Genus One is a Complex Torus	265
The Group Law on a Smooth Projective Plane Cubic	266
Problems VIII.5	267
Further Reading	267

Chapter IX. Sheaves and Čech Cohomology — 269

1. Presheaves and Sheaves — 269

Presheaves	269
Examples of Presheaves	269
The Sheaf Axiom	272
Locally Constant Sheaves	273
Skyscraper Sheaves	273
Global Sections on Compact Riemann Surfaces	275
Restriction to an Open Subset	276
Problems IX.1	276
2. Sheaf Maps	278
Definition of a Map between Sheaves	278
Inclusion Maps	278
Differentiation Maps	279
Restriction or Evaluation Maps	279
Multiplication Maps	280
Truncation Maps	281
The Exponential Map	281
The Kernel of a Sheaf Map	282
1-1 and Onto Sheaf Maps	282
Short Exact Sequences of Sheaves	284
Exact Sequences of Sheaves	286
Sheaf Isomorphisms	286
Using Sheaves to Define the Category	288
Problems IX.2	289
3. Čech Cohomology of Sheaves	290
Čech Cochains	291
Čech Cochain Complexes	291
Cohomology with respect to a Cover	292
Refinements	293
Čech Cohomology Groups	295
The Connecting Homomorphism	297
The Long Exact Sequence of Cohomology	298
Problems IX.3	300
4. Cohomology Computations	301
The Vanishing of $\check{H}^1$ for $\mathcal{C}^\infty$ Sheaves	302
The Vanishing of $\check{H}^1$ for Skyscraper Sheaves	303
Cohomology of Locally Constant Sheaves	304
The Vanishing of $\check{H}^2(X, \mathcal{O}_X[D])$	304
De Rham Cohomology	305
Dolbeault Cohomology	306
Problems IX.4	308
Further Reading	308
Chapter X. Algebraic Sheaves	309

1. Algebraic Sheaves of Functions and Forms	309
Algebraic Curves	309
Algebraic Sheaves of Functions	309
Algebraic Sheaves of Forms	310
The Zariski Topology	311
Problems X.1	312
2. Zariski Cohomology	312
The Vanishing of $\check{H}^1(X_{Zar}, \mathcal{F})$ for a Constant Sheaf $\mathcal{F}$	313
The Interpretation of $H^1(D)$	314
GAGA Theorems	315
Further Computations	316
The Zero Mean Theorem	317
The High Road to Abel's Theorem	319
Problems X.2	320
Further Reading	321
Chapter XI. Invertible Sheaves, Line Bundles, and $\check{H}^1$	323
1. Invertible Sheaves	323
Sheaves of $\mathcal{O}$-Modules	323
Definition of an Invertible Sheaf	324
Invertible Sheaves associated to Divisors	325
The Tensor Product of Invertible Sheaves	326
The Inverse of an Invertible Sheaf	328
The Group of Isomorphism Classes of Invertible Sheaves	329
Problems XI.1	330
2. Line Bundles	331
The Definition of a Line Bundle	331
The Tautological Line Bundle for a Map to $\mathbb{P}^n$	333
Line Bundle Homomorphisms	334
Defining a Line Bundle via Transition Functions	335
The Invertible Sheaf of Regular Sections of a Line Bundle	337
Sections of the Tangent Bundle and Tangent Vector Fields	340
Rational Sections of a Line Bundle	342
The Divisor of a Rational Section	343
Problems XI.2	344
3. Avatars of the Picard Group	345
Divisors Modulo Linear Equivalence and Cocycles	345
Invertible Sheaves Modulo Isomorphism	348
Line Bundles Modulo Isomorphism	351
The Jacobian	356
Problems XI.3	357
4. $\check{H}^1$ as a Classifying Space	357
Why $\check{H}^1(\mathcal{O}^*)$ Classifies Invertible Sheaves and Line Bundles	357

Locally Trivial Structures	359
A General Principle Regarding $\check{H}^1$	360
Cyclic Unbranched Coverings	360
Extensions of Invertible Sheaves	361
First-Order Deformations	364
Problems XI.4	368
Further Reading	369
References	371
Index of Notation	377

Preface

This text has evolved from lecture notes for a one-semester course which I have taught 5 times in the last 8 years as an introduction to the ideas of algebraic geometry using the theory of algebraic curves as a foundation.

There are two broad aims for the book: to keep the prerequisites to a bare minimum while still treating the major theorems seriously; and to begin to convey to the reader some of the language of modern algebraic geometry.

In order to present the material of Algebraic Curves to an initially relatively unsophisticated audience I have taken the approach that Algebraic Curves are best encountered for the first time over the complex numbers. Therefore the book starts out as a primer on Riemann surfaces, with complex charts and meromorphic functions taking center stage. In particular, one semester of graduate complex analysis should be sufficient preparation, and it is not assumed that the reader has any serious background in either algebraic topology or commutative algebra. But I try to stress that the main examples (from the point of view of algebraic geometry) come from projective curves, and slowly but surely the text evolves to the algebraic category, culminating in an algebraic proof of the Riemann-Roch theorem. After returning to the analytic side of things for Abel's theorem, the progression is repeated again when sheaves and cohomology are discussed: first the analytic, then the algebraic category.

The proof of Riemann-Roch presented here is an adaptation of the adelic proof, expressed completely in terms of solving a Mittag-Leffler problem. This is a very concrete approach, and in particular no cohomology or sheaf theory is used. However, cohomology groups clandestinely appear (as obstruction spaces to solving Mittag-Leffler problems), motivating their explicit introduction later on.

The other goal is to begin to convey, as much as possible, the language of modern algebraic geometry to the student. This language is that of rational functions, divisors, bundles, sheaves, cohomology, and the Zariski topology, to name some of the highlights presented here. I hope that a student who has read the later chapters of this book will be prepared to understand at least the first few minutes of a modern colloquium talk discussing algebraic curves and algebraic geometry. I consider the treatment of sheaves and cohomology given here to be rather gentle; for example, most of the sheaves which are used as the initial examples were introduced in a natural way much earlier in the text. Hence by the time a sheaf is even defined the reader will actually have a decent understanding of what the technicalities entail. In addition, the zero-th and first cohomology groups will already have been seen in the proof of the Riemann-Roch theorem (without them being called that of course).

The first three chapters are introductory, discussing the basic definitions of Riemann surfaces and holomorphic maps between them. Of the 12 sections in these chapters, 5 are devoted entirely to examples of one sort or another. The main theorems here are that the sum of the orders of a meromorphic function on a compact Riemann surface is zero, and Hurwitz's Formula relating the genera of compact Riemann surfaces given a map between them. The fourth chapter on integration is meant to get to the Residue Theorem in a direct manner.

Chapters 5-8 form the technical heart of the text. Divisors and how they are used to organize forms, functions, and maps are introduced in Chapter 5, and in Chapter 6 the Riemann-Roch Theorem and Serre Duality are proved, after introducing the concept of an algebraic curve, which is defined here as a compact Riemann surface whose field of global meromorphic functions separates points and tangents. Chapter 7 is devoted to applications of Riemann-Roch. Here is found the classification of curves of low genus, Clifford's Theorem, the analysis of the canonical map, and Riemann's count of $3g - 3$ parameters for curves of genus at least two. A section on the degree of a projective curve culminates in Castelnuovo's bound on the genus. It is here most of all that the reader will feel an urge to learn more algebraic geometry, and in particular some higher-dimensional theory. The final section concerns inflection points of linear systems and Weierstrass points in particular. In Chapter 8 Abel's Theorem is proved; along the way the algebraic proof of the Residue Theorem is indicated. The final section discusses the group law on a smooth cubic curve.

The last three chapters introduce sheaves and Čech cohomology. Initially the classical topology is used, focusing in on the standard sheaves of holomorphic and meromorphic functions and forms. The Zariski topology and the algebraic sheaves are brought into the picture next, and the obstruction space for solving a Mittag-Leffler problem (seen in the proof of the Riemann-Roch Theorem) is here realized as an $\check{H}^1$ of an algebraic sheaf.

The last chapter is organized around the Picard group of an algebraic curve and its many manifestations: as the group of divisors modulo linear equivalence, as the group of line bundles modulo isomorphism, as the group of invertible sheaves modulo isomorphism, as the first cohomology group with values in the nowhere zero regular functions, and as the Jacobian (extended by $\mathbb{Z}$). Here there is an opportunity to explain why $\check{H}^1$ is useful to classify locally trivial objects in general, and the text closes with first-order deformations, with Riemann's count of $3g - 3$ parameters enjoying a reprise.

At the end of each chapter I have included some suggestions for further reading. These are not meant to be completely comprehensive, but simply indicate some of the sources that I am aware of which I have found illuminating.

I would like to thank Bruce Crauder, David Hahn, Luisa Paoluzzi, John Symms, and Caryn Werner for commenting on various sections; also I am greatly indebted to Ciro Ciliberto and Peter Stiller who each read through substantial portions of the text and offered many valuable suggestions.

It has been my great privilege to have been given the opportunity to study algebraic geometry in my professional life. There is no doubt that the theory of algebraic curves is the richest and deepest of the field's various roots, and I hope I have conveyed some of the special pleasure obtained in visiting this material, which serves simultaneously as one of the great jewels of classical mathematics and one of the most vital areas of modern research.

Rick Miranda
October 1994
Fort Collins, Colorado

Chapter I. Riemann Surfaces: Basic Definitions

1. Complex Charts and Complex Structures

The basic idea of a Riemann surface is that it is a space which, locally, looks just like an open set in the complex plane. In this section we make this precise.

Complex Charts. Let X be a topological space. In order to make X look, locally, like an open set in the complex plane, we want to have a local complex coordinate at every point of the space; this local coordinate can then be used to define all the local notions of functions of one complex variable. Now a coordinate on a space is simply a function from the space to the standard space, in this case the complex plane. This leads to the following definition.

DEFINITION 1.1. A *complex chart*, or simply *chart*, on X is a homeomorphism $\phi : U \to V$, where $U \subset X$ is an open set in X, and $V \subset \mathbb{C}$ is an open set in the complex plane. The open subset U is called the *domain* of the chart ϕ. The chart ϕ is said to be *centered at* $p \in U$ if $\phi(p) = 0$.

We think of a chart on X as giving a local (complex) coordinate on its domain, namely $z = \phi(x)$ for $x \in U$.

EXAMPLE 1.2. Let $X = \mathbb{R}^2$, and let U be any open subset. Define $\phi_U(x,y) = x + iy$ from U (considered as a subset of $\mathbb{R}^2$) to the complex plane. This is a complex chart on $\mathbb{R}^2$.

EXAMPLE 1.3. Again let $X = \mathbb{R}^2$. For any open subset U, define

$$\phi_U(x,y) = \frac{x}{1 + \sqrt{x^2 + y^2}} + i \frac{y}{1 + \sqrt{x^2 + y^2}}.$$

These are also complex charts on $\mathbb{R}^2$.

EXAMPLE 1.4. Let $\phi : U \to V$ be a complex chart on X. Suppose that $U_1 \subset U$ is an open subset of U. Then $\phi|_{U_1} : U_1 \to \phi(U_1)$ is a complex chart on X. This restriction of ϕ is called a *sub-chart* of ϕ.

EXAMPLE 1.5. Let $\phi : U \to V$ be a complex chart on X. Suppose that $\psi : V \to W$ is a holomorphic bijection between two open sets of the complex plane. Then the composition $\psi \circ \phi : U \to W$ is a complex chart on X. If we think of ϕ as giving a complex coordinate on U, we can view this operation as a change of coordinates.

We do not want to think of a simple change of coordinates as imposing an essentially different structure on the open set in question. In other words, with the above notation, the two charts ϕ and $\psi \circ \phi$ should not produce different answers when we get around to asking questions about local functions and forms on the domain. A careful analysis of the "difference" between these two charts leads us to the following definition.

DEFINITION 1.6. Let $\phi_1 : U_1 \to V_1$ and $\phi_2 : U_2 \to V_2$ be two complex charts on X. We say that ϕ_1 and ϕ_2 are *compatible* if either $U_1 \cap U_2 = \emptyset$, or
$$\phi_2 \circ \phi_1^{-1} : \phi_1(U_1 \cap U_2) \to \phi_2(U_1 \cap U_2)$$
is holomorphic.

Note that the definition is symmetric: if $\phi_2 \circ \phi_1^{-1}$ is holomorphic on $\phi_1(U_1 \cap U_2)$, then $\phi_1 \circ \phi_2^{-1}$ will be holomorhic on $\phi_2(U_1 \cap U_2)$. The function $T = \phi_2 \circ \phi_1^{-1}$ is called the *transition function* between the two charts; it is a bijection in any case. Transition functions enjoy the following property.

LEMMA 1.7. *Let T be a transition function between two compatible charts. Then the derivative T' is never zero on the domain of T.*

PROOF. Let S denote the inverse to T, so that $S \circ T$ is the identity on the domain of T, i.e., $S(T(w)) = w$ for all w. Taking the derivative of this equation gives $S'(T(w))T'(w) = 1$, so that $T'(w)$ cannot be zero. □

Suppose that T is the transition function between the charts ϕ and ψ, with a point p in their common domain. Denote by $z = \phi(x)$ and $w = \psi(x)$ the two local coordinates, with $z_0 = \phi(p)$ and $w_0 = \psi(p)$. The above lemma implies that the power series expansion of the transition function $T = \phi \circ \psi^{-1}$ (which expresses z as a power series in w) must be of the form
$$z = T(w) = z_0 + \sum_{n \geq 1} a_n (w - w_0)^n,$$
with $a_1 \neq 0$.

EXAMPLE 1.8. Referring to the situation of Example 1.5, let $\phi : U \to V$ be a complex chart on X, and let $\psi : V \to W$ be a holomorphic bijection between two open sets of the complex plane. Then the charts ϕ and $\psi \circ \phi$ are compatible. Moreover, $\psi \circ \phi$ will be compatible with any chart which is compatible with ϕ.

EXAMPLE 1.9. Any two sub-charts of a complex chart are compatible.

1. COMPLEX CHARTS AND COMPLEX STRUCTURES

EXAMPLE 1.10. Any two charts in Example 1.2 are compatible.

EXAMPLE 1.11. Any two charts in Example 1.3 are compatible.

EXAMPLE 1.12. No chart of Example 1.2 is compatible with any chart of Example 1.3 (unless the domains are disjoint).

A more serious example is given by the following.

EXAMPLE 1.13. Let S^2 denote the unit 2-sphere inside $\mathbb{R}^3$, i.e.,

$$S^2 = \{(x, y, w) \in \mathbb{R}^3 \mid x^2 + y^2 + w^2 = 1\}.$$

Consider the $w = 0$ plane as a copy of the complex plane $\mathbb{C}$, with $(x, y, 0)$ being identified with $z = x + iy$. Let $\phi_1 : S^2 - \{(0,0,1)\} \to \mathbb{C}$ be defined by projection from $(0, 0, 1)$. Specifically,

$$\phi_1(x, y, w) = \frac{x}{1-w} + i\frac{y}{1-w}.$$

The inverse to ϕ_1 is

$$\phi_1^{-1}(z) = \left(\frac{2\operatorname{Re}(z)}{|z|^2+1}, \frac{2\operatorname{Im}(z)}{|z|^2+1}, \frac{|z|^2-1}{|z|^2+1}\right).$$

Define $\phi_2 : S^2 - \{(0, 0, -1)\} \to \mathbb{C}$ by projection from $(0, 0, -1)$ followed by a complex conjugation:

$$\phi_2(x, y, w) = \frac{x}{1+w} - i\frac{y}{1+w}.$$

The inverse to ϕ_2 is

$$\phi_2^{-1}(z) = \left(\frac{2\operatorname{Re}(z)}{|z|^2+1}, \frac{-2\operatorname{Im}(z)}{|z|^2+1}, \frac{1-|z|^2}{|z|^2+1}\right).$$

The common domain is $S^2 - \{(0, 0, \pm 1)\}$, and is mapped by both ϕ_1 and ϕ_2 bijectively onto $\mathbb{C}^* = \mathbb{C} - \{0\}$. The composition $\phi_2 \circ \phi_1^{-1}(z) = 1/z$, which is holomorphic. Thus the two charts are compatible.

Complex Atlases. Note that in Example 1.13, every point of the sphere lies in at least one of the two complex charts. Therefore we have a local complex coordinate at each point of the sphere. This, of course, is our ultimate goal.

For X to look locally like the complex plane everywhere, we must have complex charts around every point of X. Moreover, we want these charts to be compatible. This is the notion of a *complex atlas*.

DEFINITION 1.14. A *complex atlas* (or simply *atlas*) $\mathcal{A}$ on X is a collection $\mathcal{A} = \{\psi_\alpha : U_\alpha \to V_\alpha\}$ of pairwise compatible complex charts whose domains cover X, i.e., $X = \bigcup_\alpha U_\alpha$.

Note that the charts defined in Example 1.2 form a complex atlas on $\mathbb{R}^2$, as do the charts in Example 1.3. Also, the two charts defined on the 2-sphere in Example 1.13 define a complex atlas on S^2.

EXAMPLE 1.15. If $\mathcal{A} = \{\phi_\alpha : U_\alpha \to V_\alpha\}$ is an atlas on X, and $Y \subset X$ is any open subset, then the collection of sub-charts $\mathcal{A}_Y = \{\phi_\alpha|_{Y \cap U_\alpha} : Y \cap U_\alpha \to \phi_\alpha(Y \cap U_\alpha)\}$ is an atlas on Y.

It may well be the case that two different atlases give the same local notions of complex analysis on a Riemann surface; in particular, this will happen when every chart of one atlas is compatible with every chart of the other atlas. This notion gives an equivalence relation on atlases:

DEFINITION 1.16. Two complex atlases $\mathcal{A}$ and $\mathcal{B}$ are *equivalent* if every chart of one is compatible with every chart of the other.

Note that two complex atlases are equivalent if and only if their union is also a complex atlas. An easy Zorn's lemma argument will show that every complex atlas is contained in a unique maximal complex atlas; moreover, two atlases are equivalent if and only if they are contained in the same maximal complex atlas.

DEFINITION 1.17. A *complex structure* on X is a maximal complex atlas on X, or, equivalently, an equivalence class of complex atlases on X.

Note that *any* atlas on X determines a unique complex structure. This is the usual way that complex structures are defined: by giving an atlas.

The Definition of a Riemann Surface. Recall that a topological space X is said to be *Hausdorff* if, for every two distinct points x, y in X, there are disjoint neighborhoods U and V of x and y, respectively. X is said to be *second countable* if there is a countable basis for its topology.

DEFINITION 1.18. A *Riemann surface* is a second countable connected Hausdorff topological space X together with a complex structure.

The second countability condition is a technical one, meant to exclude certain pathological examples; any Riemann surface found "in nature" (i.e., as a subset of $\mathbb{C}^n$ for example) will be second countable. In particular, if the complex structure may be defined by a countable atlas, then X must be second countable.

EXAMPLE 1.19. Let X be $\mathbb{C}$ itself, considered topologically as $\mathbb{R}^2$, with the complex structure induced by the atlas of Example 1.2. This Riemann surface is called the *complex plane*.

EXAMPLE 1.20. Let X be the 2-sphere, with complex structure given by the two-chart atlas of Example 1.13. Note that the sphere is Hausdorff and connected. This Riemann surface is called the *Riemann Sphere*. Note that if one chart of the Riemann Sphere has as coordinate z, then the other chart has the coordinate $1/z$, and there is only one point which is not in the z-chart. The

Riemann Sphere is often written as $\mathbb{C}_\infty$ or $\mathbb{C} \cup \infty$, with the complex plane $\mathbb{C}$ representing one chart, with the "point at infinity" ∞ being the single extra point. The Riemann Sphere is a *compact* Riemann surface.

EXAMPLE 1.21. Any connected open subset of a Riemann surface is a Riemann surface; use the atlas on the subset as described in Example 1.15.

Real 2-Manifolds. The reader who has seen some of the theory of manifolds will recognize in every detail the definitions and constructions. In essence, a Riemann surface is simply a connected complex manifold of dimension one (this is one *complex* dimension, remember).

It is convenient to sometimes "forget" the complex structure of a Riemann surface, and to consider it simply as a 2-manifold. Let us then briefly recall the relevant definitions. Let X be a Hausdorff topological space.

DEFINITION 1.22. An *n-dimensional real chart* on X is a homeomorphism $\phi : U \to V$, where $U \subset X$ is an open set in X, and $V \subset \mathbb{R}^n$ is an open set in $\mathbb{R}^n$. Two such real charts ϕ_1 and ϕ_2 are $\mathcal{C}^\infty$-*compatible* if either the intersection of their domains is empty, or
$$\phi_2 \circ \phi_1^{-1} : \phi_1(U_1 \cap U_2) \to \phi_2(U_1 \cap U_2)$$
is a $\mathcal{C}^\infty$ diffeomorphism, i.e., it and its inverse have partial derivatives of all orders at every point. A $\mathcal{C}^\infty$ *atlas* on X is a collection of real charts on X, which are pairwise $\mathcal{C}^\infty$-compatible, and whose domains cover X. Two such atlases are *equivalent* if their union is an atlas. A $\mathcal{C}^\infty$ *structure* on X is an equivalence class of $\mathcal{C}^\infty$ atlases. A $\mathcal{C}^\infty$ *real manifold* is a second countable connected Hausdorff space X together with a $\mathcal{C}^\infty$ structure.

Since holomorphic maps of a complex variable $z = x + iy$ are $\mathcal{C}^\infty$ in the real variables x and y, we immediately see that every Riemann surface is a 2-dimensional $\mathcal{C}^\infty$ real manifold (which we often abbreviate and refer to simply as a "2-manifold").

Let us make a few remarks concerning the topology of Riemann surfaces. Firstly, for manifolds, connectedness and path-connectedness are equivalent; thus we have that every Riemann surface is path-connected.

Next, note that a holomorphic map between two subsets of the complex plane preserves the orientation of the plane. Indeed, the familiar conformal property of holomorphic functions implies that all local angles are preserved by holomorphic maps, and in particular right angles are preserved; therefore the local notions of "clockwise" and "counterclockwise" for small circles are preserved. Since giving an orientation on a surface can be viewed as equivalent to having consistent local choices for "clockwise", holomorphic maps preserve the orientation of the plane.

Therefore, if we induce a local orientation at each point of a Riemann surface by "pulling back" the orientation via some complex chart containing that point, this local orientation is well defined, independent of the choice of complex chart.

These local orientations induce a global orientation on the Riemann surface; hence we have that every Riemann surface is orientable. The reader may consult [**Armstrong83**], [**Munkres84**], or [**Massey67**] for further details concerning orientation.

The Genus of a Compact Riemann Surface. These remarks are enough to completely determine *compact* Riemann surfaces, as far as their C^∞ structure goes. For this we appeal to the classification of compact orientable 2-manifolds; each of these is a g-holed torus for some unique integer $g \geq 0$. (See [**Armstrong83**], [**Massey67**], or [**Sieradski92**] for example.) When $g = 0$, we have no holes and the surface is topologically the 2-sphere. When $g = 1$, there is one hole, and the surface is a simple torus, topologically homeomorphic to $S^1 \times S^1$. For $g \geq 2$, the surface is obtained by attaching g "handles" to a 2-sphere. This integer g is called the *topological genus* of the compact Riemann surface, and is a fundamental invariant. Thus:

PROPOSITION 1.23. *Every Riemann surface is an orientable path-connected 2-dimensional C^∞ real manifold. Every compact Riemann surface is diffeomorphic to the g-holed torus, for some unique integer $g \geq 0$.*

We have only seen one example so far of a compact Riemann surface, namely the Riemann Sphere (Example 1.20). It has topological genus 0.

Complex Manifolds. As was seen above, the definition of a Riemann surface and the definition of a C^∞ real manifold are in all ways parallel. The reader should also be aware that higher-dimensional analogues of Riemann surfaces also exist, defined in exactly the same spirit. Here we just give the definitions, since we will rarely need to work with complex manifolds of higher dimension.

DEFINITION 1.24. Let X be a Hausdorff topological space. An *n-dimensional complex chart* on X is a homeomorphism $\phi : U \to V$, where $U \subset X$ is an open set in X, and $V \subset \mathbb{C}^n$ is an open set in $\mathbb{C}^n$. Two such n-dimensional complex charts ϕ_1 and ϕ_2 are *compatible* if either the intersection of their domains is empty, or

$$\phi_2 \circ \phi_1^{-1} : \phi_1(U_1 \cap U_2) \to \phi_2(U_1 \cap U_2)$$

is *holomorphic*, i.e., is holomorphic in each of the n variables separately at every point. An *n-dimensional complex atlas* on X is a collection of n-dimensional complex charts on X, which are pairwise compatible, and whose domains cover X. Two such atlases are *equivalent* if their union is an atlas. An *n-dimensional complex structure* on X is an equivalence class of n-dimensional complex atlases. An *n-dimensional complex manifold* is a connected Hausdorff space X together with an n-dimensional complex structure.

Problems I.1

A. Let $\phi_i : U_i \to V_i$, $i = 1, 2$, be complex charts on X with $U_1 \cap U_2 \neq \emptyset$. Suppose that $\phi_2 \circ \phi_1^{-1} : \phi_1(U_1 \cap U_2) \to \phi_2(U_1 \cap U_2)$ is holomorphic. Show that it is bijective, with inverse $\phi_1 \circ \phi_2^{-1} : \phi_2(U_1 \cap U_2) \to \phi_1(U_1 \cap U_2)$, proving that $\phi_1 \circ \phi_2^{-1}$ is also holomorphic.

B. Let $\phi : U \to V$ be a complex chart on X, and let $\psi : V \to W$ be a holomorphic bijection between two open sets in $\mathbb{C}$. Show that $\psi \circ \phi : U \to W$ is a complex chart on X. Show that $\psi \circ \phi$ is compatible with any chart on X which is compatible with ϕ.

C. Verify that any two sub-charts of a complex chart are compatible (Example 1.9 of the text).

D. Verify that any two charts in Example 1.2 are compatible.

E. Verify that any two charts in Example 1.3 are compatible.

F. Check that no chart of Example 1.2 is compatible with any chart of Example 1.3 of the notes if their domains intersect.

G. In Example 1.13, where an atlas of the Riemann Sphere is defined, check that indeed $\phi_2 \circ \phi_1^{-1}$ sends z to $1/z$ as stated.

H. Show that equivalence of complex atlases is an equivalence relation.

I. Equivalent atlases may be partially ordered by inclusion. Show that any atlas is equivalent to a unique maximal atlas.

J. Show that holomorphic bijections between open sets in the complex plane preserve the local orientation.

2. First Examples of Riemann Surfaces

In this section we'll present some easy examples of Riemann surfaces, especially of compact Riemann surfaces. These include the projective line, complex tori, and smooth plane curves.

A Remark on Defining Riemann Surfaces. To define a Riemann surface, it would appear that one needs to start with a topological space X, second countable, connected and Hausdorff, and then define a complex atlas on it; in other words, one needs to have the topology first, and then one can impose the complex structure. This is not completely accurate; one can often use the data defining an atlas to also define the topology.

This observation is based on the following remark: if an open cover $\{U_\alpha\}$ of a topological space X is given, then a subset $U \subset X$ is open in X if and only if each intersection $U \cap U_\alpha$ is open in U_α.

More generally, if any collection $\{U_\alpha\}$ of subsets of a set X is given, and topologies are given for each subset U_α, then one can define a topology on X by declaring a set U to be open if and only if each intersection $U \cap U_\alpha$ is open in U_α.

Now suppose we are given a collection of subsets $\{U_\alpha\}$ of a set X, which cover X (so that $X = \bigcup U_\alpha$), and a set of bijections $\phi_\alpha : U_\alpha \to V_\alpha$ where each V_α is an open subset of $\mathbb{C}$. Each V_α has its topology as a subset of $\mathbb{C}$, and so using the ϕ_α, we can transport this topology to every U_α: we simply declare a subset U of U_α to be open if and only if $\phi_\alpha(U)$ is open in V_α (or, equivalently, open in $\mathbb{C}$).

Now we can define a topology on all of X, again by declaring a set U to be open in X if and only if each intersection $U \cap U_\alpha$ is open in U_α.

This prescription gives a topology on X such that each U_α is an open set if and only if for each α and β, the subset $U_\alpha \cap U_\beta$ is open in U_α. From the definition of the topology on the U_α's, this condition is equivalent to asking that $\phi_\alpha(U_\alpha \cap U_\beta)$ is open in V_α (or, equivalently, open in $\mathbb{C}$).

Thus we may take the following route to define a Riemann surface:

- Start with a set X.
- Find a countable collection of subsets $\{U_\alpha\}$ of X, which cover X.
- For each α, find a bijection ϕ_α from U_α to an open subset V_α of the complex plane.
- Check that for every α and β, $\phi_\alpha(U_\alpha \cap U_\beta)$ is open in V_α. At this point we have, by the above remarks, a topology defined on X, such that each U_α is open; moreover by definition, each ϕ_α is a complex chart on X.
- Check that the complex charts ϕ_α are pairwise compatible.
- Check that X is connected and Hausdorff.

The Projective Line. Let $\mathbb{CP}^1$ denote the complex projective line, that is, the set of 1-dimensional subspaces of $\mathbb{C}^2$. If (z, w) is a nonzero vector in $\mathbb{C}^2$, its span is a point in $\mathbb{CP}^1$; we will denote the span of (z, w) by $[z : w]$. Note that every point of $\mathbb{CP}^1$ can be written in this form, as $[z : w]$, with z and w not both zero; moreover,

$$[z : w] = [\lambda z : \lambda w]$$

for any nonzero $\lambda \in \mathbb{C}^*$.

We will use the method outlined above for defining a complex structure on $\mathbb{CP}^1$.

Let $U_0 = \{[z : w] \mid z \neq 0\}$, and $U_1 = \{[z : w] \mid w \neq 0\}$. Note that U_0 and U_1 cover $\mathbb{CP}^1$. Define $\phi_0 : U_0 \to \mathbb{C}$ by $\phi_0[z : w] = w/z$; similarly define $\phi_1 : U_1 \to \mathbb{C}$ by $\phi_1[z : w] = z/w$. Both ϕ_0 and ϕ_1 are bijections, so we have the data required above. Note that $\phi_i(U_0 \cap U_1) = \mathbb{C}^*$, which is open in $\mathbb{C}$. The composition $\phi_1 \circ \phi_0^{-1}$ sends s to $1/s$, and therefore these two charts are compatible. Since both U_0 and U_1 are connected, and have nonempty intersection, their union $\mathbb{CP}^1$ is connected. Finally we show $\mathbb{CP}^1$ is Hausdorff. Take two points p and q in $\mathbb{CP}^1$. If both p and q are in either U_0 or U_1, we can separate them by open sets, since the U_i are Hausdorff. Therefore we may assume that $p \in U_0 - U_1$ and $q \in U_1 - U_0$; this forces $p = [1 : 0]$ and $q = [0 : 1]$. These are separated by $\phi_0^{-1}(D)$ and $\phi_1^{-1}(D)$, where D is the open unit disc in $\mathbb{C}$.

We will usually denote $\mathbb{CP}^1$ by simply $\mathbb{P}^1$; it is called the *complex projective line*. Note that $\mathbb{P}^1$ is the union of the two *closed* sets $\phi_0^{-1}(\bar{D})$ and $\phi_1^{-1}(\bar{D})$, where

$\bar{D}$ is the closed unit disc in $\mathbb{C}$. Since $\bar{D}$ is compact, we see that the projective line is compact.

Complex Tori. Fix ω_1 and ω_2 to be two complex numbers which are linearly independent over $\mathbb{R}$. Define L to be the lattice

$$L = \mathbb{Z}\omega_1 + \mathbb{Z}\omega_2 = \{m_1\omega_1 + m_2\omega_2 \mid m_1, m_2 \in \mathbb{Z}\}.$$

The lattice L is a subgroup of the additive group of $\mathbb{C}$. Let $X = \mathbb{C}/L$ be the quotient group, with projection map $\pi : \mathbb{C} \to X$. Note that via π, we can impose the quotient topology on X, namely, a set $U \subset X$ is open if and only if $\pi^{-1}(U)$ is open in $\mathbb{C}$. This definition makes π continuous, and since $\mathbb{C}$ is connected, so is X.

Every open set in X is the image of an open set in $\mathbb{C}$, since if U is open in X, $U = \pi(\pi^{-1}(U))$. A more serious remark is that π is an open mapping, that is, π takes any open set of $\mathbb{C}$ onto an open set in X. Indeed, if V is open in $\mathbb{C}$, then to check that $\pi(V)$ is open in X we must show that $\pi^{-1}(\pi(V))$ is open in $\mathbb{C}$; but

$$\pi^{-1}(\pi(V)) = \bigcup_{\omega \in L} (\omega + V)$$

is a union of translates of V, which are all open sets in $\mathbb{C}$.

For any $z \in \mathbb{C}$, define the closed parallelogram

$$P_z = \{z + \lambda_1\omega_1 + \lambda_2\omega_2 \mid \lambda_i \in [0,1]\}.$$

Note that any point of $\mathbb{C}$ is congruent modulo L to a point of P_z. Therefore the projection map π maps P_z onto X. Since P_z is compact, so is X.

The lattice L is a discrete subset of $\mathbb{C}$, so there is an $\epsilon > 0$ such that $|\omega| > 2\epsilon$ for every nonzero $\omega \in L$. Fix such an ϵ, and fix a point $z_0 \in \mathbb{C}$. Consider the open disc $D = D(z_0, \epsilon)$ of radius ϵ about z_0. This choice of ϵ insures that no two points of $D(z_0, \epsilon)$ can differ by an element of the lattice L.

We claim that for any z_0, and for any such ϵ, the restriction of the projection π to the open disc D maps D homeomorphically onto the open set $\pi(D)$. Clearly $\pi|_D : D \to \pi(D)$ is onto, continuous, and open (since π is). Therefore we need only check that it is 1-1; this follows from the choice of ϵ.

We are now ready to define a complex atlas on X. Again fix ϵ as above. For each $z_0 \in \mathbb{C}$, let $D_{z_0} = D(z_0, \epsilon)$, and define $\phi_{z_0} : \pi(D_{z_0}) \to D_{z_0}$ to be the inverse of the map $\pi|_{D_{z_0}}$. By the above claim, these ϕ's are complex charts on X.

To finish the construction, we must check that these charts are pairwise compatible. Choose two points z_1 and z_2, and consider the two charts $\phi_1 = \phi_{z_1} : \pi(D_{z_1}) \to D_{z_1}$ and $\phi_2 = \phi_{z_2} : \pi(D_{z_2}) \to D_{z_2}$. Let $U = \pi(D_{z_1}) \cap \pi(D_{z_2})$. If U is empty, there is nothing to prove. If U is not empty, let $T(z) = \phi_2(\phi_1^{-1}(z)) = \phi_2(\pi(z))$ for $z \in \phi_1(U)$; we must check that T is holomorphic on $\phi_1(U)$. Note that $\pi(T(z)) = \pi(z)$ for all $z \in \phi_1(U)$, so $T(z) - z = \omega(z) \in L$ for all $z \in \phi_1(U)$. This function $\omega : \phi_1(U) \to L$ is continuous, and L is discrete; hence ω is locally constant on $\phi_1(U)$. (It is constant on the connected components of U.) Thus, locally, $T(z) = z + \omega$ for some fixed $\omega \in L$, and is therefore holomorphic. Hence the two charts ϕ_1 and ϕ_2 are compatible, and the collection of charts $\{\phi_z \mid z \in \mathbb{C}\}$ is a complex atlas on X.

Hence X is a compact Riemann surface. In fact it has topological genus one; topologically, X is a simple torus. This is most easily seen by considering X as the image of the parallelogram P_0; under the map $\pi|_{P_0}$, the opposite sides are identified together, and no other identifications are made, giving the familiar construction of the torus. These Riemann surfaces (which depend of course on the lattice L) are called *complex tori*.

Graphs of Holomorphic Functions. Let $V \subset \mathbb{C}$ be a connected open subset of the complex plane, and let g be a holomorphic function defined on all of V. Consider the graph X of g, as a subset of $\mathbb{C}^2$:

$$X = \{(z, g(z)) \mid z \in V\}.$$

Give X the subspace topology, and let $\pi : X \to V$ be the first projection; note that π is a homeomorphism, whose inverse simply sends the point $z \in V$ to the ordered pair $(z, g(z))$. Thus π is a complex chart on X, whose domain covers all of X. Hence we have a complex atlas on X, composed of a single chart; this gives X the structure of a Riemann surface.

This example can be immediately generalized to any finite collection of holomorphic functions $g_1, \ldots, g_n$ on V; simply take X to be the graph in $\mathbb{C}^{n+1}$:

$$X = \{(z, g_1(z), \ldots, g_n(z)) \mid z \in V\}.$$

Smooth Affine Plane Curves. This is a further generalization of the graph construction introduced above. We would like to consider a locus $X \subset \mathbb{C}^2$ which is *locally* a graph, but perhaps not globally. The most natural way to do this is to define a locus X by requiring a complex polynomial of two variables $f(z, w)$ to vanish. Morally speaking, this should cut the complex dimension down by one, and we have a chance of producing a Riemann surface this way.

One needs a mild condition on the polynomial f for this to work, essentially insuring that X is locally a graph. This condition is based on the Implicit Function Theorem:

THEOREM 2.1 (THE IMPLICIT FUNCTION THEOREM). *Let $f(z, w) \in \mathbb{C}[z, w]$ be a polynomial, and let $X = \{(z, w) \in \mathbb{C}^2 \mid f(z, w) = 0\}$ be its zero locus. Let $p = (z_0, w_0)$ be a point of X, i.e., p is a root of f. Suppose that $\partial f / \partial w(p) \neq 0$. Then there exists a function $g(z)$ defined and holomorphic in a neighborhood of z_0, such that, near p, X is equal to the graph $w = g(z)$. Moreover $g' = -\frac{\partial f}{\partial z} / \frac{\partial f}{\partial w}$ near z_0.*

Of course, if $\partial f / \partial w(p) = 0$, it may still be true that $\partial f / \partial z(p) \neq 0$, and X will still be, locally, a graph near p, using the other variable. This motivates the following.

DEFINITION 2.2. An *affine plane curve* is the locus of zeroes in $\mathbb{C}^2$ of a polynomial $f(z,w)$. A polynomial $f(z,w)$ is *nonsingular* at a root p if either partial derivative $\partial f/\partial z$ or $\partial f/\partial w$ is not zero at p. The affine plane curve X of roots of f is *nonsingular at p* if f is nonsingular at p. The curve X is *nonsingular*, or *smooth*, if it is nonsingular at each of its points.

We can obtain complex charts on a smooth affine plane curve by using the Implicit Function Theorem to conclude that the curve is locally a graph, and then making the construction analogously to that for a graph.

Specifically, let X be a smooth affine plane curve, defined by a polynomial $f(z,w)$. Let $p = (z_0, w_0) \in X$. If $\partial f/\partial w(p) \neq 0$, find a holomorphic function $g_p(z)$ such that in a neighborhood U of p, X is the graph $w = g_p(z)$. Thus the projection $\pi_z : U \to \mathbb{C}$ (mapping (z,w) to z) is a homeomorphism from U to its image V, which is open in $\mathbb{C}$. This gives a complex chart on X.

If instead $\partial f/\partial z(p) \neq 0$, then we make the identical construction using the other projection π_w, sending (z,w) to w near p.

Since X is smooth, at least one of these partials must be nonzero at each point, and so the domains of these complex charts cover X.

Let us check that any two of these charts are compatible. Suppose first that both charts are obtained using π_z. Then, if there is nonempty intersection with their domains, the composition of the inverse of one with the other is the identity, which is certainly holomorphic. The same holds if both charts are obtained using π_w.

Therefore assume that one chart is π_z and the other is π_w. Choose a point $p = (z_0, w_0)$ in their common domain U. Assume that near p, X is locally of the form $w = g(z)$ for some holomorphic function g. Then on $\pi_z(U)$ near z_0, the inverse of π_z sends z to $(z, g(z))$. Thus the composition $\pi_w \circ \pi_z^{-1}$ of π_w with the inverse of π_z sends z to $g(z)$, which is holomorphic.

This completes the proof that any two of the charts are compatible, and gives a complex atlas on X.

The space X is certainly second countable and Hausdorff, as a subspace of $\mathbb{C}^2$. Thus to see that X is a Riemann surface, we must only check that it is connected. This is not automatic; for example, if the polynomial f defining X is the product of two linear factors with the same slope (e.g., $f(z,w) = (z+w)(z+w-1)$) then X is the union of two complex lines which do not meet; each line is a Riemann surface itself (being a graph), but the union is not connected.

One possible assumption on the polynomial f for X to be connected is that $f(z,w)$ be an irreducible polynomial; that is, that f cannot be factored nontrivially as $f = g(z,w)h(z,w)$, where both g and h are nonconstant polynomials:

THEOREM 2.3. *If $f(z,w)$ is an irreducible polynomial, then its locus of roots X is connected. Hence if f is nonsingular and irreducible, X is a Riemann surface.*

The locus of roots of an irreducible polynomial $f(z,w)$ is called an *irreducible affine plane curve*.

The proof of the connectedness of X if f is irreducible is not elementary, but requires some of the machinery of algebraic geometry. We will not present a proof here; see [**Shafarevich77**], for example. Granting this, we see that: every smooth irreducible affine plane curve is a Riemann surface.

EXAMPLE 2.4. Let $h(z)$ be a polynomial in one variable which is not a perfect square. Then the polynomial $f(z,w) = w^2 - h(z)$ is irreducible. Moreover, if $h(z)$ has distinct roots, then f is nonsingular, and its locus of roots X is a Riemann surface. (Prove this for yourself: Problem G below.)

A slight generalization will be useful later. If $f(z,w)$ is an irreducible polynomial, then the points on its locus of roots X where f is singular forms a finite set. (This is nontrivial! But let's go on.) If we delete these points, then the resulting open subset of X is a Riemann surface, using the same charts as given above. This is referred to as the *smooth part* of the affine plane curve X, and in general, if f is an irreducible polynomial, the smooth part of its zero locus is a Riemann surface.

No affine plane curve is compact: as a subset of $\mathbb{C}^2 = \mathbb{R}^4$, it is not a bounded set, since for any fixed z_0, there will be roots w to the polynomial $f(z_0, w) = 0$.

Problems I.2

A. Verify that if any collection of subsets $\{U_\alpha\}$ of a set X are given, and topologies are given for each subset U_α, then a topology can be defined on X by declaring that a subset $U \subseteq X$ is open in X if and only if $U \cap U_\alpha$ is open in U_α for every α.

B. Suppose, in problem A, that each U_α is connected. Form a graph with one vertex (called v_α) for each U_α, and with vertex v_α connected by an edge to v_β if and only if $U_\alpha \cap U_\beta \neq \emptyset$. Prove or disprove: X is connected if and only if the graph is connected.

C. Check that the function from $\mathbb{P}^1$ to S^2 sending $[z:w]$ to

$$(2\operatorname{Re}(w\bar{z}), 2\operatorname{Im}(w\bar{z}), |w|^2 - |z|^2)/(|w|^2 + |z|^2)$$

is a homeomorphism onto the unit sphere in $\mathbb{R}^3$. Therefore the projective line is a compact Riemann surface of genus zero.

D. Show that any lattice $L = \mathbb{Z}\omega_1 + \mathbb{Z}\omega_2$ in $\mathbb{C}$ with ω_1 and ω_2 linearly independent over $\mathbb{R}$ is a discrete subset of $\mathbb{C}$.

E. Show that a complex torus has topological genus one by constructing an explicit homeomorphism to the product $S^1 \times S^1$ of two circles.

F. Show that the group law of a complex torus X is divisible: for any point $p \in X$ and any integer $n \geq 1$ there is a point $q \in X$ with $n \cdot q = p$. Indeed, show that there are exactly n^2 such points q.

G. Show that the polynomial $f(z,w) = w^2 - h(z)$ is an irreducible polynomial if and only if $h(z)$ is a polynomial which is not a perfect square. Show that $f(z,w)$ is a nonsingular polynomial if and only if $h(z)$ has distinct roots.

H. Let X be an affine plane curve of degree 2, that is, defined by a quadratic polynomial $f(z,w)$. (Such a curve is called an *affine conic*.) Suppose that $f(z,w)$ is singular. Show that in fact f factors as the product of two linear polynomials, so that X is therefore the union of two intersecting lines. Give an example of a smooth affine plane conic.

I. Give an example of a smooth irreducible affine plane curve of arbitrary degree. Make sure you check the irreducibility!

J. Let ϕ be holomorphic in a neighborhood of $p \in \mathbb{C}$. Assume that $\phi'(p) \neq 0$. Prove (using the Implicit Function Theorem) that there exists a neighborhood U of p such that $\phi|_U$ is a chart on $\mathbb{C}$.

3. Projective Curves

The Projective Line $\mathbb{P}^1$ is the first in a series of examples which encompass the most important and interesting compact Riemann surfaces. These are surfaces which are embedded in *projective space*. We first discuss the case of projective *plane* curves.

The Projective Plane $\mathbb{P}^2$. We will make a construction very similar to that made for the projective line $\mathbb{P}^1$.

DEFINITION 3.1. The *projective plane* $\mathbb{P}^2$ is the set of 1-dimensional subspaces of $\mathbb{C}^3$.

If (x,y,z) is a nonzero vector in $\mathbb{C}^3$, its span is denoted by $[x:y:z]$ and is a point in the projective plane; every point in the projective plane may be written in this way. Note that

$$[x:y:z] = [\lambda x : \lambda y : \lambda z]$$

for any nonzero number λ; indeed, $\mathbb{P}^2$ can be viewed as the quotient space of $\mathbb{C}^3 - \{0\}$ by the multiplicative action of $\mathbb{C}^*$. In this way it inherits a Hausdorff topology, which is the quotient topology coming from the natural map from $\mathbb{C}^3 - \{0\}$ onto $\mathbb{P}^2$.

The entries in the notation $[x:y:z]$ are called the *homogeneous coordinates* of the corresponding point in the projective plane. The homogeneous coordinates are not unique, as noted above; however whether they are zero or not is well defined.

The space $\mathbb{P}^2$ can be covered by the three open sets

$$U_0 = \{[x:y:z] \mid x \neq 0\}; U_1 = \{[x:y:z] \mid y \neq 0\}; U_2 = \{[x:y:z] \mid z \neq 0\}.$$

Each open set U_i is homeomorphic to the affine plane $\mathbb{C}^2$. The homeomorphism on U_0 is given by sending $[x:y:z] \in \mathbb{P}^2$ to $(y/x, z/x) \in \mathbb{C}^2$; its inverse sends

$(a, b) \in \mathbb{C}^2$ to $[1 : a : b] \in \mathbb{P}^2$. On the other two open sets the map is similar, dividing by y for U_1 and by z for U_2.

We note here that the projective plane is compact: it may be covered by three compact sets, namely the closed unit poly-disks in the three open sets U_i above.

Smooth Projective Plane Curves. A polynomial F is *homogeneous* if every term has the same degree in the variables; this degree is the *degree* of the homogeneous polynomial. For example, $x^2y - 2xyz + 3z^3$ is homogeneous of degree 3 in the variables x, y, z.

Let $F(x, y, z)$ be a homogeneous polynomial of degree d. It does not make sense to evaluate F at a point of the projective plane; if $[x_0 : y_0 : z_0] \in \mathbb{P}^2$, then $F(x_0, y_0, z_0)$ is not well defined, because the homogeneous coordinates x_0, y_0, and z_0 are themselves not well defined. In particular, one sees easily that

$$F(\lambda x_0, \lambda y_0, \lambda z_0) = \lambda^d F(x_0, y_0, z_0)$$

but as noted above $[\lambda x_0 : \lambda y_0 : \lambda z_0]$ and $[x_0 : y_0 : z_0]$ are the same point in the projective plane. However this computation shows that *whether F is zero or not does make sense*. Therefore the locus

$$X = \{[x : y : z] \in \mathbb{P}^2 \mid F(x, y, z) = 0\}$$

is well defined. Moreover it is a closed subset of $\mathbb{P}^2$. The intersection X_i of X with the open sets U_i is exactly an affine plane curve when transported to $\mathbb{C}^2$. For example, in U_0 where $x \neq 0$, we have after transporting to $\mathbb{C}^2$ that

$$X_0 = X \cap U_0 \cong \{(a, b) \in \mathbb{C}^2 \mid F(1, a, b) = 0\}$$

which is the affine plane curve described by the polynomial $f(a, b) = 0$, where $f(a, b) = F(1, a, b)$.

We want to show that under a nonsingularity assumption on F, the locus X is a Riemann surface. In any case X is called the *projective plane curve* defined by F.

DEFINITION 3.2. A homogeneous polynomial $F(x, y, z)$ is *nonsingular* if there are no common solutions to the system of equations

$$(3.3) \qquad F = \frac{\partial F}{\partial x} = \frac{\partial F}{\partial y} = \frac{\partial F}{\partial z} = 0$$

in the projective plane $\mathbb{P}^2$.

This condition is equivalent to requiring that there be no *nonzero* solutions to the above system in $\mathbb{C}^3$.

Before proceeding, we note that any homogeneous polynomial F (in any number of variables x_i) satisfies *Euler's Formula*:

$$(3.4) \qquad F = \frac{1}{d} \sum_i x_i \frac{\partial F}{\partial x_i},$$

where d is the degree of F. To see this, it suffices to prove it when F is a monomial, since both sides are additive; for a monomial it is trivial.

LEMMA 3.5. *Suppose that $F(x,y,z)$ is a homogeneous polynomial of degree d. Then F is nonsingular if and only if each X_i is a smooth affine plane curve in $\mathbb{C}^2$.*

PROOF. Suppose first that one of the X_i is not smooth; we may assume by symmetry that X_0 is not smooth. Define $f(u,v) = F(1,u,v)$, so that X_0 is defined by $f = 0$ in $\mathbb{C}^2$. Since X_0 is not smooth, there is a common solution $(u_0, v_0) \in \mathbb{C}^2$ to the set of equations

$$f = \frac{\partial f}{\partial u} = \frac{\partial f}{\partial v} = 0.$$

We claim then that $[1 : u_0 : v_0]$ is a common solution to the system (3.3), and thus F is singular. For this we note that

$$\begin{aligned}
F[1 : u_0 : v_0] &= f(u_0, v_0) = 0, \\
\frac{\partial F}{\partial y}[1 : u_0 : v_0] &= \frac{\partial f}{\partial u}(u_0, v_0) = 0, \\
\frac{\partial F}{\partial z}[1 : u_0 : v_0] &= \frac{\partial f}{\partial v}(u_0, v_0) = 0, \text{ and} \\
\frac{\partial F}{\partial x}[1 : u_0 : v_0] &= (dF - u_0\frac{\partial F}{\partial y} - v_0\frac{\partial F}{\partial z})[1 : u_0 : v_0] = 0,
\end{aligned}$$

where the last computation of $\partial F/\partial x$ uses Euler's formula (3.4).

We leave the converse, which follows the same lines of computation, to the reader. □

Now suppose we do have that $F(x,y,z)$ is a nonsingular homogeneous polynomial, defining the projective plane curve X. It is a basic theorem, again a little deeper than what we can do here, that a nonsingular homogeneous polynomial is automatically irreducible. Let us simply accept this, and then note that each of the three open subsets X_i of X are smooth irreducible affine plane curves, and hence are Riemann surfaces by Theorem 2.3. Recall that the coordinate charts on the X_i are simply the projections, which in our case are easy to describe: they are the functions y/x and z/x for X_0, and are ratios of the other variables for the other pieces.

Thus to see that the complex structures given on the X_i separately are compatible, one needs to check statements like the following. Consider a point $p \in X$ which is in both X_0 and X_1: $p = [x : y : z]$ with $x, y \neq 0$. Suppose that $\phi_0 = y/x$ is a chart near p for X_0, and $\phi_1 = z/y$ is a chart near p for X_1. We must show that $\phi_1 \circ \phi_0^{-1}$ is holomorphic. Now $\phi_0^{-1}(w) = [1 : w : h(w)]$ for some holomorphic function h (locally, X is the graph of h). Hence $\phi_1 \circ \phi_0^{-1}(w) = h(w)/w$ which is holomorphic since $w \neq 0$ (p is in X_1).

Similar checks with all other possible chart combinations show that the complex structures on the X_i are all compatible, and thus induce a complex structure on X.

PROPOSITION 3.6. *Let $F(x, y, z)$ be a nonsingular homogeneous polynomial. Then the projective plane curve X which is its zero locus in $\mathbb{P}^2$ is a compact Riemann surface. Moreover at every point of X one can take as a local coordinate a ratio of the homogeneous coordinates.*

We have indicated above why X is a Riemann surface; we need only show that it is compact. However it is a closed subset of $\mathbb{P}^2$, which is compact. Such a Riemann surface is called a *smooth projective plane curve*; its *degree* is the degree of the defining polynomial.

Higher-Dimensional Projective Spaces. The opportunity exists to find Riemann surfaces in higher-dimensional projective space, which we now briefly describe.

DEFINITION 3.7. *The set of 1-dimensional subspaces of $\mathbb{C}^{n+1}$ is called projective n-space and is denoted by $\mathbb{P}^n$.*

The span of the vector $(x_0, x_1, \ldots, x_n) \in \mathbb{C}^{n+1}$ is denoted by $[\underline{x}] = [x_0 : x_1 : \cdots : x_n]$; these are the homogeneous coordinates of the corresponding point of $\mathbb{P}^n$. We have

$$\mathbb{P}^n = (\mathbb{C}^{n+1} - \{0\})/\mathbb{C}^*$$

which induces a Hausdorff topology on projective space.

Projective n-space is covered by the $n + 1$ open sets

$$U_i = \{[\underline{x}] \mid x_i \neq 0\}$$

for $i = 0, \ldots, n$. Each U_i is isomorphic to $\mathbb{C}^n$, via the map sending the $n + 1$ homogeneous coordinates $[x_0 : x_1 : \cdots : x_n]$ to the n-tuple $(x_0/x_i, x_1/x_i, \ldots, x_n/x_i)$ (with x_i/x_i deleted). These maps from U_i to $\mathbb{C}^n$ are n-dimensional complex charts on $\mathbb{P}^n$, and together they form an n-dimensional complex atlas, inducing an n-dimensional complex structure on $\mathbb{P}^n$. Therefore $\mathbb{P}^n$ is an n-dimensional complex manifold.

It is easy to see that $\mathbb{P}^n$ is compact, either by mapping the unit sphere in $\mathbb{C}^{n+1}$ onto it, or by writing it as the union of the $n + 1$ compact sets in each U_i where all of the coordinates are at most 1.

If $F(x_0, \ldots, x_n)$ is a homogeneous polynomial, then its values in $\mathbb{P}^n$ are not well defined, but whether F is zero or not is; the locus of zeroes of F is called a *hypersurface* in $\mathbb{P}^n$.

Complete Intersections. Since $\mathbb{P}^n$ is a complex manifold of dimension n, it is locally isomorphic to an open set in $\mathbb{C}^n$. Every time we impose an equation, we intuitively cut down the complex dimension by one. Thus to find a Riemann surface in projective n-space $\mathbb{P}^n$ one would first look at the common zeroes of $n-1$ homogeneous polynomials, that is, one would try to intersect $n-1$ hypersurfaces. In order to obtain a Riemann surface this way, one needs to have the analogue of the nonsingularity condition.

This is based (as in the original case of plane curves) on a higher-dimensional version of the Implicit Function Theorem. Without going into the details, we will just state the final result.

DEFINITION 3.8. Let $F_1, \ldots, F_{n-1}$ be $n-1$ homogeneous polynomials in $n+1$ variables $x_0, \ldots, x_n$. Let X be their common zero locus in $\mathbb{P}^n$. We say X is a *smooth complete intersection curve* in $\mathbb{P}^n$ if the $(n-1) \times (n+1)$ matrix of partial derivatives $(\partial F_i/\partial x_j)$ has maximal rank $n-1$ at every point of X.

PROPOSITION 3.9. *A smooth complete intersection curve in $\mathbb{P}^n$ is a compact Riemann surface. Moreover at every point of X one can take as a local coordinate a ratio x_i/x_j of the homogeneous coordinates.*

The condition on the matrix of partials is the hypothesis of the multi-variable Implicit Function Theorem, which insures that X is locally the graph of a set of $n-1$ holomorphic functions. Charts on X are then afforded by the ratios of appropriate coordinates.

Local Complete Intersections. Not all Riemann surfaces which one finds in projective n-space are smooth complete intersection curves. One example is the image of the function $H : \mathbb{P}^1 \to \mathbb{P}^3$ sending $[x : y]$ to $[x^3 : x^2y : xy^2 : y^3]$. The image is a curve X in $\mathbb{P}^3$ which requires not 2 but 3 equations to cut it out. The three equations are

$$x_0 x_3 = x_1 x_2, \quad x_0 x_2 = x_1^2, \quad \text{and} \quad x_1 x_3 = x_2^2.$$

This is the *twisted cubic curve* in $\mathbb{P}^3$. It is not easy to see that it is not a complete intersection curve, but let us leave that off for the moment.

The way to see that X is a Riemann surface is to notice that at any point of the curve, only two of the three equations are actually necessary; for example, near $[1 : 0 : 0 : 0]$, the curve is cut out by the two equations

$$x_0 x_3 = x_1 x_2 \quad \text{and} \quad x_0 x_2 = x_1^2$$

since the third equation $x_1 x_3 = x_2^2$ is a consequence of these two if one assumes that $x_0 \neq 0$, which it is not near this point.

The problem is that no single pair of the three will work at every point of X. This situation then motivates the following definition.

DEFINITION 3.10. A *local complete intersection curve* in projective n-space is a locus $X \subset \mathbb{P}^n$ given by the vanishing of a set $\{F_\alpha\}$ of homogeneous polynomials, such that near each point $p \in X$, X is actually described by $n-1$ of the polynomials

$$F_{\alpha_1} = F_{\alpha_2} = \cdots = F_{\alpha_{n-1}} = 0$$

satisfying the nonsingularity condition that the $(n-1) \times (n+1)$ matrix of partial derivatives $(\partial F_{\alpha_i}/\partial x_j)$ has maximal rank $n-1$ at the point p.

Since the charts on a complete intersection curve are locally defined by the Implicit Function Theorem, this local condition on the common zeroes of a set of homogeneous polynomials is enough to insure that the curve is a Riemann surface:

PROPOSITION 3.11. *Every connected local complete intersection curve X in $\mathbb{P}^n$ is a compact Riemann surface. Moreover at every point of X one can take as a local coordinate a ratio x_i/x_j of the homogeneous coordinates.*

It is an interesting and important theorem in algebraic geometry that every Riemann surface which is holomorphically embedded in projective space is a local complete intersection curve. (We will define "holomorphically embedded" a bit later!)

Problems I.3

A. Let $\phi_i : U_i \to \mathbb{C}^2$ for $i = 0, 1, 2$ be the maps described in the text, e.g., $\phi_0[x : y : z] = (y/x, z/x)$ and similarly for ϕ_1 and ϕ_2. Show that the ϕ_i's are homeomorphisms, where U_i has its subspace topology from $\mathbb{P}^2$, whose topology is given by the quotient topology from $\mathbb{C}^3$. Show that $\mathbb{P}^2$ is Hausdorff. Show further that $\mathbb{P}^2$ is covered by the three compact sets $\phi_i^{-1}(D)$, where $D = \{(z, w) \mid \|z\| \leq 1 \text{ and } \|w\| \leq 1\}$, and is therefore compact.

B. Show that the locus of zeroes of a homogeneous polynomial $F(x, y, z)$ in the projective plane is well defined.

C. Prove Euler's formula for a homogeneous polynomial $F(\underline{x})$ of degree d in any number of variables $\underline{x} = (x_0, x_1, \ldots, x_n)$:

$$F(\underline{x}) = \frac{1}{d} \sum_{i=0}^{n} x_i \frac{\partial F}{\partial x_i}.$$

D. Prove the other half of Lemma 3.5: if a homogeneous polynomial $F(x, y, z)$ is singular, then at least one of the affine plane curves X_i is not smooth.

E. A degree one curve in the projective plane, defined by a homogeneous polynomial in x, y, z of degree one, is called a *line*. Any such polynomial F is of the form $ax + by + cz$. One may write this polynomial in vector form as

$$F(x, y, z) = ax + by + cz = RV = \begin{pmatrix} a & b & c \end{pmatrix} \begin{pmatrix} x \\ y \\ z \end{pmatrix}$$

where R is the row vector of coefficients and V is the column vector of variables. Use this description to prove that any two distinct lines in the projective plane meet at a unique point. Give a formula for the point of intersection in terms of the coefficients of the lines.

F. Show that the curve in $\mathbb{P}^3$ defined by the two equations $x_0 x_3 = 2 x_1 x_2$ and $x_0^2 + x_1^2 + x_2^2 + x_3^2 = 0$ is a smooth complete intersection curve. What is its topological genus?

G. Show that no pair of the three equations given in the text which define the twisted cubic curve X suffice to define X. Show however that near any point of X, X is defined (locally) by two equations. Hence it is a local complete intersection curve and a compact Riemann surface. What is its topological genus?

Further Reading

It will come as no surprise that the subject of Riemann surfaces goes back to Riemann [**Riemann1892**]; Klein's exposition [**Klein1894**] followed in the last century. The first modern treatment of Riemann surfaces dates from Weyl's landmark text [**Weyl55**], first published in 1913. The third edition, published in 1955, was reworked considerably, and Weyl's approach there was foreshadowed by Chevalley a few years earlier [**Chevalley51**].

The literature on Riemann surfaces is often referred to as "vast", but this word is almost an understatement. For the basic definitions [**Springer57**], [**Pfluger57**], [**Bers58**], and [**AS60**] are still useful; these are in a slightly older style but have especially strong treatments of the topological issues. More recent are [**Gunning66**], [**S-N70**], [**G-N76**], [**Gunning76**], [**FK80**], [**Forster81**], [**Griffiths89**], [**Reyssat89**], [**Yang91**], [**Buser92**], and [**Narasimhan92**], all of which are solid and relatively complete in what they do. As an excellent survey the reader may wish to consult [**Shokurov94**]. The texts [**Beardon84**], [**JS87**] and [**Kirwan92**] are somewhat more elementary. Especially delightful is Clemens' scrapbook [**Clemens80**], which is informal yet substantial.

We have downplayed the topological questions which arise in the study; the reader could consult any number of good texts for the basic material on manifolds. In the text are mentioned [**Massey67**], [**Massey91**], [**Munkres75**], [**Armstrong83**], [**Munkres84**], and [**Sieradski92**]; the analysis on manifolds is very well done in [**Munkres91**], and [**Buser92**] has a solid discussion of the topological questions arising specifically for Riemann surfaces.

As to preliminary material, namely the basics on functions of one complex variable, the author has taught or taken courses using [**Ahlfors66**], [**Conway78**],

[**Lang85**], and [**Boas87**] as texts, and all have their strengths.

Complex manifolds of higher dimension is the subject of [**K-M71**].

Chapter II. Functions and Maps

1. Functions on Riemann Surfaces

Let X be a Riemann surface, p a point of X, and f a function on X defined near p. To check whether f has any particular property at p (for example to check/define f being holomorphic at p), one would use complex charts to transport the function to the neighborhood of a point in the complex plane, and check the property there. In this section we make this precise for a variety of properties.

The only thing to be careful of is that the property one is checking must be independent of coordinate changes, so that it does not matter what chart one uses to check the property.

Holomorphic Functions. Let X be a Riemann surface, let p be a point of X, and let f be a complex-valued function defined in a neighborhood W of p.

DEFINITION 1.1. We say that f is *holomorphic at* p if there exists a chart $\phi : U \to V$ with $p \in U$, such that the composition $f \circ \phi^{-1}$ is holomorphic at $\phi(p)$. We say f is *holomorphic in* W if it is holomorphic at every point of W.

We have some immediate remarks, which are embodied in the following.

LEMMA 1.2. *Let X be a Riemann surface, let p be a point of X, and let f be a complex-valued function defined in a neighborhood W of p. Then:*
    a. *f is holomorphic at p if and only if for every chart $\phi : U \to V$ with $p \in U$, the composition $f \circ \phi^{-1}$ is holomorphic at $\phi(p)$;*
    b. *f is holomorphic in W if and only if there exists a set of charts $\{\phi_i : U_i \to V_i\}$ with $W \subseteq \bigcup_i U_i$, such that $f \circ \phi_i^{-1}$ is holomorphic on $\phi_i(W \cap U_i)$ for each i;*
    c. *if f is holomorphic at p, f is holomorphic in a neighborhood of p.*

PROOF. To prove the first statement, let ϕ_1 and ϕ_2 be two charts whose domains contain p, and suppose that $f \circ \phi_1^{-1}$ is holomorphic at $\phi_1(p)$. We must check that $f \circ \phi_2^{-1}$ is holomorphic at $\phi_2(p)$. But

$$f \circ \phi_2^{-1} = (f \circ \phi_1^{-1}) \circ (\phi_1 \circ \phi_2^{-1})$$

which shows that $f \circ \phi_2^{-1}$ is the composition of holomorphic functions, and is therefore holomorphic.

The second statement follows immediately from the first. The third statement follows from the corresponding statement for functions on open sets in $\mathbb{C}$. □

The reader should check that the following examples all give holomorphic functions as claimed.

EXAMPLE 1.3. Any complex chart, considered as a complex-valued function on its domain, is holomorphic on its domain.

EXAMPLE 1.4. Let f be a complex-valued function on an open set in $\mathbb{C}$. Then the above definition of holomorphic (considering $\mathbb{C}$ as a Riemann surface, see Example 1.19) agrees with the usual definition.

EXAMPLE 1.5. Suppose f and g are both holomorphic at $p \in X$. Then $f \pm g$ and fg are holomorphic at p. If $g(p) \neq 0$, then f/g is holomorphic at p.

EXAMPLE 1.6. Let f be a complex-valued function on the Riemann Sphere $\mathbb{C}_\infty$ defined in a neighborhood of ∞. Then f is holomorphic at ∞ if and only if $f(1/z)$ is holomorphic at $z = 0$. In particular, if f is a rational function $f(z) = p(z)/q(z)$, then f is holomorphic at ∞ if and only if $\deg(p) \leq \deg(q)$.

EXAMPLE 1.7. Consider the projective line $\mathbb{P}^1$ with homogeneous coordinates $[z : w]$. Let $p(z, w)$ and $q(z, w)$ be homogeneous polynomials of the *same* degree. Assume that $q(z_0, w_0) \neq 0$. Then $f([z : w]) = p(z, w)/q(z, w)$ is a well defined holomorphic function in a neighborhood of $[z_0 : w_0]$.

EXAMPLE 1.8. Consider a complex torus $\mathbb{C}/L$, with quotient map $\pi : \mathbb{C} \to \mathbb{C}/L$. Let $f : W \to \mathbb{C}$ be a complex-valued function on an open subset $W \subset \mathbb{C}/L$. Then f is holomorphic at a point $p \in W$ if and only if there is a preimage z of p in $\mathbb{C}$ such that $f \circ \pi$ is holomorphic at z. In addition, f is holomorphic on W if and only if $f \circ \pi$ is holomorphic on $\pi^{-1}(W)$.

EXAMPLE 1.9. Let X be an affine plane curve which is defined by a nonsingular polynomial $f(z, w) = 0$. Then the two projections (onto the z- and w-axes) are holomorphic functions on X. Any polynomial function $g(z, w)$, when restricted to the smooth affine plane curve X, is a holomorphic function.

EXAMPLE 1.10. Let X be a projective plane curve which is defined by a nonsingular polynomial $F(x, y, z) = 0$. Let $p = [x_0 : y_0 : z_0]$ be a point on X with $x_0 \neq 0$. Then the two ratios y/x and z/x are holomorphic functions on X at p. Any polynomial function $g(y/x, z/x)$, when restricted to the smooth projective plane curve X, is a holomorphic function at p. Note that such a polynomial function may be written as a ratio $G(x, y, z)/x^d$, where G is the homogenization of the polynomial g, of degree d. More generally, if $G(x, y, z)$ is a homogeneous polynomial of degree d, and $H(x, y, z)$ is a homogeneous polynomial of the same degree d which does not vanish at p, then the ratio G/H is a holomorphic function on X at p.

EXAMPLE 1.11. The previous example generalizes immediately to smooth local complete intersection curves X inside $\mathbb{P}^n$. In particular, if $G(x_0, \ldots, x_n)$ is a homogeneous polynomial of degree d, and $H(x_0, \ldots, x_n)$ is a homogeneous polynomial of the same degree d which does not vanish at $p \in X$, then the ratio G/H is a holomorphic function on X at p.

It is useful to introduce the following notation.

DEFINITION 1.12. If $W \subset X$ is an open subset of a Riemann surface X, we will denote the set of holomorphic functions on W by $\mathcal{O}_X(W)$ (or simply $\mathcal{O}(W)$):

$$\mathcal{O}_X(W) = \mathcal{O}(W) = \{f : W \to \mathbb{C} \mid f \text{ is holomorphic }\}.$$

We note that $\mathcal{O}(W)$ is a $\mathbb{C}$-algebra.

Singularities of Functions; Meromorphic Functions. Let X be a Riemann surface, let p be a point of X, and let f be a complex-valued function defined and holomorphic in a punctured neighborhood of p. (A punctured neighborhood of a point p is a set of the form $U - \{p\}$, where U is a neighborhood of p.) The concept of the type of singularity (removable, pole, essential) for functions of a single complex variable extends readily to functions on a Riemann surface.

DEFINITION 1.13. Let f be holomorphic in a punctured neighborhood of $p \in X$.
  a. We say f has a *removable singularity* at p if and only if there exists a chart $\phi : U \to V$ with $p \in U$, such that the composition $f \circ \phi^{-1}$ has a removable singularity at $\phi(p)$.
  b. We say f has a *pole* at p if and only if there exists a chart $\phi : U \to V$ with $p \in U$, such that the composition $f \circ \phi^{-1}$ has a pole at $\phi(p)$.
  c. We say f has an *essential singularity* at p if and only if there exists a chart $\phi : U \to V$ with $p \in U$, such that the composition $f \circ \phi^{-1}$ has an essential singularity at $\phi(p)$.

We have the following analogue of Lemma 1.2, which we leave to the reader.

LEMMA 1.14. *With the above notations, f has a removable singularity (respectively pole, essential singularity) if and only if for every chart $\phi : U \to V$ with $p \in U$, the composition $f \circ \phi^{-1}$ has removable singularity (resp. pole, essential singularity) at $\phi(p)$.*

We note that if f is defined and holomorphic in a punctured neighborhood of p, then one can decide which kind of singularity f has at p by investigating the behaviour of $f(x)$ for x near p.

a) If $|f(x)|$ is bounded in a neighborhood of p, then f has a removable singularity at p. Moreover, in this case the limit $\lim_{x \to p} f(x)$ exists, and if we define $f(p)$ to be this limit, f is holomorphic at p.
   b) If $|f(x)|$ approaches ∞ as x approaches p, then f has a pole at p.
   c) If $|f(x)|$ has no limit as x approaches p, then f has an essential singularity at p.

DEFINITION 1.15. A function f on X is *meromorphic* at a point $p \in X$ if it is either holomorphic, has a removable singularity, or has a pole, at p. We say f is meromorphic on an open set W if it is meromorphic at every point of W.

As was the case with the examples of holomorphic functions, we leave to the reader to check that the following examples all give meromorphic functions as claimed.

EXAMPLE 1.16. Let f be a complex-valued function on an open set in $\mathbb{C}$. Then the above definition of meromorphic (considering $\mathbb{C}$ as a Riemann surface, see Example 1.19) agrees with the usual definition.

EXAMPLE 1.17. Suppose f and g are both meromorphic at $p \in X$. Then $f \pm g$ and fg are meromorphic at p. If g is not identically zero, then f/g is meromorphic at p.

EXAMPLE 1.18. Let f be a complex-valued function on the Riemann Sphere $\mathbb{C}_\infty$ defined in a neighborhood of ∞. Then f is meromorphic at ∞ if and only if $f(1/z)$ is meromorphic at $z = 0$. In particular, any rational function $f(z) = p(z)/q(z)$ is meromorphic at ∞; indeed, any rational function is meromorphic on all of the Riemann Sphere.

EXAMPLE 1.19. Let f and g be holomorphic functions on a Riemann surface X at p. Then the ratio f/g is a meromorphic function at p, as long as g is not identically zero in a neighborhood of p. Indeed, any function h which is meromorphic at a point $p \in X$ is locally the ratio of two holomorphic functions.

EXAMPLE 1.20. Consider the projective line $\mathbb{P}^1$ with homogeneous coordinates $[z : w]$. Let $p(z, w)$ and $q(z, w)$ be homogeneous polynomials of the same degree (with q not identically zero). Then $f([z : w]) = p(z, w)/q(z, w)$ is a well defined meromorphic function on $\mathbb{P}^1$.

EXAMPLE 1.21. Consider a complex torus $\mathbb{C}/L$, with quotient map $\pi : \mathbb{C} \to \mathbb{C}/L$. Let $f : W \to \mathbb{C}$ be a complex-valued function on an open subset $W \subset \mathbb{C}/L$. Then f is meromorphic at a point $p \in W$ if and only if there is a preimage z of p in $\mathbb{C}$ such that $f \circ \pi$ is meromorphic at z. In addition, f is meromorphic on W if and only if $f \circ \pi$ is meromorphic on $\pi^{-1}(W)$. Note that $g = f \circ \pi$ is always L-periodic, that is, $g(z + \omega) = g(z)$ for every $z \in \mathbb{C}$ and every $\omega \in L$; in fact, there is a 1-1 correspondence between functions on $\mathbb{C}/L$ and L-periodic functions on $\mathbb{C}$. A meromorphic L-periodic function on $\mathbb{C}$ is called an *elliptic function*. Thus the above correspondence induces a 1-1 correspondence between elliptic functions on $\mathbb{C}$ and meromorphic functions on $\mathbb{C}/L$.

EXAMPLE 1.22. Let X be a projective plane curve which is defined by a nonsingular polynomial $F(x,y,z) = 0$. Let $G(x,y,z)$ be a homogeneous polynomial of degree d, and $H(x,y,z)$ a homogeneous polynomial of the same degree d which does not vanish identically on X. Then the ratio G/H is a meromorphic function on X.

EXAMPLE 1.23. Again the previous example generalizes to smooth local complete intersection curves X inside $\mathbb{P}^n$. In particular, if $G(x_0, \ldots, x_n)$ is a homogeneous polynomial of degree d, and $H(x_0, \ldots, x_n)$ is a homogeneous polynomial of the same degree d which does not vanish identically on X, then the ratio G/H is a meromorphic function on X.

DEFINITION 1.24. If $W \subset X$ is an open subset of a Riemann surface X, we will denote the set of meromorphic functions on W by $\mathcal{M}_X(W)$ (or simply $\mathcal{M}(W)$):

$$\mathcal{M}_X(W) = \mathcal{M}(W) = \{f : W \to \mathbb{C} \mid f \text{ is meromorphic }\}.$$

Laurent Series. Let f be defined and holomorphic in a punctured neighborhood of $p \in X$. Let $\phi : U \to V$ be a chart on X with $p \in U$. Thinking of z as the local coordinate on X near p, so that $z = \phi(x)$ for x near p, we have that $f \circ \phi^{-1}$ is holomorphic in a neighborhood of $z_0 = \phi(p)$. Therefore we may expand $f \circ \phi^{-1}$ in a Laurent series about z_0:

$$(1.25) \qquad f(\phi^{-1}(z)) = \sum_n c_n(z - z_0)^n.$$

This is called the *Laurent Series for f about p with respect to ϕ* (or with respect to the local coordinate z). The coefficients $\{c_n\}$ of the Laurent series are called the *Laurent coefficients*.

The Laurent series definitely depends on the choice of local coordinate, that is, the choice of chart ϕ.

One can use Laurent series however to check the nature of the singularity of f at p. This is just based on the usual criterion for functions of one complex variable, and we leave it to the reader:

LEMMA 1.26. *With the above notation, f has a removable singularity at p if and only if any one of its Laurent series has no negative terms. The function f has a pole at p if and only if any one of its Laurent series has finitely many (but not zero) negative terms. The function f has an essential singularity at p if and only if any one of its Laurent series has infinitely many negative terms.*

The Order of a Meromorphic Function at a Point. Not only can one decide the nature of a singularity from a Laurent series, but, for meromorphic functions, one can extract the order of the zero or pole from any Laurent series.

DEFINITION 1.27. Let f be meromorphic at p, whose Laurent series in a local coordinate z is $\sum_n c_n(z - z_0)^n$. The *order of f at p*, denoted by $\operatorname{ord}_p(f)$, is the minimum exponent actually appearing (with nonzero coefficient) in the Laurent series:
$$\operatorname{ord}_p(f) = \min\{n \mid c_n \neq 0\}.$$

We need to check that $\operatorname{ord}_p(f)$ is well defined, independent of the choice of local coordinate used to define the Laurent series.

Suppose that $\psi : U' \to V'$ is another chart with $p \in U'$, giving local coordinate $w = \psi(x)$ for x near p. Suppose further that $\psi(p) = w_0$. Consider the holomorphic transition function $T(w) = \phi \circ \psi^{-1}$, which expresses z as a holomorphic function of w. Since T is invertible at w_0, we must have $T'(w_0) \neq 0$ (Chapter I, Lemma 1.7). If we write the power series for T, it will therefore be of the form
$$z = T(w) = z_0 + \sum_{n \geq 1} a_n (w - w_0)^n,$$
with the linear term coefficient $a_1 \neq 0$.

Suppose now that $c_{n_0}(z - z_0)^{n_0} +$ (higher order terms) is the Laurent series for f at p in terms of the coordinate z, with $c_{n_0} \neq 0$, so that the order of f computed via z is n_0. To obtain the Laurent series for f in terms of w, we simply compose the above Laurent series with the above power series expression $z - z_0 = \sum_{n \geq 1} a_n (w - w_0)^n$. We see immediately that the term of lowest possible order in the variable $w - w_0$ of the composition is $c_{n_0} a_1^{n_0} (w - w_0)^{n_0}$. Since neither c_{n_0} nor a_1 is zero, this term is actually present and the order of f computed via w is also n_0. Thus the order of f at p is well defined.

We have the following, which we leave to the reader.

LEMMA 1.28. *Suppose f is meromorphic at p. Then f is holomorphic at p if and only if $\operatorname{ord}_p(f) \geq 0$. In this case $f(p) = 0$ if and only if $\operatorname{ord}_p(f) > 0$. f has a pole at p if and only if $\operatorname{ord}_p(f) < 0$. f has neither a zero nor a pole at p if and only if $\operatorname{ord}_p(f) = 0$.*

We say that f has a *zero of order n* at p if $\operatorname{ord}_p(f) = n \geq 1$. We say f has a *pole of order n* at p if $\operatorname{ord}_p(f) = -n < 0$.

The order function behaves well with respect to products, but is more unruly with respect to sums:

LEMMA 1.29. *Let f and g be nonzero meromorphic functions at $p \in X$. Then:*
  a. $\operatorname{ord}_p(fg) = \operatorname{ord}_p(f) + \operatorname{ord}_p(g)$.
  b. $\operatorname{ord}_p(f/g) = \operatorname{ord}_p(f) - \operatorname{ord}_p(g)$.
  c. $\operatorname{ord}_p(1/f) = -\operatorname{ord}_p(f)$.

d. $\text{ord}_p(f \pm g) \geq \min\{\text{ord}_p(f), \text{ord}_p(g)\}$.

Again we leave these simple computations to the reader.

Any rational function $f(z)$ can be considered as a meromorphic function on the Riemann Sphere (Example 1.18). The order of such functions at any point can be obtained quite readily; we leave the details as a problem for the reader.

EXAMPLE 1.30. Let $f(z) = p(z)/q(z)$ be a nonzero rational function of z, considered as a meromorphic function on the Riemann Sphere as in Example 1.18. We may factor p and q completely into linear factors and write f uniquely as

$$f(z) = c \prod_i (z - \lambda_i)^{e_i},$$

where c is a nonzero constant, the λ_i's are distinct complex numbers, and the exponents e_i are integers. Then $\text{ord}_{z=\lambda_i}(f) = e_i$ for each i. Moreover, $\text{ord}_\infty(f) = \deg(q) - \deg(p) = -\sum_i e_i$. Finally, $\text{ord}_x(f) = 0$ unless $x = \infty$ or x is one of the points $z = \lambda_i$. Note that

$$\sum_{x \in X} \text{ord}_x(f) = 0,$$

which as we will see is a general phenomenon for meromorphic functions on compact Riemann surfaces.

$\mathcal{C}^\infty$ **Functions.** A real-valued function of a complex variable $z = x+iy$ is $\mathcal{C}^\infty$ at a point z_0 if, as a function of x and y, it has continuous partial derivatives of all orders at z_0. A complex-valued function of z is $\mathcal{C}^\infty$ if its real and imaginary parts are. This concept transfers immediately to a Riemann surface using the same construct as for holomorphic functions: a function f defined on a Riemann surface X is $\mathcal{C}^\infty$ at a point p if there is a chart $\phi : U \to V$ on X with $p \in U$ such that $f \circ \phi^{-1}$ is $\mathcal{C}^\infty$ at $\phi(p)$. To check that a function is $\mathcal{C}^\infty$, any chart can be used. If f is defined on all of X, then for f to be $\mathcal{C}^\infty$ it suffices to check locally using any atlas of charts on X.

Harmonic Functions. Harmonic functions play a central role in the analytic theory of Riemann surfaces. Although we will not stress this point of view, it is good to know what that aspect of the theory is about.

DEFINITION 1.31. A real-valued $\mathcal{C}^\infty$ function $h(x,y)$ of two real variables x and y defined on an open set $V \subset \mathbb{R}^2$ is *harmonic* if

$$\frac{\partial^2 h}{\partial x^2} + \frac{\partial^2 h}{\partial y^2} = 0$$

identically on V. A complex-valued function is *harmonic* if and only if its real and imaginary parts are harmonic.

The real and imaginary parts of any holomorphic function of $z = x + iy$ are harmonic as functions of x and y; this follows immediately from the Cauchy-Riemann equations. Hence holomorphic functions are harmonic.

We transport harmonicity to a Riemann surface X via charts in the usual way. Suppose h is a C^∞ at a point p on X. We say that h is harmonic at p if there is a chart $\phi : U \to V$ with $p \in U$ such that $h \circ \phi^{-1}$ is harmonic near $\phi(p)$.

This is independent of the choice of chart in fact:

LEMMA 1.32. *Suppose h is a C^∞ function defined near $p \in X$. Let ϕ_1 and ϕ_2 be two charts near p. Then $h \circ \phi_1^{-1}$ is harmonic near $\phi_1(p)$ if and only if $h \circ \phi_2^{-1}$ is harmonic near $\phi_2(p)$.*

PROOF. Let $z = x + iy$ be the local coordinate for ϕ_1, and let $w = u + iv$ be the local coordinate for ϕ_2. We know that the change of coordinates function $w = T(z) = \phi_2(\phi_1^{-1}(z))$ is holomorphic; writing this as $u = u(x,y)$ and $v = v(x,y)$, we conclude that these functions satisfy the Cauchy-Riemann equations $u_x = v_y$ and $u_y = -v_x$.

Suppose $h_2 = h \circ \phi_2^{-1}$ is harmonic in u and v. We must show that $h_1 = h \circ \phi_1^{-1}$ is harmonic in x and y; this will suffice, by symmetry. But

$$h_1(x,y) = h(\phi_1^{-1}(x,y)) = h(\phi_2^{-1}(u(x,y), v(x,y))) = h_2(u(x,y), v(x,y));$$

so the chain rule gives

$$(h_1)_x = (h_2)_u u_x - (h_2)_v u_y \text{ and } (h_1)_y = (h_2)_u u_y + (h_2)_v u_x$$

using the Cauchy-Riemann equations. Then

$$(h_1)_{xx} = (h_2)_{uu} u_x^2 - (h_2)_{uv} u_x u_y + (h_2)_u u_{xx} - (h_2)_{vu} u_y u_x + (h_2)_{vv} u_y^2 - (h_2)_v u_{yx},$$

and

$$(h_1)_{yy} = (h_2)_{uu} u_y^2 + (h_2)_{uv} u_y u_x + (h_2)_u u_{yy} + (h_2)_{vu} u_x u_y + (h_2)_{vv} u_x^2 + (h_2)_v u_{xy}.$$

Therefore

$$(h_1)_{xx} + (h_1)_{yy} = ((h_2)_{uu} + (h_2)_{vv})(u_x^2 + u_y^2) + (h_2)_u (u_{xx} + u_{yy}).$$

In the above expression, the first term is zero since h_2 is harmonic as a function of u and v, and the second term is zero since u is harmonic as a function of x and y. □

Theorems Inherited from One Complex Variable. Certain theorems concerning holomorphic and meromorphic functions are inherited immediately from the corresponding theorems concerning functions defined on open sets in the complex plane. We collect some of them here.

THEOREM 1.33 (DISCRETENESS OF ZEROES AND POLES). *Let f be a meromorphic function defined on a connected open set W of a Riemann surface X. If f is not identically zero, then the zeroes and poles of f form a discrete subset of W.*

The above theorem has an immediate implication for compact surfaces.

COROLLARY 1.34. *Let f be a meromorphic function on a compact Riemann surface, which is not identically zero. Then f has a finite number of zeroes and poles.*

THEOREM 1.35 (THE IDENTITY THEOREM). *Suppose that f and g are two meromorphic functions defined on a connected open set W of a Riemann surface X. Suppose that $f = g$ on a subset $S \subset W$ which has a limit point in W. Then $f = g$ on W.*

THEOREM 1.36 (THE MAXIMUM MODULUS THEOREM). *Let f be holomorphic on a connected open set W of a Riemann surface X. Suppose that there is a point $p \in W$ such that $|f(x)| \leq |f(p)|$ for all $x \in W$. Then f is constant on W.*

We have the following corollary of the Maximum Modulus Theorem, which is a theorem truly about Riemann surfaces, in that there is no precise counterpart for functions on complex domains.

THEOREM 1.37. *Let X be a compact Riemann surface. Suppose that f is holomorphic on all of X. Then f is a constant function.*

PROOF. Since f is holomorphic, its absolute value $|f|$ is a continuous function; therefore, since X is compact, $|f|$ achieves its maximum value at some point of X. By the Maximum Modulus Theorem, f must then be constant on X, since X is connected. □

The closest thing to the above theorem for functions on complex domains is Liouville's Theorem, which states that a bounded entire function must be constant. This can in fact be reformulated in terms of functions on the Riemann Sphere.

Harmonic functions also satisfy a maximum principle; the statement is practically the same as for holomorphic functions.

THEOREM 1.38. *Suppose that f is harmonic on a connected open set W of a Riemann surface X. Suppose that there is a point $p \in W$ such that $|f(x)| \leq |f(p)|$ for all $x \in W$. Then f is constant on W. In particular, if X is a compact Riemann surface then any harmonic function on X is constant.*

Problems II.1

A. Check that all of the functions of Examples 1.3 through 1.11 are holomorphic as claimed.

B. Check that all of the functions of Examples 1.16 through 1.23 are meromorphic as claimed.

C. Let L be a lattice in $\mathbb{C}$ and let X be the torus $\mathbb{C}/L$. Let $\pi : \mathbb{C} \to X$ be the quotient map. Show that a function f on X is meromorphic if and only if the composition $f\pi$ is a meromorphic function on $\mathbb{C}$.

D. Prove Lemma 1.26.

E. Prove Lemma 1.28.

F. Prove Lemma 1.29.

G. Verify all of the statements of Example 1.30.

H. Prove Liouville's Theorem (that a bounded entire function on $\mathbb{C}$ is constant) by showing that a bounded entire function extends to a holomorphic function on the (compact) Riemann Sphere $\mathbb{C}_\infty$.

I. Prove without invoking the Maximum Modulus Theorem that any rational function which is holomorphic at every point of the Riemann Sphere $\mathbb{C}_\infty$ is in fact a constant.

2. Examples of Meromorphic Functions

Meromorphic Functions on the Riemann Sphere. We have seen in Example 1.18 that any rational function $r(z) = p(z)/q(z)$ is meromorphic on the whole Riemann Sphere. In fact, the converse is true:

THEOREM 2.1. *Any meromorphic function on the Riemann Sphere is a rational function.*

PROOF. Let f be a meromorphic function on the Riemann Sphere $\mathbb{C}_\infty$. Since $\mathbb{C}_\infty$ is compact, it has finitely many zeroes and poles. Let $\{\lambda_i\}$ be the set of zeroes and poles of f in the finite complex plane $\mathbb{C}$, and assume that $\operatorname{ord}_{z=\lambda_i}(f) = e_i$. Consider the rational function

$$r(z) = \prod_i (z - \lambda_i)^{e_i}$$

which has the same zeroes and poles, to the same orders, as f does, in the finite plane (see Example 1.30). Let $g(z) = f/r(z)$; g is a meromorphic function on $\mathbb{C}_\infty$, with no zeroes or poles in the finite plane. Therefore, as a function on $\mathbb{C}$, it is everywhere holomorphic, and has a Taylor series

$$g(z) = \sum_{n=0}^{\infty} c_n z^n$$

which converges everywhere on $\mathbb{C}$. Note however that g is also meromorphic at $z = \infty$; in terms of the coordinate $w = 1/z$ at ∞, we have

$$g(w) = \sum_{n=0}^{\infty} c_n w^{-n}$$

and so for this to be meromorphic at $w = 0$ it must be the case that g has only finitely many terms, that is, g is a polynomial in z.

If the polynomial g is not constant, then it will have a zero in $\mathbb{C}$, which is a contradiction. Hence the ratio f/r is constant, and f is a rational function. □

COROLLARY 2.2. *Let f be any meromorphic function on the Riemann Sphere. Then*

$$\sum_p \operatorname{ord}_p(f) = 0.$$

PROOF. We have already seen in Example 1.30 that this is true for rational functions. Since any meromorphic function on $\mathbb{C}_\infty$ is rational by the above theorem, we are done. □

Recall that for a meromorphic function f, the order is positive at the zeroes and negative at the poles. Therefore the statement above that the sum of the orders is zero says exactly that f has the *same* number of zeroes and poles, if we count them according to their order. This is a recurring theme in the theory: one gets very nice answers to formulas which count things (like the number of zeroes, etc.) if one "counts properly". In this case, counting properly means counting according to the order.

Meromorphic Functions on the Projective Line. Let $\mathbb{P}^1$ be the projective line. We have claimed in Example 1.20 that ratios of homogeneous polynomials of the same degree give meromorphic functions on $\mathbb{P}^1$. This example is important enough to go through the details in the text, which we will now do.

Note that we can view $\mathbb{P}^1$ as the quotient space

$$\mathbb{P}^1 = (\mathbb{C}^2 - \{0\})/\mathbb{C}^*,$$

where $\lambda \in \mathbb{C}^*$ acts on a nonzero vector $(z, w) \in \mathbb{C}^2$ by sending it to $(\lambda z, \lambda w)$. The orbit of (z, w) is exactly the point $[z : w] \in \mathbb{P}^1$. Thus to construct functions on $\mathbb{P}^1$ we try to define functions on $\mathbb{C}^2$ which are invariant under the action of $\mathbb{C}^*$; such a function will descend to the quotient $\mathbb{P}^1$, and we can check at the end whether or not it is meromorphic. One such function is the function sending (z, w) to z/w. This is the prototype for all the examples, in fact.

A polynomial $p(z, w)$ is said to be *homogeneous* if each of its terms has the same total degree; this degree is the degree of p. Thus a homogeneous polynomial of degree d can be written uniquely as

$$p(z,w) = \sum_{i=0}^{d} a_i z^i w^{d-i}.$$

Note that if $p(z,w)$ is a homogeneous polynomial of degree d, then $p(\lambda z, \lambda w) = \lambda^d p(z,w)$. Hence p is not invariant under the action of $\mathbb{C}^*$, but at least it transforms in a very controlled way.

What is now obvious is that if $p(z,w)$ and $q(z,w)$ are both homogeneous polynomials of the *same* degree, with q not identically zero, then the ratio $r(z,w) = p(z,w)/q(z,w)$ will be invariant under the action of $\mathbb{C}^*$.

Indeed, consider the special function $u = z/w$, which is $\mathbb{C}^*$-invariant. Let $r(u)$ be any rational function of u; then r is also $\mathbb{C}^*$-invariant. If we multiply the numerator and denominator of r by the appropriate power of w, we will obtain a ratio of homogeneous polynomials of the same degree.

LEMMA 2.3. *If $p(z,w)$ and $q(z,w)$ are homogeneous of the same degree, with q not identically zero, then $r(z,w) = p(z,w)/q(z,w)$ descends to a meromorphic function on $\mathbb{P}^1$.*

PROOF. Let $\phi : \{w \neq 0\} \to \mathbb{C}$ be one of the two standard charts of $\mathbb{P}^1$, so that $\phi([z:w]) = z/w$. Note that $\phi^{-1}(u) = [u:1]$. To check that the function $r([z:w]) = p(z,w)/q(z,w)$ is meromorphic on $\{w \neq 0\}$, we must show that $r \circ \phi^{-1}$ is meromorphic on $\mathbb{C}$. But

$$r(\phi^{-1}(u)) = r([u:1]) = p(u,1)/q(u,1)$$

is a rational function of u, and is certainly meromorphic. The same computation for the other chart (sending $[z:w]$ to w/z) finishes the computation. □

Every homogeneous polynomial of positive degree in z,w factors completely into linear factors; homogeneous polynomials in two variables behave like ordinary polynomials in a single variable in this respect. Therefore a ratio of homogeneous polynomials of the same degree can always be written in the form

$$(2.4) \qquad r(z,w) = \prod_i (b_i z - a_i w)^{e_i},$$

where we may assume the different factors are relatively prime. It is easy to see that with this notation, $\text{ord}_{[a_i:b_i]}(r) = e_i$ when we consider r as a meromorphic function on $\mathbb{P}^1$. With this remark, it is easy to show the following analogue of Theorem 2.1; moreover the proof is essentially the same.

THEOREM 2.5. *Every meromorphic function on $\mathbb{P}^1$ is a ratio of homogeneous polynomials in z,w of the same degree.*

PROOF. Let f be a meromorphic function on $\mathbb{P}^1$ which is not identically zero. Since $\mathbb{P}^1$ is compact, f has finitely many zeroes and poles, which we may list as $\{[a_i : b_i]\}$. Assume that $\mathrm{ord}_{[a_i:b_i]}(f) = e_i$, and consider the ratio

$$r(z,w) = w^n \prod_i (b_i z - a_i w)^{e_i}$$

where n is chosen simply to make r a ratio of homogeneous polynomials of the same degree: $n = -\sum_i e_i$. The ratio $g = f/r$ has no zeroes or poles, except possibly at the point $[1 : 0]$ where $w = 0$. We would like to show that g is constant.

If g has a pole at $[1 : 0]$, then since g has no zeroes, $1/g$ has no poles. Hence $1/g$ is constant since $\mathbb{P}^1$ is compact; but $1/g$ has a zero at $[1 : 0]$, which gives a contradiction since $1/g = r/f$ is not identically zero.

Therefore we may assume that g does not have a pole at $[1 : 0]$. Hence g is holomorphic on all of $\mathbb{P}^1$, so g is constant since $\mathbb{P}^1$ is compact. □

Note that since r is a ratio of polynomials of the same degree, when we write r as in (2.4) we have $\sum_i e_i = 0$; therefore we see that, as with rational functions on the Riemann Sphere, we have $\sum \mathrm{ord}_p(r) = 0$. By the above theorem, every meromorphic function on $\mathbb{P}^1$ is of this form. Therefore:

COROLLARY 2.6. *Let f be any nonconstant meromorphic function on $\mathbb{P}^1$. Then*

$$\sum_p \mathrm{ord}_p(f) = 0.$$

Meromorphic Functions on a Complex Torus. Fix τ in the upper half-plane, and consider the lattice $L = \mathbb{Z} + \mathbb{Z}\tau$. Form the complex torus $X = \mathbb{C}/L$.

Just like $\mathbb{P}^1$, $\mathbb{C}/L$ is a quotient space, and so one may construct meromorphic functions on $\mathbb{C}/L$ by taking L-periodic meromorphic functions on $\mathbb{C}$. One's first instinct is to build such functions by taking ratios of L-periodic holomorphic functions on $\mathbb{C}$. The problem is that there are (essentially) no such things: any L-periodic holomorphic function on $\mathbb{C}$ would descend to a holomorphic function on $\mathbb{C}/L$, which would then be constant because $\mathbb{C}/L$ is compact.

Therefore we fall back to relying on ratios of holomorphic functions which are not separately L-periodic, but which transform in a highly controlled manner upon translation by lattice points. By a careful choice of the numerator and denominator, we can arrange the extra factor to cancel and obtain a true L-periodic function.

The entire story from this point of view is similiar to the construction of meromorphic functions on $\mathbb{P}^1$, as ratios of homogeneous polynomials. The homogeneous polynomials are not invariant under the action of $\mathbb{C}^*$, but transform very nicely ($p(z,w)$ of degree d transforms under the action of $\lambda \in \mathbb{C}^*$ to $\lambda^d p(z,w)$). A homogeneous polynomial is a product of homogeneous linear polynomials, and the analogue of homogeneous linear polynomials in our torus situation is the *theta-function*.

Fix a τ with $Im(\tau) > 0$, and define

$$\theta(z) = \sum_{n=-\infty}^{\infty} e^{\pi i [n^2 \tau + 2nz]}.$$

This series converges absolutely and uniformly on compact subsets of $\mathbb{C}$. Hence $\theta(z)$ is an analytic function on all of $\mathbb{C}$.

Note that $\theta(z+1) = \theta(z)$ for every z in $\mathbb{C}$, so that θ is periodic. (The series given above is its Fourier series.) We need to investigate how θ transforms under translation by τ. An easy series computation shows that

$$\theta(z+\tau) = e^{-\pi i [\tau + 2z]} \theta(z)$$

for every z in $\mathbb{C}$.

It follows directly that z_0 is a zero of θ if and only if $z_0 + m + n\tau$ is a zero of θ for every m and n in $\mathbb{Z}$. Moreover the order of zero of θ at z_0 is the same as the order of zero at $z_0 + m + n\tau$.

An integral computation easily shows that the only zeroes of θ are at the points $(1/2) + (\tau/2) + m + n\tau$, for integers m and n, and that these zeroes are all simple.

Consider then the translate

$$\theta^{(x)}(z) = \theta(z - (1/2) - (\tau/2) - x)$$

which has simple zeroes at the points $x + L$. Note that

$$\theta^{(x)}(z+1) = \theta^{(x)}(z) \text{ and } \theta^{(x)}(z+\tau) = -e^{-2\pi i (z-x)} \theta^{(x)}(z).$$

Now consider a ratio

$$R(z) = \frac{\prod_i \theta^{(x_i)}(z)}{\prod_j \theta^{(y_j)}(z)}.$$

This function $R(z)$ is certainly meromorphic on $\mathbb{C}$, and is periodic, i.e., $R(z+1) = R(z)$. Therefore it will be L-periodic if and only if $R(z+\tau) = R(z)$. But

$$\begin{aligned} R(z+\tau) &= \frac{\prod_{i=1}^{m} \theta^{(x_i)}(z+\tau)}{\prod_{j=1}^{n} \theta^{(y_j)}(z+\tau)} \\ &= (-1)^{m-n} \frac{\prod_{i=1}^{m} e^{-2\pi i (z-x_i)} \theta^{(x_i)}(z)}{\prod_{j=1}^{n} e^{-2\pi i (z-y_j)} \theta^{(y_j)}(z)} \\ &= (-1)^{m-n} e^{-2\pi i [(m-n)z + \sum_j y_j - \sum_i x_i]} R(z). \end{aligned}$$

Thus we need the extra factor

$$(-1)^{m-n} e^{-2\pi i [(m-n)z + \sum_j y_j - \sum_i x_i]}$$

to be identically 1 for all z. This forces $m = n$, and if so, this number is 1 if and only if
$$\sum_i x_i - \sum_j y_j \in \mathbb{Z}.$$
We have therefore proved the following.

PROPOSITION 2.7. *Fix a positive integer d, and choose any two sets of d complex numbers $\{x_i\}$ and $\{y_j\}$ such that $\sum_i x_i - \sum_j y_j$ is an integer. Then the ratio of translated theta functions*
$$R(z) = \frac{\prod_i \theta^{(x_i)}(z)}{\prod_j \theta^{(y_j)}(z)}$$
is a meromorphic L-periodic function on $\mathbb{C}$, and so descends to a meromorphic function on $\mathbb{C}/L$.

We note that since $\theta^{(x)}$ has a simple zero at each of the points of $x + L$, the above ratio R has zeroes at the points $x_i + L$ and poles at the points $y_j + L$ of $\mathbb{C}/L$.

We will be able to prove later that every meromorphic function on $\mathbb{C}/L$ is of this form, namely a ratio of translated theta-functions. Moreover in the next section we will see that every lattice L may be put into the form $\mathbb{Z} + \mathbb{Z}\tau$, so this seemingly special case is in fact the general one.

Meromorphic Functions on Smooth Plane Curves. Let $f(x, y) = 0$ define a smooth affine plane curve $X \subset \mathbb{C}^2$. We have seen in Example 1.11 that the coordinate functions x and y are both holomorphic functions on X, and hence so is any polynomial $g(x, y)$. Therefore any ratio of polynomials $r(x, y) = g(x, y)/h(x, y)$ is a meromorphic function on X, as long as the denominator $h(x, y)$ does not vanish identically on X.

If the defining polynomial $f(x, y)$ divides this denominator $h(x, y)$, then clearly h will vanish everywhere on X. A basic theorem of polynomial algebra and algebraic geometry guarantees that this is the only case when h could vanish identically on X. This is Hilbert's Nullstellensatz, and for our purposes it can be stated as follows:

THEOREM 2.8 (HILBERT'S NULLSTELLENSATZ). *Suppose h is a polynomial vanishing everywhere an irreducible polynomial f vanishes. Then f divides h.*

See [**Shafarevich77**], for example. Therefore the only condition on a ratio of polynomials g/h to obtain a meromorphic function on the affine plane curve described by $f = 0$ is that f not divide the denominator h.

The situation is very similar in the projective case. Here we have homogeneous coordinates $[x : y : z]$, with the plane curve X defined by the vanishing of an irreducible nonsingular homogeneous polynomial $F(x, y, z)$.

We no longer can take ratios of holomorphic functions, since there are no nonconstant holomorphic functions on X. But we may still take ratios: if

$G(x,y,z)$ and $H(x,y,z)$ are both homogeneous of the same degree, then the ratio $R(x,y,z) = G/H$ is a well defined complex-valued function everywhere in the plane $\mathbb{P}^2$ away from the zeroes of H.

We claim that such a ratio determines a meromorphic function on the smooth projective plane curve X defined by $F = 0$, as long as the denominator H does not vanish identically. Moreover, an easy extension of Hilbert's Nullstellensatz says that this can happen only if F divides H.

Indeed, since a projective plane curve has the same charts as a smooth affine curve, we may check that such a ratio R is meromorphic by checking on the affine charts of $\mathbb{P}^2$. To check it for example on the $\mathbb{C}^2$ where $z \neq 0$, we simply set $z = 1$ in the equation F to obtain the affine equation $f(x,y) = F(x,y,1)$ for X, and also set $z = 1$ in the homogeneous polynomials G and H. Thus we see that the function R is, in this $\mathbb{C}^2$ where $z \neq 0$, equal to the ratio of ordinary polynomials $g(x,y)/h(x,y) = G(x,y,1)/H(x,y,1)$. Hence it is meromorphic at all points of X in this $\mathbb{C}^2$, since it is a ratio of holomorphic functions there.

Similar arguments in the other two $\mathbb{C}^2$'s of $\mathbb{P}^2$ show that R is meromorphic on all of X.

Therefore:

PROPOSITION 2.9. *Let X be a smooth affine plane curve defined by an irreducible nonsingular polynomial $f(x,y) = 0$. Then any ratio of polynomials $r = g(x,y)/h(x,y)$ is a meromorphic function on X as long as f does not divide the denominator h.*

In the projective case, let X be a smooth projective plane curve defined by an irreducible nonsingular homogeneous polynomial $F(x,y,z) = 0$. Then any ratio of homogeneous polynomials $R = G(x,y,z)/H(x,y,z)$ where G and H have the same degree is a meromorphic function on X as long as F does not divide the denominator H.

Smooth Projective Curves. It is time to make a proper definition of a general Riemann surface found in a higher-dimensional projective space. The idea is exactly motivated by the requirement that the above Proposition still be true, namely that the ratios of homogeneous polynomials will give meromorphic functions.

DEFINITION 2.10. *Let X be a Riemann surface, which is a subset of a projective space $\mathbb{P}^n$. We say that X is holomorphically embedded in $\mathbb{P}^n$ if for every point p on X there is a homogeneous coordinate z_j such that:*
    a. *$z_j \neq 0$ at p;*
    b. *for every k, the ratio z_k/z_j is a holomorphic function on X near p; and*
    c. *there is a homogeneous coordinate z_i such that the ratio z_i/z_j is a local coordinate on X near p.*

A Riemann surface which is holomorphically embedded in projective space is called a *smooth projective curve.*

Let X be a smooth projective curve. If we fix a point p on X, and let z_j be the homogeneous coordinate with the above properties near p, then note that any ratio of homogeneous coordinates z_i/z_k is meromorphic at p, since $z_i/z_k = (z_i/z_j)/(z_k/z_j)$ is a ratio of holomorphic functions defined near p. (This works at least when the coordinate z_k is not identically zero on X.) Since the ratios of the coordinates are meromorphic functions on X, so will any rational function of these ratios. A rational function of these ratios can always be written itself as a ratio of homogeneous polynomials of the same degree, by clearing denominators; therefore we have immediately the statement corresponding to the previous Proposition:

PROPOSITION 2.11. *Let X be a smooth projective curve in $\mathbb{P}^n$. Then any ratio of homogeneous polynomials $R = G(z_0, z_1, \ldots, z_n)/H(z_0, z_1, \ldots, z_n)$ where G and H have the same degree is a meromorphic function on X as long as the denominator H does not vanish identically on X.*

It is easy to verify that all of the examples of Riemann surfaces which we have found in projective space so far are holomorphically embedded:

PROPOSITION 2.12. *The projective line $\mathbb{P}^1$ is a smooth projective curve. Any smooth projective plane curve $X \subset \mathbb{P}^2$ is a smooth projective curve. Any complete intersection curve, and more generally any local complete intersection curve, is a smooth projective curve.*

PROOF. Suppose X is a local complete intersection curve in $\mathbb{P}^n$. Fix a point p on X. Then near p, X is locally the graph of a set of $n-1$ holomorphic functions of a complex variable z, and therefore we may write X as the locus

$$[1 : z : g_2(z) : \cdots : g_n(z)]$$

near p (after rearranging the coordinates if necessary). Here the homogeneous coordinate z_0 is nonzero at p, and the ratio $z = z_1/z_0$ is a local coordinate at p; finally all the ratios z_i/z_0 are holomorphic at p. Since this is true for all points p, X is holomorphically embedded. $\square$

For a Riemann surface X to be holomorphically embedded in $\mathbb{P}^n$ is essentially equivalent to the above local form, namely that X is locally a graph of $n-1$ holomorphic functions. Indeed, if we fix a point p, and for example assume that z_0 is the homogeneous coordinate which is nonzero at p, with the ratio $z = z_1/z_0$ being a local coordinate on X near p, then it is clear that near p, X is the graph

$$[1 : z : g_2(z) : \cdots : g_n(z)]$$

where $g_k(z) = z_k/z_0$ is holomorphic.

There are essentially two ways to find smooth projective curves X. The first we have seen: find X as the locus of common zeroes of a suitable set of homogeneous polynomials. This leads to the local complete intersection idea, which we have introduced earlier. The second we will see a bit later: take a known Riemann surface X and find a suitable map from X into projective space.

Problems II.2

A. Consider the projective line $\mathbb{P}^1$. Fix a point $p \in \mathbb{P}^1$, and a finite set $S \subset \mathbb{P}^1$ with $p \notin S$. Show that there exists a meromorphic function f on $\mathbb{P}^1$ with a simple zero at p and no zeroes or poles at any of the points of S.

B. Show that the series defining the theta-function converges absolutely and uniformly on compact subsets of $\mathbb{C}$.

C. Show that $\theta(z+1) = \theta(z)$ for every z in $\mathbb{C}$.

D. Show that $\theta(z+\tau) = e^{-\pi i [\tau + 2z]} \theta(z)$ for every z in $\mathbb{C}$.

E. Show that z_0 is a zero of θ if and only if $z_0 + m + n\tau$ is a zero of θ for every m and n in $\mathbb{Z}$. Moreover the order of zero of θ at z_0 is the same as the order of zero at $z_0 + m + n\tau$.

F. Show that the only zeroes of θ are at the points $(1/2) + (\tau/2) + m + n\tau$, for integers m and n, and that these zeroes are simple. (Hint: integrate θ'/θ around a fundamental parallelogram.)

G. Let $\{p_i\}$ and $\{q_i\}$ be two sets of d points on a complex torus $X = \mathbb{C}/L$ (repetitions are allowed). Show that there exist numbers $\{x_i\}$ and $\{y_i\}$ in $\mathbb{C}$ such that $\pi(x_i) = p_i$ and $\pi(y_i) = q_i$ for every i with $\sum_i x_i = \sum_i y_i$ if and only if $\sum_i p_i = \sum_i q_i$ in the quotient group law of X.

H. Consider the complex torus $X = \mathbb{C}/L$. Fix a point $p \in X$, and a finite set $S \subset X$ with $p \notin S$. Show that there exists a meromorphic function f on X with a simple zero at p and no zeroes or poles at any of the points of S.

3. Holomorphic Maps Between Riemann Surfaces

The Definition of a Holomorphic Map. Modern geometric philosophy holds firmly to the notion that the first thing one does after defining the objects of interest is to define the functions of interest. In our case the objects are Riemann surfaces, and we have already addressed complex-valued functions on Riemann surfaces. However "functions" are to be taken also in the sense of mappings between the objects; once we define such mappings, we will have a *category* of Riemann surfaces.

In the case of Riemann surfaces, which have local complex coordinates, the natural property of a mapping is to be holomorphic. Let X and Y be Riemann surfaces.

DEFINITION 3.1. A mapping $F : X \to Y$ is *holomorphic at* $p \in X$ if and only if there exists charts $\phi_1 : U_1 \to V_1$ on X with $p \in U_1$ and $\phi_2 : U_2 \to V_2$ on Y with $F(p) \in U_2$ such that the composition $\phi_2 \circ F \circ \phi_1^{-1}$ is holomorphic at $\phi_1(p)$. If F is defined on an open set $W \subset X$, then we say F is *holomorphic on* W if F is holomorphic at each point of W. In particular, F is a *holomorphic map* if and only if F is holomorphic on all of X.

EXAMPLE 3.2. The identity mapping id : $X \to X$ is holomorphic for any Riemann surface X.

As is the case with holomorphic functions, one can check the holomorphicity of a map with *any* pair of charts. Specifically, we have the following.

LEMMA 3.3. *Let $F : X \to Y$ be a mapping between Riemann surfaces.*
   a. *F is holomorphic at p if and only if for any pair of charts $\phi_1 : U_1 \to V_1$ on X with $p \in U_1$ and $\phi_2 : U_2 \to V_2$ on Y with $F(p) \in U_2$, the composition $\phi_2 \circ F \circ \phi_1^{-1}$ is holomorphic at $\phi_1(p)$.*
   b. *F is holomorphic on W if and only if there are two collections of charts $\{\phi_1^{(i)} : U_1^{(i)} \to V_1^{(i)}\}$ on X with $W \subset \bigcup_i U_1^{(i)}$ and $\{\phi_2^{(j)} : U_2^{(j)} \to V_2^{(j)}\}$ on Y with $F(W) \subset \bigcup_j U_2^{(j)}$ such that $\phi_2^{(j)} \circ F \circ {\phi_1^{(i)}}^{-1}$ is holomorphic for every i and j where it is defined.*

EXAMPLE 3.4. If Y is the complex plane $\mathbb{C}$, then a holomorphic map $F : X \to Y$ is simply a holomorphic function on X.

Holomorphic maps behave quite well with respect to composition. We leave the following to the reader.

LEMMA 3.5.
   a. *If F is holomorphic, then F is continuous and C^∞.*
   b. *The composition of holomorphic maps is holomorphic: if $F : X \to Y$ and $G : Y \to Z$ are holomorphic maps, then $G \circ F : X \to Z$ is a holomorphic map.*
   c. *The composition of a holomorphic map with a holomorphic function is holomorphic: if $F : X \to Y$ is holomorphic and g is a holomorphic function on an open set $W \subset Y$, then $g \circ F$ is a holomorphic function on $F^{-1}(W)$.*
   d. *The composition of a holomorphic map with a meromorphic function is meromorphic: if $F : X \to Y$ is holomorphic and g is a meromorphic function on an open set $W \subset Y$, then $g \circ F$ is a meromorphic function on $F^{-1}(W)$. (There is one mild proviso here: the image $F(X)$ must not be a subset of the set of poles of g.)*

The second property, along with Example 3.2, insures that Riemann surfaces, with holomorphic mappings, form a category.

The last properties above are often expressed as follows. Let $F : X \to Y$ be a holomorphic map between Riemann surfaces. Then for every open set $W \subset Y$, F induces a $\mathbb{C}$-algebra homomorphism

$$F^* : \mathcal{O}_Y(W) \to \mathcal{O}_X(F^{-1}(W))$$

defined by composition with F: $F^*(g) = g \circ F$. We have the same notion for meromorphic functions, and the map is also called F^*:

$$F^* : \mathcal{M}_Y(W) \to \mathcal{M}_X(F^{-1}(W))$$

is again defined as composition with F, if F is not constant.

If $F : X \to Y$ and $G : Y \to Z$ are holomorphic maps, then it is trivial that $F^* \circ G^* = (G \circ F)^*$.

Isomorphisms and Automorphisms. When are two Riemann surfaces to be considered the same? The answer is of course the natural one.

DEFINITION 3.6. An *isomorphism* (or *biholomorphism*) between Riemann surfaces is a holomorphic map $F : X \to Y$ which is bijective, and whose inverse $F^{-1} : Y \to X$ is holomorphic. A self-isomorphism $F : X \to X$ is called an *automorphism* of X. If there exists an isomorphism between X and Y, we say that X and Y are *isomorphic* (or *biholomorphic*).

LEMMA 3.7. *The Riemann Sphere $\mathbb{C}_\infty$ and the projective line $\mathbb{P}^1$ are isomorphic.*

PROOF. The function from $\mathbb{P}^1$ to the Riemann Sphere sending $[z : w]$ to

$$(2\operatorname{Re}(z\bar{w}), 2\operatorname{Im}(z\bar{w}), |z|^2 - |w|^2)/(|z|^2 + |w|^2) \in S^2$$

is an isomorphism onto $\mathbb{C}_\infty$. $\square$

Easy Theorems about Holomorphic Maps. Several theorems concerning holomorphic maps are immediate consequences of the corresponding theorems concerning holomorphic functions. We collect some of them here.

The first is the Open Mapping Theorem for holomorphic maps.

PROPOSITION 3.8 (OPEN MAPPING THEOREM). *Let $F : X \to Y$ be a nonconstant holomorphic map between Riemann surfaces. Then F is an open mapping.*

Next is the fact that the inverse of a holomorphic map is automatically holomorphic.

PROPOSITION 3.9. *Let $F : X \to Y$ be a 1-1 holomorphic map between Riemann surfaces. Then F is an isomorphism between X and its image $F(X)$.*

We have the analogue of the Identity Theorem.

PROPOSITION 3.10 (IDENTITY THEOREM). *Let F and G be two holomorphic maps between Riemann surfaces X and Y. If $F = G$ on a subset S of X with a limit point in X, then $F = G$.*

The next proposition has no analogue in the theory of holomorphic functions, since it deals with holomorphic maps with compact domain.

PROPOSITION 3.11. *Let X be a compact Riemann surface, and let $F : X \to Y$ be a nonconstant holomorphic map. Then Y is compact and F is onto.*

PROOF. Since F is holomorphic and X is open in itself, $F(X)$ is open in Y by the open mapping theorem. On the other hand, since X is compact, $F(X)$ is compact; since Y is Hausdorff, $F(X)$ must be closed in Y. Hence $F(X)$ is both open and closed in Y, and since Y is connected, it must be all of Y. Thus F is onto, and Y is compact. □

PROPOSITION 3.12 (DISCRETENESS OF PREIMAGES). *Let $F : X \to Y$ be a nonconstant holomorphic map between Riemann surfaces. Then for every $y \in Y$, the preimage $F^{-1}(y)$ is a discrete subset of X. In particular, if X and Y are compact, then $F^{-1}(y)$ is a nonempty finite set for every $y \in Y$.*

PROOF. Fix a local coordinate z centered at $y \in Y$, and for a point $x \in F^{-1}(y)$ choose a local coordinate w centered at x. Then the map F, written in terms of these local coordinates, is a nonconstant holomorphic function $z = g(w)$; moreover g has a zero at the origin, since x (which is $w = 0$) goes to y (which is $z = 0$). Since zeroes of nonconstant holomorphic functions are discrete, we see that, in some neighborhood of x, x is the only preimage of y. This proves that $F^{-1}(y)$ is a discrete subset of X. The second statement follows since F must be onto (Proposition 3.11) and discrete subsets of compact spaces are finite. □

Meromorphic Functions and Holomorphic Maps to the Riemann Sphere. We have noted above in Example 3.4 that any holomorphic function f on a Riemann surface X can be viewed as a holomorphic map to the complex plane $\mathbb{C}$. A similar construction may be made for meromorphic maps.

Let f be a meromorphic map on X. The values which f can take are complex numbers, away from the poles of f. At a pole of f, the natural "value" is ∞. To make this precise, we define a function $F : X \to \mathbb{C}_\infty$ by

$$F(x) = \begin{cases} f(x) \in \mathbb{C} & \text{if } x \text{ is not a pole of } f \\ \infty & \text{if } x \text{ is a pole of } f. \end{cases}$$

It is easy to see that this mapping F is a holomorphic map. Moreover, we have the following correspondence, which we leave to the reader to verify.

PROPOSITION 3.13. *The above construction induces a 1-1 correspondence between*

$$\left\{ \begin{array}{c} \text{meromorphic functions } f \\ \text{on } X \end{array} \right\} \text{ and } \left\{ \begin{array}{c} \text{holomorphic maps} \\ F : X \to \mathbb{C}_\infty \\ \text{which are not identically } \infty \end{array} \right\}.$$

Of course, the constant functions correspond to the constant maps.

Since $\mathbb{C}_\infty$ is isomorphic to $\mathbb{P}^1$, there is of course a correspondence between meromorphic functions and maps to $\mathbb{P}^1$; it should be clear what the precise statement is. Let us simply write down the formula.

Suppose that f is a meromorphic function on X, and consider a point $p \in X$. In a neighborhood of p, f may be written as the ratio of two holomorphic functions $f = g/h$. The corresponding map to $\mathbb{P}^1$, in this neighborhood of p, sends a point x to $[g(x) : h(x)]$.

A meromorphic function cannot be *globally* written as a ratio of holomorphic functions in general, so this representation for the map to $\mathbb{P}^1$ is generally possible only locally, in a neighborhood of each point. However we do see that any holomorphic map to $\mathbb{P}^1$ can be locally written in this form: x goes to $[g(x) : h(x)]$, where g and h are holomorphic functions.

This correspondence between meromorphic functions and holomorphic maps to the Riemann Sphere makes it possible to make geometric arguments (namely, arguments about maps between Riemann surfaces) in order to draw conclusions about holomorphic and meromorphic functions. This is in fact a central tool in the theory.

Meromorphic Functions on a Complex Torus, Again. Let us give an example of the kind of arguments which are possible by exploiting the correspondence between meromorphic functions and holomorphic maps to the Riemann Sphere. In particular, let us prove the analogue to Corollary 2.6 for meromorphic functions on a complex torus.

LEMMA 3.14. *Let f be any nonconstant meromorphic function on a complex torus $X = \mathbb{C}/L$. Then*
$$\sum_p \operatorname{ord}_p(f) = 0.$$

PROOF. We'll give a proof in the case that $L = \mathbb{Z} + \mathbb{Z}\tau$, with $\operatorname{Im}(\tau) > 0$; this is in fact the general case (see Problem K below).

Let f be a meromorphic function on X. The statement says that, counting via order, f has exactly as many zeroes as poles. Suppose this is false; by replacing f by $1/f$ if necessary we may assume that f has more poles than zeroes. Let $p_1, \ldots, p_n$ be the zeroes of f, and let $q_1, \ldots, q_m$ be the poles of f, with $n < m$; repetitions are allowed in these lists if the zero or pole is of order higher than one.

Add $p_{n+1}, \ldots, p_m$ to the list of zeroes in an arbitrary way, with the only condition that $\sum p_i = \sum q_i$ in the quotient group law of X. Lift each p_i to $x_i \in \mathbb{C}$ and each q_j to $y_j \in \mathbb{C}$, in such a way that $\sum x_i = \sum y_i$ in $\mathbb{C}$. (See Problem II.2,G.) Form the ratio of translated theta-functions as in Proposition 2.7: $R(z) = \prod_i \theta^{(x_i)}(z) / \prod_j \theta^{(y_j)}(z)$. Since R is meromorphic and L-periodic, we may consider R as a meromorphic function on $X = \mathbb{C}/L$. As such, it has zeroes exactly at the p_i's and poles at the q_i's, for every $i = 1, \ldots, m$.

Therefore the ratio $g = R/f$ has no poles, with only zeroes at the points $p_{n+1}, \ldots, p_m$. But X is compact; therefore g is constant. Since g has zeroes, it must be identically zero, which is nonsense since R is not.

This contradiction proves the lemma. □

We will be able to give a proof of the above statement for an arbitrary compact Riemann surface shortly, along slightly different lines.

Problems II.3

A. Verify Example 3.4: if Y is the complex plane $\mathbb{C}$, prove that a holomorphic map $F : X \to Y$ is simply a holomorphic function on X.
B. Prove all the statements of Lemma 3.5.
C. Show that under the isomorphism between $\mathbb{P}^1$ and the Riemann Sphere $\mathbb{C}_\infty$, the points $[z : 1]$ are sent to the finite points z, and the point $[1 : 0]$ is sent to ∞.
D. Explicitly write down the inverse holomorphic map to the isomorphism from $\mathbb{P}^1$ to $\mathbb{C}_\infty$ given in the proof of Lemma 3.7. Check everything necessary.
E. Let $\pi : \mathbb{C} \to X = \mathbb{C}/L$ be the natural projection map defining a complex torus X. Let Y be a Riemann surface. Show that a map $F : X \to Y$ is holomorphic if and only if $F \circ \pi : \mathbb{C} \to Y$ is holomorphic. Deduce that the projection map π is a holomorphic map.
F. Let $f(z,w)$ and $g(z,w)$ be homogeneous polynomials of the same degree with no common factor, and not both identically zero. Show that the map $F : \mathbb{P}^1 \to \mathbb{P}^1$ defined by sending $[z : w]$ to $[f(z,w) : g(z,w)]$ is well defined and holomorphic. What if f and g have a common factor?
G. Let $A = \begin{pmatrix} a & b \\ c & d \end{pmatrix}$ be an invertible 2-by-2 matrix over $\mathbb{C}$. Show that the map $F_A : \mathbb{P}^1 \to \mathbb{P}^1$ sending $[z : w]$ to $[az + bw : cz + dw]$ is an automorphism of $\mathbb{P}^1$. For which matrices A is F_A the identity? Show that $F_{AB} = F_A \circ F_B$.
H. Show that after identifying $\mathbb{P}^1$ with $\mathbb{C}_\infty$, the automorphism F_A defined above takes $z \in \mathbb{C}_\infty$ to $(az + b)/(cz + d)$; hence it is a linear fractional transformation.
I. Let X be a compact Riemann surface and f a nonconstant meromorphic function on X. Show that f must have a zero on X, and must have a pole on X.
J. Prove that, given a meromorphic function f on a Riemann surface X, the associated map $F : X \to \mathbb{C}_\infty$ is holomorphic. Verify the 1-1 correspondence of Proposition 3.13.
K. Recall that a *lattice* $L \subset \mathbb{C}$ is an additive subgroup generated (over $\mathbb{Z}$) by two complex numbers ω_1 and ω_2 which are linearly independent over $\mathbb{R}$. Thus $L = \{m\omega_1 + n\omega_2 \mid m, n \in \mathbb{Z}\}$.
   1. Suppose that $L \subseteq L'$ are two lattices in $\mathbb{C}$. Show that the natural map from $\mathbb{C}/L$ to $\mathbb{C}/L'$ is holomorphic, and is biholomorphic if and only if $L = L'$.
   2. Let L be a lattice in $\mathbb{C}$ and let α be a nonzero complex number. Show that αL is a lattice in $\mathbb{C}$ and that the map
   $$\phi : \mathbb{C}/L \to \mathbb{C}/(\alpha L)$$

sending the coset $z + L$ to $(\alpha z) + (\alpha L)$ is a well defined biholomorphic map.

3. Show that every torus $\mathbb{C}/L$ is isomorphic to a torus which has the form $\mathbb{C}/(\mathbb{Z} + \mathbb{Z}\tau)$, where τ is a complex number with strictly positive imaginary part.

4. Global Properties of Holomorphic Maps

Local Normal Form and Multiplicity. It may seem strange to have the first part of a section on global properties dealing with a completely local concept. However, most global properties actually state that some function of local invariants is constant. This is the case in our situation, and so we must introduce the local invariant before proceeding.

A holomorphic map between two Riemann surfaces has a standard normal form in some local coordinates: essentially, every map looks like a power map. This we now present.

PROPOSITION 4.1 (LOCAL NORMAL FORM). *Let $F : X \to Y$ be a holomorphic map defined at $p \in X$, which is not constant. Then there is a unique integer $m \geq 1$ which satisfies the following property: for every chart $\phi_2 : U_2 \to V_2$ on Y centered at $F(p)$, there exists a chart $\phi_1 : U_1 \to V_1$ on X centered at p such that $\phi_2(F(\phi_1^{-1}(z))) = z^m$.*

PROOF. Fix a chart ϕ_2 on Y centered at $F(p)$, and choose any chart $\psi : U \to V$ on X centered at p. Then the Taylor series for the function $T(w) = \phi_2(F(\psi^{-1}(w)))$ must be of the form

$$T(w) = \sum_{i=m}^{\infty} c_i w^i$$

with $c_m \neq 0$, and $m \geq 1$ since $T(0) = 0$. Thus we have $T(w) = w^m S(w)$ where $S(w)$ is a holomorphic function at $w = 0$, and $S(0) \neq 0$. In this case there exists a function $R(w)$ holomorphic near 0 such that $R(w)^m = S(w)$, so that $T(w) = (wR(w))^m$. Let $\eta(w) = wR(w)$; since $\eta'(0) \neq 0$, we see that near 0 the function η is invertible (by the Implicit Function Theorem), and of course holomorphic. Hence the composition $\phi_1 = \eta \circ \psi$ is also a chart on X defined and centered near p. If we think of η as defining a new coordinate z (via $z = \eta(w)$), we see that z and w are related by $z = wR(w)$. Thus

$$\begin{aligned}
\phi_2(F(\phi_1^{-1}(z))) &= \phi_2(F(\psi^{-1}(\eta^{-1}(z)))) \\
&= T(\eta^{-1}(z)) \\
&= T(w) \\
&= (wR(w))^m \\
&= z^m.
\end{aligned}$$

4. GLOBAL PROPERTIES OF HOLOMORPHIC MAPS

The uniqueness of m comes from noticing that, if there are local coordinates at p and $F(p)$ such that the map F has the form $z \mapsto z^m$, then near p, there are exactly m preimages of points near $F(p)$. Thus this exponent m can be detected solely by studying the topological properties of the map F near p, and is therefore independent of the choices of the local coordinates. $\square$

DEFINITION 4.2. The *multiplicity* of F at p, denoted $\mathrm{mult}_p(F)$, is the unique integer m such that there are local coordinates near p and $F(p)$ with F having the form $z \mapsto z^m$.

EXAMPLE 4.3. Let $\phi : U \to V$ be a chart map for X, considered as a holomorphic map to $\mathbb{C}$. Then ϕ has multiplicity one at every point of U.

Note that $\mathrm{mult}_p(F) \geq 1$ always. There is a simple way to compute the multiplicity without having to find local coordinates which put the map F into local normal form, or even to have local coordinates which are centered at the point in question and its image. Take any local coordinates z near p and w near $F(p)$; say that p corresponds to z_0 and $F(p)$ to w_0. In terms of these coordinates, the map F may be written as $w = h(z)$ where h is holomorphic. Then of course $w_0 = h(z_0)$.

LEMMA 4.4. *With the above notation, the multiplicity* $\mathrm{mult}_p(F)$ *of F at p is one more than the order of vanishing of the derivative $h'(z_0)$ of h at z_0:*

$$\mathrm{mult}_p(F) = 1 + \mathrm{ord}_{z_0}(dh/dz).$$

In particular, the multiplicity is the exponent of lowest strictly positive term of the power series for h: if $h(z) = h(z_0) + \sum_{i=m}^{\infty} c_i(z-z_0)^i$ with $m \geq 1$ and $c_m \neq 0$, then $\mathrm{mult}_p(F) = m$.

PROOF. We saw in the proof of the Local Normal Form Proposition 4.1 that the multiplicity was the lowest term appearing in the power series T for F when centered local coordinates are used at p and at the image point $F(p)$. With the above notation, $z - z_0$ and $w - w_0$ are such centered local coordinates; therefore since $w - w_0 = h(z) - h(z_0)$, we see that the multiplicity is the lowest term appearing in the power series expansion for $h(z) - h(z_0)$ about $z = z_0$. By Taylor's Theorem, this is one more than the order of the derivative of h at z_0, as stated. $\square$

The above lemma shows that the points of the domain where F has multiplicity at least two form a discrete set. Indeed, such points correspond to zeroes of the derivative of a local formula h for F, and since h is holomorphic, the zeroes of its derivative are discrete. One can also check this by the local normal form.

DEFINITION 4.5. Let $F : X \to Y$ be a nonconstant holomorphic map. A point $p \in X$ is a *ramification point* for F if $\mathrm{mult}_p(F) \geq 2$. A point $y \in Y$ is a *branch point* for F if it is the image of a ramification point for F.

Thus the ramification points and branch points for a holomorphic map form discrete subsets of the domain and range respectively.

Let us do an example concerning smooth plane curves. Suppose that X is a smooth affine plane curve defined by $f(x,y) = 0$. Define $\pi : X \to \mathbb{C}$ by projection onto the x-axis: $\pi(x,y) = x$. We claim that π is ramified at $p = (x_0, y_0) \in X$ if and only if $(\partial f / \partial y)(p) = 0$.

Suppose first that $(\partial f / \partial y)(p) \neq 0$. Then π is a chart map for X near p, and so certainly has multiplicity one.

Conversely, suppose that $(\partial f / \partial y)(p) = 0$. Then since X is smooth at p, we must have $(\partial f / \partial x)(p) \neq 0$, and so the function y is a chart map for X near p. By the Implicit Function Theorem, near p, X is locally the graph of a holomorphic function $g(y)$. Hence $f(g(y), y)$ is identically zero in a neighborhood of y_0. Taking the derivative with respect to y, we see that $(\partial f / \partial x) g'(y) + (\partial f / \partial y)$ is identically zero near p. By assumption the second term is zero at p, and so since $(\partial f / \partial x)(p) \neq 0$, we must have $g'(y_0) = 0$.

But $g(y)$ is exactly the local formula for the map π. Hence by the derivative criterion Lemma 4.4, π is ramified at p.

The same remark holds for a smooth projective plane curve X. Suppose that X is defined by a homogeneous polynomial $F(x, y, z) = 0$. Consider the map $G : X \to \mathbb{P}^1$ defined by projection to the $y = 0$ line: $G[x : y : z] = [x : z]$. Then G is ramified at $p \in X$ if and only if $(\partial F / \partial y)(p) = 0$. This follows directly from the above analysis, only noting that locally, X is the affine plane curve defined by $f(x, y) = F(x, y, 1) = 0$. (One has to check the chart where $x = 1$ also; one gets the same answer.)

These statements will be useful enough to collect them below:

LEMMA 4.6. *Let X be a smooth affine plane curve defined by $f(x, y) = 0$. Define $\pi : X \to \mathbb{C}$ by $\pi(x, y) = x$. Then π is ramified at $p \in X$ if and only if $(\partial f / \partial y)(p) = 0$.*

Let X be a smooth projective plane curve defined by a homogeneous polynomial $F(x, y, z) = 0$; consider the map $G : X \to \mathbb{P}^1$ defined by $G[x : y : z] = [x : z]$. Then G is ramified at $p \in X$ if and only if $(\partial F / \partial y)(p) = 0$.

Finally let us remark on a relationship between the multiplicity (which is defined for a holomorphic map between Riemann surfaces) and the order (which is defined for a meromorphic function). There ought to be some relationship, by the correspondence given in Proposition 3.13. Let f be a meromorphic function on a Riemann surface X, and let $F : X \to \mathbb{C}_\infty$ be the associated holomorphic map to the Riemann Sphere.

Suppose $p \in X$ is not a pole of f; let $z_0 = f(p)$. Then the function $f - z_0$ has a zero at p, and by Lemma 4.4, we see that $\text{mult}_p(F) = \text{ord}_p(f - f(p))$.

Suppose that p is a pole of f; then the order of f at p is negative, and p is a zero of $1/f$; we obviously have $\text{mult}_p(F) = -\text{ord}_p(f)$ in this case. Let us collect these remarks in the following.

LEMMA 4.7. *Let f be a meromorphic function on a Riemann surface X, with associated holomorphic map $F : X \to \mathbb{C}_\infty$.*
  a. *If $p \in X$ is a zero of f, then $\mathrm{mult}_p(F) = \mathrm{ord}_p(f)$.*
  b. *If p is a pole of f, then $\mathrm{mult}_p(F) = -\mathrm{ord}_p(f)$.*
  c. *If p is neither a zero nor a pole of f, then $\mathrm{mult}_p(F) = \mathrm{ord}_p(f - f(p))$.*

The Degree of a Holomorphic Map between Compact Riemann Surfaces. Holomorphic maps between compact Riemann surfaces exhibit several beautiful properties, of which the most important is the following.

PROPOSITION 4.8. *Let $F : X \to Y$ be a nonconstant holomorphic map between compact Riemann surfaces. For each $y \in Y$, define $d_y(F)$ to be the sum of the multiplicities of F at the points of X mapping to y:*

$$d_y(F) = \sum_{p \in F^{-1}(y)} \mathrm{mult}_p(F).$$

Then $d_y(F)$ is constant, independent of y.

PROOF. The idea of the proof is to show that the function $y \mapsto d_y(F)$ is a locally constant function from Y to the integers $\mathbb{Z}$. Since Y is connected, a locally constant function must be constant, and we will be done.

Before proceeding, consider the open unit disc $D = \{z \in \mathbb{C} \mid \|z\| < 1\}$ and the map $f : D \to D$ given by $f(z) = z^m$ for some integer $m \geq 1$. This map f is of course holomorphic and onto; the only ramification point for f is at $z = 0$, where the multiplicity is m. All other points have multiplicity one. For any $w \in D$, if $w \neq 0$ there are exactly m preimages (the m m^{th} roots of w), each of multiplicity one; if $w = 0$, the only preimage is $z = 0$, which has multiplicity m. Therefore this local normal form map f satisfies the constancy condition above: the sum of the multiplicities of the preimage points is constantly m.

Clearly if one has a disjoint union of such maps, that is, a map from the disjoint union of several such disks to D (each possibly with a different power m), the constancy condition is still satisfied. Our goal is then to show that for any holomorphic nonconstant map F as in the Proposition, F is locally (above a neighborhood of any point y in the target) exactly a disjoint union of these power maps.

Fix then a point $y \in Y$, and let $\{x_1, \ldots, x_n\}$ be the inverse image of y under F. Choose a local coordinate w on Y centered at y. By the Local Normal Form Proposition 4.1, we may choose coordinates $\{z_i\}$ on X, with z_i centered at x_i for each $i = 1, \ldots, n$, such that in a neighborhood of x_i the map F sends z_i to $w = z_i^{m_i}$. Therefore, if we look at these neighborhoods of the x_i, we have exactly the desired disjoint union description of F.

What is left to prove is that, near y, there are no other preimages left unaccounted for which are not in the neighborhoods of the x_i's. This is where we use the compactness of X.

Suppose that, arbitrarily close to y, there are preimages which are not in any of the above neighborhoods of the x_i's. With this assumption we may find a sequence of points of X, none of which lie in any of the neighborhoods of the x_i's, such that the images of these points under F converge to $y \in Y$. Since X is compact, we may extract a convergent subsequence, say $\{p_n\}$; this sequence then has the property that it converges (say to a point $x \in X$) and the sequence of images $F(p_n)$ converges to y. Since F is continuous, the limit point x must lie over y: $F(x) = y$. Hence by assumption x is one of the x_i's; and so we obtain a contradiction, since none of the p_n's lie in the neighborhoods of the x_i's.

This proves that there are no other unaccounted preimages in a neighborhood of y, and finishes the proof. □

The above Proposition motivates the following definition.

DEFINITION 4.9. Let $F : X \to Y$ be a nonconstant holomorphic map between compact Riemann surfaces. The *degree* of F, denoted $\deg(F)$, is the integer $d_y(F)$ for any $y \in Y$.

This is another example of how "counting properly" gives a nice formula. Here we are counting preimages for holomorphic maps, and Proposition 4.8 says that if we count with multiplicity, the number of preimages is constant, equal to the degree of the map (by definition).

Note that when F has degree one, then it is 1-1. Therefore we obtain the following immediately from Proposition 3.9:

COROLLARY 4.10. *A holomorphic map between compact Riemann surfaces is an isomorphism if and only if it has degree one.*

For example, suppose that X is a compact Riemann surface, p is a point of X, and f is a meromorphic function on X with a simple pole at p and no other poles. Then the corresponding map $F : X \to \mathbb{C}_\infty$ has multiplicity one at p, and p is the only point mapping to ∞; hence F has degree one, and is, by the above Corollary, an isomorphism. Therefore we have shown the following simple but useful fact:

PROPOSITION 4.11. *If X is a compact Riemann surface having a meromorphic function f with a single simple pole, then X is isomorphic to $\mathbb{C}_\infty$.*

Suppose that $F : X \to Y$ is a nonconstant holomorphic map between compact Riemann surfaces. If we delete the branch points (in Y) of F, and all of their preimages (in X), we obtain a map $F : U \to V$ between 2-manifolds which is a *covering map* in the sense of topology: every point of the target V has an open neighborhood $N \subset V$ such that the inverse image of N under F breaks into a disjoint union of open sets $M_i \subset U$ with the map F sending each M_i homeomorphically onto N.

4. GLOBAL PROPERTIES OF HOLOMORPHIC MAPS

Because of this, a map F as above is sometimes called a *branched covering*. It is a covering map away from finitely many points (the branch points), and over these branch points the map has a very controlled behaviour.

The Sum of the Orders of a Meromorphic Function. We are now in a position to prove the general statement concerning the sum of the orders of a nonconstant meromorphic function on a compact Riemann surface. This generalizes what we have already seen for functions on the Riemann Sphere (Corollary 2.6) and functions on a complex torus (Lemma 3.14). In these cases, the proofs were based on the ready availability of a wealth of meromorphic functions (rational functions in the case of the Riemann Sphere, and ratios of translated theta-functions in the case of a complex torus).

Now we can give a proof in general based on the theory of the degree.

PROPOSITION 4.12. *Let f be a nonconstant meromorphic function on a compact Riemann surface X. Then*

$$\sum_p \operatorname{ord}_p(f) = 0.$$

PROOF. Let $F : X \to \mathbb{C}_\infty$ be the associated holomorphic map to the Riemann Sphere. Let $\{x_i\}$ be the points of X mapping to 0, and let $\{y_j\}$ be the points of X mapping to ∞; the x_i's are exactly the zeroes of f, and the y_j's are its poles. Let d be the degree of the mapping F.

By the definition of the degree, we have that

$$d = \sum_i \operatorname{mult}_{x_i}(F) \quad \text{and} \quad d = \sum_j \operatorname{mult}_{y_j}(F).$$

Now the only points of X where f has nonzero order are at its zeroes and poles, which are these points $\{x_i\}$ and $\{y_j\}$. By Lemma 4.7, we have that

$$\operatorname{mult}_{x_i}(F) = \operatorname{ord}_{x_i}(f) \text{ and } \operatorname{mult}_{y_j}(F) = -\operatorname{ord}_{y_j}(f).$$

Hence

$$\begin{aligned}
\sum_p \operatorname{ord}_p(f) &= \sum_i \operatorname{ord}_{x_i}(f) + \sum_j \operatorname{ord}_{y_j}(f) \\
&= \sum_i \operatorname{mult}_{x_i}(F) - \sum_j \operatorname{mult}_{y_j}(F) \\
&= 0
\end{aligned}$$

since both sums are equal to the degree d. □

Meromorphic Functions on a Complex Torus, Yet Again. As an application of the degree theory for a holomorphic map, we can now characterize all meromorphic functions on a complex torus.

PROPOSITION 4.13. *Any meromorphic function on a complex torus is given by a ratio of translated theta-functions.*

PROOF. Let X be a complex torus, and f a nonconstant meromorphic function on X. We have seen that f has as many zeroes as poles (Lemma 3.14); let $\{p_i\}$, $i = 1, \ldots, n$ be the zeroes and $\{q_i\}$, $i = 1, \ldots, n$ be the poles of f, with repetitions allowed for zeroes and poles of higher order.

We will show below that in fact $\sum_i p_i = \sum_i q_i$ in the quotient group law of X. If this is true, then we may finish the argument by lifting each p_i to $x_i \in \mathbb{C}$ and each q_i to $y_i \in \mathbb{C}$ with $\sum_i x_i = \sum_i y_i$ (see Problem II.2,G). Then the ratio of translated theta-functions $R(z) = \prod_i \theta^{(x_i)}(z) / \prod_j \theta^{(y_j)}(z)$ has the same zeroes and poles as f does, to the same orders. Hence f/R has no zeroes or poles, so must be constant since X is compact.

Now suppose that $\sum_i p_i \neq \sum_i q_i$ in the quotient group law of X. Choose points p_0 and q_0 in X such that $\sum_{i=0}^n p_i = \sum_{i=0}^n q_i$ in X. Form the ratio of translated theta-functions $R(z) = \prod_{i=0}^n \theta^{(x_i)}(z) / \prod_{j=0}^n \theta^{(y_j)}(z)$ as above, and consider the meromorphic function $g = R/f$ on X. Note that g has exactly one zero (at p_0) and one pole (at q_0), both of order one, since all other zeroes and poles are cancelled away.

Let $G : X \to \mathbb{C}_\infty$ be the holomorphic map to the Riemann Sphere which corresponds to the meromorphic function g. Since g has a single simple zero, and a single simple pole, we see that as a holomorphic map, G has degree one. Hence G is an isomorphism, by Corollary 4.10. But X has genus one and the Riemann Sphere has genus zero; there certainly can be no isomorphism between them.

This contradiction shows that we must have had $\sum_i p_i = \sum_i q_i$ in the quotient group law of X after all, completing the proof. □

The Euler Number of a Compact Surface. Let S be a compact 2-manifold (possibly with boundary). A *triangulation* of S is a decomposition of S into closed subsets, each homeomorphic to a triangle, such that any two triangles are either disjoint, meet only at a single vertex, or meet only along a single edge.

DEFINITION 4.14. Let S be a compact 2-manifold, possibly with boundary. Suppose a triangulation of S is given, with v vertices, e edges, and t triangles. The *Euler number* of S (with respect to this triangulation) is the integer $e(S) = v - e + t$.

Please forgive the double use of the notation e; no confusion will arise if the reader is awake.

4. GLOBAL PROPERTIES OF HOLOMORPHIC MAPS

The main fact about Euler numbers is that they do not depend on the particular triangulation one uses to compute them. These ideas properly belong to a course in topology, and we will only sketch the proof of the following.

PROPOSITION 4.15. *The Euler number is independent of the choice of triangulation. For a compact orientable 2-manifold without boundary of topological genus g, the Euler number is $2 - 2g$.*

PROOF. [Sketch] First we must introduce the notion of a *refinement* of a triangulation. Suppose one has a triangulation of a surface; let T be one of the triangles. One obtains a "finer" triangulation by adding a vertex somewhere in the interior of T, and adding three edges from that new vertex to the three original vertices of T. This essentially replaces T with three triangles, adding a net of one vertex, three edges, and two triangles. Note that the Euler number is unchanged by this operation.

Another way to refine the triangulation is to take two neighboring triangles which meet along a common edge E. Then one adds a vertex somewhere in the interior of E, and two edges to each of the opposite vertices of the two triangles. This essentially bisects each of the two triangles, adding a net of one vertex, three edges, and two triangles again.

If the 2-manifold has a boundary, one may simply bisect a single triangle along an edge which forms part of the boundary. This adds a net of one vertex, two edges, and one triangle.

These three operations are called *elementary refinements*; note that none of them change the Euler number. A general refinement is obtained by making a sequence of elementary refinements. Therefore a triangulation and any refinement give the same Euler number for the surface.

Now comes the main theorem concerning triangulations: any two triangulations of a compact 2-manifold (even with boundary) have a common refinement. (To see this, simply superimpose both triangulations on the surface, then add lots of vertices and edges to make the union a triangulation; finally note that doing this is a refinement of either one.)

This is now enough to show that the Euler number is well defined: since the Euler number is constant under refinement, and any two triangulations have a common refinement, we see that any two triangulations give the same Euler number.

Now make a specific computation with any triangulation you wish, and discover that a sphere has Euler number 2; this is the genus zero case. Also check that a cylinder has Euler number 0, and that a closed disk has Euler number 1.

Now to increase the genus of a surface by one, one removes two disks, and attaches a cylinder along the two bounding circles. Removing the two disks drops the Euler number by one each, so by two total; adding the cylinders does not change the Euler number. Therefore the Euler number decreases by two if the genus increases by one.

Since the Euler number of a sphere (with genus zero) is 2, by induction we see that a surface of genus g has Euler number $2 - 2g$. $\square$

Note that we have swept under the rug an important theorem concerning compact Riemann surfaces: they can all be triangulated. This we leave to the reader to either look up or prove by induction.

Hurwitz's Formula. The constancy of the degree for a holomorphic map between compact Riemann surfaces, combined with the theory of the Euler number, gives an important formula relating the genera of the domain and range with the degree and ramification of the map. This is known as Hurwitz's formula, and its use is ubiquitous in the theory of compact Riemann surfaces.

THEOREM 4.16 (HURWITZ'S FORMULA). *Let $F : X \to Y$ be a nonconstant holomorphic map between compact Riemann surfaces. Then*

$$2g(X) - 2 = \deg(F)(2g(Y) - 2) + \sum_{p \in X} [\mathrm{mult}_p(F) - 1].$$

PROOF. Note that since X is compact, the set of ramification points is finite, so that the sum (which may be restricted to the ramification points of F) is a finite sum.

Take a triangulation of Y, such that each branch point of F is a vertex. Assume there are v vertices, e edges, and t triangles. Lift this triangulation to X via the map F, and assume there are v' vertices, e' edges, and t' triangles on X. Note that every ramification point of F is a vertex on X.

Since there are no ramification points over the general point of any triangle, each triangle of Y lifts to $\deg(F)$ triangles in X. Thus $t' = \deg(F)t$. Similarly $e' = \deg(F)e$. Now fix a vertex $q \in Y$. The number of preimages of q in X is $|F^{-1}(q)|$, which we can rewrite as

$$\begin{aligned}
|F^{-1}(q)| &= \sum_{p \in F^{-1}(q)} 1 \\
&= \deg(F) + \sum_{p \in F^{-1}(q)} [1 - \mathrm{mult}_p(F)].
\end{aligned}$$

Therefore the total number of preimages of vertices of Y, which is the number v' of vertices of X, is

$$\begin{aligned}
v' &= \sum_{\text{vertex } q \text{ of } Y} \left(\deg(F) + \sum_{p \in F^{-1}(q)} [1 - \mathrm{mult}_p(F)] \right) \\
&= \deg(F)v - \sum_{\text{vertex } q \text{ of } Y} \sum_{p \in F^{-1}(q)} [\mathrm{mult}_p(F) - 1] \\
&= \deg(F)v - \sum_{\text{vertex } p \text{ of } X} [\mathrm{mult}_p(F) - 1].
\end{aligned}$$

Therefore

$$\begin{aligned}
2g(X) - 2 &= -e(X) \\
&= -v' + e' - t' \\
&= -\deg(F)v + \sum_{\text{vertex } p \text{ of } X} [\text{mult}_p(F) - 1] + \deg(F)e - \deg(F)t \\
&= -\deg(F)e(Y) + \sum_{\text{vertex } p \text{ of } X} [\text{mult}_p(F) - 1] \\
&= \deg(F)(2g(Y) - 2) + \sum_{p \in X} [\text{mult}_p(F) - 1],
\end{aligned}$$

the last equality holding because every ramification point of F is a vertex of X. □

We may view this proof as resolving two different ways of computing preimages. If we "count properly", we take into account the ramification of the map and all of the multiplicities. If we count "naively", we get a computation of the Euler number. Putting these two things together gives Hurwitz's formula.

Problems II.4

A. Verify the statement in Example 4.3 that chart maps have constant multiplicity one. Is the converse true? (I.e., is every holomorphic map from an open set in X to an open set in $\mathbb{C}$ with constant multiplicity one, a chart map?)

B. Let F be a holomorphic map between Riemann surfaces. Prove that the set of points p with $\text{mult}_p(F) \geq 2$ forms a discrete subset of the domain by using the Local Normal Form.

C. Let $F : X \to Y$ and $G : Y \to Z$ be two nonconstant holomorphic maps between Riemann surfaces. Show that if $p \in X$, then $\text{mult}_p(G \circ F) = \text{mult}_p(F) \text{mult}_{F(p)}(G)$. Show that if f is a meromorphic function on Y, then $\text{ord}_p(f \circ F) = \text{mult}_p(F) \text{ord}_{F(p)}(f)$.

D. Explicitly triangulate the sphere, the disk, and the cylinder and verify that they have Euler numbers 2, 1, and 0 respectively.

E. Show that if f is a holomorphic function at p, and $\text{mult}_p(f) = 1$ (considering f as a holomorphic map locally to $\mathbb{C}$), then f is a local coordinate function at p.

F. Let f be a global meromorphic function on a compact Riemann surface X. Show that f is a local coordinate at all but finitely many points of X.

G. Let $f(z) = z^3/(1-z^2)$, considered as a meromorphic function on the Riemann Sphere $\mathbb{C}_\infty$. Find all points p such that $\text{ord}_p(f) \neq 0$. Consider the associated map $F : \mathbb{C}_\infty \to \mathbb{C}_\infty$. Show that F has degree 3 as a holomorphic map, and find all of its ramification and branch points. Verify Hurwitz's formula for this map F.

H. Let $f(z) = 4z^2(z-1)^2/(2z-1)^2$, considered as a meromorphic function on the Riemann Sphere $\mathbb{C}_\infty$. Find all points p such that $\text{ord}_p(f) \neq 0$. Consider the associated map $F : \mathbb{C}_\infty \to \mathbb{C}_\infty$. Show that F has degree 4 as a holomorphic map, and find all of its ramification and branch points. Verify Hurwitz's formula for this map F.

I. Let $F : X \to Y$ be a nonconstant holomorphic map between compact Riemann surfaces.
   1. Show that if $Y \cong \mathbb{P}^1$, and F has degree at least two, then F must be ramified.
   2. Show that if X and Y both have genus one, then F is unramified.
   3. Show that $g(Y) \leq g(X)$ always.
   4. Show that if $g(Y) = g(X) \geq 2$, then F is an isomorphism.

J. Let X be the projective plane curve of degree d defined by the homogeneous polynomial $F(x,y,z) = x^d + y^d + z^d$. This curve is called the *Fermat curve of degree d*. Let $\pi : X \to \mathbb{P}^1$ be given by $\pi[x : y : z] = [x : y]$.
   1. Check that the Fermat curve is smooth.
   2. Show that π is a well defined holomorphic map of degree d.
   3. Find all ramification and branch points of π.
   4. Use Hurwitz's formula to compute the genus of the Fermat curve: you should get
   $$g(X) = \frac{(d-1)(d-2)}{2}.$$

K. Let U be the affine plane curve defined by $x^2 = 3 + 10t^4 + 3t^8$. Let V be the affine plane curve defined by $w^2 = z^6 - 1$. Show that both curves are smooth. Show that the function $F : U \to V$ defined by $z = (1+t^2)/(1-t^2)$ and $w = 2tx/(1-t^2)^3$ is holomorphic and nowhere ramified whenever $t \neq \pm 1$.

Further Reading

The basic material on singularities of complex functions is standard fare in all texts on complex variables; each of the texts mentioned at the end of Chapter I have plenty on this, and also sections on harmonic functions, which are sometimes given short shrift in a first course.

Many authors introduce meromorphic functions on a torus (also known as *elliptic functions*) via the Weierstrass P-function; this is the approach taken for example in [**Ahlfors66**], [**JS87**], [**Lang85**], and [**Lang87**]. We have taken the approach of theta-functions, to emphasize the analogy between ratios of theta-functions (on a torus) and ratios of homogeneous polynomials (on the projective line); this is also the approach of [**Clemens80**]. For (much) more depth on theta-functions, see [**R-F74**], [**Gunning76**], and [**Mumford83**].

We have mentioned Shafarevich's text [**Shafarevich77**] for the Nullstellensatz; there are many other references, many in texts in algebra, for example, [**Z-S60**], [**AM69**], [**Hungerford74**], and [**Lang84**]; students just starting out may find the treatment in [**Artin91**] less steep. The Nullstellensatz is at the heart of algebraic geometry, and it is mentioned in all elementary texts, but often not proved; see [**Fulton69**], [**Mumford76**], [**Hartshorne77**], [**Kendig77**], [**Reid88**], [**Harris92**], and [**C-L-O92**] for variety.

The Hurwitz Formula is a fundamental result, surely the centerpiece of this chapter. Another proof may be had without the topological arguments, using differential forms; with this approach the formula is a consequence of Riemann-Roch. It is sometimes referred to as the Riemann-Hurwitz Formula. The approach we have taken is similar to that in [**Reyssat89**], [**JS87**], and [**Kirwan92**], while [**Narasimhan92**] and [**Forster81**] take the differential forms route.

Chapter III. More Examples of Riemann Surfaces

1. More Elementary Examples of Riemann Surfaces

Lines and Conics. Special examples of smooth projective plane curves are the curves of low degree. Curves of degree one are *lines*, curves of degree two are *conics*, curves of degree three are *cubics*, etc.

Lines are relatively easy to understand:

LEMMA 1.1. *Any line in $\mathbb{P}^2$ is nonsingular and is isomorphic to $\mathbb{P}^1$.*

PROOF. Let $[x : y : z]$ be the homogeneous coordinates of $\mathbb{P}^2$. Then any line X is given by an equation $F(x, y, z) = ax + by + cz = 0$, where the coefficients a, b, c are not all zero. These coefficients are exactly the three partial derivatives of F, and therefore F is nonsingular.

To see that X is isomorphic to $\mathbb{P}^1$, we may assume that $a \neq 0$. Then an isomorphism from $\mathbb{P}^1$ to X is given by sending $[r : s]$ to $[-(br + cs)/a : r : s]$. □

Conics are more interesting, and of course have been one of the favorite objects of study for geometers for millenia. Conics are defined by quadratic equations of the form

$$F(x, y, z) = ax^2 + 2bxy + 2cxz + dy^2 + 2eyz + fz^2,$$

where a, b, c, d, e, f are complex constants, not all zero. We have inserted the factor of 2 in the coefficients so that we may write F conveniently in matrix form

$$F(x, y, z) = \begin{pmatrix} x & y & z \end{pmatrix} \begin{pmatrix} a & b & c \\ b & d & e \\ c & e & f \end{pmatrix} \begin{pmatrix} x \\ y \\ z \end{pmatrix} = V^\top A_F V,$$

where V is the column vector of variables. We see that F determines and is determined by a 3-by-3 symmetric matrix A_F.

LEMMA 1.2. *The quadratic polynomial F is nonsingular if and only if the matrix A_F is invertible.*

PROOF. The vector of 3 partial derivatives of F is exactly $2A_F V$. Hence for F to be singular there must exist a point in $\mathbb{P}^2$ (represented by a nonzero column vector V_0) such that $2A_F V_0 = 0$. This happens if and only if A_F is a singular matrix. □

Now suppose that T is a nonsingular 3-by-3 matrix. Let F_A be the quadratic equation defined by the symmetric matrix A. Note that $B = T^\top A T$ is also symmetric; let F_B be the quadratic equation defined by B.

LEMMA 1.3. *The map T, defined by sending V to TV, gives an isomorphism from the curve X_B defined by F_B to the curve X_A defined by F_A.*

PROOF. Clearly if the point V lies on X_B, so that $V^\top(T^\top A T)V = 0$, then $(TV)^\top A(TV) = 0$, so that the point TV lies on the curve X_A. Therefore T maps X_B to X_A, and by symmetry the inverse map T^{-1} maps X_A back onto X_B. One now checks that T is holomorphic to complete the proof; we leave this to the reader. □

Now we appeal to some linear algebra: over the complex numbers, any invertible symmetric matrix A may be factored as $A = T^\top T$ for some invertible T. We conclude the following:

COROLLARY 1.4. *Any smooth projective plane conic is isomorphic to the conic defined by the identity matrix, which is the conic given by $x^2 + y^2 + z^2 = 0$. In particular, any two smooth projective plane conics are isomorphic. Moreover, there is an isomorphism of the form $V \mapsto TV$, where T is an invertible 3-by-3 matrix.*

Thus to study conics up to isomorphism, we can pick any one and study it. It is convenient to pick the conic X defined by $F = xz - y^2$, which is nonsingular. Define a map $G : \mathbb{P}^1 \to X$ by sending $[r : s]$ to $[r^2 : rs : s^2]$; check that this is a holomorphic map. Moreover it is an isomorphism: the inverse map sends $[x : y : z]$ on X to the point $[x : y]$ in $\mathbb{P}^1$ (if one of x or y is nonzero) or to the point $[y : z]$ (if one of y or z is nonzero). Note that this is well defined: if both conditions are satisfied, then $[x : y] = [y : z]$ in $\mathbb{P}^1$, since $xz = y^2$. Therefore:

COROLLARY 1.5. *Any smooth projective plane conic is isomorphic to $\mathbb{P}^1$. In particular, it has topological genus zero.*

The isomorphism described above is actually geometrically inspired. Let L be the line in the plane defined by $z = 0$, and let p be the point $[0 : 0 : 1]$ which is on the conic X (but not on the line L). For any point $\ell \in L$, one can form the line M_ℓ joining ℓ to p; this line will meet X at p and at one other point x_ℓ. The map sending ℓ to x_ℓ is G.

Conversely, given $x \neq p$ on the conic X, form the line joining x to p, and intersect that with L; this gives the map H from the conic back to the line.

One usually hears this isomorphism as being given by "projection from a point on the conic".

Now let us use a bit more symmetric linear algebra; the precise statement we need is that two symmetric matrices A and B are related by an invertible matrix T with $B = T^\top A T$ if and only if they have the same rank. Therefore quadratic equations are classified (up to these changes of coordinates given by the action of the matrix T) simply by the rank, which can be 0, 1, 2, or 3. The rank zero case is a bit silly: it means that the matrix A is identically zero, and therefore so is the equation F_A. In the case of rank one, the equation can be put into the form $F = x^2$, and in general F is the square of a linear form. One calls this kind of conic a *double line*: the linear form defines a line, and the "double" structure comes from the squaring. In the case of rank two, the equation can be put into the form $F = xy$, and in general F is the product of two distinct linear forms. One says that this conic consists of *two lines*. Finally the case of rank three is the case of a smooth conic which we have described above.

Glueing Together Riemann Surfaces. The preferred method to describe a Riemann surface is to give a set or space Z and then give the charts, whose domains are then subsets of Z. The chart domains are themselves Riemann surfaces, being open sets in Z. We may a posteriori think of Z as the union of the chart domains.

In several circumstances it is convenient to be able to give the open subsets abstractly, without defining the entire set Z all at once at the beginning of the process. Such a method would then start by taking a collection of Riemann surfaces (which are intended ultimately to be open subsets of the final Riemann surface) and "glue" these individual Riemann surfaces together.

The topologists have thought about these things already for us, and have provided us with the proper notion of glueing. Let us briefly describe this, in the special case where just two subsets are glued together. Suppose that X and Y are topological spaces, with open subsets $U \subset X$ and $V \subset Y$. Suppose further that a homeomorphism $\phi : U \to V$ is given.

Form the disjoint union $X \coprod Y$, and partition this disjoint union into the following three types of subsets:
- Singleton sets $\{x\}$ where $x \in X - U$;
- Singleton sets $\{y\}$ where $y \in Y - V$;
- Doubleton sets $\{u, \phi(u)\}$ where $u \in U$.

Let Z be the set of these subsets; thus there is one point of Z for every point of $X - U$ and for every point of $Y - V$, and one point of Z for every pair of corresponding points of U and V. Clearly there is an onto map from $X \coprod Y$ to Z, sending a point (in either X or Y) to the subset which it is in. If one gives Z the quotient topology for this map π, which declares a subset $W \subset Z$ to be open if and only if $\pi^{-1}(W)$ is open in $X \coprod Y$, we obtain a topological space which is the *glueing of X and Y along U and V via ϕ*. The space Z is denoted by

$X \coprod Y/\phi$.

Note that the natural inclusions of X and Y into Z are continuous. Moreover, if $A \subset X$ is an open set, then its image in Z is also open; indeed, the image is homeomorphic to A under the inclusion.

Our interest is when all the spaces in sight are Riemann surfaces, and the map ϕ is an isomorphism; we conclude under a mild hypothesis that the glueing is a Riemann surface.

PROPOSITION 1.6. *Let X and Y be Riemann surfaces. Suppose that $U \subset X$ and $V \subset Y$ are nonempty open sets, and there is given an isomorphism $\phi : U \to V$ between them. Then there is a unique complex structure on the identification space $Z = X \coprod Y/\phi$ such that the natural inclusions of X and Y into Z are holomorphic. In particular, if Z is Hausdorff, it is a Riemann surface.*

PROOF. Define a complex structure on Z as follows. Let $j_X : X \to Z$ and $j_Y : Y \to Z$ be the natural inclusions. For every chart $\psi : U_\alpha \to \psi(U_\alpha)$ on X, take the open set $j_X(U_\alpha) \subset Z$, and define a chart map on $j_X(U_\alpha)$ by using $\psi \circ j_X^{-1}$. Make similar charts on Z for the charts of Y. This gives a set of charts whose domains cover Z, which are easily checked to be pairwise compatible. Hence we have an induced complex structure on Z.

If we desire the natural inclusions of X and Y into Z to be holomorphic, these charts are forced on us; hence the complex structure is unique with this condition.

Finally, since X and Y are both connected, so is Z; therefore Z is a Riemann surface if it is Hausdorff. □

One can construct the Riemann Sphere $\mathbb{C}_\infty$ by glueing together two copies of the complex plane $\mathbb{C}$. We let $X = Y = \mathbb{C}$, $U = V = \mathbb{C}^*$, and the map ϕ is defined by $\phi(z) = 1/z$.

Hyperelliptic Riemann Surfaces. Let $h(x)$ be a polynomial of degree $2g + 1 + \epsilon$, where ϵ is either 0 or 1, and assume that $h(x)$ has distinct roots. Form the smooth affine plane curve X by the equation $y^2 = h(x)$. Let $U = \{(x,y) \in X \mid x \neq 0\}$; U is an open subset of X.

Let $k(z) = z^{2g+2}h(1/z)$; note that $k(z)$ is a polynomial in z, and also has distinct roots since h does. Form the smooth affine plane curve Y by the equation $w^2 = k(z)$. Let $V = \{(z,w) \in Y \mid z \neq 0\}$; V is an open subset of Y.

Define an isomorphism $\phi : U \to V$ by

$$\phi(x,y) = (z,w) = (1/x, y/x^{g+1}).$$

Let Z be the Riemann surface obtained by glueing X and Y together along U and V via ϕ.

LEMMA 1.7. *With the above construction, Z is a compact Riemann surface of genus g. The meromorphic function x on X extends to a holomorphic map*

$\pi : Z \to \mathbb{C}_\infty$ which has degree 2. The branch points of π are the roots of h (and the point ∞ if h has odd degree).

PROOF. One checks readily that Z is Hausdorff, and hence is a Riemann surface. Z is compact, since it is the union of the two compact sets

$$\{(x,y) \in X \mid \|x\| \leq 1\} \text{ and } \{(z,w) \in Y \mid \|z\| \leq 1\}.$$

The map π obviously has degree 2, and so the inverse image of any point under π is either two points with multiplicity one, or one point with multiplicity two. The latter type gives a ramification point, and occurs exactly over the $2g+2$ roots of h (if h has even degree), or over the $2g+1$ roots of h and over ∞ (if h has odd degree). Such a point contributes 1 to the sum in the Hurwitz formula; thus the total contribution to the sum in the Hurwitz formula is $2g+2$. Therefore we see that

$$\begin{aligned} 2g(Z) - 2 &= \deg(\pi)(2g(\mathbb{C}_\infty) - 2) + (2g+2) \\ &= 2(-2) + 2g + 2 = 2g - 2. \end{aligned}$$

Thus $g(Z) = g$ as claimed. $\square$

DEFINITION 1.8. A compact Riemann surface constructed in this way is called a *hyperelliptic* Riemann surface.

Note that any hyperelliptic surface Z defined by $y^2 = h(x)$ has an automorphism $\sigma : Z \to Z$, namely

$$\sigma(x,y) = (x, -y).$$

Note that σ is an *involution*, that is, $\sigma \circ \sigma = \text{id}$. This involution is called the *hyperelliptic involution* on X. It commutes with the projection map $\pi : X \to \mathbb{C}_\infty$ is the sense that $\pi \circ \sigma = \pi$.

Meromorphic Functions on Hyperelliptic Riemann Surfaces. Using the hyperelliptic involution σ, we can describe all meromorphic functions on a hyperelliptic Riemann surface X, defined by an equation $y^2 = h(x)$.

For any meromorphic function f on X, the *pullback* function $\sigma^* f = f \circ \sigma$ is also meromorphic on X, since σ is a holomorphic map. Since $\sigma^2 = \text{id}$, the sum $f + \sigma^* f$ is σ^*-invariant: $\sigma^*(f + \sigma^* f) = f + \sigma^* f$.

Now the basic example of a σ^*-invariant function is one which is pulled back from $\mathbb{C}_\infty$. This is a function g of the form $g = \pi^* r = r \circ \pi$ for some meromorphic function r on $\mathbb{C}_\infty$. The next lemma shows that these are in fact all of the σ^*-invariant functions on X.

LEMMA 1.9. *Let g be a meromorphic function on X such that $\sigma^* g = g$. Then there is a unique meromorphic function r on $\mathbb{C}_\infty$ such that $g = \pi^* r = r \circ \pi$.*

PROOF. One simply defines $r(p)$ for $p \in \mathbb{C}_\infty$ by choosing a preimage $q \in X$ for p (so that $\pi(q) = p$) and setting $r(p) = g(q)$. The σ^*-invariance of g implies that r is well defined; then one needs to check that r is meromorphic, which is straightforward, and is left as an exercise. □

Therefore, given any meromorphic function f on X, the σ^*-invariant part $f^+ = (1/2)(f + \sigma^* f)$ is pulled back from a function r on $\mathbb{C}_\infty$: $f^+ = r \circ \pi$.

Note that of the two coordinate functions x and y on X, the function x is σ^*-invariant, and the function y is not. However y does enjoy an anti-invariance: $\sigma^* y = -y$. This anti-invariance holds for the function $f^- = (1/2)(f - \sigma^* f)$ also, for any meromorphic function f on X. Therefore the ratio f^-/y is again σ^*-invariant, and we conclude that there exists a meromorphic function s on $\mathbb{C}_\infty$ such that $f^- = ys$.

Since $f = f^+ + f^-$, and all meromorphic functions on $\mathbb{C}_\infty$ are rational by Theorem 2.1 of Chapter II, we have the following.

PROPOSITION 1.10. *Every meromorphic function f on a hyperelliptic Riemann surface X defined by $y^2 = h(x)$ can be written uniquely as*

$$f = r(x) + ys(x),$$

where $r(x)$ and $s(x)$ are rational functions of x.

Maps Between Complex Tori. Suppose that L and M are lattices in $\mathbb{C}$, defining complex tori $X = \mathbb{C}/L$ and $Y = \mathbb{C}/M$. Fix any complex number $a \in \mathbb{C}$, and consider the translation map $z \mapsto z + a$. This map descends to a holomorphic map $T_a : Y \to Y$; moreover, T_a depends only on $a \mod M$, which is a point $q \in Y$, and T_a is an automorphism of Y with inverse T_{-a}. Such an automorphism, which is usually denoted simply by T_q, is called a *translation* of Y: it sends $y \in Y$ to $y + q$ (where the sum is understood to be that of the quotient group law in Y).

Now let $F : X \to Y$ be a holomorphic map. By composing F with a suitable translation on Y we may assume that $F(0) = 0$.

Note that F is unramified by Hurwitz's formula. Hence $F : X \to Y$ is a covering map in the sense of topology, and hence so is the composition $F \circ \pi : \mathbb{C} \to X \to Y$. Since the domain is simply connected, this must be isomorphic as a covering to the universal covering of Y, which is $\pi : \mathbb{C} \to Y$. Therefore there is a map $G : \mathbb{C} \to \mathbb{C}$ and a commutative diagram

$$\begin{array}{ccc} \mathbb{C} & \xrightarrow{G} & \mathbb{C} \\ \pi \downarrow & & \downarrow \pi \\ X & \xrightarrow{F} & Y \end{array}.$$

Note that the map G must be holomorphic, since all other maps in the diagram are holomorphic and unramified. Moreover, since $F(0) = 0$, G must send 0 to a lattice point; we may assume in fact that $G(0) = 0$, since composing with translation by a lattice point does not affect the projection map π.

Now G is a holomorphic map which must send the lattice L to the lattice M. Indeed, for any complex number z, and any element $\ell \in L$, we must have $G(z + \ell) \equiv G(z) \mod M$; hence there is a lattice point $\omega(z, \ell) \in M$ such that $\omega(z, \ell) = G(z+\ell) - G(z)$. But M is a discrete set, and $\mathbb{C}$ is connected; therefore for fixed ℓ we see that $\omega(z, \ell)$ is independent of z.

Taking the derivative with respect to z, we have that $\omega'(z, \ell) \equiv 0$. But this says that $G'(z + \ell) = G'(z)$, so that the derivative G' is invariant under translation by lattice points. Hence the values of G' all occur in a fundamental parallelogram for L; since such a parallelogram is compact, the values of G' are bounded. Therefore G' is a bounded entire function; hence it is constant, and so G is *linear*. Since $G(0) = 0$, there is a complex constant γ such that $G(z) = \gamma z$.

Since G sends the lattice L into the lattice M, we must have that $\gamma L \subseteq M$. This implies in particular that the induced map F is a group homomorphism. Therefore we have shown the following:

PROPOSITION 1.11. *Let X and Y be two complex tori given by lattices L and M respectively. Then any holomorphic map $F : X \to Y$ is induced by a linear map $G : \mathbb{C} \to \mathbb{C}$ of the form $G(z) = \gamma z + a$, where γ is a constant such that $\gamma L \subseteq M$. The constant a may be taken to be zero if and only if F sends 0 to 0; in this case the map F is a homomorphism of groups. The holomorphic map F is an isomorphism if and only if $\gamma L = M$. In general, the degree of F is the index $|M/\gamma L|$ of γL inside M.*

Only the last two statements require any more argument, but the first is rather obvious; if $\gamma L = M$, then $\gamma^{-1} M = L$, and so the map $H(z) = \gamma^{-1}(z - a)$ induces a holomorphic map from Y to X which is an inverse for F. We leave the final statement to the reader as an exercise.

Using these ideas we may easily determine all of the *automorphisms* of a complex torus. Again we assume that an automorphism $F : X \to X$ sends 0 to 0 (else we may compose with a translation to achieve this). F is then induced by a linear map G of the form $G(z) = \gamma z$, for some γ such that $\gamma L = L$.

This forces $\|\gamma\| = 1$, and in fact γ must be a root of unity. We see the obvious values $\gamma = \pm 1$ as possibilities; these correspond to F being the identity map and the inverse map, respectively. Every complex torus has these two automorphisms.

Assume then that γ is not real. Let ℓ be a number of minimal length in $L - \{0\}$; then so is $\gamma\ell$, and ℓ and $\gamma\ell$ must generate L over $\mathbb{Z}$.

Now $\gamma^2 \ell$ is also in L, and so we may write $\gamma^2 \ell = m\gamma\ell + n\ell$ for some integers m and n. Dividing by ℓ we see that γ satisfies the quadratic equation $z^2 - mz - n = 0$. The only roots of unity which satisfy quadratic equations are the 4th and 6th roots of unity. Therefore we may assume that $\gamma = i$ or that $\gamma = \exp(\pi i/3)$. In the first case the lattice L is a square lattice (which has orthogonal generators of the same length) and in the second case L is an hexagonal lattice (with generators of the same length separated by an angle of $\pi/3$).

Let us summarize:

PROPOSITION 1.12. *Let $X = \mathbb{C}/L$ be a complex torus. Then any holomorphic map $F : X \to X$ fixing 0 is induced by multiplication by some $\gamma \in \mathbb{C}$, and is therefore a homomorphism of the group structure on X. Moreover if F is an automorphism, then either:*
  a. *L is a square lattice and γ is a 4th root of unity;*
  b. *L is an hexagonal lattice and γ is a 6th root of unity; or*
  c. *L is neither square nor hexagonal and $\gamma = \pm 1$.*

Therefore if we set $\text{Aut}_0(X)$ to be the automorphisms of X fixing 0, we have that

$$\begin{aligned} \text{Aut}_0(X) &\cong \mathbb{Z}/4 \quad \text{if } L \text{ is square;} \\ \text{Aut}_0(X) &\cong \mathbb{Z}/6 \quad \text{if } L \text{ is hexagonal;} \\ \text{Aut}_0(X) &\cong \mathbb{Z}/2 \quad \text{otherwise.} \end{aligned}$$

In particular: the complex torus defined using a square lattice is *not* isomorphic to a complex torus defined using an hexagonal lattice. Thus there are nonisomorphic complex tori (and hence nonisomorphic Riemann surfaces of genus one)! Of course two Riemann surfaces with different genera cannot be isomorphic, but the complex tori have given us the first example of nonisomorphic Riemann surfaces with the same genus.

We will be able to show later that every Riemann surface of genus zero is isomorphic to $\mathbb{P}^1$, so our first chance at finding this phenomenon is in genus one. In fact for every genus $g \geq 1$ there are nonisomorphic Riemann surfaces (and lots of them).

We can be a bit more precise in the case of complex tori using the methods above. First we note that every complex torus is isomorphic to a complex torus X_τ defined by a lattice L_τ generated by 1 and τ, where τ is a complex number with positive imaginary part. Indeed, if L is generated by ω_1 and ω_2, then using $\gamma = 1/\omega_1$ maps L into the lattice generated by 1 and ω_2/ω_1. If this ratio is in the upper half-plane, then this is τ; otherwise we may take $\tau = -\omega_2/\omega_1$ equally well as a generator.

Now we ask the question: when are X_τ and $X_{\tau'}$ isomorphic? For this we must have a complex number γ such that $\gamma L_\tau = L_{\tau'}$; this is equivalent to having the two numbers γ and $\gamma\tau$ generating $L_{\tau'}$. In order that they lie in $L_{\tau'}$, there must be integers a, b, c, and d such that $\gamma = c\tau' + d$ and $\gamma\tau = a\tau' + b$. Eliminating γ from these equations gives that $\tau = (a\tau' + b)/(c\tau' + d)$. Moreover for γ and $\gamma\tau$ to *generate* $L_{\tau'}$, we must have the determinant $ad - bc$ equal to ± 1. In fact it must equal 1, since both τ τ' lie in the upper half-plane. These conditions are also clearly sufficient, and we have proven the following:

1. MORE ELEMENTARY EXAMPLES OF RIEMANN SURFACES

PROPOSITION 1.13. *Two complex tori X_τ and $X_{\tau'}$ are isomorphic if and only if there is a matrix $\begin{pmatrix} a & b \\ c & d \end{pmatrix}$ in $\mathrm{SL}_2(\mathbb{Z})$ such that $\tau = (a\tau' + b)/(c\tau' + d)$.*

The group $\mathrm{SL}_2(\mathbb{Z})$ acts on the upper half-plane $\mathbb{H}$ (the matrix sends τ to $(a\tau + b)/(c\tau + d)$) and so we see that isomorphism classes of complex tori are in 1-1 correspondence with points of the orbit space $\mathbb{H}/\mathrm{SL}_2(\mathbb{Z})$. This orbit space is in fact isomorphic to the complex numbers, via the so-called j-function. The interested reader should consult [**Serre73**] or [**Lang87**] for rather complete treatments. But in any case we see that there are uncountably many isomorphism classes of complex tori, and that they vary with essentially one parameter (the lattice generator τ).

Problems III.1

A. Verify that the isomorphism T between two conics described in the text is indeed a holomorphic map. Verify that the map from $\mathbb{P}^1$ to the conic $xz = y^2$ sending $[r : s]$ to $[r^2 : rs : s^2]$ is a holomorphic map.

B. Check that the charts on the glueing space $Z = X \coprod Y/\phi$ defined in the proof of Proposition 1.6 are pairwise compatible.

C. Show that if one glues together $\mathbb{C}$ and $\mathbb{C}$ along $\mathbb{C}^*$ and $\mathbb{C}^*$ via the glueing map $\phi(z) = z$, the resulting space is not Hausdorff.

D. Let $h(x)$ be a polynomial of degree $2g+1+\epsilon$ (with $\epsilon \in \{0,1\}$) having distinct roots and let $U = \{(x,y) \in \mathbb{C}^2 \mid y^2 = h(x) \text{ and } x \neq 0\}$. As in the text let $k(z) = z^{2g+2}h(1/z)$ and let $V = \{(z,w) \in \mathbb{C}^2 \mid w^2 = k(z) \text{ and } z \neq 0\}$. Show that the mapping $\phi : U \to V$ defined by $(z,w) = (1/x, y/x^{g+1})$ is an isomorphism of Riemann surfaces.

E. Check that the function r defined in the proof of Lemma 1.9 is meromorphic.

F. Let X be the compact hyperelliptic curve defined by $x^2 = 3 + 10t^4 + 3t^8$. Let Y be the compact hyperelliptic curve defined by $w^2 = z^6 - 1$. Let U and V be the corresponding affine plane curves, which are the complements in X and Y respectively of the points at infinity. Show that the function $F : U \to V$ defined by $z = (1+t^2)/(1-t^2)$ and $w = 2tx/(1-t^2)^3$ extends to a holomorphic map from X to Y of degree 2, which is nowhere ramified. What is the genus of X and of Y?

G. Let X be a complex torus. Show that any translation map of X, which is induced from a translation in the complex plane, is a holomorphic map.

H. Let X be a complex torus. Show that the full group of automorphisms of X is a semidirect product of the group of translations with the group $\mathrm{Aut}_0(X)$ of automorphisms fixing 0.

I. Let X be a complex torus, and let F be a nontrivial automorphism of X. Show that if F is not a translation, then F has a fixed point.

J. Let X and Y be complex tori defined by lattices L and M respectively, and $F : X \to Y$ be a holomorphic map induced by a linear map $G(z) = \gamma z + a$ with $\gamma L \subset M$. Show that the degree of F is the index of γL inside M.

2. Less Elementary Examples of Riemann Surfaces

Plugging Holes in Riemann Surfaces. If one takes a Riemann surface and deletes one point, one still has a Riemann surface, albeit with a "hole" in it. The process can be reversed if we define a "hole" properly. Defining something which isn't supposed to be there requires some care, but is not too troublesome after all.

DEFINITION 2.1. Let X be a Riemann surface. A *hole chart* on X is a complex chart $\phi : U \to V$ on X such that V contains an open punctured disc $D_0 = \{z \mid 0 < \|z - z_0\| < \epsilon\}$ with the closure in X of $\phi^{-1}(D_0)$ inside U, and this closure is transported via ϕ to the punctured closed disc $D_1 = \{z \mid 0 < \|z - z_0\| \leq \epsilon\}$.

In other words, a hole chart has a hole in it: the closure of D_0 in $\mathbb{C}$ has z_0 in it, but the closure of the corresponding open set $\phi^{-1}(D_0)$ in X does not have any point corresponding to z_0.

Now suppose that X is a Riemann surface with a hole chart $\phi : U \to V$ on it. Let D_0 be the open punctured disc as above, and let D be simply the open disc $D = \{z \mid \|z - z_0\| < \epsilon\}$. Note that D is a Riemann surface in its own right, and D_0 is an open subset of D, which is isomorphic to the open subset $\phi^{-1}(D_0) \subset X$ via the chart map ϕ suitably restricted. Form the identification space $Z = X \coprod D / \phi$; the assumption on the closure of $\phi^{-1}(D_0)$ exactly implies that Z is Hausdorff. Thus Z is a Riemann surface, which we refer to as the surface obtained from X by *plugging the hole in the hole chart* ϕ.

Compactifications of certain Riemann surfaces may be effected by means of plugging holes. Suppose that X is a Riemann surface with a finite number of disjoint hole charts $\phi_i : U_i \to V_i$. Let G_i be the open subset $\phi_i^{-1}(D_0)$ in X. Suppose that $X - \cup_i G_i$ is compact. Then the surface obtained from X by plugging the holes in these hole charts is compact, since it can be decomposed as the union of finitely many compact sets (namely $X - \cup_i G_i$ and the closures of the discs which are glued in to plug the holes).

The simplest example of this is the compactification of $\mathbb{C}$ to the Riemann Sphere $\mathbb{C}_\infty$. The hole chart on $\mathbb{C}$ is the function $\phi(z) = 1/z$, defined for $z \neq 0$.

A more sophisticated example is the compactification of the smooth affine plane curve given by the hyperelliptic equation $y^2 = h(x)$, where h is a polynomial with distinct roots. We have already produced a compactification above by glueing together two such Riemann surfaces. However we can also obtain the same compact Riemann surfaces by plugging the holes.

Assume first that h has odd degree $2g + 1$. Then the chart ϕ defined by $\phi(x, y) = y/x^{g+1}$ is defined for $\|x\|$ large, and is a hole chart on X; the "hole" is the point at infinity. Plugging this hole gives a compact Riemann surface.

If h has even degree $2g + 2$, then we know that X has two points at infinity. As x approaches ∞, y/x^{g+1} approaches one of the two square roots of $\alpha = \lim_{x \to \infty} h(x)/x^{2g+2}$. (The number α is just the top coefficient of h.) The two hole charts are ϕ_i $i = 1, 2$, where $\phi_i(x,y) = 1/x$ for both i, but ϕ_1 is defined for $\|x\|$ large and y/x^{g+1} near $+\sqrt{\alpha}$, and ϕ_2 is defined for $\|x\|$ large and y/x^{g+1} near $-\sqrt{\alpha}$. Plugging these two holes gives a compact Riemann surface.

Nodes of a Plane Curve. Certain types of plane curves which are not smooth everywhere (and hence are not Riemann surfaces) can give rise to Riemann surfaces with the following construction. First assume that X is an affine plane curve given by $f(z,w) = 0$, such that at all but finitely many points of X, at least one of the partials $\partial f/\partial z$ or $\partial f/\partial w$ is nonzero. Therefore if we delete these finitely many points, where $f = \partial f/\partial z = \partial f/\partial w = 0$, we obtain a Riemann surface using charts afforded by the Implicit Function Theorem as in Chapter I, Section 2.

The deletion of these points gives a Riemann surface with holes in it, and under some mild hypotheses it is not hard to discover the hole charts.

DEFINITION 2.2. A point p on an affine plane curve X defined by $f(z,w) = 0$ is called a *node* of the plane curve X if p is a singular point of X (i.e., $f(p) = \partial f/\partial z(p) = \partial f/\partial w(p) = 0$), but the Hessian matrix of second partials

$$\begin{pmatrix} \partial^2 f/\partial z^2 & \partial^2 f/\partial z \partial w \\ \partial^2 f/\partial w \partial z & \partial^2 f/\partial w^2 \end{pmatrix}$$

is nonsingular at p, i.e.,

$$\frac{\partial^2 f}{\partial z^2}(p) \frac{\partial^2 f}{\partial w^2}(p) \neq \left(\frac{\partial^2 f}{\partial z \partial w}(p)\right)^2.$$

In terms of the coefficients for f, this condition means that if we expand f about the point $p = (z_0, w_0)$, the constant term is zero (since $f(p) = 0$), the linear terms are zero (since $\partial f/\partial z(p) = \partial f/\partial w(p) = 0$), and the quadratic terms are of the form

$$a(z - z_0)^2 + b(z - z_0)(w - w_0) + c(w - w_0)^2$$

where the homogeneous quadratic equation $ax^2 + bxy + cy^2$ factors into distinct homogeneous linear factors $\ell_1(x,y)\ell_2(x,y)$.

The Implicit Function Theorem applied to a smooth point of $f(z,w) = 0$ can be interpreted as saying that near a smooth point, the locus of roots X of f looks very much like the tangent line to X at p. In other words, if $f(p) = 0$ and one of the derivatives of f is not zero at p, then X is locally the graph of a function, which of course is locally like its tangent line. Note that the tangent line at a point is exactly the zeroes of the linear part of f, expanded about that point.

The same principle can be applied here, to one higher order: if X has a node at p, then locally near p the curve should look like the zeroes of its quadratic part. We can make this precise as follows:

LEMMA 2.3. *Suppose the locus of roots X of $f(z,w)$ has a node at $p = (z_0, w_0)$. Factor the quadratic term of f as above, writing*

$$f(z,w) = \ell_1(z - z_0, w - w_0)\ell_2(z - z_0, w - w_0) + \text{ higher order terms,}$$

where the ℓ_i are distinct homogeneous linear polynomials. Then as a power series, f itself factors as $f = gh$, where

$$g(z,w) = \ell_1(z - z_0, w - w_0) + \text{ higher order terms, and}$$

$$h(z,w) = \ell_2(z - z_0, w - w_0) + \text{ higher order terms.}$$

PROOF. This is a simple version of a general factoring principle known as Hensel's Lemma: if the lowest order terms of a power series factors into distinct factors, then the entire power series factors compatibly.

In this special case the lemma is easy to see. For sanity change coordinates to $x = \ell_1(z - z_0, w - w_0)$ and $y = \ell_2(z - z_0, w - w_0)$, and write

$$f(x,y) = xy + \sum_{i=3}^{\infty} f_i(x,y),$$

where f_i is homogeneous of degree i in x and y. We seek power series $g = x + \sum_{i \geq 2} g_i$ and $h = y + \sum_{i \geq 2} h_i$ such that $f = gh$, where we have g_i and h_i homogeneous of degree i. We note first that imposing $f = gh$ forces

(2.4) $$f_i = xh_{i-1} + yg_{i-1} + \sum_{j=2}^{i-2} g_j h_{i-j}$$

for each $i \geq 3$. For $i = 3$, this requires simply that $f_3 = xh_2 + yg_2$, and clearly for any f_3 of degree 3 one can solve this for g_2 and h_2.

One now proceeds by induction on i. Suppose that all g_j and h_j have been found for $j < i-1$, and we want to determine g_{i-1} and h_{i-1}. Then the constraint (2.4) gives the condition that

$$xh_{i-1} + yg_{i-1} = f_i - \sum_{j=2}^{i-2} g_j h_{i-j},$$

and the right-hand side is, by induction, a known homogeneous polynomial of degree i. Clearly one can solve for g_{i-1} and h_{i-1} in this case. This recursive procedure produces the power series g and h, factoring f. □

2. LESS ELEMENTARY EXAMPLES OF RIEMANN SURFACES

Resolving a Node of a Plane Curve. It is an easy exercise to show that these power series g and h must converge near the node point p, since f does (f is a polynomial, after all). Thus near p, the locus X of zeroes of f is the locus of zeroes of gh, which is simply the union of the locus X_g of zeros of g and the locus X_h of zeroes of h.

These separate loci X_g and X_h are, near p, Riemann surfaces! Using the change of coordinates as in the proof above, we see for example that

$$g(x,y) = x + \text{ (higher order terms in } x \text{ and } y\text{)}$$

and so $\partial g/\partial x(p) = 1 \neq 0$. Therefore the Implicit Function Theorem gives that near p, X_g is the graph of a function of y, and so is a Riemann surface. The same remarks hold of course for X_h.

Now let us return to the singular curve X defined by $f = 0$ at p, and delete the point p, producing a Riemann surface Y (at least near p). This surface Y, near p, is equal to the union of $X_g - \{p\}$ and $X_h - \{p\}$. Let U_g and U_h be the open sets on Y which are equal separately to $X_g - \{p\}$ and $X_h - \{p\}$, respectively. Then Y has two obvious hole charts on it: one is the composition of the isomorphism of U_g with $X_g - \{p\}$ with a chart on X_g near p, and the other is the same for U_h. Plugging these two hole charts is called *resolving the node of X at p*.

This entire process is really local to the singularity at p. It can be performed equally well on a projective plane curve with nodes; after all, a projective plane curve is locally an affine plane curve, and the concept of a node transfers immediately. Since a projective plane curve, whether singular or not, is certainly compact (it is a closed subset of the projective plane, which is compact), the result of resolving the nodes of a projective plane curve is a compact Riemann surface, if it is connected. As with affine plane curves, the resolution is connected if and only if the homogeneous polynomial defining the projective plane curve is irreducible. Therefore:

PROPOSITION 2.5. *Let $F(x,y,z)$ be an irreducible homogeneous polynomial of degree d, defining the locus of roots $X \subset \mathbb{P}^2$. Assume that at all but finitely many points of X, F is a nonsingular polynomial, that is, at least one of its first partials is nonzero. Assume further that these finitely many singular points are nodes of X. Then the Riemann surface obtained by resolving these nodes of X is a compact Riemann surface.*

The Genus of a Projective Plane Curve with Nodes. We have seen in previous examples that smooth projective plane curves of degree either one or two (i.e., lines and conics) have genus zero. Moreover, the Fermat curve of degree d (defined by $x^d + y^d + z^d = 0$) has genus $g = (d-1)(d-2)/2$. (Problem II.4, G.) This is indeed the formula in general for a smooth projective plane curve of degree d, although it is a bit beyond us now to prove this.

There are two approaches to the proof which can be outlined now, although neither can be executed just yet. The first is to write down a suitable meromorphic function on the curve X, say a ratio of linear homogeneous polynomials. Considering this function as a holomorphic map F to $\mathbb{P}^1$, one then tries to compute the ramification and degree of F, and then apply the Hurwitz formula. In other words, one does the same computation as was done in the special case of the Fermat curve, for the general smooth curve.

The second is to show that the genus does not change when one varies the coefficients of the homogeneous polynomial: essentially one wants to show that the genus is locally constant as a function of the coefficients. Then one shows that the space of coefficients for smooth plane curves is connected. Since the formula is true for the Fermat curve, the result follows.

There are other approaches, which involve the theory of algebraic surfaces, but these require more theory than can be stated concisely at this point.

What about a projective plane curve whose only singularities are nodes? Let us argue in the spirit of the second approach, and consider "nearby" curves whose coefficients differ only slightly from that of the nodal curve. At a node, we locally have a curve of the form $xy = 0$; the nearby curve looks locally like $xy = t$ for some small parameter t. Topologically, as t approaches zero, a small circle (homeomorphic to S^1) is becoming contracted to the node point. Therefore the Euler number of the nearby smooth curve and the Euler number of the nodal curve differs by exactly one, which is the difference between the Euler number of a circle (0) and of the nodal point (1). Thus after resolving the node (which replaces one point by two), we see that the resolution curve has an Euler number which is two greater than that of the nearby smooth curve.

This same analysis holds at each node; hence if there are n nodes to the curve, the Euler number increases by $2n$ in going from the nearby smooth curve to the resolution of the nodes. Since the Euler number is equal to $2 - 2g$, an increase of $2n$ in the Euler number implies a decrease of n in the genus. We therefore arrive at the formula for the genus of a projective plane curve with nodes, which is called *Plücker's formula*:

PROPOSITION 2.6 (PLÜCKER'S FORMULA). *Let X be a projective plane curve of degree d with n nodes and no other singularities. Then the genus g of X is*

$$g = (d-1)(d-2)/2 - n.$$

We will return to Plücker's formula and give a proper proof later, along the lines of the first approach described above.

The point of bringing this all up now is simply to point out that *every* Riemann surface can be obtained as a projective plane curve with nodes. Indeed, if a Riemann surface of genus g is not hyperelliptic, then it can be obtained as a projective plane curve of degree $2g - 2$ with exactly $2g^2 - 8g + 6$ nodes! This too is beyond us now, but it feels good to have at least a minimal understanding of every Riemann surface.

Resolving Monomial Singularities. It is a basic fact of plane curve geometry that *any* type of singularity can be suitably resolved. Although a general analysis of plane curve singularities is not appropriate now, there is a type of singularity which is very similar to the node, whose resolution involves really no further ideas, and which will come up shortly. These are singularities which are locally of the form $z^n = w^m$, for positive integers n and m.

What does it mean for a singularity to be locally of this form? Consider the plane curve $f(x, y) = 0$, and assume that it is singular at the origin.

DEFINITION 2.7. The singularity of $f(x,y) = 0$ at the origin is said to be (n,m)-*monomial* if there are power series $g(x,y)$ and $h(x,y)$, each having no constant term, and having linearly independent linear terms, such that $f(x,y) = g(x,y)^n - h(x,y)^m$ as power series in two variables.

First note that if either n or m is 1, then the curve is not singular at the origin. Hence we may assume that both n and m are at least two.

Note that a node is a $(2,2)$-monomial singularity; if it is given by $xy = 0$ locally, when we set $g = (x+y)/2$ and $h = (x-y)/2$, we have $xy = g^2 - h^2$.

Now let us turn to resolving a monomial singularity. The existence of the power series g and h gives a second pair of local analytic coordinates $z = g(x,y)$ and $w = h(x,y)$ on the plane. Therefore we may consider the equation $z^n = w^m$ as the prototype.

First assume that n and m are relatively prime; choose integers a and b such that $an + bm = 1$. Consider the function $r(t) = (z, w) = (t^m, t^n)$; define $s(z,w) = t = z^b w^a$. Note that r and s are inverse maps between a neighborhood of $t = 0$ and a neighborhood of the monomial singularity on the singular curve. This function s then gives a hole chart on the curve with the singular point deleted; plugging this one hole resolves the singularity.

Next assume that $n = m$. If we let $\zeta = \exp(2\pi i/n)$ be a primitive n^{th} root of unity, then the equation $z^n - w^n$ factors completely into linear factors:

$$z^n - w^n = \prod_{i=0}^{n-1}(z - \zeta^i w).$$

Each of the factors obviously defines a smooth curve; they just all pass through the origin, which is of course the singular point. Therefore removing the origin gives a space which decomposes into n smooth curves, each with a hole in it. Plugging these n holes resolves the singularity in this case.

Finally assume that $n < m$ and $(n, m) = k$ with $1 < k \leq n$. The resolution of this singularity simply combines features of the previous two cases. Write $n = ka$ and $m = kb$; then $(a, b) = 1$. Note now that if $\zeta = \exp(2\pi i/k)$, the equation now factors as

$$z^n - w^n = (z^a)^k - (w^b)^k = \prod_{i=0}^{k-1}(z^a - \zeta^i w^b).$$

Each of the factors we know how to resolve; this was our first case, up to a harmless constant, and we have seen that there is one hole to plug for each factor. Doing this for each factor gives a resolution of the curve with k plugged holes.

Summarizing, we have:

LEMMA 2.8. *A plane curve singularity which is (n,m)-monomial is resolved by the above process, which involves removing the singular point and plugging the resulting $k = (n,m)$ holes.*

Special names are traditionally given to certain monomial singularities. We have already seen that a $(2,2)$-monomial singularity is a *node*. A $(2,3)$-monomial singularity is an *ordinary cusp*. A $(2,4)$-monomial singularity is a *tacnode*. A $(2,5)$-monomial singularity is called a *higher-order cusp*, as are all $(2,m)$-monomial singularities with $m \geq 5$ and odd. A $(2,6)$-monomial singularity is called a *higher-order tacnode*, as are all $(2,m)$-monomial singularities with $m \geq 6$ and even. In general, a $(2,m)$-monomial singularity is said to be *of type a_{m-1}*. A $(3,3)$-monomial singularity is an *ordinary triple point*, or *of type d_4*. A $(3,4)$-monomial singularity is said to be *of type e_6*. A $(3,5)$-monomial singularity is said to be *of type e_8*. A $(3,6)$-monomial singularity is an *infinitely near triple point*.

A (n,n)-monomial singularity is an *ordinary n-fold point*. It is only a special type of one, however:

DEFINITION 2.9. *A plane curve singularity $f(x,y) = 0$ at the origin is an ordinary n-fold point if the lowest term of f is the degree n term, and this term (which is a homogeneous polynomial of degree n in x and y) factors completely into distinct linear factors.*

The n-fold analogue of Hensel's Lemma then insures that the entire polynomial $f(x,y)$ factors compatibly into n power series, each of which is smooth at the origin. Therefore we can resolve an ordinary n-fold point by removing the singular point and plugging the n holes in the resulting n factors.

Finally, suppose that the polynomial $f(x,y)$ locally factors as $f = gh$, with each of the curves $g = 0$ and $h = 0$ having a monomial singularity. Then we see immediately how to resolve the singularity of f: remove the singular point, and separately plug the holes in the $g = 0$ and $h = 0$ curves. Examples of this are the singularities *of type d_n*, $n \geq 4$, which have local equations of the form $f(x,y) = x(y^2 - x^{n-2})$. (Note that when $n = 3$ this is just a tacnode and when $n = 4$ this is an ordinary triple point.) Also the missing singularity *of type e_7* has the equation $x(x^2 - y^3)$.

The singularities of types a_n ($n \geq 1$), d_n ($n \geq 4$), and e_n ($n = 6, 7, 8$) are called *simple plane curve singularities*; they are important especially in the theory of algebraic surfaces.

Cyclic Coverings of the Line. There is a construction which is very similar in spirit to the construction of hyperelliptic surfaces, and gives many new examples of Riemann surfaces with especially simple equations describing them. These are the *cyclic coverings of a line*. Choose an integer d and a polynomial $h(x)$ of degree k with distinct roots. Consider the affine plane curve X defined by the equation

$$y^d = h(x);$$

the assumption that h has distinct roots implies that X is smooth.

Let us show that X has finitely many holes at infinity; when these holes are plugged, we will obtain a compact Riemann surface Y.

We attempt the same sort of change of coordinates which was used in the hyperelliptic analysis. Let $x = 1/z$, so that $y^d = h(1/z)$. Write $k = d\ell - \epsilon$, where $0 \leq \epsilon < d$, and multiply the equation through by $z^{d\ell}$ to obtain $(yz^\ell)^d = z^{d\ell}h(1/z) = z^\epsilon(z^k h(1/z))$.

Now we let $w = yz^\ell$ and $g(z) = z^k h(1/z)$; note that g is a polynomial with a nonzero constant term c, and near $x = \infty$ the curve X is described by the equation $w^d = z^\epsilon g(z)$ with z near 0.

If $\epsilon = 0$, as z approaches 0 we see that w can approach any of the d d^{th} roots of the constant term c. Moreover the projection to z gives d hole charts on X near $x = \infty$; plugging these gives a compact Riemann surface Y.

If $\epsilon \neq 0$ then as z approaches 0, so does w. By choosing an ϵth root of $g(z)$ we may absorb the function g into the z^ϵ, and note that the curve is then described by a monomial singularity equation $w^d = z^\epsilon$. This we have seen how to resolve to produce a Riemann surface in the previous section: we must remove the singular point (if $\epsilon \geq 2$) and plug the resulting (d, ϵ) holes.

This completes the analysis; we have compactified X to a compact Riemann surface Y by plugging certain holes at $x = \infty$; moreover we have been led to resolving certain monomial singularities.

Now however we see that the assumption that $h(x)$ has distinct roots was unnecessary: if h has a multiple root at $x = x_0$, then one simply obtains a monomial singularity, which we are prepared to resolve. To see this, assume for simplicity that $x_0 = 0$, so that h has a root of order n at 0. Then the equation for X is $y^d = x^n r(x)$, where r has a nonzero constant term. Taking the nth root of r and absorbing this into the x^n factor, we see the monomial form $y^d = x^n$. We simply resolve this, and any other singularities of X coming from multiple roots of h in a similar way, to produce the compact Riemann surface Y (after plugging the holes at $x = \infty$).

Any Riemann surface obtained this way, by resolving and compactifying a plane curve defined by $y^d = h(x)$, is called a *cyclic covering of the line*.

There is a natural projection map $\pi : Y \to \mathbb{P}^1$ induced by sending (x, y) to x; this is of course a holomorphic map.

The "cyclicity" of these curves comes from the existence of an automorphism

σ generalizing the hyperelliptic involution. Choose a dth root of unity ζ, and consider the map $\sigma : X \to X$ sending (x, y) to $(x, \zeta y)$. This map is of order d, and extends to an automorphism of order d of the compact Riemann surface Y. Moreover σ commutes with the projection map π, in the sense that $\pi \circ \sigma = \pi$.

Problems III.2

A. Let X be the smooth affine plane curve defined by the equation $y^2 = h(x)$, where $h(x)$ is a polynomial of degree $2g+1$ with distinct roots. Show that the map $\phi(x, y) = y/x^{g+1}$ defines a hole chart on X for $\|x\|$ large.

B. Convince yourself that the "difference" (topologically speaking) between the locus $xy = t$ and the locus $xy = 0$ for small t, near the origin, is exactly that the node of $xy = 0$ is deforming into a circle. (Remember that this is all happening in $\mathbb{C}^2 \cong \mathbb{R}^4$!)

C. Use Plücker's formula to show that a projective plane curve of degree $2g-2$ with exactly $2g^2 - 8g + 6$ nodes has a resolution of genus g.

D. Let Y be the Riemann surface defined by the equation $y^d = h(x)$, a cyclic covering of the line. Let π be the projection map and σ the cyclic automorphism.

  1. Show that π is a holomorphic map of degree d from Y to $\mathbb{P}^1$.
  2. Check that the cyclic map σ of a cyclic covering of the line is an automorphism of the compact Riemann surface Y as claimed. Show that every fiber of the projection map $\pi : Y \to \mathbb{P}^1$ is an orbit of σ.
  3. Show that above a root of h of order n, there are (d, n) points of Y, each of multiplicity $d/(d, n)$ for the projection map π.
  4. Given the degree of h and the orders of its roots, give a formula for the genus of Y using Hurwitz's formula.

E. Let Y be the Riemann surface defined by the equation $y^d = h(x)$, a cyclic covering of the line. Let π be the projection map and σ the cyclic automorphism; let $\zeta = \exp(2\pi i/d)$ be a primitive d^{th} root of unity.
   Note that given any meromorphic function f on Y, the composition $\sigma^* f = f \circ \sigma$ is also meromorphic. For each $i = 0, \ldots, d-1$, let $\mathcal{M}_i$ be the space of those meromorphic functions f on Y such that $\sigma^* f = \zeta^i f$.

  1. Show that $x \in \mathcal{M}_0$ and $y \in \mathcal{M}_1$.
  2. Show that every f in $\mathcal{M}_0$ is of the form $\pi^* r$, for some meromorphic function r on $\mathbb{C}_\infty$.
  3. Show that every f in $\mathcal{M}_i$ is of the form $y^i \pi^* r$, for some meromorphic function r on $\mathbb{C}_\infty$.
  4. Show that every meromorphic function f can be written uniquely in the form $f = \sum_i f_i$, where $f_i \in \mathcal{M}_i$ for each i.
  5. Conclude that the field of meromorphic functions on Y is the field of all functions of the form
  $$\sum_{i=0}^{d-1} r_i(x) y^i,$$

where $r_i(x)$ is a rational function of x for each i.

3. Group Actions on Riemann Surfaces

A basic construction for Riemann surfaces is to take a known Riemann surface and divide it by the action of a group. In this section we develop the first ideas of this theory.

Finite Group Actions. Let G be a group and X a Riemann surface. We will assume that G is a finite group for most of this section; at the end we will make some remarks about the infinite case.

An *action* of G on X is a map $G \times X \to X$, which we will denote by $(g, p) \mapsto g \cdot p$, which satisfies
  a. $(gh) \cdot p = g \cdot (h \cdot p)$ for $g, h \in G$ and $p \in X$, and
  b. $e \cdot p = p$ for $p \in X$, where $e \in G$ is the identity.

Technically, this is called a *left action* of G on X. To denote that G acts on X, we write $G : X$.

Note that if we fix $g \in G$, the map sending p to $g \cdot p$ is a bijection; its inverse is the map sending p to $g^{-1} \cdot p$.

The *orbit* of a point $p \in X$ is the set $G \cdot p = \{g \cdot p \mid g \in G\}$. If $A \subset X$ is any subset, we denote by $G \cdot A$ the set of orbits of points in A: $G \cdot A = \{g \cdot a \mid g \in G \text{ and } a \in A\}$.

The *stabilizer* of a point $p \in X$ is the subgroup $G_p = \{g \in G \mid g \cdot p = p\}$. The stabilizer is often called the *isotropy subgroup* of p.

Note that points in the same orbit have conjugate stabilizers: indeed, $G_{g \cdot p} = gG_pg^{-1}$. Moreover if G is a finite group, then the order of the orbit times the order of the stabilizer equals the order of the group:

$$|G \cdot p| \, |G_p| = |G|.$$

The *kernel* of an action of G on X is the subgroup $K = \{g \in G \mid g \cdot p = p \text{ for all } p \in X\}$. It is the intersection of all stabilizer subgroups. It is not hard to see that the kernel is a normal subgroup of G, and that the quotient group G/K acts on X with trivial kernel and identical orbits to the G action. Therefore we usually may assume that the kernel is trivial; this is called an *effective* action.

The action is *continuous*, respectively *holomorphic*, if for every $g \in G$, the bijection sending p to $g \cdot p$ is a continuous, respectively holomorphic, map from X to itself. If it is holomorphic, it will necessarily be an automorphism of X.

The *quotient space* X/G is the set of orbits. There is a natural quotient map $\pi : X \to X/G$ sending a point to its orbit. We give a topology to X/G by declaring a subset $U \subset X/G$ to be open if and only if $\pi^{-1}(U)$ is open in X; this is the *quotient topology* on X/G. Clearly the quotient map π is continuous; it is an open mapping if the action is continuous, in particular, if it is holomorphic.

Our goal is to put a complex structure on X/G so that the quotient map π is a holomorphic map.

Stabilizer Subgroups. The first step in the process is to understand the stabilizers more precisely.

PROPOSITION 3.1. *Let G be a group acting holomorphically and effectively on a Riemann surface X, and fix a point $p \in X$. Suppose that the stabilizer subgroup G_p is finite. Then in fact G_p is a finite cyclic group.*

In particular, if G is finite, all stabilizer subgroups are finite cyclic subgroups.

PROOF. Fix a local coordinate z centered at p. For any $g \in G_p$, write $g(z) = \sum_{n=1}^{\infty} a_n(g) z^n$; this power series has no constant term since $g(p) = p$. Moreover note that $a_1(g) \neq 0$, since g is an automorphism of X and hence has multiplicity one at every point, in particular at p.

Consider the function $a_1 : G_p \to \mathbb{C}^\times$. Note that it is a homomorphism of groups: $a_1(gh)$ is calculated by computing the power series for $g(h(z))$, and this is

$$\begin{aligned} g(h(z)) &= g(\sum_{n=1}^{\infty} a_n(h) z^n) \\ &= \sum_{m=1}^{\infty} a_m(g) [\sum_{n=1}^{\infty} a_n(h) z^n)]^m \\ &= a_1(g) a_1(h) z + \text{ higher order terms} \end{aligned}$$

so that $a_1(gh) = a_1(g) a_1(h)$.

To finish the proof, we will show that this homomorphism is 1-1. This suffices, since the only finite subgroups of $\mathbb{C}^\times$ are cyclic.

To see that a_1 is 1-1, consider a group element g in the kernel of a_1. This means that $g(z) = z + $ higher order terms. In order to show that the kernel is trivial, we must show that in fact $g(z) = z$, i.e., that all higher order terms of g are zero.

Suppose not; let $m \geq 2$ be the exponent of the first nonzero higher order term of g. Therefore $g(z) = z + az^m \mod z^{m+1}$ with $a \neq 0$.

Now it is elementary to check, by induction, that $g^k(z) = z + kaz^m \mod z^{m+1}$. But since the stabilizer subgroup is finite, this element g must have finite order. Hence for some k, g^k is the identity, i.e., $g^k(z) = z$. Therefore for some k, ka must be zero, forcing $a = 0$. This contradiction shows that in fact g is the identity, and completes the proof. □

PROPOSITION 3.2. *Let G be a finite group acting holomorphically and effectively on a Riemann surface X. Then the points of X with nontrivial stabilizers are discrete.*

PROOF. Suppose that there is a sequence $\{p_n\}$ converging to p such that each p_i has a nontrivial element g_i fixing it. Since G is finite, we may pass to a subsequence and assume that each p_i is fixed by the *same* nontrivial element g. Since g is continuous, it must fix the limit point p also. However, since g is a holomorphic automorphism of X, the Identity Theorem implies that g is the identity. This contradiction proves that points with nontrivial stabilizers cannot accumulate, and in particular they form a discrete set. □

The Quotient Riemann Surface. In order to put a complex structure on the quotient surface X/G, we must find complex charts. The following proposition is fundamental.

PROPOSITION 3.3. *Let G be a finite group acting holomorphically and effectively on a Riemann surface X. Fix a point $p \in X$. Then there is an open neighborhood U of p such that:*
  (a) *U is invariant under the stabilizer G_p, i.e., $g \cdot u \in U$ for every $g \in G_p$ and $u \in U$;*
  (b) *$U \cap (g \cdot U) = \emptyset$ for every $g \notin G_p$;*
  (c) *the natural map $\alpha : U/G_p \to X/G$, induced by sending a point in U to its orbit, is a homeomorphism onto an open subset of X/G;*
  (d) *no point of U except p is fixed by any element of G_p.*

PROOF. Let $G - G_p = \{g_1, \ldots, g_n\}$ be the elements of G not fixing p. Since X is Hausdorff, for each i, we may find open neighborhoods V_i of p and W_i of $g_i \cdot p$ with $V_i \cap W_i = \emptyset$. Note that $g_i^{-1} \cdot W_i$ is an open neighborhood of p for each i. Let $R_i = V_i \cap (g_i^{-1} \cdot W_i)$, let $R = \bigcap_i R_i$, and let

$$U = \bigcap_{g \in G_p} g \cdot R.$$

Clearly each R_i is an open neighborhood of p, and hence so is R and U. Moreover $g \cdot U = U$ for $g \in G_p$; the terms of the intersection defining U are simply permuted upon applying g. This proves (a).

To prove (b), note that $R_i \cap (g_i \cdot R_i) \subset V_i \cap W_i = \emptyset$; hence $R \cap (g_i \cdot R) = \emptyset$ and $U \cap (g_i \cdot U) = \emptyset$ for each i.

Finally, the map $\alpha : U/G_p \to X/G$ is obviously 1-1. It is continuous and open since the composition with the quotient map from U to U/G_p gives the quotient map $\pi|_U$, which is continuous and open. Hence it is a homeomorphism onto its image in X/G.

Finally, (d) follows by the discreteness of the set of points with nontrivial isotropy: simply shrink U if necessary. □

The above Proposition points the way towards defining charts on X/G: we define charts on U/G_p and transport these to X/G via the map α.

Choose a point $\bar{p} \in X/G$, and suppose that $\bar{p}$ is the orbit of a point $p \in X$. Suppose first that $|G_p| = 1$, so that the stabilizer of p is trivial. Then Proposition 3.3 implies that there is a neighborhood U of p such that $\pi|_U : U \to W \subset X/G$ is a homeomorphism onto a neighborhood W of $\bar{p}$. By shrinking U if necessary, we may assume that U is the domain of a chart $\phi : U \to V$ on X. We take as a chart on X/G the composition $\psi = \phi \circ \pi|_U^{-1} : W \to V$. Since both ϕ and $\pi|_U$ are homeomorphisms, this is a chart on X/G.

In order to form a chart near a point $\bar{p}$ with $m = |G_p| \geq 2$, we must find an appropriate function from a neighborhood of $\bar{p}$ to $\mathbb{C}$. Again using Proposition 3.3, choose a G_p-invariant neighborhood U of p such that the natural map $\alpha : U/G_p \to W \subset X/G$ is a homeomorphism onto a neighborhood W of $\bar{p}$. Moreover we may assume that the map $U \to U/G_p$ is exactly m-to-1 away from the point p.

We seek a mapping $\phi : W \to \mathbb{C}$ to serve as a chart near $\bar{p}$. The composition of such a map with α and the quotient map from U to U/G_p would be a G_p-invariant function $h : U \to U/G_p \xrightarrow{\alpha} W \xrightarrow{\phi} \mathbb{C}$ on a neighborhood of p. We will find ϕ by first finding this function h.

Let z be a local coordinate centered at p. For each $g \in G_p$, we have the function $g(z)$, which has multiplicity one at p. Define

$$h(z) = \prod_{g \in G_p} g(z).$$

Note that h has multiplicity $m = |G_p|$ at p, and is defined in some G_p-invariant neighborhood of p; we may shrink U to this neighborhood if necessary, and assume that h is defined on U.

Clearly h is holomorphic and G_p-invariant: applying an element of G_p simply permutes the factors in the definition of h. Therefore h descends to a continuous function $\bar{h} : U/G_p \to \mathbb{C}$. Moreover, since h is open, so is $\bar{h}$.

Finally we claim that $\bar{h}$ is 1-1. This is simply because the holomorphic map h has multiplicity m, and hence is m-to-1 near p; so is the map from U to U/G_p away from p. Therefore $\bar{h}$ is 1-1.

Since $\bar{h}$ is 1-1, continuous, and open, it is a homeomorphism; composing it with the inverse of $\alpha : U/G_p \to W$ gives a chart map ϕ on W:

$$\phi : W \xrightarrow{\alpha^{-1}} U/G_p \xrightarrow{\bar{h}} V \subset \mathbb{C}.$$

Note that the first case of multiplicity one is really a special case of the second case: if $m = 1$, then $h(z) = z$, and we recover the charts described in the first case.

THEOREM 3.4. *Let G be a finite group acting holomorphically and effectively on a Riemann surface X. Then the above construction of complex charts on X/G makes X/G into a Riemann surface. Moreover the quotient map $\pi : X \to X/G$ is holomorphic of degree $|G|$, and $\mathrm{mult}_p(\pi) = |G_p|$ for any point $p \in X$.*

PROOF. These complex charts certainly cover X/G. We must check that they are all compatible, and give a complex atlas on X, and hence a complex structure. Since the points with nontrivial stabilizers are discrete, we may assume that no two chart domains, constructed in the $m \geq 2$ case, meet; hence there is nothing to check there.

Suppose next that the two charts are both constructed in the $m = 1$ case. Then they are compatible, since the original charts on X are compatible.

Finally suppose that we have one chart $\phi_1 : \overline{U}_1 \to V_1$ constructed in the $m = 1$ case, and one $\phi_2 : \overline{U}_2 \to V_2$ constructed in the $m \geq 2$ case. Let U_1 and U_2 be the open sets in X used to construct these charts. Choose a point $\bar{r}$ in the intersection $\overline{U}_1 \cap \overline{U}_2$ of the domains of the two charts; lift $\bar{r}$ to r in $U_1 \cap U_2$. (If U_1 and U_2 do not intersect, replace U_1 by a translate under the group which does intersect U_2.) Let w be the local coordinate in U_1 and z the local coordinate in U_2. The local coordinate in $\overline{U}_1$ is also w, and the local coordinate in $\overline{U}_2$ is $h(z)$, constructed as above. Since h is a holomorphic function, and since z and w are themselves compatible, we see that ϕ_1 and ϕ_2 are compatible.

Since G is finite and X is Hausdorff, so is X/G; since X is connected and $\pi : X \to X/G$ is onto, X/G is also connected. Therefore these charts make X/G into a Riemann surface.

That π is holomorphic is immediate from the definitions of the charts on X/G. Clearly the degree of π is the order of the group $|G|$. Finally, the multiplicity of π at a point p is exactly the multiplicity of the function $h(z)$ constructed above, and this is precisely $|G_p|$. □

The above analysis gives the following interesting Corollary for the way a finite group can act on a Riemann surface, locally. It may be thought of as a version of the Local Normal Form.

COROLLARY 3.5 (LINEARIZATION OF THE ACTION). *Let G be a finite group acting holomorphically and effectively on a Riemann surface X. Fix a point $p \in X$ with nontrivial stabilizer of order m. Let $g \in G_p$ generate the stabilizer subgroup. Then there is a local coordinate z on X centered at p such that $g(z) = \lambda z$, where λ is a primitive m^{th} root of unity. (By replacing g by a different generator of G_p, we may obtain $\lambda = \exp(2\pi i/m)$.)*

PROOF. Choose a local coordinate w on X/G near $G \cdot p$. The Local Normal Form Proposition 4.1 gives the existence of a local coordinate z on X near p such that $w = z^m$ is the formula for π in these coordinates. The preimages of points corresponding to small nonzero values of w exactly differ by m^{th} roots of unity in the z-coordinate. However these preimages are also orbits under the action of elements of the stabilizer subgroup G_p. Therefore, for small z, this G_p-orbit consists of exactly the points $\{\exp(2\pi ik/m)z \mid 0 \leq k \leq m-1\}$. This forces $g(z) = \lambda z$ for some $\lambda = \exp(2\pi ik/m)$ as stated. □

Ramification of the Quotient Map. Let G be a finite group acting holomorphically and effectively on a compact Riemann surface X, with quotient $Y = X/G$. Suppose that $y \in Y$ is a branch point of the quotient map $\pi : X \to Y$. Let $x_1, \ldots, x_s$ be the points of X lying above y; they form a single orbit for the action of G on X. Since the x_i's are all in the same orbit, they all have conjugate stabilizer subgroups, and in particular each stabilizer subgroup is of the same order, say r. Moreover the number s of points in this orbit is the index of the stabilizer, and so is equal to $|G|/r$. These remarks prove the following.

LEMMA 3.6. *Let G be a finite group acting holomorphically and effectively on a compact Riemann surface X, with quotient map $\pi : X \to Y = X/G$. Then for every branch point $y \in Y$ there is an integer $r \geq 2$ such that $\pi^{-1}(y)$ consists of exactly $|G|/r$ points of X, and at each of these preimage points π has multiplicity r.*

We therefore have the following, applying Hurwitz's formula (Theorem 4.16) for the genus:

COROLLARY 3.7. *Let G be a finite group acting holomorphically and effectively on a compact Riemann surface X, with quotient map $\pi : X \to Y = X/G$. Suppose that there are k branch points $y_1, \ldots, y_k$ in Y, with π having multiplicity r_i at the $|G|/r_i$ points above y_i. Then*

$$\begin{aligned} 2g(X) - 2 &= |G|(2g(X/G) - 2) + \sum_{i=1}^{k} \frac{|G|}{r_i}(r_i - 1) \\ &= |G|[2g(X/G) - 2 + \sum_{i=1}^{k}(1 - \frac{1}{r_i})]. \end{aligned}$$

The quantity $\sum_{i=1}^{k}(1 - \frac{1}{r_i})$ is clearly of some importance in studying actions of finite groups on compact Riemann surfaces. In particular, the value of 2 is interesting, given the above formula. The following Lemma is completely elementary, and we leave it to the reader.

LEMMA 3.8. *Suppose that k integers $r_1, \ldots, r_k$ with $r_i \geq 2$ for each i are given. Let $R = \sum_{i=1}^{k}(1 - \frac{1}{r_i})$.*

(a) $R < 2 \iff k, \{r_i\} = \begin{cases} k = 1, \text{ any } r_1; \\ k = 2, \text{ any } r_1, r_2; \text{ or} \\ k = 3, \{r_i\} = \{2, 2, \text{ any } r_3\}; \text{ or} \\ k = 3, \{r_i\} = \{2, 3, 3\}, \{2, 3, 4\}, \text{ or } \{2, 3, 5\}. \end{cases}$

(b) $R = 2 \iff k, \{r_i\} = \begin{cases} k = 3, \{r_i\} = \{2, 3, 6\}, \{2, 4, 4\}, \text{ or } \{3, 3, 3\}; \text{ or} \\ k = 4, \{r_i\} = \{2, 2, 2, 2\}. \end{cases}$

(c) *If $R > 2$ then in fact $R \geq 2\frac{1}{42}$.*

Let us apply these results towards computing the possible finite groups which can act on the Riemann Sphere. Suppose then that G is a finite group acting holomorphically and effectively on $\mathbb{C}_\infty$. Since $\mathbb{C}_\infty$ has genus zero, so must $\mathbb{C}_\infty/G$, and so the Hurwitz formula in this case says that

$$-2 = |G|[-2 + R],$$

where as above $R = \sum_{i=1}^{k}(1 - \frac{1}{r_i})$. In particular, we see that if $G \neq \{1\}$ then $R \neq 0$ and there must be ramification, i.e., $k \geq 1$; in addition we must have $R < 2$, and solving for $|G|$ we see that

$$|G| = \frac{2}{2 - R}.$$

Note that we cannot have $k = 1$ in fact. Just numerically, this makes $R = 1 - 1/r$ for some $r \geq 2$, so $0 < R < 1$ and $2 > 2 - R > 1$; hence $|G| = 2/(2 - R)$ will not be an integer.

A more topological argument is that if $k = 1$, there is only one branch point for the quotient map π. Hence π is unramified over $\mathbb{C}_\infty$ − one point, which is simply connected and has no nontrivial coverings. Therefore π would have to have degree one, which it does not.

Not all of the possibilities of Lemma 3.8(a) in case $k = 2$ can occur, either. In fact if $k = 2$ then r_1 and r_2 must be equal. To see this suppose that the branch points are at y_1 and y_2. Consider a small loop γ in $\mathbb{C}_\infty/G$ around y_1, which starts and ends at a point y_0. This loop γ may be lifted to a curve in $\mathbb{C}_\infty$ starting at any of the $|G|$ points in the fiber of π over y_0. The permutation of this fiber of π given by sending a point p in the fiber to the endpoint of the lift of γ which starts at p, is of order r_1.

Similar considerations apply to a small loop around y_2, giving a permutation which is of order r_2. However since $\mathbb{C}_\infty/G \cong \mathbb{C}_\infty$, these two loops are homotopic; hence the permutations must have the same order, so $r_1 = r_2 = r$, say. Note that $|G| = 2/(2 - R) = r$ in this case. Indeed, this case is achieved by a cyclic group of order r, acting on $\mathbb{C}_\infty$ by multiplying the coordinate z by r^{th} roots of unity.

In case $k = 3$, we see that:

$$\begin{aligned}
\text{if } \{r_i\} &= 2,2,r, & \text{then} && |G| &= 2r; \\
\text{if } \{r_i\} &= 2,3,3, & \text{then} && |G| &= 12; \\
\text{if } \{r_i\} &= 2,3,4, & \text{then} && |G| &= 24; \\
\text{if } \{r_i\} &= 2,3,5, & \text{then} && |G| &= 60.
\end{aligned}$$

The first case is achieved by the action of a dihedral group. The latter cases are achieved by actions of A_4, S_4, and A_5. These are the famous "platonic solid actions", which are groups acting on the sphere leaving either a tetrahedron (the $2,3,3$ case), a cube and an octahedron (the $2,3,4$ case), or a dodecahedron and an icosahedron (the $2,3,5$ case) invariant.

Let us finish this subsection by briefly mentioning finite group actions on Riemann surfaces of genus one. Suppose X has genus one, and G is a finite group acting holomorphically and effectively on X. Then X/G has genus at most one. If X/G has genus one, then we see from Corollary 3.7 that $0 = |G|R$, so $R = 0$

and there is no ramification for the map π. Hence none of the automorphisms of X given by the action of group elements of G have any fixed points. This forces them all to be translations of X, and so G is a finite abelian group of translations of X.

If X/G has genus 0, then we see that $0 = |G|(-2+R)$, so $R = 2$ and we have the four cases of Lemma 3.8(b) for the ramification possibilities.

Hurwitz's Theorem on Automorphisms. For Riemann surfaces of genus 2 or more, Corollary 3.7 leads to a bound on the order of the group G which can act holomorphically and effectively. This was first proved by Hurwitz, and is known as Hurwitz' Theorem.

THEOREM 3.9 (HURWITZ' THEOREM). *Let G be a finite group acting holomorphically and effectively on a compact Riemann surface X of genus $g \geq 2$. Then*
$$|G| \leq 84(g-1).$$

PROOF. Corollary 3.7 gives that
$$2g - 2 = |G|[2g(X/G) - 2 + R],$$
where as above $R = \sum_i (1 - 1/r_i)$.

Suppose first that $g(X/G) \geq 1$. If $R = 0$, so there is no ramification to the quotient map, then $g(X/G) \geq 2$, which implies that $|G| \leq g - 1$. If $R \neq 0$, this forces $R \geq 1/2$. Then $2g(x/G) - 2 + R \geq 1/2$, so we have $|G| \leq 4(g-1)$. This finishes the case that $g(X/G) \geq 1$.

Assume then that $g(X/G) = 0$. Then the above reduces to
$$2g - 2 = |G|[-2 + R],$$
which forces $R > 2$. Lemma 3.8(c) then implies that $R - 2 \geq 1/42$. Therefore $|G| \leq 84(g-1)$ as claimed. □

In fact, the group of all automorphisms of a compact Riemann surface of genus at least two is a finite group. This is a bit beyond us now; but it implies that for such a Riemann surface, we have
$$|\operatorname{Aut}(X)| \leq 84(g(X) - 1)$$
since the full group $\operatorname{Aut}(X)$ of automorphisms certainly acts holomorphically and effectively on X. We will prove the finiteness in Chapter VII.

Infinite Groups. In the above discussion we have concentrated on the actions of finite groups; however the reader should be aware that the construction of the complex structure on X/G can be easily made for a certain class of actions of infinite groups.

DEFINITION 3.10. Let G be a discrete group acting effectively on a Hausdorff space X. We say that G acts *properly discontinuously* if for each pair of points (p,q) in X there exist neighborhoods U and V of x and y respectively such that $\{g \in G \mid (g \cdot U) \cap V \neq \emptyset\}$ is finite.

This forces the quotient space to be Hausdorff. Moreover, if X is a Riemann surface and G acts properly discontinuously on X, then points of X with non-trivial stabilizers form a discrete set, and all stabilizers are finite cyclic groups.

Indeed the analogue of Proposition 3.3 holds verbatim. (See [**tomDieck87**, Chapter I, Section 3] for the basic theory.) This allows one to put a complex structure on X/G in the same manner as outlined above.

The first example of the case of an infinite group action is the action of $\mathbb{Z} \times \mathbb{Z}$ on $\mathbb{C}$ given by translation in two linearly independent directions. The quotient space is a complex torus.

The second and primary example is a discrete group of automorphisms of the complex disc. This is of fundamental importance, since the universal covering of any compact Riemann surface of genus at least two is the disc, so the deck transformations of the universal covering give a holomorphic and effective action of a discrete group, with quotient the given compact Riemann surface. A recent introduction to this can be found in [**JS87**, Chapter 5] and [**FK80**, Chapter IV].

Problems III.3

A. Let G be a finite group, acting on a set X. For $p \in X$, show that the order of the orbit of p times the order of the stabilizer subgroup of p equals the order of the group G:
$$|G \cdot p| \ |G_p| = |G|.$$

B. Show that the kernel K of an action of G on X is a normal subgroup of G, and that the quotient group G/K acts on X with trivial kernel and identical orbits to the G action.

C. Assume that G acts continuously on X. Show that the quotient map $\pi : X \to X/G$ is an open mapping.

D. Suppose that $g(z) = z + az^m \mod z^{m+1}$ with $a \neq 0$. Check that $g^k(z) = z + kaz^m \mod z^{m+1}$.

E. Let $G \subset \mathbb{C}^\times$ be a finite subgroup of order n. Show that G consists of exactly the n n^{th}-roots of unity: $G = \{\exp(2\pi i k/n) \mid 0 \leq k \leq n-1\}$.

F. Let G act continuously on a topological space X, and let Y be a topological space. Show that a map $\alpha : X/G \to Y$ is continuous if and only if $\alpha \circ \pi : X \to Y$ is continuous and G-invariant. Show that there is a 1-1-correspondence between

$$\left\{ \begin{array}{c} \text{continuous maps} \\ \alpha : X/G \to Y \end{array} \right\} \quad \text{and} \quad \left\{ \begin{array}{c} G\text{-invariant} \\ \text{continuous maps} \\ \beta : X \to Y \end{array} \right\}$$

which associates to α the map $\beta = \alpha \circ \pi$.

G. Prove Lemma 3.8. With the notation of that Lemma, show that $R = 2\frac{1}{42}$ if and only if $k = 3$ and $\{r_i\} = \{2, 3, 7\}$.

H. Show that the group of automorphisms of $\mathbb{C}_\infty$ generated by the two automorphisms sending z to $\exp(2\pi i/r)z$ and sending z to $1/z$ is a dihedral group of order $2r$, which acts holomorphically and effectively on $\mathbb{C}_\infty$. Show that there are three branch points to the quotient map, with ramification numbers $2, 2, r$.

I. Define holomorphic and effective actions of A_4, S_4, and A_5 on $\mathbb{C}_\infty$ such that the quotient map has 3 branch points with ramification numbers $2, 3, 3$, $2, 3, 4$, and $2, 3, 5$ respectively. Compute the ramification points, and show that when they are represented as points on the two-sphere $S^2 \subset \mathbb{R}^3$, one of the three orbits form the vertices of a regular solid.

J. Define holomorphic and effective actions of finite groups on Riemann surfaces of genus one which have genus zero quotient, and realize the four cases of Lemma 3.8(b).

K. Show that the "Klein curve" X defined by $xy^3 + yz^3 + zx^3 = 0$ is a smooth projective plane curve. Since it has degree 4, X has genus 3. Show that it realizes the Hurwitz bound by finding 168 automorphisms of X.

4. Monodromy

In this section we will introduce the concept of *monodromy* for a holomorphic map between compact Riemann surfaces, and show how the monodromy may be used to recover the map itself. We will assume that the reader is familiar with the basic ideas about the fundamental group of a real manifold, and the relationship between the fundamental group and covering spaces. There are many good references for this elementary material; see for example [**Munkres75**], [**Massey67**], or [**Armstrong83**].

We will at least define everything so there is no confusion about terminology.

Covering Spaces and the Fundamental Group. Let V be a connected real manifold, and fix a base point $q \in V$. A *path* on V is a continuous map $\gamma : [0, 1] \to V$. A *loop* based at q is a path on V such that $\gamma(0) = \gamma(1) = q$. Two loops γ_1 and γ_2 are said to be *homotopic* if there is a continuous map $G : [0, 1] \times [0, 1] \to V$ such that $G(0, t) = \gamma_1(t)$ and $G(1, t) = \gamma_2(t)$ for all t, and $G(s, 0) = G(s, 1) = q$ for all s. Homotopy is an equivalence relation on the set of all loops based at q. The *fundamental group* of V is the set of homotopy classes of loops based at q, and is denoted by $\pi_1(V, q)$. The operation of concatenation of loops gives a group structure to $\pi_1(V, q)$. A connected space is *simply connected* if its fundamental group is trivial.

A *covering space* of V is a continuous map $F : U \to V$ such that F is onto, and for each point $v \in V$ there is a neighborhood W of v in V such that $F^{-1}(W)$ consists of a disjoint union of open sets U_α, each mapping via F homeomorphically onto W.

A covering space $F: U \to V$ enjoys the *path-lifting property*: for any path $\gamma: [0,1] \to V$ and any preimage p of $\gamma(0)$ there is a path $\tilde{\gamma}$ on U such that $\tilde{\gamma}(0) = p$ and $F \circ \tilde{\gamma} = \gamma$. In other words, one can lift the path γ to a path on U, starting at any preimage of the starting point of γ.

There is a straightforward notion of *isomorphism* of covers: two covers $F_1: U_1 \to V$ and $F_2: U_2 \to V$ are *isomorphic* if there is a homeomorphism $G: U_1 \to U_2$ such that $F_2 \circ G = F_1$.

There exists a *universal covering space* $F_0: U_0 \to V$ such that U_0 is simply connected; moreover $F_0: U_0 \to V$ is unique up to isomorphism. The universal property of the universal cover is that if $F: U \to V$ is any other connected covering space of V, then F factors through F_0 uniquely, in the sense that there is a unique covering map $G: U_0 \to U$ such that $F_0 = F \circ G$.

The fundamental group $\pi_1(V, q)$ acts on the universal cover $F_0: U_0 \to V$ as follows. Fix a point $p \in U_0$ which maps to the base point $q \in V$. Choose a loop γ on V based at q, and a point $u \in U$. Choose a path α on U starting at u and ending at p. Then $F_0 \circ \alpha$ is a path on V, starting at $F_0(u)$ and ending at q. Its reverse, $-F_0 \circ \alpha$, starts at q and ends at $F_0(u)$. Consider the unique lift $\tilde{\gamma}$ of the loop γ which starts at p, and the unique lift β of the reverse path $-F_0 \circ \alpha$ which starts at the endpoint $\tilde{\gamma}(1)$ of $\tilde{\gamma}$. The endpoint $\beta(1)$ of this last path β lies over the point $F_0(u)$.

There is a lot to check here, but the bottom line is that this point $\beta(1)$ depends only on the point u and the homotopy class $[\gamma]$ of the loop γ; call the point $[\gamma] \cdot u$. This gives an action of $\pi_1(V, q)$ on the universal cover $F_0: U_0 \to V$, and the action preserves the fibers of the covering map F_0. Moreover, the orbit space $U_0/\pi_1(V, q)$ is naturally homeomorphic to the original space V.

Given any subgroup $H \subseteq \pi_1(V, q)$ of the fundamental group, the above action may be restricted to an action of H on the universal cover. The orbit space U_0/H maps to V, and is a covering space of V. Moreover every connected covering space of V occurs this way; two such orbit spaces are isomorphic (as coverings of V) if and only if the subgroups are conjugate subgroups of the fundamental group. Therefore there is a 1-1 correspondence

$$\left\{ \begin{array}{c} \text{isomorphism classes of} \\ \text{connected coverings} \\ F: U \to V \end{array} \right\} \leftrightarrow \left\{ \begin{array}{c} \text{conjugacy classes} \\ \text{of subgroups} \\ H \subseteq \pi_1(V, q) \end{array} \right\}.$$

As noted above, given the subgroup $H \subseteq \pi_1(V, q)$, the covering is obtained by taking the orbit space U_0/H, where U_0 is the universal covering space. Conversely, given a connected covering $F: U \to V$, choose a point $p \in U$ lying over the base point q, and take the subgroup $H \subseteq \pi_1(V, q)$ to be those homotopy classes $[\gamma]$ such that $[\gamma] \cdot p = p$. This subgroup depends on the point p, but its conjugacy class does not. The degree of the covering (that is, the number of preimages of a point of V) is exactly the index of the subgroup H inside the fundamental group.

EXAMPLE 4.1. Let $X = \mathbb{C}/L$ be a complex torus. Then the natural quotient map $\pi : \mathbb{C} \to X$ is the universal cover of X. The fundamental group of X is a free abelian group on two generators, isomorphic to the lattice L. The action of L on the universal cover $\mathbb{C}$ is by translation.

EXAMPLE 4.2. Let V be the punctured unit disc $\{z \in \mathbb{C} \mid 0 < |z| < 1\}$. Let $\mathbb{H}$ be the upper half plane $\{z \in \mathbb{C} \mid \text{Im}(z) > 0\}$. The map $F : \mathbb{H} \to V$ defined by $F(z) = \exp(2\pi i z)$ is the universal cover of V. The fundamental group of V is an infinite cyclic group, generated by any loop in V with winding number one about the origin. The action of the fundamental group on the universal cover $\mathbb{H}$ is by translation by integers: z is sent to $z + n$ for an integer n.

If we identify the fundamental group $\pi_1(V, q)$ with $\mathbb{Z}$, we see that the only subgroups are those generated by a nonnegative integer $N \geq 0$: $N\mathbb{Z}$. When $N = 0$, we have the trivial subgroup, and this corresponds to the universal cover $\mathbb{H} \to V$, which has infinite degree. When $N = 1$, we have the entire fundamental group, and this corresponds to the trivial covering of V by itself (via the identity). For $N \geq 2$, the covering space corresponds to the quotient of the universal cover $\mathbb{H}$ by the translation $z \mapsto z + N$; this quotient is also a punctured disc D_N, and the quotient map $\pi_N : \mathbb{H} \to D_N$ sends z to $\exp(2\pi i z/N)$. If we denote the coordinate in the disc D_N by w_N, we see that the covering map may be expressed as $w_N = \exp(2\pi i z/N)$, where z is the coordinate in $\mathbb{H}$. In particular the original coordinate in the space V is w_1, and the covering $F_N : D_N \to V$ is given by $w_1 = w_N^N$. Therefore these intermediate coverings are simply the punctured disc again, and the covering map is a power map, of degree N.

The Monodromy of a Finite Covering. Let $F : U \to V$ be a connected covering space of finite degree d, so that all points have exactly d preimages. If F corresponds to a subgroup $H \subseteq \pi_1(V, q)$, then the degree d is the index of the subgroup H.

Consider the fiber $F^{-1}(q)$ over q. Denote the d points in this fiber by $\{x_1, \ldots, x_d\}$. Every loop γ in V based at q can be lifted to d paths $\tilde{\gamma}_1, \ldots, \tilde{\gamma}_d$, where $\tilde{\gamma}_i$ is the unique lift of γ which starts at x_i. In other words, $\tilde{\gamma}_i(0) = x_i$ for every i.

Next consider the endpoints $\tilde{\gamma}_i(1)$; these also lie over q, and indeed form the entire preimage set $F^{-1}(q)$. Hence each is an x_j for some j; we denote $\tilde{\gamma}_i(1)$ by $x_{\sigma(i)}$.

This function σ is a *permutation* of the indices $\{1, \ldots, d\}$, and it is easy to see that it depends only on the homotopy class of the loop γ. Therefore we have a group homomorphism

$$\rho : \pi_1(V, q) \to S_d$$

where S_d denotes the symmetric group of all permutations on d indices.

DEFINITION 4.3. The *monodromy representation* of a covering map $F : U \to V$ of finite degree d is the group homomorphism $\rho : \pi_1(V, q) \to S_d$ defined above.

The connectedness of the domain U gives the following property to the monodromy representation. We say that a subgroup $H \subseteq S_d$ is *transitive* if for any pair of indices i and j there is a permutation σ in the subgroup H which sends i to j: $\sigma(i) = j$.

LEMMA 4.4. *Let $\rho : \pi_1(V, q) \to S_d$ be the monodromy representation of a covering map $F : U \to V$ of finite degree, with U connected. Then the image of ρ is a transitive subgroup of S_d.*

PROOF. With the notation introduced above, fix two indices i and j, and consider the two points x_i and x_j in the fiber of F over q. Since U is connected, we may find a path $\tilde\gamma$ on U starting at x_i and ending at x_j. Let $\gamma = F \circ \tilde\gamma$ be the image of $\tilde\gamma$ in V; note that γ is a loop in V based at q, since both x_i and x_j map to q under F. Then by construction we have that $\rho([\gamma])$ is a permutation which sends i to j. □

EXAMPLE 4.5. Let $\pi_N : D_N \to V = D_1$ be the covering map of punctured unit discs given by the Nth power map: if w_i is the coordinate in D_i then the map is given by $w_1 = w_n^N$. Let $q = 1/2^N$ be the base point in D_1. If we let $\zeta = \exp(2\pi i/N)$ be a primitive N^{th} root of unity, then the preimages of q are the points $x_i = \zeta^i/2$, for $i = 1, \ldots, N$. The generator γ for the fundamental group $\pi_1(V, q)$ is given by the loop $w_1(t) = \exp(2\pi it)/2^N$ for $t \in [0, 1]$. This loop lifts to the loops $\tilde\gamma_i$ given by $w_N(t) = \zeta^i \exp(2\pi it/N)/2$ for $t \in [0, 1]$, whose starting point is at $\zeta^i/2$ and whose ending point is at $\zeta^{i+1}/2$. Therefore the monodromy representation ρ for this covering sends the generator $[\gamma]$ of the fundamental group to the cyclic permutation which sends i to $i + 1$ (modulo N) for each i.

The Monodromy of a Holomorphic Map. Let us apply this theory of covering spaces, the fundamental group, and monodromy representations to the case of a holomorphic nonconstant map $F : X \to Y$ between compact Riemann surfaces. Because of ramification, F is *not* in general a covering map. Let $R \subset X$ be the finite set of ramification points of F, and let $B = F(R) \subset Y$ be the finite set of branch points. Let $V = Y - B$ and let $U = X - F^{-1}(B)$. Note that we are removing all of the branch points from Y, and all of the ramification points from X; but in addition from X we are also removing any point which maps to a branch point, that is, any point in the same fiber of F as a ramification point. These need not all be ramification points.

Note that for any $v \in V$, the preimage set $F^{-1}(v)$ consists of d distinct points, each having multiplicity one for the holomorphic map F. Therefore the restriction $F|_U : U \to V$ is a true covering map, of degree d.

This covering therefore has a monodromy representation $\rho : \pi_1(V, q) \to S_d$; it is called the *monodromy representation of the holomorphic map F*. Since X is connected, so is the open subset U, and hence the image is a transitive subgroup of S_d.

For each branch point $b \in Y$, choose a small open neighborhood W of b in Y; the punctured open set $W - \{b\}$ is an open subset of V, isomorphic to a small punctured disc. Denote by $u_1, \ldots, u_k$ the k preimages of b in X; the number k will be less than the degree d of F since b is a branch point, so that at least one of the u_j's are ramification points.

We choose W small enough so that $F^{-1}(W)$ decomposes as a disjoint union of open neighborhoods $U_1, \ldots, U_k$ of the points $u_1, \ldots, u_k$ respectively. Set $m_j = \text{mult}_{u_j}(F)$ to be the multiplicity of F at these preimage points; by the Local Normal Form, there are coordinates z_j on the U_j's and z on W so that the map F has the form $z = z_j^{m_j}$ on U_j.

Now consider $U_j - \{u_j\}$; $U_j - \{u_j\}$ is isomorphic to a punctured disc, and the map F sends $U_j - \{u_j\}$ to $W - \{b\}$ via the m_j^{th} power map. Choose a path α from the base point q to a point q_0 in $W - \{b\}$, and a loop β in $W - \{b\}$ based at q_0 with winding number one around the branch point b. Then the path $\alpha^{-1}\beta\alpha$ (composing paths from right to left) is a loop γ on V based on q. We call such a loop on V a *small loop on V around b*.

In analyzing how the small loop γ lifts to the covering $F : U \to V$, it is clear that traversing the path α simply gives an identification of the fiber of F over q with the fiber of F over q_0; following the reverse path α^{-1} gives the inverse identification. Therefore the permutation σ of the fiber of F over q which is induced (via the monodromy representation ρ) by the small loop γ around b is actually determined up to this identification by the loop β around the branch point b.

Above the open set $W - \{b\}$ we have k punctured discs $U_j - \{u_j\}$, each mapping to $W - \{b\}$ via a power map. This situation was analyzed in Example 4.5; the monodromy for each cover $U_j - \{u_j\} \to W - \{b\}$ is a cyclic permutation of those m_j preimages of q_0 which lie in U_j. In fact the loop β induces a cyclic permutation of these points, and therefore the loop γ also maps to a cyclic permutation of the corresponding identified points in the fiber above the base point q. Therefore we know the cycle structure of the permutation σ, and we have proved the following.

LEMMA 4.6. *Suppose that above the branch point $b \in Y$ there are k preimages $u_1, \ldots, u_k$, with $\text{mult}_{u_j}(F) = m_j$. Then with the above notation the cycle structure of the permutation σ representing a small loop around b (after the identification via the path α) is $(m_1, \ldots, m_k)$.*

Coverings via Monodromy Representations. Suppose a connected real manifold V is given, with a chosen base point q. Suppose further that we have a group homomorphism $\rho : \pi_1(V, q) \to S_d$, from the fundamental group of V to a symmetric group S_d, with a transitive image. Fix an index, say 1. Let $H \subseteq \pi_1(V, q)$ be the subgroup consisting of those homotopy classes $[\gamma]$ such that

$\rho([\gamma])$ fixes the index 1:

$$H = \{[\gamma] \in \pi_1(V,q) \mid \rho([\gamma])(1) = 1\}.$$

Then H has index d in $\pi_1(V,q)$, and by the general theory induces a connected covering space $F_\rho : U_\rho \to V$. Moreover this covering has the property that its monodromy representation is exactly the given homomorphism ρ of course.

This process essentially gives an inverse to the mapping which sends a covering to its monodromy representation; the only caveat is that this only works for coverings of finite degree. Hence we have the following: for a connected real manifold V, there is a 1-1 correspondence

$$\left\{\begin{array}{c} \text{isomorphism classes of} \\ \text{connected coverings} \\ F: U \to V \\ \text{of degree } d \end{array}\right\} \leftrightarrow \left\{\begin{array}{c} \text{group homomorphisms} \\ \rho : \pi_1(V,q) \to S_d \\ \text{with transitive image} \\ (\text{up to conjugacy in } S_d) \end{array}\right\}.$$

The reason for the conjugacy in S_d is easy to see: this simply reflects a relabeling of the points in the fiber of the covering over the base point.

Now further assume that V is a Riemann surface. It is a general principle that *coverings of Riemann surfaces are Riemann surfaces*. We have seen this principle at work before in discussions concerning complex tori; chart maps on the covering space are given by composing the covering map with chart functions on the target Riemann surface. Moreover this way of putting charts on the covering space is forced if you want the covering map to be holomorphic. Therefore:

LEMMA 4.7. *Let $F : U \to V$ be a connected covering map of a Riemann surface V. Then there is a unique complex structure on U such that F is a holomorphic map.*

If the reader is interested he or she may supply the details of the proof of the lemma quite easily. In particular the universal cover of any Riemann surface is a Riemann surface.

Combining this with the previous 1-1 correspondence gives us the following.

COROLLARY 4.8. *Let V be a Riemann surface. Then there is a 1-1 correspondence*

$$\left\{\begin{array}{c} \textit{isomorphism classes of} \\ \textit{unramified holomorphic maps} \\ F: U \to V \\ \textit{of degree } d \end{array}\right\} \leftrightarrow \left\{\begin{array}{c} \textit{group homomorphisms} \\ \rho : \pi_1(V,q) \to S_d \\ \textit{with transitive image} \\ (\textit{up to conjugacy in } S_d) \end{array}\right\}.$$

Holomorphic Maps via Monodromy Representations. We want to apply the constructions given above to construct *branched* coverings of Riemann surfaces, that is, holomorphic maps with ramification. Of course such a map is not a covering map in the sense of topology, so we must finesse this somehow.

Let Y be a compact Riemann surface, and let $B \subset Y$ be a finite subset. Let $V = Y - B$ be the complement of B, which is an open subset of Y and is also a Riemann surface. Fix a base point $q \in V$. Suppose that one has a group homomorphism $\rho : \pi_1(V,q) \to S_d$, with transitive image. Let $F_\rho : U_\rho \to V$ be the covering map induced by ρ; the space U_ρ is a Riemann surface and the map F_ρ is a holomorphic map of degree d, by the previous discussion.

Let us focus attention on a point $b \in B$ which has been removed from Y to create the open set V. Let W be a small open neighborhood of b in Y, so that $W - \{b\}$ is isomorphic to a punctured disc. If W is small enough, the preimage $F_\rho^{-1}(W - \{b\})$ will decompose into a disjoint collection $\tilde{U}_j$ of covers of $W - \{b\}$. Now recall that finite degree covers of a punctured disc have been classified: they are all punctured discs, and the covering map is a power map. We may therefore suppose that each $\tilde{U}_j$ is a punctured disc also, and that the map F_ρ restricted to $\tilde{U}_j$ is a power map; say that the power for the domain $\tilde{U}_j$ is m_j.

We may further shrink W if necessary and assume that W is completely contained in a chart domain for Y. Then we see that on each $\tilde{U}_j$ we have a *hole chart* for U_ρ, since $\tilde{U}_j$ is isomorphic to a punctured disc. Hence we may plug each of these holes in U_ρ; moreover when we do this (for each branch point and for each hole chart) the resulting surface X_ρ maps holomorphically to Y, extending the covering map $F_\rho : U_\rho \to V$. The reason for the existence of the extension is simple: on each $\tilde{U}_j$ the map F_ρ is the m_j^{th} power map from $\tilde{U}_j$ to $W - \{b\}$, and the power map from one punctured disc to another extends to the unpunctured discs. Therefore if the disc which plugs the hole of $\tilde{U}_j$ is denoted by U_j, we have a unique extension of the holomorphic map F_ρ from U_j to W. These all combine to extend the map F_ρ to a map $F_\rho : X_\rho \to Y$.

Note lastly that the Riemann surface X_ρ obtained by plugging all these holes is compact. Indeed, if we delete each W from Y we obtain a compact subset, and its preimage Z in X_ρ is also compact (a finite covering of a compact set is compact). Since X_ρ is the union of Z and the closures of all of the U_j's (over all of the branch points), we see that X_ρ is a union of finitely many compact sets and is therefore compact.

Finally note that if we remove B from Y and its preimage from X_ρ, we obtain a covering map with monodromy representation ρ.

Note that $F_\rho : X_\rho \to Y$ has as its branch points at most the finite set B: at all other points F_ρ is unramified. We may compute the multiplicities of F_ρ at the points lying above a point $b \in B$ by considering the cycle structure of a small loop γ in V around b constructed as in the proof of Lemma 4.6. Such a loop γ must be of the form $\alpha^{-1}\beta\alpha$, where α is a path from the base point q to a point near b, and β is a small loop winding once around b. Then if the cycle structure of $\rho([\gamma])$ is $(m_1, \ldots, m_k)$, then there are k preimages $u_1, \ldots, u_k$ of b in X_ρ, and $\mathrm{mult}_{u_j}(F_\rho) = m_j$.

We see that it is not necessary that each point b of B be a branch point for the map F_ρ: in particular, if the cycle structure above is $(1, 1, \ldots, 1)$, then above b we will have d preimages, each having multiplicity one, and so there will be no ramification above b.

Summarizing, we have the following.

PROPOSITION 4.9. *Let Y be a compact Riemann surface, let B be a finite subset of Y, and let q be a base point of $Y - B$. Then there is a 1-1 correspondence*

$$\left\{ \begin{array}{c} \text{isomorphism classes of} \\ \text{holomorphic maps} \\ F : X \to Y \\ \text{of degree } d \\ \text{whose branch points lie in } B \end{array} \right\} \leftrightarrow \left\{ \begin{array}{c} \text{group homomorphisms} \\ \rho : \pi_1(Y - B, q) \to S_d \\ \text{with transitive image} \\ \text{(up to conjugacy in } S_d) \end{array} \right\}.$$

Moreover at a point $b \in B$, if γ is a small loop in $Y - B$ around b based at q, and if $\rho([\gamma])$ has cycle structure $(m_1, \ldots, m_k)$, then there are k preimages $u_1, \ldots, u_k$ of b in the corresponding cover $F_\rho : X_\rho \to Y$, with $\mathrm{mult}_{u_j}(F_\rho) = m_j$ for each j.

Holomorphic Maps to $\mathbb{P}^1$. The previous proposition is especially useful in constructing Riemann surfaces together with holomorphic maps to the projective line $\mathbb{P}^1$. Fix n points $b_1, \ldots, b_n$ in $\mathbb{P}^1$ and a base point q (which is not one of the b_i's). Let $V = \mathbb{P}^1 - \{b_1, \ldots, b_n\}$; V is a Riemann surface, and since $\mathbb{P}^1$ is topologically a sphere, we have that the fundamental group of V is a free group on n generators $[\gamma_1], \ldots, [\gamma_n]$, subject to the single relation that

$$[\gamma_1][\gamma_2] \cdots [\gamma_n] = 1$$

in $\pi_1(V, q)$. Indeed, each $[\gamma_i]$ is the homotopy class of a small loop on V around b_i.

Therefore a group homomorphism $\rho : \pi_1(V, q) \to S_d$ is determined by choosing n permutations $\sigma_i = \rho([\gamma_i])$, subject only to the condition that

$$\sigma_1 \sigma_2 \cdots \sigma_n = 1$$

in S_d. The image of ρ will be the subgroup generated by the σ_i's.

Applying Proposition 4.9, we have the following.

COROLLARY 4.10. *Fix a finite set $B = \{b_1, \ldots, b_n\} \subset \mathbb{P}^1$. Then there is a 1-1 correspondence*

$$\left\{ \begin{array}{c} \textit{isomorphism classes of} \\ \textit{holomorphic maps} \\ F : X \to \mathbb{P}^1 \\ \textit{of degree } d \\ \textit{whose branch points} \\ \textit{lie in } B \end{array} \right\} \leftrightarrow \left\{ \begin{array}{c} \textit{conjugacy classes of n-tuples} \\ (\sigma_1, \ldots, \sigma_n) \textit{ of permutations in } S_d \\ \textit{such that } \sigma_1 \cdots \sigma_n = 1 \\ \textit{and the subgroup} \\ \textit{generated by the } \sigma_i\textit{'s} \\ \textit{is transitive} \end{array} \right\}.$$

Moreover if σ_i has cycle structure $(m_1, \ldots, m_k)$, then there are k preimages $u_1, \ldots, u_k$ of b_i in the corresponding cover $F : X \to Y$, with $\mathrm{mult}_{u_j}(F) = m_j$ for each j.

Hyperelliptic Surfaces. Recall that a hyperelliptic curve is a compact Riemann surface X defined by an equation of the form $y^2 = h(x)$, where h is a polynomial with distinct roots. The coordinate function x induces a holomorphic map $F : X \to \mathbb{P}^1$ which has degree 2.

Using the monodromy theory developed above, we can prove a sort of converse to this statement:

PROPOSITION 4.11. *Let X be a compact Riemann surface. Suppose that $F : X \to \mathbb{P}^1$ is a holomorphic map of degree 2. Then X is a hyperelliptic curve.*

PROOF. Let g be the genus of X; by Hurwitz's formula we have that the number of branch points of F is $2g+2$, and since the degree of F is 2, each branch point has as its preimage a single ramification point with multiplicity two. Let $B = \{b_1, \ldots, b_{2g+2}\}$ be the set of branch points of F, and let $V = \mathbb{P}^1 - B$; the monodromy representation for F is a group homomorphism $\rho : \pi_1(V) \to S_2$ and we denote by σ_i the image under ρ of the homotopy class of a small loop in V around b_i.

Since there is one point of multiplicity two lying above b_i for each i, the cycle structure of σ_i must be (2) for each i; this forces σ_i to be the transposition $(12) \in S_2$ for each i. Hence once the branch points $\{b_1, \ldots, b_{2g+2}\}$ are chosen, there is no choice for the permutations σ_i, and the Riemann surface X is determined up to isomorphism by the branch points alone.

On the other hand, if z is an affine coordinate on $\mathbb{P}^1$, and none of the branch points b_i is the point at infinity, then the hyperelliptic curve Y defined by the equation $y^2 = \prod_{i=1}^{2g+2}(z - b_i)$ also covers $\mathbb{P}^1$ with the same branch points by a map of degree 2 (given by the function z). Therefore this covering $z : Y \to \mathbb{P}^1$ has the same branch points and the same monodromy as does the given map $F : X \to \mathbb{P}^1$. By the 1-1 correspondence, we must have that these coverings are isomorphic, and so $X \cong Y$.

If one of the branch points, say b_{2g+2}, is the point at infinity, simply consider instead the hyperelliptic curve defined by the equation $y^2 = \prod_{i=1}^{2g+1}(z - b_i)$; the argument proceeds in the same manner. $\square$

This Proposition is often used as the defining property of a hyperelliptic curve: a hyperelliptic curve is one which has a map of degree two onto the projective line.

Problems III.4

A. Let L be a lattice in $\mathbb{C}$ and let $\pi : \mathbb{C} \to X = \mathbb{C}/L$ be the natural quotient map. Show that if $M \subset L$ is a sublattice of L, then the covering of X corresponding to the subgroup M is the complex torus $\mathbb{C}/M$, and the covering map is the natural map sending $\mathbb{C}/M$ to $\mathbb{C}/L$.

B. Let L be a lattice generated by 1 and τ, with $\text{Im}(\tau) > 0$. Let H be the subgroup of L generated by 1; that is, H is the integers. Show that the covering of $X = \mathbb{C}/L$ corresponding to the subgroup H is isomorphic to $\mathbb{C}^*$, and write down the covering maps.

C. Let $V = \mathbb{C}^*$. Show that the universal covering of V is $\mathbb{C}$, and find the universal covering map. Show that the fundamental group of V is infinite cyclic. Determine all connected coverings of V up to isomorphism.

D. Suppose that a connected covering $F_\rho : U_\rho \to V$ is defined via a group homomorphism $\rho : \pi_1(V, q) \to S_d$ with transitive image as in the text. Show that the monodromy representation of F_ρ is ρ.

E. Prove Lemma 4.7.

F. Suppose that a holomorphic map $F : X \to \mathbb{P}^1$ of degree d is defined by the correspondence of Corollary 4.10, that is, a set of branch points $\{b_1, \ldots, b_n\}$ in $\mathbb{P}^1$ are chosen, and a set of corresponding permutations $\sigma_1, \ldots, \sigma_n$ in S_d are given, which generate a transitive subgroup of S_d and whose product is 1. Suppose that the permutation σ_i is a product of k_i disjoint cycles. Show that the genus g of the compact Riemann surface X is

$$g = 1 + \frac{(n-2)d - \sum_{i=1}^n k_i}{2}.$$

G. Let $f(z) = z^3/(1-z^2)$ define a holomorphic map of degree 3 from $\mathbb{P}^1$ to itself. Find all of the branch points, and the corresponding permutations in S_3.

H. Let $f(z) = 4z^2(z-1)^2/(2z-1)^2$ define a holomorphic map of degree 4 from $\mathbb{P}^1$ to itself. Show that there are three branch points, and that the three corresponding permutations in S_4 are $\sigma_1 = (12)(34)$, $\sigma_2 = (13)(24)$, and $\sigma_3 = (14)(23)$ up to conjugacy.

I. Let X denote the Fermat curve of degree d in $\mathbb{P}^2$, defined by the homogeneous polynomial $x^d + y^d + z^d = 0$. Let $F : X \to \mathbb{P}^1$ be defined by $F([x : y : z]) = [x : y]$. Show that F has d branch points, and find the d corresponding permutations.

J. Let G be the dihedral group of order $2r$ acting on $\mathbb{P}^1$, with three branch points b_1, b_2, b_3 for the quotient map $\pi : \mathbb{P}^1 \to \mathbb{P}^1$; moreover assume that for each $i = 1, 2, 3$ the map π has multiplicity r_i at each of $2r/r_i$ points lying above b_i, with $\{r_i\} = \{2, 2, r\}$. Find the three corresponding permutations in S_{2r}.

K. Do the same computation as above, for the groups of order 12, 24, and 60 which act on $\mathbb{P}^1$; here the $\{r_i\}$ numbers are $\{2,3,3\}$, $\{2,3,4\}$, and $\{2,3,5\}$ respectively.

L. Let Y be a Riemann surface of genus $g \geq 1$. The fundamental group of Y is a free group on $2g$ generators $a_1, \ldots, a_g, b_1, \ldots, b_g$ subject to the single relation that

$$a_1 b_1 a_1^{-1} b_1^{-1} a_2 b_2 a_2^{-1} b_2^{-1} \cdots a_g b_g a_g^{-1} b_g^{-1} = 1.$$

Therefore an unramified covering $F: X \to Y$ of degree two is determined by giving $2g$ permutations in S_2 satisfying the above relation, which generate a transitive subgroup. For the permutations to generate a transitive subgroup is easy: not all of the permutations should be the identity. Show that the number of nonisomorphic unramified coverings of Y is $2^{2g} - 1$. In the case of $g = 1$, assume that Y is a complex torus given by a lattice L in $\mathbb{C}$; find the three sublattices of L corresponding to the three nonisomorphic covers.

5. Basic Projective Geometry

In this section we will develop somewhat further the basic notions of projective n-space $\mathbb{P}^n$.

Homogeneous Coordinates and Polynomials. Recall that $\mathbb{P}^n$ is the set of 1-dimensional subspaces of $\mathbb{C}^{n+1}$. If $(x_0, \ldots, x_n)$ is a nonzero vector in $\mathbb{C}^{n+1}$, its span, which is a 1-dimensional subspace, is denoted by $[x_0 : \cdots : x_n] \in \mathbb{P}^n$. Every point of $\mathbb{P}^n$ may be written in this way; moreover

$$[x_0 : \cdots : x_n] = [\lambda x_0 : \cdots : \lambda x_n] \text{ for any } \lambda \in \mathbb{C}, \lambda \neq 0$$

and if $[x_0 : \cdots : x_n] = [y_0 : \cdots : y_n]$ then there is a nonzero complex number λ such that $y_i = \lambda x_i$ for each i.

The x_i's are called the *homogeneous coordinates* on $\mathbb{P}^n$. We note that their values are not determined at a point $p \in \mathbb{P}^n$, but whether x_i is zero or not does make sense.

Similarly, suppose that $F(x_0, \ldots, x_n)$ is a homogeneous polynomial. Then we cannot *evaluate* F at a point $p \in \mathbb{P}^n$ (by writing $p = [a_0 : \cdots : a_n]$ and forming the number $F(p) = F(a_0, \ldots, a_n)$) but again whether this number $F(p)$ is zero or not does make sense.

A projective space may be constructed from any finite-dimensional complex vector space V, by taking the 1-dimensional subspaces of V. This is called the *projectivization* of V, and is denoted by $\mathbb{P}V$. If $v \in V$ is a nonzero vector, then its span is a point of $\mathbb{P}V$, denoted by $[v]$. If V has dimension $n+1$, and one chooses a basis for V, (which essentially gives an explicit isomorphism of V with $\mathbb{C}^{n+1}$), then we see that $\mathbb{P}V$ is "isomorphic" to $\mathbb{P}^n$.

Projective Algebraic Sets. The subsets of projective space which we are most interested in are the smooth projective curves. We have mentioned previously that every such subset is a local complete intersection curve, defined by the vanishing of a set of homogeneous polynomials (with the extra Jacobian condition). We take this idea as the definition of an "algebraic" subset of $\mathbb{P}^n$:

DEFINITION 5.1. A subset $Z \subset \mathbb{P}^n$ is an *algebraic set* if there is a set of homogeneous polynomials $\{F_\alpha\}$ such that $Z = \{p \in \mathbb{P}^n \mid F_\alpha(p) = 0 \text{ for every } \alpha\}$.

Denote by $k[\underline{x}]$ the ring of polynomials $k[x_0, \ldots, x_n]$. If S is a set of homogeneous polynomials, we will denote by $Z(S)$ the set of common zeroes in $\mathbb{P}^n$ of the polynomials in S.

The two "extreme" cases of projective algebraic sets are the largest ones and the smallest ones. Intuitively speaking, the largest ones should be the common zeroes of the smallest sets of polynomials: the singletons. A *hypersurface* in $\mathbb{P}^n$ is an algebraic subset which is the zeroes of a single polynomial F, i.e., an algebraic subset of the form $Z(\{F\})$. It is obvious that every algebraic set is an intersection of hypersurfaces, and indeed that a subset of $\mathbb{P}^n$ is an algebraic subset if and only if it is an intersection of hypersurfaces.

At the other extreme, the smallest possible algebraic subset would be a single point, and it is true that a single point is an algebraic set. To see this, suppose that $p = [a_0 : \cdots : a_n] \in \mathbb{P}^n$; then p is the only common zero of the set of linear polynomials $F_{ij} = a_i x_j - a_j x_i$. Alternatively, suppose that $a_0 = 1$ (which we may assume by reordering the variables and scaling the coordinates); then p is the only common zero of the set of linear polynomials $G_j = x_j - a_j x_0$.

In fact, any finite subset of $\mathbb{P}^n$ is algebraic; this is a consequence of the following lemma, which we leave to the reader.

LEMMA 5.2. *Any intersection of algebraic subsets of $\mathbb{P}^n$ is an algebraic subset. Any finite union of algebraic subsets of $\mathbb{P}^n$ is an algebraic subset.*

We see therefore that one has a topology on $\mathbb{P}^n$ whose closed sets are the algebraic subsets; this topology is called the *Zariski topology* on $\mathbb{P}^n$.

Linear Subspaces. Probably the most important algebraic subsets other than the hypersurfaces and the finite sets are the *linear subspaces* of $\mathbb{P}^n$. These are exactly the subsets described by a set of homogeneous polynomials, which all have degree *one*.

An alternate way of viewing a linear subspace is afforded by considering the original vector space $\mathbb{C}^{n+1}$. Suppose that $W \subset \mathbb{C}^{n+1}$ is a vector subspace. Then the 1-dimensional subspaces of W forms a linear subspace of $\mathbb{P}^n$, and every linear subspace of $\mathbb{P}^n$ is obtained in this way, for a unique vector subspace.

This is nicely expressed without coordinates: if $\mathbb{P}V$ is the projectivization of a vector space V, and $W \subset V$ is a vector subspace, then $\mathbb{P}W \subset \mathbb{P}V$ is a linear subspace of $\mathbb{P}V$.

Note that the intersection of any collection of linear subspaces is a linear subspace.

The *dimension* of a linear subspace $L \subset \mathbb{P}^n$ is defined to be one less than the dimension of the vector subspace W to which $L = \mathbb{P}W$ corresponds:

$$\dim \mathbb{P}W = \dim W - 1.$$

Note that with this convention, the empty set (which is $\mathbb{P}\{0\}$) has dimension -1.

Linear subspaces of dimension zero are the points; a linear subspace of dimension one is called a *line*. In general, a linear subspace of dimension k is called a *k-plane*. A *hyperplane* is a linear subspace of codimension one, that is, of dimension $n-1$ in $\mathbb{P}^n$.

Suppose that $Z \subset \mathbb{P}^n$ is any subset. We define the *span of Z*, denoted by $\mathrm{span}(Z)$, to be the intersection of all linear subspaces containing Z. If $L = \mathrm{span}(Z)$, we might also say that Z *spans* L. We say that Z is *nondegenerate* if Z spans all of $\mathbb{P}^n$.

If Z is a finite set of points $Z = \{p_1, \ldots, p_r\}$, then we say that Z is *linearly independent* if the dimension of the span of Z is maximal, i.e., if $\dim \mathrm{span}(Z) = \#(Z) - 1$. The finite set Z is *dependent* if not.

Thus two distinct points are always independent and span a line. Three points either are independent (and span a 2-plane) or are dependent and span a line. Points lying on a line are said to be *collinear*.

We have the following dimension formula, which follows easily from the corresponding formula for vector subspaces of a vector space:

LEMMA 5.3. *If L and M are two linear subspaces of $\mathbb{P}^n$, then*

$$\dim(\mathrm{span}(L \cup M)) = \dim(L) + \dim(M) - \dim(L \cap M).$$

We leave the proof to the reader.

The Ideal of a Projective Algebraic Set. Suppose that $Z \subset \mathbb{P}^n$ is a subset. Since homogeneous polynomials are the only "functions" which we have available to work with in $\mathbb{P}^n$, it is natural to ask which ones vanish at all the points of Z. We define

$$I(Z) = \text{the ideal of } k[\underline{x}]$$
generated by all homogeneous polynomials F vanishing on Z;

$I(Z)$ is called the *homogeneous ideal* of the subset Z.

The study of projective algebraic sets and their ideals is the main topic of the field of algebraic geometry. We will not delve too deeply into this in this text, but the reader should be aware of some of the language.

5. BASIC PROJECTIVE GEOMETRY

Linear Automorphisms and Changing Coordinates. Suppose that $T : \mathbb{C}^{n+1} \to \mathbb{C}^{n+1}$ is a $\mathbb{C}$-linear isomorphism. Then T transports subspaces to subspaces, preserving dimension; in particular it sends each 1-dimensional subspace to another. Hence T induces a map $T : \mathbb{P}^n \to \mathbb{P}^n$; such a map is called a *linear automorphism* of $\mathbb{P}^n$.

In terms of homogeneous coordinates, suppose that we think of $\mathbb{C}^{n+1}$ as column vectors in the usual way, so that applying the map T is equal to multiplication by an invertible square matrix $A_T = (a_{ij})$ of size $n+1$:

$$T \begin{pmatrix} x_0 \\ x_1 \\ \vdots \\ x_n \end{pmatrix} = A_T \begin{pmatrix} x_0 \\ x_1 \\ \vdots \\ x_n \end{pmatrix} = \begin{pmatrix} \sum_{j=0}^n a_{0j} x_j \\ \sum_{j=0}^n a_{1j} x_j \\ \vdots \\ \sum_{j=0}^n a_{nj} x_j \end{pmatrix}.$$

Hence the same formulas are used for transforming the homogeneous coordinates of points in $\mathbb{P}^n$ under T:

$$T[x_0 : \cdots : x_n] = [\sum_{j=0}^n a_{0j} x_j : \cdots : \sum_{j=0}^n a_{nj} x_j].$$

Often the application of an invertible linear transformation T on projective space is called *changing the coordinates*, or *choosing coordinates*. This is perhaps more apt when thinking about the projectivization $\mathbb{P}V$ of a vector space V of dimension $n+1$ over $\mathbb{C}$. A choice of basis $v_0, \ldots, v_n$ for V gives "coordinates" on V: the coordinates of $\sum_j c_j v_j$ are the c_j's. This choice of basis is equivalent to giving a $\mathbb{C}$-linear isomorphism $\phi : \mathbb{C}^{n+1} \to V$, which send the standard i^{th} basis vector of $\mathbb{C}^{n+1}$ to v_i; using ϕ we obtain a corresponding isomorphism from $\mathbb{P}^n$ to $\mathbb{P}V$, putting homogeneous coordinates on $\mathbb{P}V$. The formalism is the same: the homogeneous coordinates of the point $[\sum_j c_j v_j] \in \mathbb{P}V$ are $[c_0 : \cdots : c_n]$. Choosing another basis gives a different isomorphism, and different homogeneous coordinates.

More generally, if V and W are two vector spaces, and $T : V \to W$ is an isomorphism between them, then T induces a map $T : \mathbb{P}V \to \mathbb{P}W$; such a map is called a *linear isomorphism* of the projective spaces.

One of the most common uses of changing coordinates is to take some collection of subsets of $\mathbb{P}^n$ and choose coordinates so that the subsets are described either by simple equations or by simple coordinates. Some examples are given in the lemma below, which we leave to the reader to check.

LEMMA 5.4. *Let $\mathbb{P}^n$ be projective n-space.*

   a. *Given any point $p \in \mathbb{P}^n$, there are coordinates so that*

$$p = [1 : 0 : 0 : \cdots : 0].$$

b. *Given any point $p \in \mathbb{P}^n$, and any hyperplane $H \subset \mathbb{P}^n$ not containing p, there are coordinates so that $p = [1:0:0:\cdots:0]$ and H is described by $x_0 = 0$.*

c. *Given any $n+1$ linearly independent points $\{p_0, p_1, \ldots, p_n\}$ of $\mathbb{P}^n$, there are coordinates so that*

$$\begin{aligned} p_0 &= [1:0:0:\cdots:0] \\ p_1 &= [0:1:0:\cdots:0] \\ &\vdots \\ p_n &= [0:0:0:\cdots:1]. \end{aligned}$$

d. *Given any $n+2$ points $\{p_0, p_1, \ldots, p_n, p_{n+1}\}$ of $\mathbb{P}^n$, such that any $n+1$ of them are linearly independent, there are coordinates so that $p_0, \ldots, p_n$ are as above, and $p_{n+1} = [1:1:\cdots:1]$.*

e. *Given a k-plane $L \subset \mathbb{P}^n$, there are coordinates so that L is described by $x_{k+1} = x_{k+2} = \cdots = x_n = 0$.*

f. *Given a k-plane $L \subset \mathbb{P}^n$, and an $(n-k-1)$-plane $M \subset \mathbb{P}^n$, which are disjoint, there are coordinates so that L is described by $x_{k+1} = x_{k+2} = \cdots = x_n = 0$ and M is described by $x_0 = x_1 = \cdots = x_k = 0$.*

Two disjoint linear subspaces L and M as in Lemma 5.4.f above are said to be *complementary* linear subspaces.

Projections. Let $L \subset \mathbb{P}^n$ be a k-plane and $M \subset \mathbb{P}^n$ be an $(n-k-1)$-plane which are disjoint (and hence complementary) subspaces. Note that L and M together span all of $\mathbb{P}^n$.

Suppose p is a point not in L. Then the span of $L \cup \{p\}$ is a linear subspace L_1 which has dimension one more than that of L: L_1 is a $(k+1)$-plane. Hence by the dimension formula (Lemma 5.3), we see that

$$\begin{aligned} \dim(L_1 \cap M) &= \dim(L_1) + \dim(M) - \dim \operatorname{span}(L_1 \cup M) \\ &= (k+1) + (n-k-1) - (n) = 0, \end{aligned}$$

so that $L_1 \cap M$ is a single point, in M of course.

DEFINITION 5.5. The *projection* from L to M is the mapping

$$\pi : (\mathbb{P}^n - L) \to M$$

defined by sending a point $p \notin L$ to the intersection point of $\operatorname{span}(L \cup \{p\})$ with M:

$$\pi(p) = \operatorname{span}(L \cup \{p\}) \cap M.$$

The subspace L is called the *center of projection*.

Note the odd use of the word: the projection from L is exactly *not* defined on L!

It is easy to see that if L is defined by $x_{k+1} = x_{k+2} = \cdots = x_n = 0$ and M is described by $x_0 = x_1 = \cdots = x_k = 0$, then

$$\pi[x_0 : \cdots : x_n] = [0 : 0 : \cdots : 0 : x_{k+1} : x_{k+2} : \cdots : x_n].$$

One often suppresses the choice of the target subspace M in the language, and refers to π simply as "the projection from L". The reason for this is that if M_1 and M_2 are two complementary subspaces to L, with projections π_1 from L to M_1 and π_2 from L to M_2, then the restriction of π_2 to M_1 is a linear isomorphism $\phi : M_1 \to M_2$, and

$$\phi \circ \pi_1 = \pi_2.$$

So for most purposes it doesn't matter which subspace one is projecting to.

In abstract terms, projections may be defined as follows. Suppose that W is a vector subspace of V, and $L = \mathbb{P}W$ is the corresponding linear subspace of $\mathbb{P}V$. Then the quotient space V/W is a vector space, and the quotient map $\pi : V \to V/W$ induces the projection map $\pi : (\mathbb{P}V - L) \to \mathbb{P}(V/W)$.

This point of view makes it even clearer that the target space of a projection is not so important: in the above formulation, it is not even a subspace!

Probably the most common use of projections is when the center of projection is a single point p. Then the projection from p maps $\mathbb{P}^n - \{p\}$ to a hyperplane, isomorphic to a $\mathbb{P}^{n-1}$.

Projections send linear subspaces to linear subspaces; the dimension of the image depends on how much the subspace meets the center of the projection.

Projections compose nicely: if $L_1 \subset L_2 \subset \mathbb{P}^n$ are linear subspaces, and if π_i is the projection from L_i, then π_2 is the composition of π_1 with projection from the image of L_2. In particular, any projection may be viewed as a composition of projections, each of which is a projection from a single point.

Projection maps are always onto, and never 1-1. It is an exercise to check that if π is the projection from L, and $p \neq q$ are distinct points not in L, then $\pi(p) = \pi(q)$ if and only if the line joining p and q meets L.

Secant and Tangent Lines. Let us return now to a smooth projective curve $X \subset \mathbb{P}^n$. Suppose that p and q are distinct points on X. The line joining p and q is called a *secant line* to X, and in general any line of $\mathbb{P}^n$ which meets X in at least two distinct points is called a secant line to X. The line through two points p and q is often denoted by $\overline{pq}$.

Let $L \subset \mathbb{P}^n$ be a linear subspace, with a complementary space M, and suppose that L is disjoint from X. If π is the projection with center L, then $\pi|_X$ maps X to the lower-dimensional linear space M.

If p and q are distinct points of X, then the projection $\pi(p) = \pi(q)$ if and only if the secant line through p and q meets the center of projection L. Hence $\pi|_X$ will be 1-1 if L is disjoint from the union of the secant lines.

More interesting, and slightly more difficult to define, are the tangent lines to X. Fix a point $p \in X$. Since X is holomorphically embedded, we may choose coordinates in $\mathbb{P}^n$ such that $p = [1:0:0:\cdots:0]$ and X is, near p, defined by the locus
$$[1:z:g_2(z):\cdots:g_n(z)],$$
where z is a local coordinate centered at p and $g_2, \ldots, g_n$ are holomorphic functions of z, with $g_i(0) = 0$ for every i.

Define a point q by taking the derivative of the above local parametrization of X:
$$q = [0:1:\frac{dg_2}{dz}(0):\cdots:\frac{dg_n}{dz}(0)].$$

DEFINITION 5.6. The *tangent line* to X at p is the line joining p and q.

It is an exercise to check that the tangent line is well defined, independent of the choices made in the local parametrization of X.

If X is a straight line in $\mathbb{P}^n$, then X is its own tangent line at any of its points.

Returning to the situation of a projection π from a center L, restricted to the Riemann surface X, we ask the question: given a point p, when does $\pi|_X$ map a neighborhood of p isomorphically onto a Riemann surface in the target projective space? We have already seen above that in order for $\pi|_X$ to be 1-1, the center L must be disjoint from all of the secant lines.

This is not enough for the image to be a holomorphically embedded Riemann surface, however. Consider the twisted cubic curve $X \subset \mathbb{P}^3$, which is the image of the mapping from $\mathbb{P}^1$ to $\mathbb{P}^3$ which locally sends z to $[1:z:z^2:z^3]$. Let L be the single point $[0:1:0:0]$; projection from L, restricted to X, locally sends z to $[1:z^2:z^3] \in \mathbb{P}^2$. This is *not* a holomorphically embedded Riemann surface near $z = 0$.

The problem with the above example is that the tangent line to X at the point $[1:0:0:0]$ (corresponding to $z = 0$) is the line joining $[1:0:0:0]$ to $[0:1:0:0]$, and this line then meets the center of projection L. If this does not happen, then the image is locally a Riemann surface:

PROPOSITION 5.7. *Let $X \subset \mathbb{P}^n$ be a smooth projective curve. Let $L \subset \mathbb{P}^n$ be a linear space disjoint from X. Suppose that L does not meet any secant line to X, so that the projection π from L, when restricted to X, is 1-1. Fix a point $p \in X$. Then there is a neighborhood U of p such that $\pi(U)$ is a holomorphically embedded Riemann surface (in the complementary space to L) if and only if the tangent line to X at p does not meet L.*

PROOF. We may choose coordinates so that the k-plane L is defined by $x_0 = x_1 = \cdots = x_{n-k-1} = 0$, and that $p = [1:0:\cdots:0]$. In this case, since X is holomorphically embedded, there is a local coordinate z centered at p such that X is locally parametrized near p by $[1:g_1(z):g_2(z):\cdots:g_n(z)]$, where the

g_i's are holomorphic functions of z, $g_i(0) = 0$ for each i, and at least one g_i has nonvanishing derivative at 0 (so that this g_i is a local coordinate also at p).

The projection π, in terms of the local coordinate z, sends z to $[1 : g_1(z) : \cdots : g_{n-k-1}(z)]$. This is also a holomorphically embedded Riemann surface if and only if one of the g_i's with $i \leq n - k - 1$ is a local coordinate at the image of p, i.e., one of the g_i's with $i \leq n - k - 1$ has a nonvanishing derivative at 0.

Now the tangent line to X at p is the line joining p to the point $[0 : g_1'(0) : g_2'(0) : \cdots : g_n'(0)]$. This line meets L if and only if this point is in L, and so the tangent line meets L if and only if $g_i'(0) = 0$ for every $i \leq n - k - 1$.

Hence the tangent line to X at p meets L if and only if the projection $\pi(X)$, near $\pi(p)$, is not a holomorphically embedded Riemann surface. $\square$

Projecting Projective Curves. The discussion above immediately gives us the following.

COROLLARY 5.8. *Suppose that $X \subset \mathbb{P}^n$ is a smooth projective curve. Let L be a linear space disjoint from X, which is the center of the projection π. Then $\pi|_X$ is 1-1 and $\pi(X)$ is a smooth projective curve if and only if L does not meet any secant or tangent line to X. In this case $\pi|_X : X \to \pi(X)$ is an isomorphism of Riemann surfaces.*

Can we find a linear subspace L disjoint from any secant or tangent line? This is basically a matter of determining dimensions. Fix a smooth projective curve X and consider the space

$$\mathcal{I} = \{(p, q, r) \mid p \in X, q \in X, p \neq q, r \in \overline{pq}\}$$

of triples of points whose first and second point are different, and lie on X, and whose third point lies on the secant line through the first two points. The space $\mathcal{I}$ is clearly a 3-dimensional complex manifold: If z is a local coordinate near p, and w is a local coordinate near q, and $r = p + \lambda q$, then $\mathcal{I}$ is parametrized locally near (p, q, r) by (z, w, λ).

The function $\alpha : \mathcal{I} \to \mathbb{P}^n$ sending (p, q, r) to r is a continuous map, and has image equal to the union of all the secant lines to X. We conclude that if $n \geq 4$, then α cannot be an onto map; therefore there is a point $p_0 \in \mathbb{P}^n$ which does not lie on any secant line.

Similarly consider the space

$$\mathcal{J} = \{(p, r) \mid r \text{ lies on the tangent line to } X \text{ at } p\}.$$

Again, it is easy to see that $\mathcal{J}$ is a 2-dimensional complex manifold: if z is a local coordinate near p, and if we choose coordinates so that X is described near p as the locus $[1 : z : g_2(z) : \cdots : g_n(z)]$, then we may write $r = p + \lambda[0 : 1 : g_2'(0) : \cdots : g_n'(0)]$, and we see that (z, λ) parametrizes $\mathcal{J}$ near the point (p, r).

The function $\beta : \mathcal{J} \to \mathbb{P}^n$ sending (p, r) to r is a continuous map, and has image equal to the union of all the tangent lines to X. We conclude that if $n \geq 3$, then β cannot be onto; therefore there is a point $p_0 \in \mathbb{P}^n$ which does not lie on any tangent line.

Putting the two constructions together, we see that if $n \geq 4$, there is a point p_0 which does not lie on any secant or tangent line to X. Therefore, by Corollary 5.8, the projection from p_0 maps X isomorphically onto a smooth projective curve in $\mathbb{P}^{n-1}$.

We may therefore proceed recursively, continuing to project the curve X until we reach $\mathbb{P}^3$. Therefore:

PROPOSITION 5.9. *Let X be a smooth projective curve in $\mathbb{P}^n$ with $n \geq 4$. Then there is a projection to $\mathbb{P}^3$ which maps X isomorphically onto a smooth projective curve in $\mathbb{P}^3$.*

One can refine this argument and show the following.

PROPOSITION 5.10. *Let X be a smooth projective curve in $\mathbb{P}^n$ with $n \geq 3$. Then there is a projection to $\mathbb{P}^2$ which maps X isomorphically onto a smooth projective plane curve with nodes.*

We leave the details to the reader; the idea is that since there are points not on any tangent lines, we can project to $\mathbb{P}^2$, and *locally* we have an isomorphism onto the image, which is a Riemann surface. But globally, since the projecting point may lie on some secants, two different points may be mapped to the same point in the plane, creating nodes on the image.

Problems III.5

A. Let $p = [x_0 : \cdots : x_n]$ and $q = [y_0 : \cdots : y_n]$ be points of $\mathbb{P}^n$ given by homogeneous coordinates. Show that $p = q$ if and only if for every i and j, $x_i y_j = x_j y_i$.
B. Show that if $S_1 \subset S_2 \subset k[\underline{x}]$ then $Z(S_2) \subset Z(S_1) \subset \mathbb{P}^n$.
C. Show that if $S \subset k[\underline{x}]$ generates the ideal $I \subset k[\underline{x}]$, then $Z(S) = Z(I)$.
D. Prove Lemma 5.2.
E. If L and M are two linear subspaces of $\mathbb{P}^n$, show that

$$\dim(\operatorname{span}(L \cup M)) = \dim(L) + \dim(M) - \dim(L \cap M).$$

F. Show that any four distinct points on the twisted cubic curve in $\mathbb{P}^3$ are linearly independent.
G. Show that the homogeneous ideal $I(X)$ of the plane conic curve X defined by $F(x, y, z) = xz - y^2 = 0$ is the principal ideal generated by F.
H. Show that the homogeneous ideal of the twisted cubic is generated by the three quadratic equations $F_1 = x_1^2 - x_0 x_2$, $F_2 = x_2^2 - x_1 x_3$, and $F_3 = x_0 x_3 - x_1 x_2$ which cut it out.
I. Let p and q be distinct points in $\mathbb{P}^n$. Find a map $F : \mathbb{P}^1 \to \mathbb{P}^n$ which sends 0 to p, ∞ to q, and has image equal to the line joining p and q.

J. Show that given any degree d, and any finite set of points of $\mathbb{P}^n$, there is a hypersurface of degree d not containing any of the points of the set.

K. Prove Lemma 5.4.

L. Suppose that W_1 and W_2 are vector subspaces of V, so that $\mathbb{P}W_1$ and $\mathbb{P}W_2$ are linear subspaces of $\mathbb{P}V$. Show that $\mathbb{P}W_1$ and $\mathbb{P}W_2$ are complementary (i.e., they are disjoint and the sum of their dimensions is one less than the dimension of $\mathbb{P}V$) if and only if the vector space V is the internal direct sum of W_1 and W_2.

M. Check that if L is defined by $x_{k+1} = x_{k+2} = \cdots = x_n = 0$ and M is described by $x_0 = x_1 = \cdots = x_k = 0$, then the projection π from L to M has the formula
$$\pi[x_0 : \cdots : x_n] = [0 : 0 : \cdots : 0 : x_{k+1} : x_{k+2} : \cdots : x_n].$$

N. Show that if $\pi : \mathbb{P}^n - L \to M$ is the projection from L to M, and $p \neq q$ are distinct points not in L, then $\pi(p) = \pi(q)$ if and only if the line joining p and q meets L.

O. Suppose that π is a projection with center L, and that L' is another linear subspace. Show that the image $\pi(L')$ is a linear subspace of the target space, and that
$$\dim \pi(L') = \dim(L') - \dim(L \cap L') - 1.$$

P. Let X be a smooth projective plane curve defined by $F(x, y, z) = 0$, where F is a nonsingular homogeneous polynomial. Show that if $p = [x_0 : y_0 : z_0]$ is a point on X, then the tangent line to X at p is the line defined by
$$\frac{\partial F}{\partial x}(x_0, y_0, z_0)x + \frac{\partial F}{\partial y}(x_0, y_0, z_0)y + \frac{\partial F}{\partial z}(x_0, y_0, z_0)z = 0.$$

Q. Show that the complement of an algebraic set in $\mathbb{P}^n$ is path-connected.

Further Reading

The discussion about lines and conics is de rigueur for any book on curve geometry; [**Reid88**] and [**Clemens80**] are recent books with sections devoted to conics in particular. For further reading on maps between complex tori, [**JS87**] and [**Serre73**] are fine; all books devoted to elliptic curves treat this, in particular [**Lang87**], [**Silverman86**], and [**Husemoller87**] among many others take off from here.

The singularities of projective plane curves are discussed in [**Walker50**], [**S-K59**], [**Seidenberg68**], [**Fulton69**], [**Samuel69**] [**O-O81**], [**Brieskorn86**], and [**Kirwan92**]; an older viewpoint is taken in [**Coolidge31**]. This is a classical subject, and its literature may be as large as that on Riemann surfaces themselves.

Forming quotients of manifolds and algebraic varieties by actions of groups is also a subject unto itself. Eventually an expert will want to read [**Mumford65**], but not right away.

The brief section above on Projective Geometry is meant only to scratch the surface and whet the reader's appetite for more algebraic details, leading to Algebraic Geometry. A sampler of relatively recent books might include [**Mumford76**], [**Kendig77**], [**Shafarevich77**], [**Hartshorne77**], [**G-H78**], [**Iitaka82**], [**Namba84**], [**Reid88**], [**C-L-O92**], and [**Harris92**]. For the previous generation [**H-P47**] and [**S-R49**] were widely read and are still valuable.

Chapter IV. Integration on Riemann Surfaces

1. Differential Forms

As you know from a first course in one complex variable, the basic tool and, indeed, the motivation for much of the subject is contour integration. In order to transport the theory of integration to Riemann surfaces, we need to have objects to integrate. These objects are called *forms*, and they come in various flavors.

Holomorphic 1-Forms.

DEFINITION 1.1. A *holomorphic 1-form* on an open set $V \subset \mathbb{C}$ is an expression ω of the form
$$\omega = f(z)\mathrm{d}z$$
where f is a holomorphic function on V. We say that ω is a holomorphic 1-form *in the coordinate z*.

This is the basic object which we would like to transport up to a general Riemann surface via complex charts. When we do this, we will require some compatibility condition whenever two charts have overlapping domains. This motivates the following.

DEFINITION 1.2. Suppose that $\omega_1 = f(z)\mathrm{d}z$ is a holomorphic 1-form in the coordinate z, defined on an open set V_1. Also suppose that $\omega_2 = g(w)\mathrm{d}w$ is a holomorphic 1-form in the coordinate w, defined on an open set V_2. Let $z = T(w)$ define a holomorphic mapping from the open set V_2 to V_1. We say that ω_1 *transforms to* ω_2 *under* T if $g(w) = f(T(w))T'(w)$.

Note that the above definition is cooked up exactly so that the expression for ω_1 transforms into the expression for ω_2 when one sets $\mathrm{d}z = T'(w)\mathrm{d}w$ (as one should!).

Also note that if T is invertible with inverse function S, then ω_1 transforms to ω_2 under T if and only if ω_2 transforms to ω_1 under S.

Given the above notation, we are ready to transport this construct to a Riemann surface:

DEFINITION 1.3. Let X be a Riemann surface. A *holomorphic 1-form* on X is a collection of holomorphic 1-forms $\{\omega_\phi\}$, one for each chart $\phi: U \to V$ in the coordinate of the target V, such that if two charts $\phi_i : U_i \to V_i$ (for $i = 1, 2$) have overlapping domains, then the associated holomorphic 1-form ω_{ϕ_1} transforms to ω_{ϕ_2} under the change of coordinate mapping $T = \phi_1 \circ \phi_2^{-1}$.

To define a holomorphic 1-form on a Riemann surface, one does not need to actually give a holomorphic 1-form on every chart, but only the charts of some atlas:

LEMMA 1.4. *Let X be a Riemann surface and $\mathcal{A}$ a complex atlas on X. Suppose that holomorphic 1-forms are given for each chart of $\mathcal{A}$, which transform to each other on their common domains. Then there exists a unique holomorphic 1-form on X extending these holomorphic 1-forms on each of the charts of $\mathcal{A}$.*

PROOF. Let ψ be a chart of X not in the atlas; our task is to define the holomorphic 1-form with respect to ψ or, equivalently, in terms of the local coordinate w of ψ. Fix a point p in the domain of ψ, and choose chart ϕ in the atlas containing p in its domain; let z be the associated local variable. Let $f(z)dz$ be the holomorphic 1-form with respect to ϕ. Then simply define the holomorphic 1-form with respect to ψ as $f(T(w))T'(w)dw$, where $z = T(w)$ describes the change of coordinates $\phi \circ \psi^{-1}$.

Now one checks that this definition is independent of the choice of ϕ, and gives a 1-form with respect to ψ at every point of the domain. Next one checks that all of these holomorphic 1-forms transform to each other, and thus define a holomorphic 1-form on X. This 1-form is obviously unique. □

Meromorphic 1-Forms. In the same spirit as above we may define meromorphic 1-forms, as expressions which are locally of the form $f(z)dz$ where f is meromorphic:

DEFINITION 1.5. A *meromorphic 1-form* on an open set $V \subset \mathbb{C}$ is an expression ω of the form
$$\omega = f(z)dz$$
where f is a meromorphic function on V. We say that ω is a meromorphic 1-form *in the coordinate z*.

The compatibility condition for meromorphic 1-forms is identical to that for holomorphic 1-forms:

DEFINITION 1.6. Suppose that $\omega_1 = f(z)dz$ is a meromorphic 1-form in the coordinate z, defined on an open set V_1. Also suppose that $\omega_2 = g(w)dw$ is a meromorphic 1-form in the coordinate w, defined on an open set V_2. Let $z = T(w)$ define a holomorphic mapping from the open set V_2 to V_1. We say that ω_1 *transforms to* ω_2 *under* T if $g(w) = f(T(w))T'(w)$.

1. DIFFERENTIAL FORMS

Transporting the notion of meromorphic 1-forms from the complex plane to a Riemann surface is now done in the same way also:

DEFINITION 1.7. Let X be a Riemann surface. A *meromorphic 1-form* on X is a collection of meromorphic 1-forms $\{\omega_\phi\}$, one for each chart $\phi : U \to V$ in the variable of the target V, such that if two charts $\phi_i : U_i \to V_i$ (for $i = 1, 2$) have overlapping domains, then the associated meromorphic 1-form ω_{ϕ_1} transforms to ω_{ϕ_2} under the change of coordinate mapping $T = \phi_1 \circ \phi_2^{-1}$.

As is the case for holomorphic 1-forms, we may define a meromorphic 1-form using only the charts in a given atlas; we leave this to the reader.

LEMMA 1.8. *Let X be a Riemann surface and $\mathcal{A}$ a complex atlas on X. Suppose that meromorphic 1-forms are given for each chart of $\mathcal{A}$, which transform to each other on their common domains. Then there exists a unique meromorphic 1-form on X extending these meromorphic 1-forms on each of the charts of $\mathcal{A}$.*

Let ω be a meromorphic 1-form defined in a neighborhood of a point p. Choosing a local coordinate centered at p, we may write $\omega = f(z)\mathrm{d}z$ where f is a meromorphic function at $z = 0$.

DEFINITION 1.9. The *order of ω at p*, denoted by $\mathrm{ord}_p(\omega)$, is the order of the function f at 0.

It is easy to see that $\mathrm{ord}_p(\omega)$ is well defined, independent of the choice of local coordinate. A meromorphic 1-form ω is holomorphic at p if and only if $\mathrm{ord}_p(\omega) \geq 0$.

We say p is a *zero of ω of order n* if $\mathrm{ord}_p(\omega) = n > 0$. We say p is a *pole of ω of order n* if $\mathrm{ord}_p(\omega) = -n < 0$. The set of zeroes and poles of a meromorphic 1-form is a discrete set.

Defining Meromorphic Functions and Forms with a Formula. The definition of a meromorphic or holomorphic 1-form ω suggests that in order to define ω on a Riemann surface X, one must give local expressions for ω (of the form $f(z)\mathrm{d}z$) in each chart of an atlas for X. In fact, one can define ω by giving a *single* formula in a *single* chart. This is sufficient to determine ω by the Identity Theorem for meromorphic functions and forms: if two meromorphic 1-forms agree on an open set, they must be identical.

Of course, this way of defining a form does not guarantee that the form actually exists on all of X. It may well happen that if one has a meromorphic local expression $f(z)\mathrm{d}z$ in one chart, then when one transforms this local expression to another chart it may fail to be meromorphic. For example, the meromorphic 1-form $\exp(z)\mathrm{d}z$ on the finite chart $\mathbb{C}$ of $\mathbb{C}_\infty$ does not extend to a meromorphic 1-form in a neighborhood of ∞.

A second problem may arise, namely that the local expression does not transform uniquely to the other points of X. For example, consider the meromorphic 1-form $\sqrt{z}\mathrm{d}z$ defined on the complex plane with the negative real axis removed, where the branch of the square root is chosen so that $\sqrt{1} = 1$. This can be extended to the negative real axis, but not uniquely. Hence we do not obtain a meromorphic 1-form on all of $\mathbb{C}^*$.

However, it is very convenient to simply use a single formula in one specified chart to define a meromorphic 1-form ω, and to let the burden fall to the reader to check that the formula transforms uniquely to give a meromorphic 1-form on all of X. This way of defining meromorphic 1-forms is employed systematically.

The same remarks hold also for meromorphic functions: they can be determined by a single formula in a single chart.

Using dz and $d\bar{z}$. We can relax the holomorphic or meromorphic conditions for 1-forms and obtain a notion of C^∞ 1-forms. These should locally be expressions of the form $f(x,y)dx + g(x,y)dy$, where x and y are the local real variables (i.e., $z = x + iy$).

However it is useful to abandon completely the use of the real and imaginary parts x and y of the complex variable z, and instead depend solely on z and its complex conjugate $\bar{z}$. This is possible, since

$$x = (z + \bar{z})/2 \text{ and } y = (z - \bar{z})/2i,$$

and

$$z = x + iy \text{ and } \bar{z} = x - iy,$$

so that any function expressible in terms of x and y is expressible in terms of z and $\bar{z}$, and vice-versa. Furthermore, the same holds for the differentials:

$$dx = (dz + d\bar{z})/2 \text{ and } dy = (dz - d\bar{z})/2i,$$

since

$$dz = dx + idy \text{ and } d\bar{z} = dx - idy.$$

Thus any expression one would like to construct of the form $f(x,y)dx + g(x,y)dy$ can be written instead in the form $r(z,\bar{z})dz + s(z,\bar{z})d\bar{z}$. This we will do religiously.

This principle is carried over to partial derivatives also. Given a C^∞ function $f(x,y)$, we have

$$\begin{aligned}\frac{\partial f}{\partial z} &= \frac{\partial f}{\partial x}\frac{\partial x}{\partial z} + \frac{\partial f}{\partial y}\frac{\partial y}{\partial z} \\ &= \frac{1}{2}\frac{\partial f}{\partial x} + \frac{1}{2i}\frac{\partial f}{\partial y},\end{aligned}$$

and

$$\begin{aligned}\frac{\partial f}{\partial \bar{z}} &= \frac{\partial f}{\partial x}\frac{\partial x}{\partial \bar{z}} + \frac{\partial f}{\partial y}\frac{\partial y}{\partial \bar{z}} \\ &= \frac{1}{2}\frac{\partial f}{\partial x} - \frac{1}{2i}\frac{\partial f}{\partial y}.\end{aligned}$$

1. DIFFERENTIAL FORMS

Thus we can define the differential operators $\partial/\partial z$ and $\partial/\partial \bar{z}$ by

$$\frac{\partial}{\partial z} = \frac{1}{2}(\frac{\partial}{\partial x} - i\frac{\partial}{\partial y})$$

and

$$\frac{\partial}{\partial \bar{z}} = \frac{1}{2}(\frac{\partial}{\partial x} + i\frac{\partial}{\partial y}).$$

With this notation, we note that a $\mathcal{C}^\infty$ function f is holomorphic on an open set V if and only if

$$\frac{\partial f}{\partial \bar{z}} = 0,$$

since this condition is exactly the Cauchy-Riemann equations for f.

$\mathcal{C}^\infty$ 1-Forms. With the dz and $d\bar{z}$ notation at hand, we can easily develop the notion of $\mathcal{C}^\infty$ 1-forms.

DEFINITION 1.10. A $\mathcal{C}^\infty$ *1-form* on an open set $V \subset \mathbb{C}$ is an expression ω of the form

$$\omega = f(z,\bar{z})dz + g(z,\bar{z})d\bar{z}$$

where f and g are $\mathcal{C}^\infty$ functions on V. We say that ω is a $\mathcal{C}^\infty$ 1-form *in the coordinate z*.

The transformation rule is the following:

DEFINITION 1.11. Suppose that $\omega_1 = f_1(z,\bar{z})dz + g_1(z,\bar{z})d\bar{z}$ is a $\mathcal{C}^\infty$ 1-form in the coordinate z, defined on an open set V_1. Also suppose that $\omega_2 = f_2(w,\bar{w})dw + g_2(w,\bar{w})d\bar{w}$ is a $\mathcal{C}^\infty$ 1-form in the coordinate w, defined on an open set V_2. Let $z = T(w)$ define a holomorphic mapping from the open set V_2 to V_1. We say that ω_1 *transforms to* ω_2 *under* T if $f_2(w,\bar{w}) = f_1(T(w), \overline{T(w)})T'(w)$ and $g_2(w,\bar{w}) = g_1(T(w), \overline{T(w)})\overline{T'(w)}$.

Note that the definition is made in this way because of the differential formula for the chain rule: if $z = T(w)$, then $dz = T'(w)dw$, and $d\bar{z} = \overline{T'(w)}d\bar{w}$. Also note that the dz part of the expression transforms into the dw part, and the $d\bar{z}$ part into the $d\bar{w}$: there is no "mixing" of the two halves of the expression upon changes of coordinates. This is the real reason to use z and $\bar{z}$ instead of x and y here; in an x,y formulation, there are cross-terms everywhere.

We use the same method as before to transport these ideas to a Riemann surface:

DEFINITION 1.12. Let X be a Riemann surface. A $\mathcal{C}^\infty$ *1-form* on X is a collection of $\mathcal{C}^\infty$ 1-forms $\{\omega_\phi\}$, one for each chart $\phi: U \to V$ in the variable of the target V, such that if two charts $\phi_i : U_i \to V_i$ (for $i = 1,2$) have overlapping domains, then the associated $\mathcal{C}^\infty$ 1-form ω_{ϕ_1} transforms to ω_{ϕ_2} under the change of coordinate mapping $T = \phi_1 \circ \phi_2^{-1}$.

We have the same remark concerning defining a $\mathcal{C}^\infty$ 1-form only on the charts of an atlas, which we again leave to the reader:

LEMMA 1.13. *Let X be a Riemann surface and $\mathcal{A}$ a complex atlas on X. Suppose that $\mathcal{C}^\infty$ 1-forms are given for each chart of $\mathcal{A}$, which transform to each other on their common domains. Then there exists a unique $\mathcal{C}^\infty$ 1-form on X extending these $\mathcal{C}^\infty$ 1-forms on each of the charts of $\mathcal{A}$.*

1-Forms of Type $(1,0)$ and $(0,1)$. Since, under transformation by holomorphic changes of coordinates, the $\mathrm{d}z$ and $\mathrm{d}\bar{z}$ parts of a $\mathcal{C}^\infty$ 1-form are preserved, we may split the definition of a $\mathcal{C}^\infty$ 1-form into two separate definitions, namely of $\mathcal{C}^\infty$ 1-forms with only $\mathrm{d}z$ parts, and ones with only $\mathrm{d}\bar{z}$ parts.

DEFINITION 1.14. A $\mathcal{C}^\infty$ 1-form is *of type $(1,0)$* if it is locally of the form $f(z,\bar{z})\mathrm{d}z$. It is *of type $(0,1)$* if it is locally of the form $g(z,\bar{z})\mathrm{d}\bar{z}$.

Since the transformation rules for $\mathcal{C}^\infty$ 1-forms preserve the $\mathrm{d}z$ part and the $\mathrm{d}\bar{z}$ part, this definition is well defined: if a form looks like a form of type $(1,0)$ in one chart, it will in any other chart on the common domain.

Note that any holomorphic 1-form is of type $(1,0)$. A meromorphic 1-form would be of type $(1,0)$ if it was $\mathcal{C}^\infty$ (which it is not at its poles).

$\mathcal{C}^\infty$ 2-Forms. One introduces 1-forms in order to have something to integrate around paths, which we will see later. Similarly, one often has a desire to perform a surface integral over a suitable 2-dimensional piece of a Riemann surface. The appropriate integrand in this case is a 2-form.

DEFINITION 1.15. A $\mathcal{C}^\infty$ *2-form* on an open set $V \subset \mathbb{C}$ is an expression η of the form
$$\eta = f(z,\bar{z})\mathrm{d}z \wedge \mathrm{d}\bar{z}$$
where f is a $\mathcal{C}^\infty$ function on V. We say that η is a $\mathcal{C}^\infty$ 2-form *in the coordinate z*.

These types of differentials for surface integrals behave formally as follows. Firstly, one has
$$\mathrm{d}z \wedge \mathrm{d}\bar{z} = -\mathrm{d}\bar{z} \wedge \mathrm{d}z$$
since changing the order in the wedge product corresponds to reversing the orientation of the surface over which the integration is being performed (thus changing the sign of the integral). Secondly,
$$\mathrm{d}z \wedge \mathrm{d}z = \mathrm{d}\bar{z} \wedge \mathrm{d}\bar{z} = 0$$
since one cannot have a surface integral using only one variable!

The transformation rule is the following:

DEFINITION 1.16. Suppose that $\eta_1 = f(z,\bar{z})dz \wedge d\bar{z}$ is a $\mathcal{C}^\infty$ 2-form in the coordinate z, defined on an open set V_1. Also suppose that $\eta_2 = g(w,\bar{w})dw \wedge d\bar{w}$ is a $\mathcal{C}^\infty$ 2-form in the coordinate w, defined on an open set V_2. Let $z = T(w)$ define a holomorphic mapping from the open set V_2 to V_1. We say that η_1 transforms to η_2 under T if $g(w,\bar{w}) = f(T(w), \overline{T(w)})\|T'(w)\|^2$.

The above definition comes exactly from making the change of coordinates both in the function parts and the dz and $d\bar{z}$ parts of the expression, and then using the rules given above for simplifying and cancelling, noting that $\|T'(w)\|^2 = T'(w)\overline{T'(w)}$.

Again the same method is used to transport these ideas to a Riemann surface:

DEFINITION 1.17. Let X be a Riemann surface. A $\mathcal{C}^\infty$ *2-form* on X is a collection of $\mathcal{C}^\infty$ 2-forms $\{\eta_\phi\}$, one for each chart $\phi : U \to V$ in the variable of the target V, such that if two charts $\phi_i : U_i \to V_i$ (for $i = 1, 2$) have overlapping domains, then the associated $\mathcal{C}^\infty$ 2-form η_{ϕ_1} transforms to η_{ϕ_2} under the change of coordinate mapping $T = \phi_1 \circ \phi_2^{-1}$.

Finally the same atlas remark holds again:

LEMMA 1.18. *Let X be a Riemann surface and $\mathcal{A}$ a complex atlas on X. Suppose that $\mathcal{C}^\infty$ 2-forms are given for each chart of $\mathcal{A}$, which transform to each other on their common domains. Then there exists a unique $\mathcal{C}^\infty$ 2-form on X extending these $\mathcal{C}^\infty$ 2-forms on each of the charts of $\mathcal{A}$.*

Problems IV.1

A. Let X be the Riemann Sphere $\mathbb{C}_\infty$, with local coordinate z in one chart and $w = 1/z$ in the other chart. Let ω be a meromorphic 1-form on X. Show that if $\omega = f(z)dz$ in the coordinate z, then f must be a rational function of z. Show further that there are no nonzero holomorphic 1-forms on $\mathbb{C}_\infty$. Where are the zeroes and poles, and the orders, of the meromorphic 1-form defined by dz? Of the 1-form dz/z?

B. Let L be a lattice in $\mathbb{C}$, and let $\pi : \mathbb{C} \to X = \mathbb{C}/L$ be the natural quotient map. Show that the local formula dz in every chart of $\mathbb{C}/L$ is a well defined holomorphic 1-form on $\mathbb{C}/L$. Show that this 1-form has no zeroes. Show that the local formula $d\bar{z}$ in every chart of $\mathbb{C}/L$ is a well defined $\mathcal{C}^\infty$ 1-form on $\mathbb{C}/L$.

C. Let X be a smooth affine plane curve defined by $f(u,v) = 0$. Show that du and dv define holomorphic 1-forms on X, as do $p(u,v)du$ and $p(u,v)dv$ for any polynomial $p(u,v)$. Show that if $r(u,v)$ is any rational function, then $r(u,v)du$ and $r(u,v)dv$ are meromorphic 1-forms on X. Show that $(\partial f/\partial u)du = -(\partial f/\partial v)dv$ as holomorphic 1-forms on X.

D. Let X be a smooth projective plane curve defined by a homogeneous polynomial $F(x,y,z) = 0$. Let $f(u,v) = F(u,v,1)$ define the associated smooth affine plane curve. Show that du and dv define meromorphic 1-forms on all of X, as do $r(u,v)du$ and $r(u,v)dv$ for any rational function r. Show that $(\partial f/\partial u)du = -(\partial f/\partial v)dv$ as meromorphic 1-forms on X.

E. With the notation of the previous problem, suppose that $F(x,y,z)$ has degree $d \geq 3$. Show that if $p(u,v)$ is any polynomial of degree at most $d-3$, then

$$p(u,v)\frac{du}{\partial f/\partial v}$$

defines a holomorphic 1-form on the compact Riemann surface X.

F. Suppose that X is a projective plane curve of degree d with nodes, defined by the affine equation $f(u,v) = 0$. Show that if $p(u,v)$ is any polynomial of degree at most $d-3$, which vanishes at the nodes of X, then

$$p(u,v)\frac{du}{\partial f/\partial v}$$

defines a holomorphic 1-form on the resolution $\tilde{X}$ of the nodes.

G. Let X be a compact hyperelliptic Riemann surface defined by $y^2 = h(x)$, where h has degree $2g+1$ or $2g+2$ (so that X has genus g). Show that dx/y is a holomorphic 1-form on X if $g \geq 1$. Show that $p(x)dx/y$ is a holomorphic 1-form on X if $p(x)$ is a polynomial in x of degree at most $g-1$.

H. Let X be a cyclic cover of the line defined by $y^d = h(x)$. Show that $r(x,y)dx$ defines a meromorphic 1-form on X. Give criteria for when $r(x,y)dx$ is a holomorphic 1-form.

I. Let L be a lattice in $\mathbb{C}$, and let $\pi : \mathbb{C} \to X = \mathbb{C}/L$ be the natural quotient map. Show that $dz \wedge d\bar{z}$ is a well defined C^∞ 2-form on $\mathbb{C}/L$.

J. Prove Lemma 1.8.

2. Operations on Differential Forms

There are several operations which one can perform with forms to produce other forms. We briefly describe them here, and we will leave the details of most of the constructions to the reader.

Multiplication of 1-Forms by Functions. Suppose that h is a C^∞ function on a Riemann surface X, and ω is a C^∞ 1-form on X. We may define a C^∞ 1-form $h\omega$ locally, by writing $\omega = fdz + gd\bar{z}$ and declaring $h\omega$ to be $hfdz + hgd\bar{z}$. It is an immediate check that this gives a well defined 1-form $h\omega$ on X. The properties listed below are all obvious:

- If ω is of type $(1,0)$, then so is $h\omega$.
- If ω is of type $(0,1)$, then so is $h\omega$.
- If ω is holomorphic and h is holomorphic, so is $h\omega$.
- If ω is meromorphic and h is meromorphic, so is $h\omega$.
- If h and ω are meromorphic at p then $\text{ord}_p(h\omega) = \text{ord}_p(h) + \text{ord}_p(\omega)$.

One can also multiply a C^∞ 2-form η by a function h, obtaining a C^∞ 2-form $h\eta$ defined locally in the obvious way: if $\eta = f(z,\bar{z})dz \wedge d\bar{z}$ with respect to a coordinate z, then $h\eta = h(z,\bar{z})f(z,\bar{z})\mathrm{d}z \wedge \mathrm{d}\bar{z}$ with respect to that coordinate.

Differentials of Functions. Let f be a C^∞ function defined on a Riemann surface. Then one can define the C^∞ 1-forms $\mathrm{d}f$, ∂f, and $\bar{\partial}f$ on X by the following rule. Let $\phi : U \to V$ be a chart on X giving a local coordinate z. Write f on U in terms of the local coordinate as $f(z,\bar{z})$. Define

$$\partial f = \frac{\partial f}{\partial z}\mathrm{d}z, \quad \bar{\partial}f = \frac{\partial f}{\partial \bar{z}}\mathrm{d}\bar{z},$$

and

$$\mathrm{d}f = \partial f + \bar{\partial}f = \frac{\partial f}{\partial z}\mathrm{d}z + \frac{\partial f}{\partial \bar{z}}\mathrm{d}\bar{z}.$$

LEMMA 2.1. *The above local recipe gives well defined C^∞ 1-forms $\mathrm{d}f$, ∂f, and $\bar{\partial}f$ on X. A C^∞ function f is holomorphic if and only if $\bar{\partial}f = 0$. The operators d, ∂, and $\bar{\partial}$ are $\mathbb{C}$-linear and satisfy the product rules*

$$d(fg) = f\,dg + g\,df; \quad \partial(fg) = f\,\partial g + g\partial f; \quad \bar{\partial}(fg) = f\bar{\partial}g + g\bar{\partial}f.$$

A C^∞ 1-form ω is said to be *exact* on an open set U if there is a C^∞ function f defined on U such that $df = \omega$ on U.

Recall that every meromorphic function f on a Riemann surface can be used as a local coordinate at a point p where f is holomorphic and $\mathrm{ord}_p(f - f(p)) = 1$; moreover this is the case at all but a discrete set of points p. Therefore if such an f is given, we may write *any* meromorphic 1-form ω with an expression $g(z)df$ for a suitable meromorphic function g. This is a convenient method for giving formulas for meromorphic 1-forms without having to be too explicit about where the formula is valid.

The Wedge Product of Two 1-Forms. The formalism used in the definition of 2-forms can be extended, by the use of linearity, to define a wedge product of two 1-forms. The ω_1 and ω_2 be two C^∞ 1-forms on X. Choosing a local variable z we may write $\omega_1 = f_1\mathrm{d}z + g_1\mathrm{d}\bar{z}$ and $\omega_2 = f_2\mathrm{d}z + g_2\mathrm{d}\bar{z}$. Define with respect to this local variable the C^∞ 2-form $\omega_1 \wedge \omega_2$ by

$$\omega_1 \wedge \omega_2 = (f_1g_2 - f_2g_1)\mathrm{d}z \wedge \mathrm{d}\bar{z}.$$

LEMMA 2.2. *The above definition gives a well defined C^∞ 2-form on X.*

Differentiating 1-Forms. Let ω be a $\mathcal{C}^\infty$ 1-form on a Riemann surface X. Then one can define the $\mathcal{C}^\infty$ 2-forms $d\omega$, $\partial\omega$, and $\overline{\partial}\omega$ on X by the following rule. Let $\phi : U \to V$ be a chart on X giving a local coordinate z. Write ω on U in terms of the local coordinate as $f(z, \overline{z})dz + g(z, \overline{z})d\overline{z}$. Define

$$\partial \omega = \frac{\partial g}{\partial z}dz \wedge d\overline{z}, \quad \overline{\partial}\omega = -\frac{\partial f}{\partial \overline{z}}dz \wedge d\overline{z},$$

and

$$d\omega = \partial\omega + \overline{\partial}\omega = (\frac{\partial g}{\partial z} - \frac{\partial f}{\partial \overline{z}})dz \wedge d\overline{z}.$$

LEMMA 2.3. *The above local recipe gives well defined $\mathcal{C}^\infty$ 2-forms $d\omega$, $\partial\omega$, and $\overline{\partial}\omega$ on X. A $\mathcal{C}^\infty$ 1-form ω of type $(1,0)$ is holomorphic if and only if $\overline{\partial}\omega = 0$. The operators d, ∂, and $\overline{\partial}$ are $\mathbb{C}$-linear and satisfy the product rules*

$$d(f\omega) = df \wedge \omega + fd\omega; \quad \partial(f\omega) = \partial f \wedge \omega + f\partial\omega; \quad \overline{\partial}(f\omega) = \overline{\partial}f \wedge \omega + f\overline{\partial}\omega$$

if f is a $\mathcal{C}^\infty$ function and ω a $\mathcal{C}^\infty$ 1-form. In addition, we have

$$ddf = \partial\partial f = \overline{\partial}\,\overline{\partial}f = 0$$

for any $\mathcal{C}^\infty$ function f.

Note also that

$$\partial\overline{\partial}f = -\overline{\partial}\partial f$$

for a $\mathcal{C}^\infty$ function f.

A $\mathcal{C}^\infty$ function f is said to be *harmonic* on an open set U if $\partial\overline{\partial}f = 0$ on U.

A $\mathcal{C}^\infty$ 1-form ω is said to be *d-closed* (or simply *closed*) if $d\omega = 0$; it is ∂-*closed* if $\partial\omega = 0$ and $\overline{\partial}$-*closed* if $\overline{\partial}\omega = 0$.

Note that since $ddf = 0$, every exact form is closed; the converse is not generally true. Similar remarks hold for ∂-exact and $\overline{\partial}$ exact forms.

The following is a simple consequence of applying the Cauchy-Riemann equations.

LEMMA 2.4. *If ω is a holomorphic 1-form, then ω is d-closed: $d\omega = 0$. Conversely, if ω is of type $(1,0)$ and is d-closed, then ω is holomorphic.*

Pulling Back Differential Forms. Let $F : X \to Y$ be a nonconstant holomorphic map between two Riemann surfaces. Let ω be a $\mathcal{C}^\infty$ 1-form on Y. We can define a $\mathcal{C}^\infty$ 1-form $F^*\omega$ on X using the following rule. Fix a chart $\phi : U \to V$ on X such that $F(U)$ is contained in the domain U' of a chart $\psi : U' \to V'$ on Y. This gives local coordinates z on U' and w on U, and in terms of these local coordinates the holomorphic map F has the form $z = h(w)$ for some holomorphic function h.

Assume that ω is equal to $f(z,\overline{z})dz + g(z,\overline{z})d\overline{z}$ in the variable z. We define the 1-form $F^*\omega$ with respect to the variable w by setting

$$F^*\omega = f(h(w), \overline{h(w)})h'(w)dw + g(h(w), \overline{h(w)})\overline{h'(w)}d\overline{w}.$$

2. OPERATIONS ON DIFFERENTIAL FORMS

LEMMA 2.5. *The above prescription gives a well defined C^∞ 1-form $F^*\omega$ on X.*

The form $F^*\omega$ is called the *pullback* of ω via F. The following are immediate:
- If ω is holomorphic, so is $F^*\omega$.
- If ω is meromorphic, so is $F^*\omega$.
- If ω is of type $(1,0)$, so is $F^*\omega$.
- If ω is of type $(0,1)$, so is $F^*\omega$.

Recall that we may also pull back functions: if f is a function on Y, then F^*f is simply the function $f \circ F$.

A completely analogous idea allows us to pull back 2-forms also, using the local formula

$$F^*(f(z,\bar{z})\mathrm{d}z \wedge \mathrm{d}\bar{z}) = f(h(w),\overline{h(w)})\|h'(w)\|^2 \mathrm{d}w \wedge \mathrm{d}\bar{w}.$$

The operation of F^* commutes with all three types of differentiation, at all levels. Specifically, if f is a C^∞ function and ω is a C^∞ 1-form, we have

- $F^*(df) = d(F^*f)$ and $F^*(d\omega) = d(F^*\omega)$.
- $F^*(\partial f) = \partial(F^*f)$ and $F^*(\partial\omega) = \partial(F^*\omega)$.
- $F^*(\bar{\partial}f) = \bar{\partial}(F^*f)$ and $F^*(\bar{\partial}\omega) = \bar{\partial}(F^*\omega)$.

The pullback of a meromorphic 1-form enjoys an order formula relating the order of the form and the multiplicity of the map to the order of the pullback:

LEMMA 2.6. *Suppose that $F: X \to Y$ is a holomorphic map between Riemann surfaces, and ω is a meromorphic 1-form on Y. Fix a point $p \in X$. Then*

$$\mathrm{ord}_p(F^*\omega) = (1 + \mathrm{ord}_{F(p)}(\omega))\,\mathrm{mult}_p(F) - 1.$$

PROOF. We may choose local coordinates w at p and z at $F(p)$ such that near p, F has the form $z = w^n$, where $n = \mathrm{mult}_p(F)$. With respect to the variable z, the form ω equals $(cz^k + \text{higher order terms in } z)dz$, where $k = \mathrm{ord}_{F(p)}(\omega)$. Thus the form $F^*\omega$ equals $(cw^{nk} + \text{higher order terms in } w)(nw^{n-1})dw$ with respect to this variable w. We see immediately then that the order of $F^*\omega$ is $nk + n - 1$ as claimed. □

Some Notation. Fix an open set U on a Riemann surface X. It is convenient to be able to speak of the space of k-forms of various types defined on U alone.

We employ the following notation, most of which is quite standard.

$$\begin{aligned}
\mathcal{E}(U) = \mathcal{E}^{(0)}(U) &= \{\mathcal{C}^\infty \text{ functions } f : U \to \mathbb{C}\}. \\
\mathcal{E}^{(1)}(U) &= \{\mathcal{C}^\infty \text{ 1-forms defined on } U\}. \\
\mathcal{E}^{(1,0)}(U) &= \{\mathcal{C}^\infty \text{ 1-forms of type } (1,0) \text{ defined on } U\}. \\
\mathcal{E}^{(0,1)}(U) &= \{\mathcal{C}^\infty \text{ 1-forms of type } (0,1) \text{ defined on } U\}. \\
\mathcal{E}^{(2)}(U) &= \{\mathcal{C}^\infty \text{ 2-forms defined on } U\}. \\
\mathcal{O}(U) &= \{\text{ holomorphic functions } f : U \to \mathbb{C}\}. \\
\Omega^1(U) &= \{\text{ holomorphic 1-forms defined on } U\}. \\
\mathcal{M}(U) = \mathcal{M}^{(0)}(U) &= \{\text{ meromorphic functions } f \text{ defined on } U\}. \\
\mathcal{M}^{(1)}(U) &= \{\text{ meromorphic 1-forms defined on } U\}.
\end{aligned}$$

All of these sets are complex vector spaces. Moreover, the spaces $\mathcal{E}(U)$, $\mathcal{O}(U)$, and $\mathcal{M}(U)$ are rings (in fact, $\mathbb{C}$-algebras); if U is connected then $\mathcal{O}(U)$ is an integral domain and $\mathcal{M}(U)$ is a field. The usual multiplication makes the spaces $\mathcal{E}^{(1)}(U)$, $\mathcal{E}^{(1,0)}(U)$, $\mathcal{E}^{(1,0)}(U)$, and $\mathcal{E}^{(2)}(U)$ into modules over the ring $\mathcal{E}(U)$; similarly the spaces $\Omega^1(U)$ and $\mathcal{M}^{(1)}(U)$ are modules over $\mathcal{O}(U)$ and if U is connected $\mathcal{M}^{(1)}(U)$ is a vector space over $\mathcal{M}(U)$.

We have the obvious relationships

$$\begin{aligned}
\mathcal{O}(U) &\subset \mathcal{E}(U), \\
\mathcal{O}(U) &\subset \mathcal{M}(U), \\
\Omega^1(U) &\subset \mathcal{E}^{(1,0)}(U), \\
\Omega^1(U) &\subset \mathcal{M}^{(1)}(U), \text{ and} \\
\mathcal{E}^{(1)}(U) &= \mathcal{E}^{(1,0)}(U) \oplus \mathcal{E}^{(0,1)}(U).
\end{aligned}$$

Note that if $V \subset U$ are open sets, then for all of these spaces there are natural "restriction" maps from the space of forms over U to the corresponding space over V. All such maps are denoted by ρ_V^U. We always have

$$\rho_U^U = \text{id} \quad \text{and} \quad \rho_W^V \circ \rho_V^U = \rho_W^U \text{ if } W \subset V \subset U.$$

If $F : X \to Y$ is a holomorphic map, and $V \subset Y$ is an open set, then we have

$$F^* : \mathcal{E}^{(i)}(V) \to \mathcal{E}^{(i)}(F^{-1}(V))$$

for each $i = 0, 1, 2$, and similarly for all of the other spaces mentioned above. The fact that F^* commutes with the various forms of differentiation can be expressed by the commutativity of the obvious squares.

F^* also commutes with the restriction maps, as do all forms of differentiation.

The Poincaré and Dolbeault Lemmas. The Poincaré and Dolbeault Lemmas address the question: when is a function equal to the derivative of another function, at least locally? More precisely, when is a 1-form ω equal to df or $\bar\partial f$, locally? Clearly since $ddf = 0$, a necessary condition for $\omega = df$ is that $d\omega = 0$; since $\bar\partial f$ has type $(0,1)$, a necessary condition for $\omega = \bar\partial f$ is that ω be of type $(0,1)$.

It turns out that these conditions are sufficient as well. We will not use these results in an important way, and so will not give proofs; they can be found in many texts.

PROPOSITION 2.7 (POINCARÉ'S LEMMA). *Let ω be a C^∞ 1-form on a Riemann surface X. Suppose that $d\omega = 0$ identically in a neighborhood of a point p in X. Then on some neighborhood U of p there is a C^∞ function f defined on U with $\omega = df$ on U.*

A proof can be found in [**Munkres91**]; the idea is to use path integration (which we will discuss in the next section) and show that the function $f(z) = \int_p^z \omega$ is well defined (using $d\omega = 0$) and satisfies $df = \omega$ (by the fundamental theorem of calculus).

Dolbeault's Lemma is not as elementary.

PROPOSITION 2.8 (DOLBEAULT'S LEMMA). *Let ω be a C^∞ $(0,1)$-form on a Riemann surface X. Then on some neighborhood U of p there is a C^∞ function f defined on U with $\omega = \bar\partial f$ on U.*

In the real analytic category a proof is elementary, and goes as follows. Write $\omega = g(z,\bar z)\mathrm{d}\bar z$. We seek a function f such that $\partial f/\partial \bar z = g$. If g is real analytic, then it can be expanded in a series and we may write $g = \sum_{i,j} c_{ij} z^i \bar z^j$. Then we may integrate term-by-term, and set $f = \sum_{i,j} c_{ij} z^i \bar z^{j+1}/(j+1)$.

See for example [**Forster81**] for a general proof.

Problems IV.2

A. Check that if ω is a C^∞ 1-form and h is a C^∞ function, then $h\omega$ defined as in the text is a C^∞ 1-form.
B. Prove Lemma 2.1.
C. Prove Lemma 2.2, i.e., that the wedge product of two 1-forms is a well defined 2-form.
D. Prove Lemma 2.3.
E. Prove Lemma 2.4.
F. Prove Lemma 2.5, i.e., that the pullback of a 1-form is well defined.
G. Prove that the pullback of a 2-form is well defined.
H. Let a holomorphic map $F: \mathbb{C}_\infty \to \mathbb{C}_\infty$ be defined by the formula $w = z^N$ for some integer $N \geq 2$, where we use z as an affine coordinate in the domain and w as an affine coordinate in the range. Compute the pullback

$F^*((1/w)dw)$ of the form $(1/w)dw$. Compute the orders of $F^*((1/w)dw)$ at all of its zeroes and poles.

I. Let X be a hyperelliptic curve defined by $y^2 = h(x)$. Let $\pi : X \to \mathbb{P}^1$ be the double covering map sending (x,y) to x. Let $\omega = \pi^*(dx/h(x))$. Compute the orders of ω at all of its zeroes and poles.

3. Integration on a Riemann Surface

We are now in a position to describe contour integration for a Riemann surface.

Paths. The concept of a 1-form is specifically designed to provide an integrand for a "contour integral" on a Riemann surface. The other ingredient of such an integral is the contour itself. This we now develop briefly; these ideas should be quite well known.

DEFINITION 3.1. A *path* on a Riemann surface X is a continuous and piecewise $\mathcal{C}^\infty$ function $\gamma : [a,b] \to X$ from a closed interval in $\mathbb{R}$ to X. The points $\gamma(a)$ and $\gamma(b)$ are the *endpoints* of the path ($\gamma(a)$ is sometimes called the *initial point*). We say the path γ is *closed* if $\gamma(a) = \gamma(b)$.

There are several obvious remarks to make.

EXAMPLE 3.2. Let $\gamma : [a,b] \to X$ be a path on X. Suppose that $\alpha : [c,d] \to [a,b]$ is a continuous and piecewise $\mathcal{C}^\infty$ function sending c to a and d to b. Then $\gamma \circ \alpha$ is a path on X. This is referred to as a *reparametrization* of the path γ. Any path γ may be reparametrized so that its domain is $[0,1]$.

EXAMPLE 3.3. Let $\gamma : [a,b] \to X$ be a path on X. The *reversal* of γ, denoted by $-\gamma$, is the path defined by sending $t \in [a,b]$ to $\gamma(a+b-t)$. Its initial point is the endpoint of γ, and its endpoint is the initial point of γ.

EXAMPLE 3.4. If $F : X \to Y$ is a $\mathcal{C}^\infty$ map (in particular if it is a holomorphic map), then $F \circ \gamma$ is a path on Y. The path $F \circ \gamma$ is often denoted by $F_*\gamma$.

EXAMPLE 3.5. Let p be a point of a Riemann surface X, and let S be a subset of X whose closure does not contain the given point p. Then there is a closed path γ on X with the following properties:
- γ is 1-1 and the image of γ lies completely inside the domain U of a chart $\phi : U \to V$ on X.
- The closed path $\phi \circ \gamma$ on V has winding number 1 about the point $\phi(p)$.
- No point of S which lies in the domain U is mapped to the interior of $\phi \circ \gamma$, i.e., for every $s \in S \cap U$, the winding number of $\phi \circ \gamma$ about $\phi(s)$ is zero.

We say that such a path is a *small path enclosing p and not enclosing any point of S*.

We note that this definition is independent of which coordinate chart is used. One can also arrange, by suitable choice of the coordinate chart, that

- the chart ϕ is centered at p,
- the domain of γ is $[0, 2\pi]$, and
- the closed path $\phi \circ \gamma$ on V is exactly the path $z(t) = r \exp(it)$ for some real number $r > 0$, in the local coordinate z of V.

Finally we note that the interior of a small path enclosing p is well defined: it is the connected component of $X - \text{image}(\gamma)$ containing p.

EXAMPLE 3.6. Suppose γ_1 and γ_2 are two paths on X with the endpoint of γ_1 being the same as the initial point of γ_2. Then there is a path γ on X with domain $[0, 1]$ such that $\gamma|_{[0,1/2]}$ and $\gamma|_{[1/2,1]}$ are reparametrizations of γ_1 and γ_2 respectively. This is the process of *concatenation* of the two paths. It can be extended in the obvious way to any finite number of paths.

The above construction can be trivially reversed: if γ is a path on X with domain $[a, b]$, then any partition $a = a_0 < a_1 < \ldots < a_n = b$ of the interval gives a decomposition of γ into n paths, of which γ is the concatenation. One calls this a *partitioning* of the path γ.

The following is immediate using the compactness of a closed interval.

LEMMA 3.7. *Let γ be a path on a Riemann surface X. Then γ may be partitioned into a finite number of paths $\{\gamma_i\}$, such that each γ_i is $\mathcal{C}^\infty$, with image contained in a single chart domain of X.*

Note that any two such partitionings have a common refinement. Thus any quantity defined via a partition of a path which is invariant under refinement is actually a function of the path itself, not the partition.

Integration of 1-Forms Along Paths. We are now prepared to define the integral of a $\mathcal{C}^\infty$ 1-form along a path. Let ω be a $\mathcal{C}^\infty$ 1-form on a Riemann surface X. Let γ be a path on X. Choose a partition $\{\gamma_i\}$ of γ so that each γ_i is $\mathcal{C}^\infty$ on its domain $[a_{i-1}, a_i]$ and has image contained in the domain U_i of a chart ϕ_i. With respect to each chart ϕ_i, write the 1-form ω as $\omega = f_i(z, \bar{z})dz + g_i(z, \bar{z})d\bar{z}$. Consider the composition $\phi_i \circ \gamma_i$ as defining the function $z = z(t)$ for t in the domain of γ_i.

DEFINITION 3.8. With the above notation, we define *the integral of ω along γ* to be the complex number

$$\int_\gamma \omega = \sum_i \int_{t=a_{i-1}}^{a_i} [f_i(z(t), \overline{z(t)})z'(t) + g_i(z(t), \overline{z(t)})\overline{z'(t)}]dt.$$

Note that if the image of γ is contained in the domain of a single chart $\phi: U \to V$, and if $\omega = fdz + gd\bar{z}$ in this chart, then

$$\int_\gamma \omega = \int_{\phi\gamma} fdz + gd\bar{z}$$

where the integral on the right is the usual contour integral of the path $\phi\gamma$ in V.

It is an immediate check that the above definition is independent of the choice of coordinate charts; this is exactly the motivation for the definition of how a 1-form transforms under change of coordinates. Moreover, it is invariant under a refinement of the partition. Therefore, as noted above, the integral is well defined, depending only on the path γ and the 1-form ω.

The following lemma contains some immediate remarks, which we leave to the reader.

LEMMA 3.9.   (a) *The integral is independent of the choice of parametrization. In other words,*

$$\int_{\gamma\alpha} \omega = \int_{\gamma} \omega$$

if α is any reparametrization of the domain of the path γ.
(b) *The integral is $\mathbb{C}$-linear in ω:*

$$\int_{\gamma} (\lambda\omega_1 + \mu\omega_2) = \lambda \int_{\gamma} \omega_1 + \mu \int_{\gamma} \omega_2.$$

(c) *The fundamental theorem of calculus holds: if f is a C^∞ function defined in a neighborhood of the image of $\gamma : [a, b] \to X$, then*

$$\int_{\gamma} df = f(\gamma(b)) - f(\gamma(a)).$$

(d) *The integral is linear under partition of the path, i.e., if γ is partitioned into paths $\{\gamma_i\}$, then*

$$\int_{\gamma} \omega = \sum_{i} \int_{\gamma_i} \omega.$$

(e) *If one reverses the direction of a path, the sign of the integral changes:*

$$\int_{-\gamma} \omega = - \int_{\gamma} \omega.$$

(f) *If $F : X \to Y$ is a holomorphic map between Riemann surfaces, then the operation of F_* on paths is adjoint to the operation of F^* on 1-forms. In other words, if γ is a path on X and ω is a 1-form on Y, then*

$$\int_{F_*\gamma} \omega = \int_{\gamma} F^*\omega.$$

Chains and Integration Along Chains. It is useful to employ a summation notation for the partitioning of a path, and also in other situations. The proper setting for this is the notion of a *chain*.

DEFINITION 3.10. A *chain* on a Riemann surface X is a finite formal sum of paths, with integer coefficients.

The set of all chains on X forms a free abelian group $\mathrm{CH}(X)$, with basis the set of paths on X. Every chain can be uniquely written in the form $\gamma = \sum_j n_j \gamma_j$ where the n_j's are integers (positive or negative) and the γ_j's are paths on X.

Given a chain $\gamma = \sum_j n_j \gamma_j$, and a C^∞ 1-form ω, we can define the integral of ω over γ by extending the path integrals by linearity:

$$\int_\gamma \omega = \sum_j n_j \int_{\gamma_j} \omega.$$

Note that if γ is partitioned into paths $\{\gamma_i\}$, we may write $\gamma = \sum_i \gamma_i$, and not get into trouble with the integration conventions. Similarly, the notation $-\gamma$ for a reversal of a path γ now has two meanings (as the reversed path and also the chain $(-1) \cdot \gamma$), but integration cannot see the difference between these two meanings, so we will not fuss about it.

With this notation, we now have that integration is a *bilinear* operation, $\mathbb{C}$-linear in the 1-forms as has been mentioned above, and now $\mathbb{Z}$-linear in the chains.

The Residue of a Meromorphic 1-Form. Let ω be a 1-form on a Riemann surface X which is meromorphic at a point $p \in X$. Choosing a local coordinate z centered at p, we may write ω via a Laurent series as

$$\omega = f(z) \mathrm{d}z = \Big(\sum_{n=-M}^\infty c_n z^n \Big) \mathrm{d}z$$

where $c_{-M} \neq 0$, so that $\mathrm{ord}_p(\omega) = -M$.

DEFINITION 3.11. The *residue* of ω at p, denoted by $\mathrm{Res}_p(\omega)$, is the coefficient c_{-1} in a Laurent series for ω at p.

We note that a Laurent series is certainly not well defined; it is our task to show that at least this one coefficient c_{-1} is, however. This will follow from the next lemma.

LEMMA 3.12. *Let ω be a meromorphic 1-form defined in a neighborhood of $p \in X$. Let γ be a small path on X enclosing p and not enclosing any other pole of ω. Then*

$$\mathrm{Res}_p(\omega) = \frac{1}{2\pi i} \int_\gamma \omega.$$

PROOF. Let $\phi : U \to V$ be a chart on X centered at p containing the image of γ, so that γ satisfies the conditions of the definition of small path enclosing p with respect to this chart. Write $\omega = f(z) \mathrm{d}z$ in the local coordinate z on V, and assume that $f(z)$ has a Laurent series $\sum_n c_n z^n$. Then

$$\int_\gamma \omega = \int_{\phi\gamma} f(z) \mathrm{d}z,$$

which is equal to $2\pi i c_{-1}$ by the ordinary Residue Theorem in the complex plane. □

COROLLARY 3.13. *The residue of a meromorphic 1-form is a well defined complex number.*

This follows from the previous lemma, since the integral is independent of the chart, and hence of the local coordinate used to expand the 1-form in a Laurent series.

LEMMA 3.14. *Suppose f is a meromorphic function at $p \in X$. Then df/f is a meromorphic 1-form at p, and*

$$\mathrm{Res}_p(df/f) = \mathrm{ord}_p(f).$$

PROOF. Choose a chart centered at p, giving a local coordinate z, and assume that $\mathrm{ord}_p(f) = n$. Then we may write $f = cz^n +$ higher order terms near p, with $c \neq 0$. Note that then $1/f = c^{-1}z^{-n} +$ higher order terms near p. In this case $df = (ncz^{n-1} +$ higher order terms $)dz$ near p, so that $df/f = (n/z +$ higher order terms $)dz$; this clearly has residue $n = \mathrm{ord}_p(f)$ at p. □

Integration of 2-Forms. Let T be a triangle on a Riemann surface X, that is, the homeomorphic image of a triangle in $\mathbb{C}$. Suppose that T is contained completely inside the domain of a chart $\phi : U \to V$. Then if η is a $\mathcal{C}^\infty$ 2-form on X, we may write $\eta = f(z, \bar{z})dz \wedge d\bar{z}$ in this chart. With this set-up, we may define

$$\iint_T \eta = \iint_{\phi(T)} f(z, \bar{z})dz \wedge d\bar{z}$$
$$= \iint_{\phi(T)} (-2i)f(x+iy, x-iy)dx \wedge dy$$

where this last integral is the usual surface integral in $\mathbb{C} = \mathbb{R}^2$.

Note that if η is contained in the domain of two charts, then the integral is well defined: this amounts to simply a change of variable in the double integral.

In general, suppose that $D \subseteq X$ is a triangulable closed set. Then we may define $\iint_D \eta$ by first triangulating D so that each triangle is contained in a single chart domain, and then adding the separate integrals over the triangles together.

Since any two triangulations of D will have a common refinement, one need only show that the definition is well defined under a refinement of a triangulation. This boils down to simply showing that if a single triangle is subdivided, the integral does not change. But this is simply the addition formula for integrating over the union of two closed sets. Thus:

LEMMA 3.15. *The above prescription gives a well defined integral $\iint_D \eta$ whenever D is a triangulable closed set of X and η is a $\mathcal{C}^\infty$ 2-form on X.*

We note here a useful construction. If T is any triangle on X completely contained in some chart domain, we can construct a path ∂T by traversing the boundary of T counterclockwise, parametrized by arc-length. (The initial point can be taken to be any one of the vertices, fix one to be specific, it will never matter.) This gives a closed path ∂T on X. If D is any triangulable closed set on X, we may decompose D into triangles $\{T_i\}$, and set $\partial D = \sum_i \partial T_i$, which is a *chain* on X, called the *boundary chain of D*. This chain depends on the triangulation, but only up to some mild transformations, essentially replacing paths by partitions and reparametrizations. Since we only use this construction of ∂D in order to integrate over ∂D, and since integration is unaffected by partitioning and reparametrizing, we do not need to pay too much attention to the choices made.

Stoke's Theorem. We now have all the ingredients to write down the Riemann surface version of Stoke's Theorem:

THEOREM 3.16 (STOKE'S THEOREM). *Let D be a triangulable closed set on a Riemann surface X, and let ω be a C^∞ 1-form on X. Then*

$$\int_{\partial D} \omega = \iint_D d\omega.$$

PROOF. Since both sides are additive with respect to the triangles composing a triangulation of D, we may assume D is a triangle which is contained inside some chart domain. At this point we may transfer both integrals to the complex plane via the chart map, and then notice that the theorem is simply Green's Theorem in the plane. □

The Residue Theorem. In the standard first course in complex variables, one inevitably comes across the Residue Theorem, which states that the sum of the residues is equal to some integral. The Riemann surface version is even simpler:

THEOREM 3.17 (THE RESIDUE THEOREM). *Let ω be a meromorphic 1-form on a compact Riemann surface X. Then*

$$\sum_{p \in X} \mathrm{Res}_p(\omega) = 0.$$

PROOF. Note of course that since the poles of ω form a discrete set in X, the sum is actually finite since X is compact. Let $p_1, p_2, \ldots, p_n$ be the poles of ω. For each pole p_i, choose a small path γ_i on X enclosing p_i and no other pole of ω, and let U_i be the interior of γ_i. Note that by the usual residue theorem in the plane, we have

$$\int_{\gamma_i} \omega = 2\pi i \, \mathrm{Res}_{p_i}(\omega)$$

by Lemma 3.12.

Let $D = X - \cup_i U_i$; then D is triangulable, and $\partial D = -\sum_i \gamma_i$ as a chain on X. Therefore

$$\begin{aligned}
\sum_i \operatorname{Res}_{p_i}(\omega) &= \frac{1}{2\pi i} \sum_i \int_{\gamma_i} \omega \\
&= \frac{-1}{2\pi i} \int_{-\sum_i \gamma_i} \omega \\
&= \frac{-1}{2\pi i} \int_{\partial D} \omega \\
&= \frac{-1}{2\pi i} \iint_D d\omega \quad \text{by Stoke's Theorem} \\
&= 0
\end{aligned}$$

since $d\omega = 0$ in a neighborhood of D, where ω is holomorphic. □

One cannot stress too much the importance of the Residue Theorem in the theory of Riemann Surfaces. The Residue Theorem can be taken to be the basis for the proof of the Riemann-Roch Theorem, which describes rather precisely the space of meromorphic functions with prescribed poles on a compact Riemann surface.

There is also an algebraic proof of the Residue Theorem which avoids the use of integration; this will be described later.

As a first application, applying the Residue Theorem to df/f, and using Lemma 3.14, we have the following.

COROLLARY 3.18. *Let f be a nonconstant meromorphic function on a compact Riemann surface X. Then*

$$\sum_{p \in X} \operatorname{ord}_p(f) = 0.$$

Recall that we have previously proved this statement (as Proposition 4.12 of Chapter II) using the theory of the degree of a holomorphic map.

Homotopy. The concept of homotopic paths extends readily to Riemann surfaces. Let $\Gamma : [a,b] \times [0,1] \to X$ be a continuous function. For each $s \in [0,1]$, define $\gamma_s : [a,b] \to X$ by $\gamma_s(t) = \Gamma(t,s)$. Assume that each γ_s is a path on X. Assume further that all of these paths have the same initial point and the same endpoint; in other words, the map Γ is constant on the two sets $\{a\} \times [0,1]$ and $\{b\} \times [0,1]$.

DEFINITION 3.19. A map Γ as above defines a *homotopy* between the paths γ_0 and γ_1 on X. We say that the two paths γ_0 and γ_1 are *homotopic*, or *homotopic via Γ*.

Note that homotopic paths necessarily have the same initial points and the same endpoints.

The basic theorem concerning homotopy of paths carries over verbatim from the theory of contour integration in the complex plane. In our context, it is the following.

PROPOSITION 3.20. *Suppose γ_0 and γ_1 are homotopic paths on a Riemann surface X. Then if ω is any closed 1-form on X (i.e., $d\omega = 0$), then*

$$\int_{\gamma_0} \omega = \int_{\gamma_1} \omega.$$

PROOF. The point is that if D is the image of the rectangle under the homotopy, then D is triangulable and $\partial D = \gamma_1 - \gamma_0$ up to partitioning and reparametrization. Therefore

$$\int_{\gamma_1 - \gamma_0} \omega = \int_{\partial D} \omega = \iint_D d\omega = 0$$

since ω is closed. □

Note that any holomorphic 1-form is closed, so the above proposition applies immediately to integrals of holomorphic 1-forms: the integrals depend only on the homotopy class of the path of integration, not on the path itself.

Let $\pi_1(X, p)$ be the fundamental group of X, consisting of homotopy classes of closed paths starting and ending at $p \in X$. The above proposition implies that for any closed 1-form ω, the map

$$\int_{-} \omega : \pi_1(X, p) \to \mathbb{C},$$

defined by sending the homotopy class of the closed path γ to $\int_\gamma \omega$, is well defined, independent of the choice of particular path γ in the homotopy class.

Moreover, this map, for fixed ω, is a group homomorphism from the fundamental group to $\mathbb{C}$. Since $\mathbb{C}$ is an abelian group, this group homomorphism must factor through the abelianization of $\pi_1(X, p)$.

In other words, we note that every commutator $aba^{-1}b^{-1}$ of $\pi_1(X, p)$ is sent to zero by this group homomorphism, and thus the commutator subgroup $[\pi_1, \pi_1]$ (which is generated by such commutators) is in the kernel of this integration map. Thus a fundamental theorem for group homomorphisms implies that integration of ω induces a well defined group homomorphism from the abelianization $\pi_1(X, p)/[\pi_1, \pi_1]$ to $\mathbb{C}$.

The quotient group $\pi_1(X, p)/[\pi_1, \pi_1]$ is denoted by $H_1(X)$, and is called the *first homology group* of X. If X is a compact orientable 2-manifold of genus g, which is the case for a compact Riemann surface, then $H_1(X)$ is a free abelian group of rank $2g$.

Homology. There is another viewpoint on homology of which it is useful to be aware. Consider the group CH(X) of chains on X. Each chain is a finite formal sum

$$\sum_i n_i \gamma_i,$$

where each n_i is an integer and γ_i is a path on X. To each chain we can associate a finite formal sum of points on X, by mapping each path γ_i to the formal difference of its endpoints, and extending by linearity. This gives a group homomorphism from the group of all chains CH(X) to the free abelian group on the set of points of X. The kernel of this homomorphism is the set of chains which has every endpoint of a path γ_i, canceled by an initial point of another. We denote this kernel by CLCH(X), the set of *closed chains* on X.

Now it is trivial that if D is a triangulable closed set in X, then the chain ∂D is a closed chain; this follows since the boundary ∂T of any triangle is closed. Such a closed chain is called a *boundary chain* on X. The subgroup of CLCH(X) generated by all boundary chains ∂D is denoted by BCH(X).

DEFINITION 3.21. The quotient group CLCH(X)/BCH(X) is called the *first homology group* of X, and is denoted by $H_1(X)$.

It is a basic theorem in homotopy and homology theory for manifolds that the definition given above for $H_1(X)$ in terms of closed chains modulo boundary chains gives the same answer as that given in the previous subsection, as the abelianization of $\pi_1(X)$. The precise statement is that the natural map from the set of based paths on X to CLCH(X) (sending a path γ to itself) induces an isomorphism between $\pi_1(X)/[\pi_1, \pi_1]$ and CLCH(X)/BCH(X).

With respect to integration, suppose that ω is a closed 1-form. Then integration of ω gives a group homomorphism from the group of closed chains CLCH(X) to $\mathbb{C}$. By Stoke's theorem,

$$\int_{\partial D} \omega = 0$$

for any boundary chain ∂D on X; therefore this group homomorphism from CLCH(X) to $\mathbb{C}$ has all of BCH(X) in its kernel, and so induces a homomorphism

$$\int_- \omega : H_1(X) \to \mathbb{C}.$$

This homomorphism $\int_- \omega$ associated to ω is called the *period mapping* for ω. One can consider its domain to be either the homology group $H_1(X)$ or the fundamental group $\pi_1(X)$, as needs arise.

Problems IV.3

A. Check the assertions made in Examples 3.2 - 3.6.

B. Prove Lemma 3.7.

C. Check that the definition of the integral of a C^∞ 1-form along a path is independent of the choice of coordinate charts and is invariant under refinement of the chosen partition.
D. Prove Lemma 3.9.
E. Let L be a lattice in $\mathbb{C}$, and let $\pi : \mathbb{C} \to X = \mathbb{C}/L$ be the natural quotient map.
   a. Let $z_0 \in L$ be a lattice point. Define the curve $\gamma : [0,1] \to \mathbb{C}$ by $\gamma(t) = tz_0$. Show that $\pi\gamma$ is a closed path on $\mathbb{C}/L$.
   b. Compute $\int_{\pi\gamma} dz$.
   c. Compute $\iint_X dz \wedge d\bar{z}$.
F. Let τ be a complex number with strictly positive imaginary part. Let h be a meromorphic function on $\mathbb{C}$ which is $(\mathbb{Z} + \mathbb{Z}\tau)$-periodic; in other words, $h(z+1) = h(z+\tau) = h(z)$ for all z. For any point p in $\mathbb{C}$, let γ_p be the path which is the counterclockwise boundary of the parallelogram with vertices $p, p+1, p+\tau+1, p+\tau, p$ (in that order). Assume p is chosen so that there are no zeroes or poles of h on γ_p. Show that

$$\frac{1}{2\pi i} \int_{\gamma_p} z \frac{h'(z)}{h(z)} dz$$

is an element of the lattice $(\mathbb{Z} + \mathbb{Z}\tau)$.

G. Check by direct computation that if $r(z)$ is a rational function of z, then the meromorphic 1-form $r(z)dz$ on the Riemann Sphere $\mathbb{C}_\infty$ satisfies the Residue Theorem. (Hint: write $r(z)$ in partial fractions.)
H. Check that if L is a lattice in $\mathbb{C}$ and $h(z)$ is an L-periodic meromorphic function, then the meromorphic 1-form $\omega = h(z)dz$, considered as a form on the complex torus $\mathbb{C}/L$, satisfies the Residue Theorem.

Further Reading

We have taken what might be called a "low road" approach to differential forms; the "high road" is to define a form as a section of a bundle, or a sheaf. For an introduction to forms on real manifolds, see [**B-T82**].

The most important, and maybe the only, result in this chapter is the Residue Theorem; the rest is mainly definitions of what should be familiar objects, in the Riemann surface setting. The proof we have given is the standard analytic one, found in many texts, e.g., [**Forster81**], [**Narasimhan92**], [**G-H78**]. There is an algebraic proof, which we will discuss later; see [**Serre59**] for this approach.

Chapter V. Divisors and Meromorphic Functions

1. Divisors

Divisors are, at first, a way of organizing into one package the zeroes and poles of a meromorphic function or 1-form. It turns out that a seemingly simple idea has many other applications, however.

The Definition of a Divisor. Let X be a Riemann surface. We will denote by $\mathbb{Z}^X$ the group of all functions from X to the integers, which is a group under pointwise addition. Given a function $D : X \to \mathbb{Z}$, the *support* of D is the set of points $p \in X$ where $D(p) \neq 0$.

DEFINITION 1.1. A *divisor* on X is a function $D : X \to \mathbb{Z}$ whose support is a discrete subset of X. The divisors on X form a group under pointwise addition, denoted by $\mathrm{Div}(X)$.

It follows immediately that if X is a compact Riemann surface, then a function $D : X \to \mathbb{Z}$ is a divisor if and only if it has finite support; therefore the group $\mathrm{Div}(X)$ for compact X is exactly the free abelian group on the set of points of X.

We usually denote a divisor D by using a summation notation, and write

$$D = \sum_{p \in X} D(p) \cdot p,$$

where the set of points p such that $D(p) \neq 0$ is discrete.

The Degree of a Divisor on a Compact Riemann Surface. The finiteness of the support of a divisor on a compact Riemann surface allows us to take the formal sum in the notation for a divisor and make an actual sum:

DEFINITION 1.2. The *degree* of a divisor D on a compact Riemann surface is the sum of the values of D:

$$\deg(D) = \sum_{p \in X} D(p).$$

The degree function deg : $\mathrm{Div}(X) \to \mathbb{Z}$ is a group homomorphism. Its kernel is the subgroup $\mathrm{Div}_0(X)$ consisting of divisors of degree 0.

The Divisor of a Meromorphic Function: Principal Divisors. Let X be a Riemann surface and let f be a meromorphic function on X which is not identically zero.

DEFINITION 1.3. The *divisor of f*, denoted by $\mathrm{div}(f)$, is the divisor defined by the order function:
$$\mathrm{div}(f) = \sum_p \mathrm{ord}_p(f) \cdot p.$$
Any divisor of this form is called a *principal divisor* on X. The set of principal divisors on X is denoted by $\mathrm{PDiv}(X)$.

We note that by Lemma 1.29 of Chapter II, we have the following:

LEMMA 1.4. *Let f and g be nonzero meromorphic functions on X. Then:*
 (a) $\mathrm{div}(fg) = \mathrm{div}(f) + \mathrm{div}(g)$.
 (b) $\mathrm{div}(f/g) = \mathrm{div}(f) - \mathrm{div}(g)$.
 (c) $\mathrm{div}(1/f) = -\mathrm{div}(f)$.

The above lemma shows that the set $\mathrm{PDiv}(X)$ of principal divisors on X forms a subgroup of $\mathrm{Div}(X)$. In fact it is a subgroup of $\mathrm{Div}_0(X)$ when X is compact:

LEMMA 1.5. *If f is a nonzero meromorphic function on a compact Riemann surface, then $\deg(\mathrm{div}(f)) = 0$.*

This statement is exactly Proposition 4.12 of Chapter II (and Corollary 3.18 of Chapter III): the sum of the orders of a meromorphic function on a compact Riemann surface is zero.

EXAMPLE 1.6. Let X be the Riemann Sphere $\mathbb{C}_\infty$, with coordinate z in the finite plane $\mathbb{C}$. Let $f(z)$ be any rational function, which we can then factor completely and write as
$$f(z) = c \prod_{i=1}^n (z - \lambda_i)^{e_i}$$
where the e_i are integers and the λ_i are distinct complex numbers. Then
$$\mathrm{div}(f) = \sum_{i=1}^n e_i \cdot \lambda_i - (\sum_{i=1}^n e_i) \cdot \infty.$$

EXAMPLE 1.7. Let $\theta(z)$ be the standard theta-function, which is holomorphic on all of $\mathbb{C}$, and has simple zeroes at the points $(1/2) + (\tau/2) + \ell$, for all lattice points $\ell \in \mathbb{Z} + \mathbb{Z}\tau$. Then
$$\mathrm{div}(\theta) = \sum_{m,n \in \mathbb{Z}} 1 \cdot (1/2) + (\tau/2) + m + n\tau.$$

This divisor on $\mathbb{C}$ does not have finite support.

1. DIVISORS

Occasionally it is useful to focus on only the zeroes or only the poles of a meromorphic function f.

DEFINITION 1.8. The *divisor of zeroes of f*, denoted by $\text{div}_0(f)$, is the divisor

$$\text{div}_0(f) = \sum_{p \text{ with } \text{ord}_p(f)>0} \text{ord}_p(f) \cdot p.$$

Similarly, the *divisor of poles of f*, denoted by $\text{div}_\infty(f)$, is the divisor

$$\text{div}_\infty(f) = \sum_{p \text{ with } \text{ord}_p(f)<0} (-\text{ord}_p(f)) \cdot p.$$

Note that both of these divisors are nonnegative functions, with disjoint support, and

(1.9) $$\text{div}(f) = \text{div}_0(f) - \text{div}_\infty(f).$$

The Divisor of a Meromorphic 1-Form: Canonical Divisors. Let X be a Riemann surface and let ω be a meromorphic 1-form on X which is not identically zero.

DEFINITION 1.10. The *divisor of ω*, denoted by $\text{div}(\omega)$, is the divisor defined by the order function:

$$\text{div}(\omega) = \sum_p \text{ord}_p(\omega) \cdot p.$$

Any divisor of this form is called a *canonical divisor* on X. The set of canonical divisors on X is denoted by $\text{KDiv}(X)$.

EXAMPLE 1.11. Let ω be the 1-form dz on the Riemann Sphere $\mathbb{C}_\infty$. Then $\text{div}(\omega) = -2 \cdot \infty$, since ω has no zeroes, and has a double pole at ∞. More generally, if $\omega = f(z)dz$, where $f = c\sum_i (z - \lambda_i)^{e_i}$ is a rational function of z, then

$$\text{div}(\omega) = \sum_i e_i \cdot \lambda_i - (2 + \sum_i e_i) \cdot \infty.$$

In particular, all such meromorphic 1-forms on $\mathbb{C}_\infty$ have degree -2.

We have the formula

$$\text{div}(f\omega) = \text{div}(f) + \text{div}(\omega)$$

when f is a nonzero meromorphic function and ω is a nonzero meromorphic 1-form on X.

The above formula shows that if one adds a principal divisor to a canonical divisor, the result is a canonical divisor. There is a stronger version of this, based on the following lemma.

LEMMA 1.12. *Let ω_1 and ω_2 be two meromorphic 1-forms on a Riemann surface X, with ω_1 not identically zero. Then there is a unique meromorphic function f on X with $\omega_2 = f\omega_1$.*

PROOF. Choose a chart $\phi : U \to V$ on X giving local coordinate z. Write $\omega_i = g_i(z)dz$ for meromorphic functions g_i on V. Let $h = g_2/g_1$ be the ratio of these functions, which is also a meromorphic function on V. Now define $f = h \circ \phi$, a meromorphic function on U.

It is easy to check that f is well defined, independent of the choice of coordinate chart. This is the desired function. □

COROLLARY 1.13. *The set* $\mathrm{KDiv}(X)$ *of canonical divisors is exactly a coset of the subgroup* $\mathrm{PDiv}(X)$ *of principal divisors. In other words, the difference of any two canonical divisors is principal.*

Therefore we have that

$$\mathrm{KDiv}(X) = \mathrm{div}(\omega) + \mathrm{PDiv}(X)$$

for any nonzero meromorphic 1-form ω.

Finally we note that we also have the concept of the divisor of zeroes $\mathrm{div}_0(\omega)$ and the divisor of poles $\mathrm{div}_\infty(\omega)$ of a meromorphic 1-form, defined in exactly the same way as for a meromorphic function.

The Degree of a Canonical Divisor on a Compact Riemann Surface.
Let X be a compact Riemann surface of genus g. Suppose that f is a meromorphic function on X; consider f as a holomorphic map $F : X \to \mathbb{C}_\infty$. Let us assume F has degree d. Then by Hurwitz's formula, we see that

$$\sum_p [\mathrm{mult}_p(F) - 1] = 2g - 2 + 2\deg(F).$$

Consider the meromorphic 1-form ω on $\mathbb{C}_\infty$ of degree -2, defined by $\omega = dz$; it has a double pole at ∞, and no other poles or zeroes. Let $\eta = F^*(\omega)$ be the pullback of ω to X. It is not hard to see, using Hurwitz's formula and Lemma 2.6 of Chapter IV, that the degree of $\mathrm{div}(\eta)$ is $2g - 2$:

$$\begin{aligned}
\deg(\operatorname{div}(\eta)) &= \sum_{p \in X} \operatorname{ord}_p(\eta) \\
&= \sum_{p \in X} \operatorname{ord}_p(F^*(\omega)) \\
&= \sum_{p \in X} [(1 + \operatorname{ord}_{F(p)}(\omega)) \operatorname{mult}_p(F) - 1] \\
&= \sum_{\substack{q \neq \infty \\ p \in F^{-1}(q)}} [\operatorname{mult}_p(F) - 1] + \sum_{p \in F^{-1}(\infty)} (-\operatorname{mult}_p(F) - 1) \\
&= \sum_{p \in X} [\operatorname{mult}_p(F) - 1] - \sum_{p \in F^{-1}(\infty)} 2\operatorname{mult}_p(F) \\
&= 2g - 2 + 2\deg(F) - 2\deg(F) \\
&= 2g - 2.
\end{aligned}$$

This computation shows the following:

PROPOSITION 1.14. *If X is a compact Riemann surface which has a nonconstant meromorphic function, then there is a canonical divisor on X of degree $2g - 2$.*

The assumption that X has a nonconstant meromorphic function will be dispensed with later: every compact Riemann surface has one, and in fact has many. However this is highly nontrivial!

The Boundary Divisor of a Chain. Suppose $\gamma = \sum_i n_i \gamma_i$ is a chain on X. Assume for simplicity that each of the paths γ_i is defined on $[0,1]$. Since the sum is finite, we see that the boundary

$$\partial \gamma = \sum_i n_i [\gamma_i(1) - \gamma_i(0)]$$

is also a finite sum, and we may consider it then as a divisor on X. This divisor $\partial \gamma$, which was briefly introduced in Section 3 of Chapter IV, is called the *boundary divisor* of the chain γ. It obviously has degree 0, and it is easy to see since X is path-connected that any divisor of degree 0 is the boundary divisor of some chain on X.

Note that a chain is closed if and only if its boundary divisor is zero.

The Inverse Image Divisor of a Holomorphic Map. Let $F : X \to Y$ be a nonconstant holomorphic map between Riemann surfaces.

DEFINITION 1.15. Let q be a point of Y. The *inverse image divisor of q*, denoted by $F^*(q)$, is the divisor

$$F^*(q) = \sum_{p \in F^{-1}(q)} \operatorname{mult}_p(F) \cdot p.$$

Note that if X and Y are compact, then the degree of the inverse image divisor is independent of the point q and is the degree of the map F.

More generally, we can extend the above construction to any divisor D on Y.

DEFINITION 1.16. Let $D = \sum_{q \in Y} n_q \cdot q$ be a divisor on Y. The *pullback of D to X*, denoted by $F^*(D)$, is the divisor

$$F^*(D) = \sum_{q \in Y} n_q F^*(q).$$

In other words, thinking of divisors as functions, we have

$$F^*(D)(p) = \mathrm{mult}_p(F) D(F(p)).$$

Pullbacks behave very nicely with respect to most operations on divisors.

LEMMA 1.17. *Let $F : X \to Y$ be a nonconstant holomorphic map between Riemann surfaces. Then:*
  (a) *The pullback is a group homomorphism $F^* : \mathrm{Div}(Y) \to \mathrm{Div}(X)$.*
  (b) *The pullback of a principal divisor is principal. Indeed, if f is a meromorphic function on Y, then $F^*(\mathrm{div}(f)) = \mathrm{div}(F^*(f)) = \mathrm{div}(f \circ F)$.*
  (c) *If X and Y are compact, so that divisors have degrees, we have*

$$\deg(F^*(D)) = \deg(F) \deg(D).$$

Warning: the pullback of a canonical divisor is not necessarily canonical; we shall see a bit later what the difference is.

PROOF. Statement (a) follows simply from the definition of F^*, since it is extended by linearity from the pullback of a point.

To see (b), suppose that f is a meromorphic function on Y, and let $p \in X$. Then, using functional notation for divisors, we have $F^*(\mathrm{div}(f))(p) = \mathrm{mult}_p(F)(\mathrm{div}(f)(F(p))) = \mathrm{mult}_p(F) \mathrm{ord}_{F(p)}(f)$. On the other hand $\mathrm{ord}_p(f \circ F)$ is also the product $\mathrm{mult}_p(F) \mathrm{ord}_{F(p)}(f)$.

Statement (c) follows immediately from the definition if D is a single point on Y; it then follows in general since both sides of the equality are linear in D. □

The Ramification and Branch Divisor of a Holomorphic Map. Let $F : X \to Y$ be a nonconstant holomorphic map between Riemann surfaces.

DEFINITION 1.18. The *ramification divisor* of F, denoted by R_F, is the divisor on X defined by

$$R_F = \sum_{p \in X} [\mathrm{mult}_p(F) - 1] \cdot p.$$

The *branch divisor* of F, denoted by B_F, is the divisor on Y defined by

$$B_F = \sum_{y \in Y} [\sum_{p \in F^{-1}(y)} (\mathrm{mult}_p(F) - 1)] \cdot y.$$

Note that if X and Y are compact, then these sums are finite, and the ramification divisor has the same degree as the branch divisor. The degree of these divisors in this case is exactly the error term in Hurwitz's formula relating the genus of X and Y; this formula can therefore be written as

$$2g(X) - 2 = \deg(F)(2g(Y) - 2) + \deg(R_F).$$

A more precise version of the Hurwitz formula relates the pullback of a canonical divisor on Y to a canonical divisor on X. The result is the following, whose proof we leave as an exercise.

LEMMA 1.19. *Let $F : X \to Y$ be a nonconstant holomorphic map between Riemann surfaces. Let ω be a meromorphic 1-form on Y, not identically zero. Then the difference between the pullback of the divisor of ω and the divisor of the pullback of ω is the ramification divisor of the map F:*

$$\mathrm{div}(F^*\omega) = F^*(\mathrm{div}(\omega)) + R_F.$$

If X and Y are compact, and one takes the degree of both sides of this equation, one recovers the Hurwitz formula.

Intersection Divisors on a Smooth Projective Curve. Let X be a smooth projective curve, that is, a Riemann surface holomorphically embedded in projective space $\mathbb{P}^n$. We will write the homogeneous coordinates in $\mathbb{P}^n$ as $[x_0 : x_1 : \cdots : x_n]$. Fix a homogeneous polynomial $G(x_0, \ldots, x_n)$ which is not identically zero on X.

We want to define the intersection divisor $\mathrm{div}(G)$ on X, which records the points where $G = 0$ on X. Of course there are multiplicities (i.e., orders of vanishing) and we must take these into account.

Fix a point $p \in X$ where G vanishes, and choose a homogeneous polynomial H of the same degree as G, which does not vanish at p. (One way to do this is to choose a coordinate x_i which is not zero at p, and use $H = x_i^d$.)

In this case the ratio G/H is a meromorphic function on X, which vanishes at p. We define the integer $\mathrm{div}(G)(p)$ to be the order of this meromorphic function at p. Note that since G vanishes at p and H does not, this order is strictly positive.

At points q where $G \neq 0$ we set $\mathrm{div}(G)(q) = 0$.

LEMMA 1.20. *This divisor $\mathrm{div}(G)$ does not depend on the choice of the nonvanishing homogeneous polynomial H, and is therefore well defined.*

PROOF. If another polynomial H' is used, then the meromorphic function G/H changes to G/H', which is just G/H multiplied by the nonzero function H/H'. Since multiplication by a meromorphic function having order 0 does not change the order, we see that the order of G/H and of G/H' is the same, and is determined only by G. □

DEFINITION 1.21. The divisor div(G) is called the *intersection divisor* of G on X.

Note that

(1.22) $$\operatorname{div}(G_1 G_2) = \operatorname{div}(G_1) + \operatorname{div}(G_2)$$

if G_1 and G_2 are both homogeneous polynomials.

Of particular importance is when G has degree one. In this case the intersection divisor is called a *hyperplane divisor*.

There is a nice relationship between intersection divisors and principal divisors. Suppose that G_1 and G_2 are two homogeneous polynomials *of the same degree*. Then we may form the meromorphic function $f = G_1/G_2$ on the smooth projective curve X.

LEMMA 1.23. *With the above notation, if G_1 and G_2 are homogeneous polynomials of the same degree, then the divisor of $f = G_1/G_2$ is the difference of the two intersection divisors:*

$$\operatorname{div}(f) = \operatorname{div}(G_1) - \operatorname{div}(G_2).$$

PROOF. Given a point $p \in X$, choose a homogeneous polynomial H of the same degree as G_1 and G_2 which does not vanish at p. Then $\operatorname{div}(G_1)(p)$ and $\operatorname{div}(G_2)(p)$ are equal to the order of the functions G_1/H and G_2/H at p. Since $f = G_1/G_2 = (G_1/H)/(G_2/H)$, we have $\operatorname{ord}_p(f) = \operatorname{ord}_p(G_1/H) - \operatorname{ord}_p(G_2/H)$ as required. □

In particular, we see that the difference between any two hyperplane divisors is a principal divisor.

The Partial Ordering on Divisors. Let D be a divisor on a Riemann surface. We write $D \geq 0$ if $D(p) \geq 0$ for all p (thinking of D as a function). We write $D > 0$ if $D \geq 0$ and $D \neq 0$. We write $D_1 \geq D_2$ if $D_1 - D_2 \geq 0$, and similarly for $>$. Similarly we have the notion of $\leq$ and $<$ for divisors. This puts a partial ordering on the set $\operatorname{Div}(X)$ of divisors on X.

Note that every divisor D can be uniquely written in the form

$$D = P - N,$$

where P and N are nonnegative divisors with disjoint support. We have already seen an example of this decomposition in (1.9) for the divisor of a meromorphic function.

If f is a meromorphic function on X, then f is holomorphic if and only if $\operatorname{div}(f) \geq 0$. The same remark applies to divisors of meromorphic 1-forms.

There is also the notion of the *minimum* of a (finite) set of divisors, which is taken to be the function which is the minimum value among all the values of the given divisors at each point:

$$\min\{D_1, \ldots, D_n\}(p) = \min\{D_1(p), \ldots, D_n(p)\}.$$

2. LINEAR EQUIVALENCE OF DIVISORS

Note that if f and g are nonzero meromorphic functions such that $f + g$ is nonzero, then

$$\mathrm{div}(f+g) \geq \min\{\mathrm{div}(f), \mathrm{div}(g)\}$$

since the same holds true for the order function.

Problems V.1

A. Let X be the hyperelliptic surface defined by $y^2 = x^5 - x$. Note that x and y are meromorphic functions on X. Compute the principal divisors $\mathrm{div}(x)$ and $\mathrm{div}(y)$.

B. Show that the ratio of two meromorphic 1-forms on a Riemann surface is a well defined meromorphic function, independent of the coordinate chart used to define it as in Lemma 1.12.

C. Let $X = \mathbb{C}/L$ be a complex torus. Show that the form dz on X is a well defined nowhere zero holomorphic 1-form on X. Conclude that 0 is a canonical divisor on X. Conclude that on a complex torus, every canonical divisor is principal and vice versa.

D. Let f be a nonconstant meromorphic function on a Riemann surface X, and let $F : X \to \mathbb{C}_\infty$ be the associated map to the Riemann Sphere. Show that the divisor of zeroes $\mathrm{div}_0(f)$ of f is the same as the inverse image divisor $F^*(0)$. Similarly show that $\mathrm{div}_\infty(f) = F^*(\infty)$ as divisors on X.

E. Let X be the hyperelliptic surface defined by $y^2 = h(x)$, where $h(x)$ is a polynomial in x with distinct roots of even degree. Let $\pi : X \to \mathbb{C}_\infty$ be the double covering map sending (x, y) to x. Show that the ramification divisor R_π of π is the divisor of zeroes $\mathrm{div}_0(y)$ of the meromorphic function y. What goes wrong if h has odd degree?

Compute the branch divisor B_π. Show that the pullback of the branch divisor $\pi^*(B_\pi)$ is equal to twice the ramification divisor R_π: $\pi^*(B_\pi) = 2R_\pi$.

F. Prove Lemma 1.19, using Lemma 2.6 of Chapter IV.

G. Show that if X is a smooth projective curve, then $\mathrm{div}(G_1 G_2) = \mathrm{div}(G_1) + \mathrm{div}(G_2)$ if G_1 and G_2 are homogeneous polynomials.

H. Let X be the smooth projective plane cubic curve defined by $y^2 z = x^3 - xz^2$. Compute the intersection divisors of the lines defined by $x = 0$, $y = 0$, and $z = 0$ with X.

I. Show that if X is a line in the projective plane, then the intersection divisor of any other line with X has degree one. In general, show that the intersection divisor of a homogenous polynomial G of degree d with a line X has degree d.

J. Let X be the projective plane conic defined by $xy = z^2$. Then if $G = ax + by + cz$ is a homogeneous polynomial of degree one, the intersection divisor $\mathrm{div}(G)$ on X has degree two. Give criteria (in terms of the coefficients of G) for this divisor $\mathrm{div}(G)$ to be of the form $2 \cdot p$ for some point $p \in X$.

2. Linear Equivalence of Divisors

One notices that in many of the natural constructions of divisors, it is often the case that any two of the divisors differ by a principal divisor. For example, the difference between any two canonical divisors is a principal divisor. This seemingly harmless idea will become the primary way in which divisors are organized.

The Definition of Linear Equivalence. The relationship of "differing by a principal divisor" is important enough to be extracted and given a name:

DEFINITION 2.1. Two divisors on a Riemann surface X are said to be *linearly equivalent*, written $D_1 \sim D_2$, if their difference is a principal divisor, i.e., if their difference is the divisor of a meromorphic function.

There are several elementary remarks to be made:

LEMMA 2.2. *Let X be a Riemann surface. Then:*
  (a) *Linear equivalence is an equivalence relation on the set $\mathrm{Div}(X)$ of divisors on X.*
  (b) *A divisor is linearly equivalent to 0 if and only if it is a principal divisor.*
  (c) *If X is compact, then linearly equivalent divisors have the same degree: if $D_1 \sim D_2$ then $\deg(D_1) = \deg(D_2)$.*

PROOF. Statement (b) is practically the definition of linear equivalence: $D \sim 0$ if and only if $D - 0 = D$ is a principal divisor. Statement (a) then follows immediately, since we see that linear equivalence is simply the relation of being in the same coset for the subgroup $\mathrm{PDiv}(X)$ of principal divisors. A linear equivalence class is therefore exactly a coset for $\mathrm{PDiv}(X)$.

If X is compact, then principal divisors have degree 0 (Lemma 1.5). Therefore if $D_1 = \mathrm{div}(f) + D_2$, then $\deg(D_1) = \deg(\mathrm{div}(f)) + \deg(D_2) = \deg(D_2)$, which proves (c). □

We have the following examples of linearly equivalent divisors, all taken from the examples of the last section.

LEMMA 2.3. *Let X be a Riemann surface. Then:*
  (a) *If f is a meromorphic function on X which is not identically zero, then the divisor of zeroes of f is linearly equivalent to the divisor of poles of f: $\mathrm{div}_0(f) \sim \mathrm{div}_\infty(f)$.*
  (b) *Any two canonical divisors on X are linearly equivalent, and any divisor linearly equivalent to a canonical divisor is a canonical divisor.*
  (c) *If X is the Riemann Sphere $\mathbb{C}_\infty$, then any two points on X are linearly equivalent.*
  (d) *If $F : X \to Y$ is a holomorphic map, and D_1 and D_2 are linearly equivalent divisors on Y, then the pullbacks $F^*(D_1)$ and $F^*(D_2)$ are linearly equivalent divisors on X.*

(e) If $F : X \to \mathbb{C}_\infty$ is a holomorphic map, then the inverse image divisors $F^*(\lambda)$ are all linearly equivalent.

(f) If X is a smooth projective curve, and G_1 and G_2 are two homogeneous polynomials in the ambient variables of the same degree, then their intersection divisors $\text{div}(G_1)$ and $\text{div}(G_2)$ are linearly equivalent. In particular, any two hyperplane divisors on X are linearly equivalent.

PROOF. Statement (a) is immediate from the equation (1.9), which says that for a meromorphic nonzero function f, $\text{div}(f) = \text{div}_0(f) - \text{div}_\infty(f)$. Statement (b) is the content of Corollary 1.13.

To see (c), let λ_1 and λ_2 be two points in $\mathbb{C}_\infty$, neither equal to ∞. Then $f(z) = (z - \lambda_1)/(z - \lambda_2)$ is a meromorphic function with $\text{div}(f) = 1 \cdot \lambda_1 - 1 \cdot \lambda_2$. If $\lambda_2 = \infty$, then simply use $f(z) = z - \lambda_1$.

To prove (d), suppose that $D_1 - D_2 = \text{div}(f)$ on Y, for some meromorphic function f on Y. Then by Lemma 1.17, $F^*(D_1) - F^*(D_2) = \text{div}(F^*(f))$, where $F^*(f) = f \circ F$ is the composition of f with the map F. Statement (e) now follows immediately from (c) and (d).

Finally (f) is immediate from Lemma 1.23. □

The linear equivalence class of the canonical divisors is called the *canonical class* of divisors.

The terminology of linear equivalence comes from property (e) above; λ is varying on a *line* (which the Riemann Sphere is considered to be for this purpose). If we have a principal divisor D, we may write $D = \text{div}(f)$ as $D = P - N$, where both P and N are nonnegative with disjoint support. Thus P is the divisor of zeroes of f and N is the divisor of poles of f. We see immediately from the definition that P and N are linearly equivalent. Now view f not as a meromorphic function but as a holomorphic map F from X to the Riemann Sphere. The divisor P is the inverse image divisor $F^*(0)$, and N is the inverse image divisor $F^*(\infty)$. One can imagine "interpolating" between P and N by the other inverse image divisors $F^*(\lambda)$ as λ passes from 0 to ∞. This gives a family of divisors on X, varying with $\lambda \in \mathbb{C}_\infty$.

If we combine these examples given in the above lemma with the remark that for a compact Riemann surface linearly equivalent divisors have the same degree, we obtain the following corollary.

COROLLARY 2.4. *Let X be a compact Riemann surface. Then:*

(a) *If f is a meromorphic function on X which is not identically zero, then $\deg(\text{div}_0(f)) = \deg(\text{div}_\infty(f))$.*

(b) *Any two canonical divisors on X have the same degree. If X has genus g and has a nonconstant meromorphic function, then the degree of any canonical divisor is $2g - 2$.*

(c) *If X is a smooth projective curve, and G_1 and G_2 are two homogeneous polynomials in the ambient variables of the same degree, then their in-*

tersection divisors $\operatorname{div}(G_1)$ and $\operatorname{div}(G_2)$ have the same degree. In particular, any two hyperplane divisors on X have the same degree.

Statement (a) of the above corollary is another restatement of the by-now-familiar property that the sum of the orders of a meromorphic function on a compact Riemann surface is zero. Statement (b) follows from the linear equivalence between any two canonical divisors, and the computation of Proposition 1.14.

Linear Equivalence for Divisors on the Riemann Sphere. On a compact Riemann surface X, any principal divisor has degree 0. For the Riemann Sphere, this turns out also to be a sufficient condition for a divisor to be principal.

PROPOSITION 2.5. *A divisor D on the Riemann Sphere is a principal divisor if and only if $\deg(D) = 0$.*

PROOF. We have already seen that the condition is necessary. For the sufficiency, suppose that $\deg(D) = 0$, and write

$$D = \sum_i e_i \cdot \lambda_i + e_\infty \cdot \infty$$

where the λ_i are points of $\mathbb{C}$ and $e_\infty = -\sum_i e_i$. Then $D = \operatorname{div}(f)$, where

$$f(z) = \prod_i (z - \lambda_i)^{e_i}.$$

□

We leave the following two corollaries to the reader.

COROLLARY 2.6. *Let D_1 and D_2 be two divisors on the Riemann Sphere. Then $D_1 \sim D_2$ if and only if $\deg(D_1) = \deg(D_2)$.*

COROLLARY 2.7. *Let D be a divisor with $\deg(D) \geq 0$ on the Riemann Sphere. Then D is linearly equivalent to a nonnegative divisor. If $\deg(D) > 0$ then for any given point p there is a strictly positive divisor E linearly equivalent to D without p in its support.*

Principal Divisors on a Complex Torus. The problem of determining the principal divisors on a complex torus $X = \mathbb{C}/L$, where L is a lattice $\mathbb{Z} + \mathbb{Z}\tau$, introduces a new element into the situation. Note that X itself is a group, with group structure inherited from the addition in $\mathbb{C}$. This allows us to define a group homomorphism

$$A : \operatorname{Div}(X) \to X$$

by sending a formal sum $\sum_i n_i \cdot p_i$ to the actual sum in the group of X. This map is called the *Abel-Jacobi* map for the complex torus X.

THEOREM 2.8 (ABEL'S THEOREM FOR A TORUS). *A divisor D on the complex torus $X = \mathbb{C}/L$ is principal if and only if $\deg(D) = 0$ and $A(D) = 0$.*

PROOF. Let us first check that the conditions are necessary. Suppose that $D = \text{div}(f)$ for some meromorphic nonzero function f on X. Of course $\deg(D) = 0$. Let $\pi : \mathbb{C} \to X$ be the quotient map, and let $h = f \circ \pi$ be the pullback of f to an L-periodic meromorphic function h on $\mathbb{C}$. For any point $p \in \mathbb{C}$, denote by γ_p the parallelogram with vertices p, $p+1$, $p+1+\tau$, and $p+\tau$. Since the zeroes and poles of h are discrete, we may choose a point p such that h has no zeroes or poles on γ_p. Therefore the zeroes and poles of f on X are in 1-1 correspondence with the zeroes and poles of h inside γ_p, with the same orders. Now the integral

$$\int_{\gamma_p} z \frac{h'(z)}{h(z)} dz$$

is easily seen by explicit computation over the four edges of γ_p to be an element of the lattice L. (See Problem IV.3F.) On the other hand, by the ordinary residue theorem in the complex plane, the value of this integral is exactly

$$\sum_{z \text{ inside } \gamma_p} \text{ord}_z(h) z.$$

Hence modding by L gives

$$\sum_{x \in X} \text{ord}_x(f) x = 0 \text{ in } X,$$

showing that $A(\text{div}(f)) = 0$.

Conversely, assume that D has degree 0 and $A(D) = 0$. Write $D = \sum_i (p_i - q_i)$, where the p_i and q_i need not be distinct, although no p_i equals any q_j. Lift each p_i to $z_i \in \mathbb{C}$, and similarly lift each q_i to w_i. Since $A(D) = 0$, we have that $\sum_i (z_i - w_i)$ is an element of the lattice L. By altering z_1, we may assume then that in fact $\sum_i (z_i - w_i) = 0$. In this case the ratio of theta-functions

$$h(z) = \frac{\prod_i \theta^{(z_i)}(z)}{\prod_i \theta^{(w_i)}(z)}$$

is an L-periodic meromorphic function on $\mathbb{C}$ which descends to a meromorphic function f on X with $\text{div}(f) = D$. □

The following is now immediate.

COROLLARY 2.9. *Let D_1 and D_2 be two divisors on a complex torus. Then $D_1 \sim D_2$ if and only if $\deg(D_1) = \deg(D_2)$ and $A(D_1) = A(D_2)$.*

We have the following analogue of Corollary 2.7. Note the differences, however; they are significant.

COROLLARY 2.10. *Let D be a divisor with $\deg(D) > 0$ on a complex torus X. Then D is linearly equivalent to a positive divisor. If $\deg(D) = 1$ then $D \sim q$ for a unique point $q \in X$. If $\deg(D) > 1$ then for any given point $x \in X$ there is a positive divisor E linearly equivalent to D without x in its support.*

PROOF. Let $\deg(D) = d > 0$, and consider the divisor $E = D - (d-1) \cdot p - q$ for p and q arbitrary points on X. E has degree 0, and by choosing the point $q = A(D - (d-1) \cdot p)$ we may arrange $A(E) = 0$. Therefore E is a principal divisor, and D is linearly equivalent to $(d-1) \cdot p + q$. If $\deg(D) = 1$, then of course the point q is determined by $q = A(D)$. If $\deg(D) > 1$, then by varying the point p we may avoid any given point of X. $\square$

The Degree of a Smooth Projective Curve. We are now in a position to define the degree of a smooth projective curve X, which is a fundamental invariant.

DEFINITION 2.11. Let X be a smooth projective curve. The *degree* of X, denoted by $\deg(X)$, is the degree of any hyperplane divisor on X.

This is well defined, since any two hyperplane divisors on X are linearly equivalent by Lemma 2.3(f); since X is compact, these hyperplane divisors will then have the same degree.

We already have a notion of degree for smooth projective *plane* curves X, defined by the vanishing of a homogeneous polynomial $F(x,y,z)$; we have taken the degree of X to be the degree of the polynomial F. Let us check that these two definitions of degree coincide.

For this, let X be defined by $F(x,y,z) = 0$, where F has degree d. Let G be a homogeneous degree one polynomial defining the hyperplane divisor $\operatorname{div}(G)$ on X.

To compute the degree of $\operatorname{div}(G)$, we may change coordinates and assume that $G(x,y,z) = x$, and that $[0:0:1]$ is not a point of X. Consider the linear polynomial y; since $[0:0:1] \notin X$, x and y never vanish simultaneously on X. Therefore the meromorphic function $h = x/y$ can be used to determine $\operatorname{div}(x)$; indeed, the intersection divisor $\operatorname{div}(x)$ is exactly the divisor of zeroes of the function h: $\operatorname{div}(x) = \operatorname{div}_0(h) = \operatorname{div}_0(x/y)$.

Let $H : X \to \mathbb{C}_\infty$ be the associated holomorphic map to h. The divisor of zeroes of the function h is exactly the inverse image divisor $H^*(0)$ (see Problem V.1D.) Therefore the degree of $\operatorname{div}(x)$ is the degree of $H^*(0)$, which is the same as the degree of the map H, by Lemma 1.17(c).

What is the degree of the map H? Fix a general $\lambda \in \mathbb{C}$. For $H(p)$ to equal λ, we must have $p = [x : y : z]$ with $x = \lambda y$; moreover p lies on the curve X, and hence satisfies $F = 0$. If $\lambda \neq 0$, then neither x nor y can be zero, again since $[0:0:1]$ is not on X. Therefore all points of $H^{-1}(\lambda)$ can be written in the form $[\lambda : 1 : w]$ with $F(\lambda, 1, w) = 0$. For a general fixed λ, this is a polynomial in w of degree d, and has d solutions. Moreover for a general λ, these solutions are distinct, and the map H has multiplicity one at all of them, since this is the case for any λ which is not a branch point of H. Hence we see that for general λ, $H^{-1}(\lambda)$ has cardinality d; this implies that H has degree exactly d. We conclude that the intersection divisor $\operatorname{div}(x)$ has degree d.

Summarizing, we have proved the following.

PROPOSITION 2.12. *Let X be a smooth projective plane curve defined by the vanishing of a homogeneous polynomial $F(x, y, z) = 0$, where F has degree d. Then X has degree d, in the sense that any hyperplane divisor on X has degree d.*

Bezout's Theorem for Smooth Projective Plane Curves. Let X be a smooth projective curve in $\mathbb{P}^n$ of degree d. Suppose that $G(x_0, \ldots, x_n)$ is a homogeneous polynomial of degree e, defining the intersection divisor $\mathrm{div}(G)$. Intuitively, this intersection divisor records the number of points of intersection between X and the hypersurface defined by $G = 0$ (counted of course with some multiplicities). Bezout's Theorem tells us the degree of this intersection divisor, and so tells us how many points of intersection there are.

THEOREM 2.13 (BEZOUT'S THEOREM). *Let X be a smooth projective curve of degree d and let G be a homogeneous polynomial of degree e which does not vanish identically on X. Then the degree of the intersection divisor $\mathrm{div}(G)$ on X is the product of the degrees of X and of G:*

$$\deg(\mathrm{div}(G)) = \deg(X)\deg(G) = de.$$

PROOF. Let H be a homogeneous polynomial of degree one, defining a hyperplane divisor $\mathrm{div}(H)$ on X. Note that H^e has degree e, which is the same as the degree of G. Therefore by Corollary 2.4(c), the intersection divisors $\mathrm{div}(H^e)$ and $\mathrm{div}(G)$ on X have the same degree since X is compact.

Since $\mathrm{div}(H^e) = e\,\mathrm{div}(H)$, we have $\deg(\mathrm{div}(H^e)) = e\deg(\mathrm{div}(H))$. Moreover $\deg(\mathrm{div}(H)) = \deg(X) = d$ by the definition of the degree of X. Hence we have that $\deg(G) = de$ as claimed. □

There are more general forms of Bezout's Theorem which apply even when X is not a smooth curve, and even when X is not a curve at all, but a higher-dimensional subset of projective space.

Plücker's Formula. Bezout's Theorem allows us to give a proof of Plücker's formula for the genus of a smooth plane curve.

The proof is based on the following, which is a more precise version of Lemma 4.6 of Chapter II.

LEMMA 2.14. *Let X be a smooth projective plane curve defined by a homogeneous polynomial $F(x, y, z) = 0$; consider the map $\pi : X \to \mathbb{P}^1$ defined by $\pi[x : y : z] = [x : z]$. Note that $\partial F/\partial y$ is also a homogeneous polynomial. In this case the intersection divisor $\mathrm{div}(\partial F/\partial y)$ on X is exactly the ramification divisor R_π of π:*

$$\mathrm{div}(\partial F/\partial y) = R_\pi.$$

PROOF. The earlier lemma simply noted that these two divisors have the same support, that is, a point $p \in X$ is ramified for π if and only if $(\partial F/\partial y)(p) = 0$. Therefore the lemma above is a quantitative version of the earlier qualitative statement.

It is sufficient to prove the statement in the open set where $z \neq 0$; in the other open sets the argument is similar. Here X is isomorphic to the affine plane curve defined by $f(x,y) = 0$, where $f(x,y) = F(x,y,1)$; moreover π is simply the projection map sending (x,y) to x. Suppose $p = (x_0, y_0)$ is a point of ramification for π, which is therefore also a zero of $\partial f/\partial y$. Then $\partial f/\partial x$ is nonzero at p, since X is smooth at p; hence y is a local coordinate for X near p.

By the Implicit Function Theorem, near p, X is locally the graph of a holomorphic function $g(y)$. Hence $f(g(y), y)$ is identically zero in a neighborhood of y_0. Taking the derivative with respect to y, we see that $(\partial f/\partial x)g'(y) + (\partial f/\partial y)$ is identically zero on X near p; so

$$\partial f/\partial y = -(\partial f/\partial x)g'(y)$$

on X near p.

Now $g(y)$ is exactly the local formula for the projection map π. Hence the order of $g(y)$ is the multiplicity of the map π. The order drops by one upon taking a derivative, so the order of $g'(y)$ is one less than the multiplicity of π. Since $(\partial f/\partial x) \neq 0$ at p, the order of $g'(y)$ is the same as the order of $\partial f/\partial y$. Hence

$$\mathrm{ord}_p(\partial f/\partial y) = \mathrm{mult}_p(\pi) - 1.$$

The number on the left is the value of the intersection divisor $\mathrm{div}(\partial F/\partial y)$ at p; the number on the right is the value of the ramification divisor R_π at p. □

Once we understand the ramification of π, we can recover the genus of X using Hurwitz's formula.

PROPOSITION 2.15 (PLÜCKER'S FORMULA). *A smooth projective plane curve of degree d has genus $g = (d-1)(d-2)/2$.*

PROOF. Let X be a smooth projective plane curve of degree d, defined by the vanishing of the homogeneous polynomial F. Consider the holomorphic map $\pi : X \to \mathbb{P}^1$ defined by $\pi[x : y : z] = [x : z]$. This map π has degree d, and has ramification divisor equal to the intersection divisor $\mathrm{div}(\partial F/\partial y)$ by Lemma 2.14. By Bezout's Theorem (Theorem 2.13) this intersection divisor has degree $d(d-1)$, since $\partial F/\partial y$ has degree $d-1$. Therefore Hurwitz's formula yields

$$2g - 2 = d(-2) + d(d-1)$$

for the genus g of X; solving for g give $g = (d-1)(d-2)/2$ as claimed. □

This method can also be extended to provide the Plücker formula for the genus of a projective plane curve with nodes. One needs to define and check several things. Firstly, one defines the intersection divisor $\mathrm{div}(G)$ in this case

3. SPACES OF FUNCTIONS AND FORMS ASSOCIATED TO A DIVISOR

and checks that Bezout's Theorem still holds. Secondly, assume for simplicity that none of the node points lift to ramification points for the projection map π; this can always be achieved after a change of coordinates. In this case one now checks that if X is defined by $F = 0$, then $\partial F/\partial y$ vanishes at all the nodes, and the intersection divisor has value exactly one at both of the points on the Riemann surface corresponding to the node. Since these points then occur in the intersection divisor but not in the ramification divisor, the ramification divisor is equal to the intersection divisor minus the divisor of points corresponding to nodes; there are two such points for each node. Therefore the degree of the ramification divisor is equal to $d(d-1) - 2n$, where n is the number of nodes. Plugging this into Hurwitz's formula gives $2g - 2 = d(-2) + d(d-1) - 2n$, and solving for g gives

$$g = (d-1)(d-2)/2 - n.$$

Problems V.2

A. Prove Corollary 2.6, that two divisors on the Riemann Sphere $\mathbb{C}_\infty$ are linearly equivalent if and only if they have the same degree.

B. Prove Corollary 2.7.

C. Let X be the projective plane cubic defined by the equation $y^2 z = x^3 - xz^2$. Let $p_0 = [0 : 1 : 0]$, $p_1 = [0 : 0 : 1]$, $p_2 = [1 : 0 : 1]$, and $p_3 = [-1 : 0 : 1]$. Show that $2p_0 \sim 2p_i$ for each i. Show that $p_1 + p_2 + p_3 \sim 3p_0$.

D. Prove the following converse to Lemma 2.3(e). Suppose E_1 and E_2 are two divisors which are both nonnegative and have disjoint support on a Riemann surface X. Show that if $E_1 \sim E_2$, then there is a holomorphic map $F: X \to \mathbb{C}_\infty$ such that $E_1 = F^*(0)$ and $E_2 = F^*(\infty)$.

E. Prove Corollary 2.9.

F. Show that the "twisted cubic curve" in $\mathbb{P}^3$ defined by $xw = yz$, $xz = y^2$, and $yw = z^2$ has degree three, by computing the (degree of the) hyperplane divisor $\text{div}(x)$. Also compute $\text{div}(y)$ on the twisted cubic.

G. Check what needs to be checked in the outline of the proof of Plücker's formula given in the text for a projective plane curve with nodes.

3. Spaces of Functions and Forms Associated to a Divisor

One of the primary uses of divisors is to organize the meromorphic functions on a Riemann surface. This is done by employing the order function, as we will see below. For this purpose it is convenient to define

$$\text{ord}_p(f) = \infty$$

if f is identically zero in a neighborhood of p. We also use the convention that $\infty > n$ for any integer n.

The Definition of the Space $L(D)$. Let D be a divisor on a Riemann surface X.

DEFINITION 3.1. The *space of meromorphic functions with poles bounded by* D, denoted by $L(D)$, is the set of meromorphic functions

$$L(D) = \{f \in \mathcal{M}(X) \mid \text{div}(f) \geq -D\}.$$

It is immediate from the definition that $L(D)$ is a complex vector space.

The reason for the terminology is the following. Suppose that $D(p) = n > 0$. Then if $f \in L(D)$, we must have $\text{ord}_p(f) \geq -n$, which means that f may have a pole of order n at p, but no worse. Similarly, if $D(p) = -n < 0$, then if $f \in L(D)$, we must have $\text{ord}_p(f) \geq n$, forcing f to have a zero of order n at p. Hence the conditions imposed on a meromorphic function f to get into a space $L(D)$ are one of two types: either poles are being allowed (to specified order and no worse), or zeroes are being required (to at least some specified order), at a discrete set of points of X.

Another way to say the above definition is to use Laurent series. Write $D = \sum_p n_p \cdot p$. For any point p, choose a local coordinate z centered at p. Then any meromorphic function f on X has a local Laurent series with respect to this local coordinate. The condition that $f \in L(D)$ is equivalent to saying that at all points p, the local Laurent series has no terms lower than z^{-n_p}.

If $D_1 \leq D_2$, then any functions with poles bounded by D_1 has poles certainly bounded by D_2; thus we see that

(3.2) \qquad if $D_1 \leq D_2$, then $L(D_1) \subseteq L(D_2)$.

Recall that a meromorphic function is holomorphic if and only if $\text{div}(f) \geq 0$; thus

(3.3) \qquad $L(0) = \mathcal{O}(X) = \{$holomorphic functions on $X\}$.

In particular, we see that

(3.4) \qquad if X is compact, then $L(0) = \{$ constant functions on $X\} \cong \mathbb{C}$

since the only holomorphic functions on a compact Riemann surface are the constant functions.

We have the following easy but important criterion, when X is compact, for when $L(D) = \{0\}$.

LEMMA 3.5. *Let X be a compact Riemann surface. If D is a divisor on X with $\deg(D) < 0$, then $L(D) = \{0\}$.*

PROOF. Suppose that $f \in L(D)$ and f is not identically zero. Consider the divisor $E = \text{div}(f) + D$. Since $f \in L(D)$, $E \geq 0$, so certainly $\deg(E) \geq 0$. However since $\deg(\text{div}(f)) = 0$, we have $\deg(E) = \deg(D) < 0$. This contradiction proves the result. □

3. SPACES OF FUNCTIONS AND FORMS ASSOCIATED TO A DIVISOR

Complete Linear Systems of Divisors. Suppose that D is a divisor on X.

DEFINITION 3.6. The *complete linear system of D*, denoted by $|D|$, is the set of all nonnegative divisors $E \geq 0$ which are linearly equivalent to D:

$$|D| = \{E \in \text{Div}(X) \mid E \sim D \text{ and } E \geq 0\}.$$

Note that any of the divisors in a complete linear system can define that linear system; they are all linearly equivalent to each other. We have the easy remark that

if X is compact and $\deg(D) < 0$, then $|D| = \emptyset$.

There is a geometric/algebraic structure to a complete linear system $|D|$ which is related to the vector space $L(D)$. Recall the projectivization $\mathbb{P}(V)$ for a complex vector space V; it is the set of 1-dimensional subspaces of V, and if V has dimension $n+1$, then $\mathbb{P}(V)$ can be put into 1-1-correspondence with projective n-space $\mathbb{P}^n$.

Take the vector space $\mathbb{P}(L(D))$. Define a function

$$S : \mathbb{P}(L(D)) \to |D|$$

by sending the span of a function $f \in L(D)$ to the divisor $\text{div}(f) + D$. Since $\text{div}(\lambda f) = \text{div}(f)$ for any constant λ, the above map S is well defined.

LEMMA 3.7. *If X is a compact Riemann surface, the map S defined above is a 1-1 correspondence.*

PROOF. Take a divisor $E \in |D|$. Since $E \sim D$, there is a meromorphic function f on X such that $E = \text{div}(f) + D$; moreover, since $E \geq 0$, the function $f \in L(D)$. Clearly $S(f) = E$, showing that S is onto.

Suppose that $S(f) = S(g)$; they are exactly the same divisor. This implies after cancelling the D's that $\text{div}(f) = \text{div}(g)$. Therefore $\text{div}(f/g) = 0$, so that f/g has no zeroes or poles on X. Since X is compact, f/g must be a nonzero constant λ; hence f and g have the same span in $L(D)$. This shows that S is 1-1. □

Thus for a compact Riemann surface, complete linear systems have a natural projective space structure.

A general linear system is a subset of a complete linear system $|D|$, which corresponds (via the map S) to a *linear subspace* of $\mathbb{P}(L(D))$. The whole space is a linear subspace obviously, so any complete linear system is a linear system. The *dimension* of a linear system is the dimension of the linear subspace of $|D|$ considered as a projective space. A linear system of dimension one is a *pencil*, a linear system of dimension two is a *net*, and of dimension three is a *web*.

Isomorphisms between $L(D)$'s under Linear Equivalence. If two divisors are linearly equivalent, then the associated spaces of meromorphic functions are naturally isomorphic.

PROPOSITION 3.8. *Suppose that D_1 and D_2 are linearly equivalent divisors on a Riemann surface X. Write $D_1 = D_2 + \operatorname{div}(h)$ for some nonzero meromorphic function h. Then multiplication by h gives an isomorphism of complex vector spaces*

$$\mu_h : L(D_1) \xrightarrow{\cong} L(D_2).$$

In particular, if $D_1 \sim D_2$, then $\dim L(D_1) = \dim L(D_2)$.

PROOF. Suppose that $f \in L(D_1)$, so that $\operatorname{div}(f) \geq -D_1$. Then $\operatorname{div}(hf) = \operatorname{div}(h) + \operatorname{div}(f) \geq \operatorname{div}(h) - D_1 = -D_2$, so that the function $hf = \mu_h(f)$ is indeed in $L(D_2)$. Thus μ_h maps $L(D_1)$ to $L(D_2)$, and by symmetry $\mu_{1/h}$ maps $L(D_2)$ back to $L(D_1)$. Since these are inverse linear maps, μ_h is an isomorphism. □

The Definition of the Space $L^{(1)}(D)$. The same constructions used above in defining spaces of functions with poles bounded by a divisor can be used to define spaces of meromorphic 1-forms.

DEFINITION 3.9. The *space of meromorphic 1-forms with poles bounded by D*, denoted by $L^{(1)}(D)$, is the set of meromorphic 1-forms

$$L^{(1)}(D) = \{\omega \in \mathcal{M}^{(1)}(X) \mid \operatorname{div}(\omega) \geq -D\}.$$

It is immediate from the definition that $L^{(1)}(D)$ is a complex vector space. We have

$$L^{(1)}(0) = \Omega^1(X),$$

the space of global holomorphic 1-forms on X.

There is the following analogue of Proposition 3.8.

PROPOSITION 3.10. *Suppose that D_1 and D_2 are linearly equivalent divisors on a Riemann surface X. Write $D_1 = D_2 + \operatorname{div}(h)$ for some nonzero meromorphic function h. Then multiplication by h gives an isomorphism of complex vector spaces*

$$\mu_h : L^{(1)}(D_1) \xrightarrow{\cong} L^{(1)}(D_2).$$

In particular, if $D_1 \sim D_2$, then $\dim L^{(1)}(D_1) = \dim L^{(1)}(D_2)$,

The same proof given above for the spaces $L(D)$ works in this setting.

The Isomorphism between $L^{(1)}(D)$ and $L(D+K)$. The construction of the spaces $L^{(1)}(D)$ can actually be directly related to the spaces $L(D)$. Fix a canonical divisor $K = \operatorname{div}(\omega)$ (where ω is a meromorphic 1-form) and another divisor D. Suppose that f is a meromorphic function in the space $L(D+K)$; this means that $\operatorname{div}(f) + D + K \geq 0$. Consider the meromorphic 1-form $f\omega$; note that $\operatorname{div}(f\omega) = \operatorname{div}(f) + \operatorname{div}(\omega) = \operatorname{div}(f) + K$. Hence $\operatorname{div}(f\omega) + D \geq 0$, so $f\omega \in L^{(1)}(D)$. Therefore multiplication by ω gives a $\mathbb{C}$-linear map

$$\mu_\omega : L(D+K) \to L^{(1)}(D).$$

LEMMA 3.11. *With the above notation, the multiplication map μ_ω is an isomorphism of vector spaces. In particular, $\dim L^{(1)}(D) = \dim L(D+K)$.*

PROOF. The map is obviously linear and injective. To see that it is surjective, choose a 1-form $\omega' \in L^{(1)}(D)$, so that $\operatorname{div}(\omega') + D \geq 0$. By Lemma 1.12, there is a meromorphic function f such that $\omega' = f\omega$. Note that

$$\operatorname{div}(f) + D + K = \operatorname{div}(f) + D + \operatorname{div}(\omega) = \operatorname{div}(f\omega) + D = \operatorname{div}(\omega') + D \geq 0,$$

so $f \in L(D+K)$. Clearly then $\mu_\omega(f) = \omega'$. □

Computation of $L(D)$ for the Riemann Sphere. Suppose that D is a divisor on the Riemann Sphere with $\deg(D) \geq 0$. Write

$$D = \sum_{i=1}^n e_i \cdot \lambda_i + e_\infty \cdot \infty$$

with λ_i distinct in $\mathbb{C}$, such that $\sum_i e_i + e_\infty \geq 0$. Consider the function

$$f_D(z) = \prod_{i=1}^n (z - \lambda_i)^{-e_i}.$$

PROPOSITION 3.12. *With the above notations, the space $L(D)$ is exactly the space*

$$L(D) = \{g(z) f_D(z) \mid g(z) \text{ is a polynomial of degree at most } \deg(D)\}.$$

PROOF. Fix a polynomial $g(z)$ of degree d; note that $\operatorname{div}(g) \geq -d \cdot \infty$. Now the divisor of f_D is exactly

$$\sum_i -e_i \cdot \lambda_i + (\sum_i e_i) \cdot \infty,$$

and so

$$\begin{aligned}
\operatorname{div}(g(z) f_D(z)) + D &= \operatorname{div}(g) + \operatorname{div}(f_D) + D \\
&\geq (\sum_i e_i + e_\infty - d) \cdot \infty = (\deg(D) - d) \cdot \infty,
\end{aligned}$$

which is at least 0 if $d \leq \deg(D)$. This proves that the given space is a subspace of $L(D)$.

Now take any nonzero $h \in L(D)$, and consider $g = h/f_D$. We have
$$\operatorname{div}(g) = \operatorname{div}(h) - \operatorname{div}(f_D) \geq -D - \operatorname{div}(f_D) = (-\sum_i e_i - e_\infty) \cdot \infty = -\operatorname{deg}(D) \cdot \infty,$$
which shows that g can have no poles in the finite part $\mathbb{C}$, and can have a pole of order at most $\deg(D)$ at ∞. This forces g to be a polynomial of degree at most $\deg(D)$. □

This explicit computation gives immediately the dimension of the space $L(D)$:

COROLLARY 3.13. *Let D be a divisor on the Riemann Sphere. Then*
$$\dim L(D) = \begin{cases} 0 & \text{if } \deg(D) < 0, \text{ and} \\ 1 + \deg(D) & \text{if } \deg(D) \geq 0. \end{cases}$$

Computation of $L(D)$ for a Complex Torus. Let $X = \mathbb{C}/L$ be a complex torus. Let us compute the dimension of $L(D)$ for any divisor D on X.

PROPOSITION 3.14. *Let $X = \mathbb{C}/L$ be a complex torus, and let D be a divisor on X.*
  a) *If $\deg(D) < 0$, then $L(D) = \{0\}$.*
  b) *If $\deg(D) = 0$ and $D \sim 0$ then $\dim L(D) = 1$.*
  c) *If $\deg(D) = 0$ and $D \not\sim 0$ then $L(D) = \{0\}$.*
  d) *If $\deg(D) > 0$ then $\dim L(D) = \deg(D)$.*

PROOF. The first statement has been noticed already. We leave statements (b) and (c) as exercises; they are in fact true for any compact Riemann surface. To prove statement d), first let us show that it is true if $\deg(D) = 1$. By Corollary 2.10, we know that D is linearly equivalent to a positive divisor, so we may assume that $D = p$ for some point $p \in X$. Clearly the constant functions are in $L(D)$, so $L(D)$ has dimension at least one. On the other hand, suppose that $L(D)$ contains a nonconstant meromorphic function f. This function f must then have a pole; however the poles of f are bounded by p, so f has a simple pole at p and no other pole. Therefore the associated map $F: X \to \mathbb{C}_\infty$ has degree one, and is therefore an isomorphism, which is absurd. Hence $L(D)$ consists of only the constant functions and has dimension one if $\deg(D) = 1$.

To finish the proof, we may proceed by induction on D; assume then that $\deg(D) = d > 1$. Write $D = D_1 + p$ for some divisor D_1 of degree $d - 1$ and some point p. By the induction step we know that $\dim L(D_1) = d - 1$.

Find a positive divisor $E \sim D$, which does not have p in its support; this is possible by Corollary 2.10. Let f be a meromorphic function on X with $\operatorname{div}(f) = E - D$; notice that $f \in L(D)$. Also we have $\operatorname{div}(f) + D_1 = E - D + D_1 = E - p$ which is not nonnegative; hence $f \notin L(D_1)$. This proves that $L(D_1) \neq L(D)$, and so $\dim L(D) \geq d$ since $L(D_1) \subset L(D)$.

To see that the dimension of $L(D)$ is exactly d, choose a local coordinate z centered at p, and suppose that $D(p) = n$. Then every $f \in L(D)$ has a Laurent series in z whose lowest possible term is the z^{-n} term. Consider the linear map $\tau : L(D) \to \mathbb{C}$ sending f to the coefficient of the z^{-n} term of its Laurent series in z. The kernel of τ is exactly $L(D - p) = L(D_1)$. Hence $L(D)$ has dimension at most one more than $\dim L(D_1)$. Since they are not equal, we must have $\dim L(D) = \dim L(D_1) + 1 = d$. $\square$

A Bound on the Dimension of $L(D)$. Part of the argument used above in the computation of $L(D)$ for a complex torus can be applied to any Riemann surface. This will lead to a bound on the dimension of $L(D)$ for a compact Riemann surface, and in particular prove that these spaces are finite-dimensional.

LEMMA 3.15. *Let X be a Riemann surface, let D be a divisor on X, and let p be a point of X. Then either $L(D - p) = L(D)$ or $L(D - p)$ has codimension one in $L(D)$.*

PROOF. Choose a local coordinate z centered at p, and let $n = -D(p)$. Then every function f in $L(D)$ has a Laurent series at p of the form $cz^n +$ higher order terms. Define a map $\alpha : L(D) \to \mathbb{C}$ by sending f to the coefficient of the z^n term in its Laurent series. Clearly α is a linear map, and the kernel of α is exactly $L(D - p)$. If α is the identically zero map, then $L(D - p) = L(D)$. Otherwise α is onto, and so $L(D - p)$ has codimension one in $L(D)$. $\square$

On a compact Riemann surface, we can use this lemma to prove the following bound on the dimension of $L(D)$:

PROPOSITION 3.16. *Let X be a compact Riemann surface, and let D be a divisor on X. Then the space of functions $L(D)$ is a finite-dimensional complex vector space. Indeed, if we write $D = P - N$, with P and N nonnegative divisors with disjoint support, then $\dim L(D) \leq 1 + \deg(P)$. In particular, if D is a nonnegative divisor, then $\dim L(D) \leq 1 + \deg(D)$.*

PROOF. Note that the statement is true for $D = 0$: on a compact Riemann surface, $L(0)$ consists of only the constant functions and therefore has dimension one. We go by induction on the degree of the positive part P of D. If $\deg(P) = 0$, then $P = 0$, so that $\dim L(P) = 1$; since $D \leq P$, we see that $L(D) \subseteq L(P)$, so that $\dim L(D) \leq \dim L(P) = 1 = 1 + \deg(P)$ as required.

Assume then that the statement is true for divisors whose positive part has degree $k-1$, and let us prove it for a divisor whose positive part has degree $k \geq 1$. Fix such a divisor D, and write $D = P - N$ as above, with $\deg(P) = k$. Choose a point p in the support of P, so that $P(p) \geq 1$. Consider the divisor $D - p$; its positive part is $P - p$, which has degree $k - 1$. Hence the induction hypothesis applies, and we have that $\dim L(D - p) \leq \deg(P - p) + 1 = \deg(P)$. Now we apply the codimension statement Lemma 3.15, and conclude that $\dim L(D) \leq 1 + \dim L(D - p)$. Hence $\dim L(D) \leq \deg(P) + 1$ as claimed. $\square$

The finite-dimensionality of the spaces $L(D)$ implies the same for the spaces $L^{(1)}(D)$, given the isomorphism between $L^{(1)}(D)$ and $L(D+K)$ for a canonical divisor K. Therefore:

COROLLARY 3.17. *Let X be a compact Riemann surface. Then for any divisor D on X, the spaces $L^{(1)}(D)$ are finite-dimensional.*

Problems V.3
A. Show that the space $L(D)$ is a complex vector space.
B. Let $F : X \to \mathbb{C}_\infty$ be a nonconstant holomorphic map. Show that the divisors $F^*(q)$ for $q \in \mathbb{C}_\infty$ form a pencil.
C. Let D be a divisor of degree 0 on a compact Riemann surface X. Show that if $D \sim 0$, then $L(D)$ is one-dimensional. Show that if $D \not\sim 0$, then $L(D) = \{0\}$.
D. Let X be a compact Riemann surface of genus g, and assume that X has a meromorphic function, so that by Proposition 1.14 canonical divisors have degree $2g - 2$. Prove that if $\deg(D) < 2 - 2g$, we must have $L^{(1)}(D) = 0$. This is the 1-form analogue of Lemma 3.5.
E. Prove Proposition 3.10.
F. Let $L = \mathbb{Z}+\mathbb{Z}\tau$, where τ is a complex number with strictly positive imaginary part. Let $X = \mathbb{C}/L$ be the associated quotient torus, and let $\pi : \mathbb{C} \to X$ be the natural quotient map. Finally let $p_0 = \pi(0)$ be the origin of the group law on X.
   1. Recall that for any meromorphic function f on a Riemann Surface, and any meromorphic 1-form ω, the product $f\omega$ is also a meromorphic 1-form. Use this to show that if h is any meromorphic function on the torus X, then hdz is a meromorphic 1-form on X.
   2. Suppose h is a meromorphic function on X in the space $L(np_0)$ for some integer n. Show that $\operatorname{Res}_{p_0}(hdz) = 0$.
   3. Let z be a local coordinate on X centered at p_0. Suppose that for some integer n, h is a meromorphic function on X in $L(np_0)$, with a Laurent series expansion $\sum_{i=-n}^{\infty} c_i z^i$. Show that if $c_i = 0$ for every $i \leq 0$, then the meromorphic function h is identically 0.
   4. Suppose that h is in $L(2p_0)$. Show that $h(x) = h(-x)$ for all x in X. (This is equivalent to the Laurent series for h (in a coordinate z centered at p_0) having only even degree terms.)
   5. Show that no nonconstant function h in $L(2p_0)$ is the square of any meromorphic function on X.
   6. Show that there exists a unique function $f \in L(2p_0)$ such that the Laurent series for f (in a coordinate z centered at p_0) has the form
   $$f(z) = \frac{1}{z^2} + a_2 z^2 + a_4 z^4 + \dots.$$

7. Show that there exists a unique function $g \in L(3p_0)$ such that the Laurent series for g (in a coordinate z centered at p_0) has the form
$$g(z) = \frac{1}{z^3} + b_1 z^1 + \dots.$$

8. Show that $g(x) = -g(-x)$ for all x in X. (This is equivalent to the Laurent series for g (in a coordinate z centered at p_0) having only odd degree terms.) Hence the Laurent series for g actually has the form
$$g(z) = \frac{1}{z^3} + b_1 z^1 + b_3 z^3 + \dots.$$

9. Find the Laurent series for f^2, f^3, and g^2 (up through the 'z' term) in terms of the above-written Laurent series for f and g.
10. Show that $g^2 = f^3 + Af + B$ for some constants A and B.
11. Show that the polynomial $w^3 + Aw + B$ has no double roots. (Hint: suppose that α is a double root. Show that the meromorphic function $g/(f-\alpha)$ is a square root of a function in $L(2p_0)$.)

G. Show that given any two meromorphic functions f and g on X, there is a divisor D such that f and g are both in $L(D)$.

H. Suppose that X is a compact Riemann surface and $D > 0$ is a strictly positive divisor on X such that $\dim L(D) = 1 + \deg(D)$. Conclude that there exists a point $p \in X$ such that $\dim L(p) = 2$. Conclude that X is isomorphic to the Riemann Sphere.

I. Let X be a Riemann surface, and let E be any divisor on X. Suppose that D is a nonnegative divisor with finite support. Show that $L(E) \subseteq L(E+D)$ has codimension at most $\deg(D)$.

4. Divisors and Maps to Projective Space

One of the primary ways of understanding Riemann surfaces is to map them into a projective space. If we can exhibit a Riemann surface X as holomorphically embedded in a projective space, that is, as a smooth projective curve, the tools of algebraic geometry can come into play, in particular the use of hyperplane divisors, etc. Therefore, via intersections, embeddings of X into projective space give rise to divisors; the converse is also true, as we will see.

Holomorphic Maps to Projective Space. The first task is to understand what is meant by a holomorphic map to $\mathbb{P}^n$. The condition is local on the domain.

DEFINITION 4.1. Let X be a Riemann surface. A map $\phi : X \to \mathbb{P}^n$ is *holomorphic* at a point $p \in X$ if there are holomorphic functions $g_0, g_1, \dots, g_n$ defined on X near p, not all zero at p, such that $\phi(x) = [g_0(x) : g_1(x) : \dots : g_n(x)]$ for x near p. We say ϕ is a *holomorphic map* if it is holomorphic at all points of X.

154   CHAPTER V. DIVISORS AND MEROMORPHIC FUNCTIONS

Note that if one of the g_i's is nonzero at p, then it will be nonzero in a neighborhood of p, and so the map $\phi(x) = [g_0(x) : g_1(x) : \cdots : g_n(x)]$ will be well defined for x near p.

Maps to Projective Space Given By Meromorphic Functions. On a compact Riemann surface, there are no nonconstant holomorphic functions. Therefore one cannot expect to use the same holomorphic functions g_i at all points of X to define a holomorphic map ϕ. In fact, one can use meromorphic functions as we now discuss.

Let X be a Riemann surface. Choose $n+1$ meromorphic functions $f = (f_0, f_1, \ldots, f_n)$ on X, not all identically zero. Define $\phi_f : X \to \mathbb{P}^n$ by setting

$$\phi_f(p) = [f_0(p) : f_1(p) : \cdots : f_n(p)].$$

Note that a priori, ϕ_f is defined at p if
- p is not a pole of any f_i, and
- p is not a zero of every f_i.

Moreover ϕ_f is a holomorphic map at all such points p where it is defined.

We claim that even at points which violate the above conditions, ϕ_f can be defined, in such a way that ϕ_f is holomorphic. This is due to the fundamental property of homogeneous coordinates of projective space, namely that

$$[x_0 : x_1 : \cdots : x_n] = [\lambda x_0 : \lambda x_1 : \cdots : \lambda x_n]$$

for any nonzero number λ.

LEMMA 4.2. *If the meromorphic functions $\{f_i\}$ are not all identically zero, then the map $\phi_f : X \to \mathbb{P}^n$ given above extends to a holomorphic map defined on all of X.*

PROOF. Fix a point $p \in X$, and let $n = \min_i \operatorname{ord}_p(f_i)$. The problem comes exactly when $n \neq 0$: if p is a pole of some f_i, then $n < 0$, and if p is a zero of every f_i, then $n > 0$.

Now in a neighborhood of p, we may assume that no f_i has a pole other than possibly at p, and there are no common zeroes to the f_i's, other than possibly at p. Hence if we choose a local coordinate z on X centered at p, then every $f_i(z)$ is holomorphic for z near 0 but $z \neq 0$, and there is no z near 0 which is a common root to every f_i. Hence for $z \neq 0$ we may multiply each $f_i(z)$ by z^{-n}, without changing the value of ϕ_f. Thus if we set $g_i(z) = z^{-n} f_i(z)$ for each i, we have

$$\begin{aligned}\phi_f(z) &= [f_0(z) : f_1(z) : \cdots : f_n(z)] \text{ for } z \neq 0 \\ &= [z^{-n} f_0(z) : z^{-n} f_1(z) : \cdots : z^{-n} f_n(z)] \text{ for } z \neq 0 \\ &= [g_0(z) : g_1(z) : \cdots : g_n(z)].\end{aligned}$$

Now this last expression for $\phi_f(z)$ has every coordinate holomorphic near 0, and has at least one coordinate nonzero at 0. Therefore the value of ϕ_f is well defined at $z = 0$, namely it is $[g_0(0) : g_1(0) : \cdots : g_n(0)]$. This process extends ϕ_f to all of X, in a holomorphic way. □

It is a basic result that every holomorphic map $\phi : X \to \mathbb{P}^n$ can be defined this way.

PROPOSITION 4.3. *Let $\phi : X \to \mathbb{P}^n$ be a holomorphic map. Then there is an $(n+1)$-tuple of meromorphic functions $f = (f_0, f_1, \ldots, f_n)$ on X such that $\phi = \phi_f$. Moreover if two $(n+1)$-tuples $f = (f_0, f_1, \ldots, f_n)$ and $g = (g_0, g_1, \ldots, g_n)$ of meromorphic functions induce the same map, so that $\phi_f = \phi_g$ as holomorphic maps to $\mathbb{P}^n$, then there is a meromorphic function λ on X such that $g_i = \lambda f_i$ for every i.*

PROOF. Fix the holomorphic map $\phi : X \to \mathbb{P}^n$. Let $[x_0 : \cdots : x_n]$ be the homogeneous coordinates of $\mathbb{P}^n$. By reordering the variables we may assume that x_0 is not identically zero on the image $\phi(X)$. Define f_i on X to be the composition of ϕ with the function x_i/x_0 on $\mathbb{P}^n$. The function f_0 is the constant function 1 in this case.

We claim that f_i is a meromorphic function on X. To see this, fix a point $p \in X$, and write ϕ in a neighborhood of p as $\phi(z) = [g_0(z) : g_1(z) : \cdots : g_n(z)]$ for holomorphic functions g_i of a local coordinate z centered at p. Note that g_0 is not identically zero near p, by assumption. Then clearly $f_i(z) = g_i/g_0$ is meromorphic at p, since it is a ratio of holomorphic functions.

Finally it is clear that $\phi = \phi_f$, where $f = (1, f_1, f_2, \ldots, f_n)$.

To prove the uniqueness statement, suppose that $\phi_f = \phi_g$ with the notation above. Let us assume for simplicity that none of the functions f_i or g_i are identically zero; if so, these must simply be omitted from the discussion. At all points p except the finitely many zeroes and poles of the functions f_i and g_i, we have $[f_0(p) : \cdots : f_n(p)] = [g_0(p) : \cdots : g_n(p)]$ as points in projective space, and none of these coordinates are zero. Therefore there is a nonzero $\lambda(p)$, depending on p, such that $g_i(p) = f_i(p)\lambda(p)$ for every i. We see that λ is a holomorphic function at these points, since it is equal to g_i/f_i for every i. Moreover this also shows that λ is meromorphic on all of X, since it is a ratio of global meromorphic functions at all but finitely many points. □

Recall that $\mathcal{M}(X)$ is the field of global meromorphic functions on X. The above proposition then gives a 1-1 correspondence between the set of holomorphic maps from X to $\mathbb{P}^n$ and the projective space $\mathbb{P}^n_{\mathcal{M}(X)}$ (which is the set of 1-dimensional subspaces of the $(n+1)$-dimensional vector space $\mathcal{M}(X)^{n+1}$ defined over the field $\mathcal{M}(X)$).

The Linear System of a Holomorphic Map. Let $\phi : X \to \mathbb{P}^n$ be a holomorphic map to projective space. To every such holomorphic map ϕ we can associate a linear system, which we now describe.

Write $\phi = [f_0 : f_1 : \cdots : f_n]$ where each f_i is a meromorphic function on X. Let $D = -\min_i\{\mathrm{div}(f_i)\}$ be the inverse of the minimum divisor of the divisors of the functions. Therefore, for $p \in X$, we have that $-D(p)$ is the minimum among the orders of the f_i at p, and so $-D(p) \leq \mathrm{ord}_p(f_i)$ for each i.

Therefore $-D \leq \mathrm{div}(f_i)$ for each i, and we have that $f_i \in L(D)$ for every i. Hence if we let V_f be the $\mathbb{C}$-linear span of the functions $\{f_i\}$, that is, the set of all linear combinations $\sum_i a_i f_i$ with $a_i \in \mathbb{C}$, we have that $V_f \subseteq L(D)$ is a linear subspace of $L(D)$.

Therefore the set of divisors $|\phi| = \{\mathrm{div}(g) + D \mid g \in V_f\}$ forms a linear system on X, a subsystem of the complete linear system $|D|$ of all positive divisors linearly equivalent to D.

Clearly the construction of D depends on the choice of the meromorphic functions $\{f_i\}$ used to define ϕ. But in fact the linear system depends only on ϕ:

LEMMA 4.4. *The linear system $|\phi|$ defined above is well defined, independent of the choice of the functions $\{f_i\}$ used to define ϕ.*

PROOF. Suppose that ϕ is also defined by $\phi = [g_0 : \cdots : g_n]$ for meromorphic functions g_i on X. By Proposition 4.3 there is a meromorphic function λ on X such that $g_i = \lambda f_i$ for each i. Since $\mathrm{div}(g_i) = \mathrm{div}(\lambda) + \mathrm{div}(f_i)$, the minimum of the divisors of the g_i's will differ from the minimum of the divisors of the f_i's by exactly the divisor $\mathrm{div}(\lambda)$. Hence if we call D the negative of the minimum for the f_i's as above, and D' the negative of the minimum for the g_i's, we have that $D' = D - \mathrm{div}(\lambda)$. In particular, $D \sim D'$ and so the complete linear systems are the same: $|D| = |D'|$.

Now it is clear also that the linear systems $|\phi_f|$ and $|\phi_g|$, defined as above using the f_i's and the g_i's respectively, are also the same. Indeed, a typical member of $|\phi_g|$ is a divisor of the form $\mathrm{div}(\sum_i a_i g_i) + D'$, and since

$$\begin{aligned}\mathrm{div}(\sum_i a_i g_i) + D' &= \mathrm{div}(\sum_i a_i \lambda f_i) + D' \\ &= \mathrm{div}(\sum_i a_i f_i) + \mathrm{div}(\lambda) + D' \\ &= \mathrm{div}(\sum_i a_i f_i) + D,\end{aligned}$$

this is a general element of $|\phi_f|$ also. Hence the two linear systems are the same and the definition of $|\phi|$ is well defined. □

DEFINITION 4.5. Given a holomorphic map $\phi : X \to \mathbb{P}^n$ with nondegenerate image, the linear system $|\phi|$ defined above is called the *linear system of the map ϕ*.

With a linear system of divisors naturally associated to a holomorphic map ϕ, one might be tempted to define the *degree* of the map ϕ to be the degree of the divisors in the associated linear system $|\phi|$. This agrees with the prior notion of the degree of a map to $\mathbb{P}^1$, but in general is too dangerous a use of the terminology. The reason for this, as we will see later, is that if ϕ maps X onto a smooth projective curve Y inside $\mathbb{P}^n$, then we have *two* different definitions of the degree of ϕ; one coming from the old definition of the degree of the map $\phi : X \to Y$, and the one coming from the degree of the divisors in the linear system $|\phi|$, and these do not agree in general.

Note also that if ϕ maps X to $\mathbb{P}^n$ with nondegenerate image (which is equivalent to having the coordinate functions $\{f_i\}$ linearly independent), then the dimension of the linear system $|\phi|$ is exactly n, since the dimension of the associated vector space of functions is $n + 1$.

A linear system of dimension n whose divisors all have degree d is often called a "g_d^n". A natural question now arises: which g_d^n's can be the linear systems of holomorphic maps? There is one property which can be singled out, that the linear system $|\phi|$ enjoys:

LEMMA 4.6. *Let $\phi : X \to \mathbb{P}^n$ be a holomorphic map. Then for every point $p \in X$ there is a divisor $E \in |\phi|$ which does not have p in its support. In other words, there is no point of X which is contained in every divisor of the linear system $|\phi|$.*

PROOF. Fix $p \in X$, and write $\phi = [f_0 : \cdots : f_n]$ for meromorphic functions f_i. Recall that we define $D = -\min_i\{\mathrm{div}(f_i)\}$. Suppose that the minimum order of the f_i's at p is k; assume that this minimum is achieved with the function f_j, so that $\mathrm{ord}_p(f_j) = k$. Then $D(p) = -k$, and $E = \mathrm{div}(f_j) + D$ is an element of the linear system $|\phi|$. But $E(p) = \mathrm{ord}_p(f_j) + D(p) = k - k = 0$, so E does not have p in its support. □

Base Points of Linear Systems. The property above will turn out to be the only restriction on a linear system, in order that it occurs as the linear system of a holomorphic map. It is important enough to discuss it on its own.

DEFINITION 4.7. Let Q be a linear system (that is, Q is a g_d^r) on a Riemann surface X. A point p is a *base point* of the linear system Q if every divisor $E \in Q$ contains p (i.e., every $E \in Q$ satisfies $E \geq p$). A linear system Q is said to be *base-point-free* (or simply *free*) if it has no base points.

The simplest example of a linear system which is base-point-free is the system $|0|$ consisting of divisors of holomorphic functions. If X is compact, this system just has the single divisor 0 in it.

We have seen above that if ϕ is a holomorphic map to $\mathbb{P}^n$ with nondegenerate image, then the associated linear system $|\phi|$ is base-point-free.

One can express the property of being a base point using spaces of functions. Suppose that $Q \subset |D|$ is a linear system, a subsystem of a complete linear system $|D|$. Let $V \subseteq L(D)$ be the vector subspace which corresponds to Q, so that the divisors in the linear system Q are exactly those of the form $\text{div}(f) + D$ for $f \in V$.

Now for every $f \in L(D)$, and for every point $p \in X$, we have $D(p) + \text{ord}_p(f) \geq 0$. So p is a base point of Q if and only if for every $f \in V$, we have $D(p) + \text{ord}_p(f) \geq 1$. Since f is already in $L(D)$, this condition on f exactly says that $f \in L(D - p)$. Hence we are led to the following criterion:

LEMMA 4.8. *A point $p \in X$ is a base point of the linear system $Q \subseteq |D|$ defined by the vector subspace $V \subseteq L(D)$ if and only if $V \subset L(D - p)$. In particular p is a base point of the complete linear system $|D|$ if and only if $L(D - p) = L(D)$.*

We adopt the convention, which is consistent with the above, that if $|D|$ is empty, then every point is a base point.

Another way to express the above is that p is *not* a base point of Q if and only if there is a function $f \in V$ with $\text{ord}_p(f) = -D(p)$ exactly.

Combining Lemma 4.8 and Proposition 3.16 we arrive at the following.

PROPOSITION 4.9. *Let D be a divisor on a compact Riemann surface X. Then a point $p \in X$ is a base point of the complete linear system $|D|$ if and only if $\dim L(D - p) = \dim L(D)$. Hence $|D|$ is base-point-free if and only if for every point $p \in X$, $\dim L(D - p) = \dim L(D) - 1$.*

The following examples come from our rather complete knowledge of the dimensions of the spaces $L(D)$ for the Riemann Sphere and for a complex torus.

EXAMPLE 4.10. Every divisor of nonnegative degree on the Riemann Sphere has a base-point-free complete linear system.

EXAMPLE 4.11. Suppose X is a complex torus. Then any divisor of degree at least 2 has a base-point-free complete linear system.

The Hyperplane Divisor of a Holomorphic Map to $\mathbb{P}^n$. Let $\phi : X \to \mathbb{P}^n$ be a holomorphic map. We have seen above that we may associate to ϕ a linear system $|\phi|$ of divisors on X, by considering a set of $n + 1$ meromorphic functions which define ϕ.

There is another, more geometric, way of obtaining a linear system from the holomorphic map ϕ, which is inspired by the construction of a hyperplane divisor for a smooth projective curve.

Suppose that $H \subset \mathbb{P}^n$ is a hyperplane, defined by the vanishing of a single homogeneous polynomial of degree one. Suppose that X does not lie entirely inside H. We will define a divisor $\phi^*(H)$ associated to this hyperplane.

Fix a point $p \in X$, and suppose that L is the homogeneous linear equation for H. Since X does not lie inside H, the equation L does not vanish identically on X.

Choose another homogeneous linear equation M which is not zero at $\phi(p)$, and consider the function $h = (L/M) \circ \phi$, defined in a neighborhood of p. This

4. DIVISORS AND MAPS TO PROJECTIVE SPACE

is a holomorphic function near p, since if we choose a local coordinate z centered at p and write ϕ near p as $\phi(z) = [g_0(z) : g_1(z) : \cdots : g_n(z)]$ for holomorphic functions g_i, not all 0 at $z = 0$, the function h is a ratio of one linear combination of g_i's divided by another, with the denominator not zero at p.

We set $\phi^*(H)(p)$ to be the order $\mathrm{ord}_p(h)$ of h at p; since h is holomorphic, this is a nonnegative integer. Moreover it is strictly positive if and only if $\phi(p) \in H$.

DEFINITION 4.12. The above construction defines a divisor $\phi^*(H)$ on X, and is called a *hyperplane divisor* for the map ϕ.

One must check that this is well defined, but this goes exactly like the argument which shows that an intersection divisor on a smooth projective curve is well defined; we leave it to the reader. Indeed, it is possible to define, for any homogeneous polynomial G in the ambient variables, a divisor $\phi^*(G)$, using the same ideas.

In any case we note that the hyperplane divisor $\phi^*(H)$ depends only on the hyperplane H, and not on its equation L: if one multiplies the equation by a constant, one does not change the order of the function which defines $\phi^*(H)(p)$.

We want to show that the set of hyperplane divisors $\{\phi^*(H)\}$ forms exactly the linear system $|\phi|$ for the map ϕ. This relies on the following simple observation:

LEMMA 4.13. *Suppose that the homogeneous coordinates of $\mathbb{P}^n$ are $[x_0 : \cdots : x_n]$, and that H is defined by the linear equation $L = \sum_i a_i x_i = 0$. Let the holomorphic map ϕ be defined by $\phi = [f_0 : \cdots : f_n]$, and set $D = -\min_i\{\mathrm{div}(f_i)\}$. If $\phi(X)$ is not contained in the hyperplane H, then*

$$\phi^*(H) = \mathrm{div}(\sum_i a_i f_i) + D.$$

PROOF. Fix a point $p \in X$, and choose j such that $\mathrm{ord}_p(f_j) = -D(p)$ is the minimum order. In this case the coordinate x_j will not vanish at p, so we may take $M = x_j$ in defining the hyperplane divisor $\phi^*(H)$. The function $h = (L/M) \circ \phi$ is then $h = (\sum_i a_i f_i)/f_j)$, and does not vanish identically near p since X does not lie inside H. Hence

$$ord_p(h) = \mathrm{ord}_p(\sum_i a_i f_i) - \mathrm{ord}_p(f_j) = \mathrm{ord}_p(\sum_i a_i f_i) + D(p)$$

as claimed. □

Now the desired statement is immediate.

COROLLARY 4.14. *Let $\phi : X \to \mathbb{P}^n$ be a holomorphic map. Then the set of hyperplane divisors $\{\phi^*(H)\}$ forms the linear system $|\phi|$ of the map.*

We see in particular another quick proof that the linear system $|\phi|$ of a holomorphic map has no base points. This is clear from the description of this linear system as the set of hyperplane divisors: a point p is in the support of a hyperplane divisor $\phi^*(H)$ if and only if $\phi(p) \in H$, and given any point of projective space, one can find a hyperplane which does not contain that point.

Defining a Holomorphic Map via a Linear System. We will now prove that the base-point-free property of the linear system of a holomorphic map in fact characterizes such systems.

PROPOSITION 4.15. *Let $Q \subset |D|$ be a base-point-free linear system of (projective) dimension n on a compact Riemann surface X. Then there is a holomorphic map $\phi : X \to \mathbb{P}^n$ such that $Q = |\phi|$. Moreover ϕ is unique up to the choice of coordinates in $\mathbb{P}^n$.*

PROOF. We have been running around these ideas enough now that the proof of the Proposition is almost easier than the statement. Suppose that the linear system Q corresponds to a vector subspace $V \subseteq L(D)$, so that the divisors of Q are those of the form $\text{div}(f) + D$, for $f \in V$. Choose a basis $f_0, \ldots, f_n$ for V. Then the holomorphic map $\phi = [f_0 : \cdots : f_n]$ has $Q = |\phi|$ as desired.

To see the uniqueness statement, suppose that $\phi' = [g_0 : \cdots : g_n]$ also has $Q = |\phi'|$. The divisors of $|\phi'|$ are then of the form $\text{div}(g) + D'$ where g is a general linear combination of the g_i's, and D' is the inverse of the minimum of the divisors of the g_i's. In any case since $|\phi| = |\phi'|$, we may change coordinates in the ϕ' map and assume that for each i, $\text{div}(f_i) + D = \text{div}(g_i) + D'$. If we set $h_i = f_i/g_i$, we see that $\text{div}(h_i) = D' - D$ is constant, independent of i; since all of these ratios have the same divisor, they must all be equal (up to a constant factor). By adjusting the g_i's further by constant factors, we may then assume that there is a single meromorphic function h on X such that $h = f_i/g_i$. At this point we have that $\phi = \phi'$, and so ϕ is unique, up to the changes of coordinates in $\mathbb{P}^n$ which were applied in the proof. □

Therefore we have a 1-1 correspondence

$$\left\{ \begin{array}{c} \text{base-point-free} \\ \text{linear systems} \\ \text{of dimension } n \text{ on } X \end{array} \right\} \leftrightarrow \left\{ \begin{array}{c} \text{holomorphic maps } \phi : X \to \mathbb{P}^n \\ \text{with nondegenerate image,} \\ \text{up to linear coordinate changes} \end{array} \right\}.$$

Removing the Base Points. The most important case of the construction of holomorphic maps via linear systems is to use complete linear systems $|D|$. One immediate problem is that in general complete linear systems may well have base points. However this is not fatal, as we now discuss.

Suppose that the complete linear system $|D|$ has base points. Let $F = \min\{E \mid E \in |D|\}$ be the minimum of all of the divisors in the linear system; the divisor F is the largest divisor that occurs in every divisor of $|D|$. It is obvious that the complete linear system $|D - F|$ then has no base points, and moreover $|D| = F + |D - F|$, that is, every divisor of $|D|$ is F plus a divisor in $|D - F|$ and conversely.

The divisor F is called the *fixed part*, or *fixed divisor*, of the linear system $|D|$. The complete linear system $|D - F|$ is called the *moving part* of $|D|$.

As far as choosing functions in order to define a holomorphic map ϕ goes, one loses nothing by passing to the moving part of $|D|$, by the following simple observation:

LEMMA 4.16. *If F is the fixed divisor of the complete linear system $|D|$, then $L(D - F) = L(D)$.*

PROOF. Clearly since $F \geq 0$, we have that $D - F \leq D$ and so $L(D - F) \subseteq L(D)$. To see the reverse inclusion, let $f \in L(D)$, so that $\operatorname{div}(f) + D \geq 0$. Therefore $\operatorname{div}(f) + D \in |D|$, and we may write $\operatorname{div}(f) + D = F + D'$ for some nonnegative divisor D'. Then $\operatorname{div}(f) + (D - F) = D' \geq 0$, so that $f \in L(D - F)$. □

In any case we see that base points of $|D|$ allow us to "shrink" the divisor without affecting the space of functions. We therefore lose nothing by restricting attention to holomorphic maps defined by complete linear systems which are base-point-free.

Given a divisor D with $|D|$ base-point-free, we denote by ϕ_D the holomorphic map associated to the complete linear system $|D|$.

Criteria for ϕ_D to be an Embedding. The results given above allow us to restrict attention to holomorphic maps ϕ_D where $|D|$ is a complete linear system without base points. We first ask the question: when is ϕ_D a 1-1 map? We need the following preliminary lemma.

LEMMA 4.17. *Let X be a compact Riemann surface, and let D be a divisor on X with $|D|$ base-point-free. Fix a point $p \in X$. Then there is a basis $f_0, f_1, \ldots, f_n$ for $L(D)$ such that $\operatorname{ord}_p(f_0) = -D(p)$ and $\operatorname{ord}_p(f_i) > -D(p)$ for $i \geq 1$.*

PROOF. Consider the codimension one subspace $L(D - p)$ of $L(D)$, and let $f_1, \ldots, f_n$ be a basis for $L(D - p)$. Extend this to a basis for $L(D)$ by adding a function f_0 in $L(D) - L(D - p)$. Then $\operatorname{ord}_p(f_i) \geq -D(p) + 1 > -D(p)$ for every $i \geq 1$.

If in addition $\operatorname{ord}_p(f_0) > -D(p)$, then $f_0 \in L(D-p)$, which is a contradiction. Therefore $\operatorname{ord}_p(f_0) = -D(p)$ as required. □

The above lemma provides a convenient tool in studying whether the map ϕ_D is 1-1. We have the following criterion for this, expressed in terms of the function spaces.

PROPOSITION 4.18. *Let X be a compact Riemann surface, and let D be a divisor on X with $|D|$ base-point-free. Fix distinct points p and q in X. Then $\phi_D(p) = \phi_D(q)$ if and only if $L(D - p - q) = L(D - p) = L(D - q)$. Hence ϕ_D is 1-1 if and only if for every pair of distinct points p and q on X, we have $\dim L(D - p - q) = \dim L(D) - 2$.*

PROOF. Since changing the basis for $L(D)$ only gives a linear change of coordinates for the map ϕ_D, we may certainly check whether $\phi_D(p) = \phi_D(q)$ using any basis. Choose the basis given by the previous lemma. With this basis, note that $\phi_D(p) = [1 : 0 : 0 : \cdots : 0]$. Therefore $\phi_D(q) = \phi_D(p)$ if and only if $\phi_D(q) = [1 : 0 : 0 : \cdots : 0]$ also, which is clearly equivalent to having $\mathrm{ord}_q(f_0) < \mathrm{ord}_q(f_i)$ for each $i \geq 1$, by the construction of the map $\phi_D = \phi_f$. Since q is not a base point of $|D|$, this happens if and only if $\mathrm{ord}_q(f_0) = -D(q)$ and $\mathrm{ord}_q(f_i) > -D(q)$ for each $i \geq 1$. This happens if and only if $\{f_1, \ldots, f_n\}$ is a basis for $L(D - q)$. However this basis was chosen exactly so that $\{f_1, \ldots, f_n\}$ was a basis for $L(D - p)$; therefore the above criterion is equivalent to having $L(D - p)$ equal to $L(D - q)$.

This says that every function f in $L(D)$ with $\mathrm{ord}_p(f) > -D(p)$ also satisfies $\mathrm{ord}_q(f) > -D(q)$. Hence $L(D - p) \subseteq L(D - p - q)$, since p and q are distinct. Thus we see that the condition is equivalent to having the three spaces $L(D-p)$, $L(D - q)$, and $L(D - p - q)$ all equal, which proves the first statement.

Since $|D|$ is base-point-free, we have that $\dim L(D - p) = \dim L(D - q) = \dim L(D) - 1$. Therefore $\dim L(D-p-q)$ is either $\dim L(D)-1$ or $\dim L(D)-2$, by Lemma 3.15. If ϕ_D is 1-1, then by the first part we see that $L(D - p - q)$ is a proper subspace of $L(D - p)$ for all p and q, and so must have dimension equal to $\dim L(D) - 2$.

Conversely, if the dimension always does drop by 2, then the tower of subspaces $L(D - p - q) \subset L(D - p) \subset L(D)$ must all be distinct for every p and q, so that ϕ_D is 1-1. $\square$

Having ϕ_D 1-1 is not completely satisfying. The problem is that the image of ϕ_D, even if it is 1-1, may not be a holomorphically embedded Riemann surface. The prototype for this phenomenon is the map from $\mathbb{C}$ to $\mathbb{P}^2$ given by sending z to $[1 : z^2 : z^3]$. In the relevant chart $U_0 \cong \mathbb{C}^2$ of $\mathbb{P}^2$ where the first coordinate is nonzero, this map sends z to (z^2, z^3). This image cannot possibly be a holomorphically embedded Riemann surface, since if it were, the composition of the map ϕ_D with a chart map near $(0,0)$ for the image would be a 1-1 holomorphic map between Riemann surfaces, hence would be a biholomorphism. But the derivative of ϕ_D is zero at the origin, and so by the chain rule the derivative of the composition would be zero, which is a contradiction.

Another way to see that the image in this example is *not* holomorphically embedded in $\mathbb{P}^2$ is to notice that at $z = 0$ none of the coordinates of the map is a local coordinate on the curve. What is necessary and sufficient is that, if we choose a basis $f_0, f_1, \ldots, f_n$ for $L(D)$ as above to use as the coordinates of the map ϕ_D, where f_0 has minimum order $-D(p)$ at p and all other f_i have order strictly greater, then we require that at least one of the f_i with $i \geq 1$ have order exactly $-D(p) + 1$ and no more. This will have the effect that, using a local coordinate z near p, and after scaling by $z^{-D(p)}$, the zeroth coordinate will not vanish at p, all other coordinates will vanish at p, and at least one of the other

4. DIVISORS AND MAPS TO PROJECTIVE SPACE

coordinates will have a simple zero at p.

This is the condition which allows us to apply the Implicit Function Theorem, and to conclude that the image is a holomorphically embedded Riemann surface. (Essentially, if f_i has order exactly $-D(p)+1$ at p, then f_i/f_0 is a local coordinate on the image.)

Hence, if we assume that ϕ_D is 1-1, then the image is a Riemann surface and ϕ_D is an isomorphism onto its image if and only if there is a function in $L(D-p)$ but not in $L(D-2p)$. This function is the desired f_i having order $-D(p)+1$ and no more. Therefore we have shown the following.

LEMMA 4.19. *Let X be a compact Riemann surface, and let D be a divisor on X with $|D|$ base-point-free. Assume that ϕ_D is 1-1. Fix a point p in X. Then the image of ϕ_D is a holomorphically embedded Riemann surface near $\phi_D(p)$ (and hence ϕ_D is an isomorphism onto its image near p) if and only if $L(D-2p) \neq L(D-p)$.*

Again, on a compact Riemann surface, we may rephrase this using dimension. The codimension of $L(D-2p)$ inside $L(D-p)$ is either 0 or 1, and we need it to be 1 for the above criterion. However we have seen that when $|D|$ is base-point-free, then $L(D-p)$ has codimension 1 in $L(D)$. Thus the above condition boils down to having $\dim L(D-2p) = \dim L(D) - 2$.

Note that this is the same condition as the 1-1 condition, simply allowing $q = p$. Therefore the whole analysis can be expressed as follows.

PROPOSITION 4.20. *Let X be a compact Riemann surface, and let D be a divisor on X whose linear system $|D|$ has no base points. Then ϕ_D is a 1-1 holomorphic map and an isomorphism onto its image (which is a holomorphically embedded Riemann surface in $\mathbb{P}^n$), if and only if for every p and q in X, we have $\dim L(D - p - q) = \dim L(D) - 2$. (The case $p = q$ is explicitly required here!)*

When the map ϕ_D is an isomorphism onto its image, we say that it is an *embedding*. One thinks of ϕ_D as just including X inside a projective space.

A divisor D such that $|D|$ has no base points and ϕ_D is an embedding is called a *very ample* divisor. This terminology is horrible but standard.

EXAMPLE 4.21. Every divisor D of positive degree on the Riemann Sphere is very ample.

EXAMPLE 4.22. Suppose X is a complex torus. Then any divisor of degree at least 3 is very ample.

The Degree of the Image and of the Map. Note that hyperplane divisors, which were introduced previously for smooth projective curves holomorphically embedded in projective space, are now defined for any Riemann surface mapping to projective space. This is a significant generalization, and illustrates a general principle in modern algebraic geometry: if you can define something for an object, you probably can (and should) define it for a map too.

In the case of hyperplane divisors, if the mapping is a holomorphic embedding, the two notions coincide.

Hyperplane divisors were used for smooth projective curves to define their degree. Suppose now that $\phi : X \to \mathbb{P}^n$ is a holomorphic map, and suppose further that the image $Y = \phi(X)$ is a smooth projective curve. Then Y has a degree, the degree of a hyperplane divisor of Y. The map $\phi : X \to Y$ has a degree also, the number of preimages of a general point of Y. Finally we have the degree of the hyperplane divisors of the map ϕ. These degrees are related as follows:

PROPOSITION 4.23. *Suppose $\phi : X \to \mathbb{P}^n$ is a holomorphic map with a smooth projective curve Y as the image. Let H be a hyperplane of $\mathbb{P}^n$. Then*

$$\deg(\phi^*(H)) = \deg(\phi) \cdot \deg(Y)$$

where in the above formula $\deg(\phi)$ denotes the degree of the map $\phi : X \to Y$. In particular, if D is a very ample divisor on X, so that ϕ_D is a holomorphic embedding of X into $\mathbb{P}^n$, then

$$\deg(\phi(X)) = \deg(D).$$

PROOF. The degree formula is based on the following: if we fix a point $p \in X$, and H is defined by $L = 0$, then

$$\phi^*(H)(p) = \mathrm{mult}_p(\phi) \cdot \mathrm{div}(L)(\phi(p)),$$

where $\mathrm{div}(L)$ is the hyperplane divisor of L on Y. The above is exactly the equation $\mathrm{ord}_p(h \circ \phi) = \mathrm{mult}_p(\phi) \cdot \mathrm{ord}_{\phi(p)}(h)$ for a meromorphic function h on Y (see Problem II.4C), applied to the meromorphic function $h = L/M$, where M is a linear homogeneous polynomial not vanishing at p.

Then

$$\begin{aligned}
\deg(\phi^*(H)) &= \sum_{p \in X} \phi^*(p) \\
&= \sum_{p \in X} \mathrm{mult}_p(\phi) \cdot \mathrm{div}(L)(\phi(p)) \\
&= \sum_{q \in Y} \sum_{p \in \phi^{-1}(q)} \mathrm{mult}_p(\phi) \cdot \mathrm{div}(L)(q) \\
&= \sum_{q \in Y} \mathrm{div}(L)(q) \sum_{p \in \phi^{-1}(q)} \mathrm{mult}_p(\phi) \\
&= \sum_{q \in Y} \mathrm{div}(L)(q) \deg(\phi) \\
&= \deg(\phi) \deg(\mathrm{div}(L)),
\end{aligned}$$

which proves the degree formula. □

EXAMPLE 4.24. Let X be the hyperelliptic curve of genus 2 defined by the equation $v^2 = u^6 - 1$. Consider the map $\phi : X \to \mathbb{P}^2$ given by $\phi = [1 : u : u^2]$. The image of this map is the smooth conic curve Y defined by $xz = y^2$. The map $\phi : X \to Y$ has degree 2. The hyperplane divisors of ϕ have degree 4.

Rational and Elliptic Normal Curves. Consider the Riemann Sphere $\mathbb{C}_\infty$, and let D be the divisor $n \cdot \infty$ for some $n \geq 1$. This divisor is very ample by the criterion given above, since $\dim L(D) = n+1$ but $\dim L(D-p-q) = n-1$ for any p and q. The embedding ϕ_D maps the Riemann Sphere to $\mathbb{P}^n$, and the image is called a *rational normal curve*.

If we choose a local coordinate z on $\mathbb{C}$, so that $z = \infty$ is the point at infinity, then a basis for $L(n \cdot \infty)$ is $\{1, z, z^2, \ldots, z^n\}$. Hence the map ϕ_D to $\mathbb{P}^n$, using this basis, sends z to $[1 : z : z^2 : \cdots : z^n]$. The point at infinity is sent to $[0 : 0 : \cdots : 0 : 1]$.

Note that when $n = 1$, the map is the standard isomorphism between $\mathbb{C}_\infty$ and $\mathbb{P}^1$.

When $n = 2$, the map sends the Riemann Sphere isomorphically onto the conic curve given by the homogeneous polynomial equation $XZ = Y^2$ in terms of the homogeneous coordinates $[X : Y : Z]$ of $\mathbb{P}^2$. When $n = 3$, the map sends the Riemann Sphere isomorphically onto the twisted cubic in $\mathbb{P}^3$.

These maps (from $\mathbb{P}^1$ or $\mathbb{C}_\infty$ to $\mathbb{P}^n$ given by a basis for all the space of all polynomials of a certain degree) are called *Veronese maps*.

Let X be a complex torus. Then any divisor D of degree $d \geq 3$ or more is very ample, and gives a holomorphic embedding. Since $\dim L(D) = \deg(D) = d$ in this case, we see that ϕ_D maps X to $\mathbb{P}^{d-1}$, onto a smooth curve of degree d. The image is called an *elliptic normal curve of degree d*.

When $d = 3$, we have an embedding of X into the plane. This embedding maps X to a smooth cubic curve in the plane.

Working Without Coordinates. Let V be a complex vector space. Recall that the *projectivization* of V, denoted by $\mathbb{P}V$, is the set of 1-dimensional subspaces of V.

There is a dual construction which is useful to be aware of.

DEFINITION 4.25. The *dual projective space* $(\mathbb{P}V)^*$ is the set of codimension one subspaces of V.

Note that any codimension one subspace W of V induces a hyperplane $\mathbb{P}W \subset \mathbb{P}V$; indeed, the dual space $(\mathbb{P}V)^*$ may be identified with the set of hyperplanes in $\mathbb{P}V$.

We leave the following to the reader.

LEMMA 4.26. *There is a natural bijection between $\mathbb{P}(V^*)$ and $(\mathbb{P}V)^*$ given by associating to the span of a nonzero functional $f : V \to \mathbb{C}$ the codimension one subspace which is the kernel of f.*

In the situation we are currently in, of analysing linear systems $|D|$ on a Riemann surface X and their induced maps ϕ_D, codimension one subspaces arise quite naturally. Suppose that $|D|$ is a base-point-free linear system on X of dimension at least one. Then for any point $p \in X$, the subset $\{E \in |D| \mid E \geq p\}$ is a hyperplane in $|D|$; indeed, under the bijection of $|D|$ with $\mathbb{P}(L(D))$, it corresponds exactly to the codimension one subspace $\mathbb{P}(L(D-p))$.

This remark allows us to give a coordinate-free description of the map ϕ_D. Proving the following is elementary, only an exercise in unraveling the notation, and we leave it to the reader.

PROPOSITION 4.27. *The map $\phi : X \to |D|^*$ sending a point $p \in X$ to $\{E \in |D| \mid E \geq p\}$ is, with suitable coordinates, the map ϕ_D.*

Problems V.4

A. Let f be a meromorphic function on a Riemann surface X. Show that the holomorphic map $\phi : X \to \mathbb{P}^1$ defined by $\phi(p) = [1 : f(p)]$ is, after identifying $\mathbb{P}^1$ with $\mathbb{C}_\infty$, exactly the holomorphic map $F : X \to \mathbb{C}_\infty$ associated to f.

B. Verify the statements of Examples 4.10 and 4.11.

C. Verify the statements of Examples 4.21 and 4.22.

D. Note that a hyperplane in $\mathbb{P}^1$ is just a single point. Show that if $q \in \mathbb{P}^1$, then the hyperplane divisor for a holomorphic map $\phi : X \to \mathbb{P}^1$ is the same as the inverse image divisor $\phi^*(q)$.

E. Assume that $\phi : X \to \mathbb{P}^n$ is an embedding onto the projective curve $Y \subset \mathbb{P}^n$. Show that the hyperplane divisors on X, defined in this Section, correspond (via the isomorphism between X and Y) to the hyperplane divisors on Y defined in Section 1.

F. We recall the notation of Problem F, Section 3. Let $D = 3 \cdot p_0$ and let $\{1, f, g\}$ be the basis of $L(D)$ discussed there. Show that with this basis,

ϕ_D is an embedding of $\mathbb{C}/L$ onto the smooth cubic curve defined by $Y^2Z = X^3 + AXZ^2 + BZ^3$.

G. Prove Lemma 4.26.
H. Prove that the subset $\{E \in |D| \mid E \geq p\}$ is exactly the set $p + |D-p|$. Prove that it is a hyperplane in $|D|$, and under the bijection of $|D|$ with $\mathbb{P}(L(D))$, it corresponds exactly to $\mathbb{P}(L(D-p))$.
I. Prove Proposition 4.27.
J. Generalize Example 4.24 to any hyperelliptic curve: show that if $v^2 = h(u)$ defines a hyperelliptic curve of genus g, then $\phi = [1 : u : u^2 : \cdots : u^{g-1}]$ defines a degree 2 map onto a rational normal curve of degree $g-1$ in $\mathbb{P}^{g-1}$, and that the hyperplane divisors of ϕ have degree $2g - 2$.
K. Let Q be a linear system without base points. Show that for any finite set of points $\{p_1, \ldots, p_n\}$ there is a divisor D in Q without any p_i in its support.

Further Reading

No book on algebraic curves or Riemann surfaces can avoid divisors; the treatment given here is the standard one.

Bezout's Theorem generalizes to many settings. For a thorough discussion of the local intersections of plane curves and their contribution to the product of the degrees, see [**Fulton69**]. More general statements are discussed in all texts in algebraic geometry, e.g., [**Mumford76**], [**Hartshorne77**], [**G-H78**], and [**Shafarevich77**].

Plücker's Formula also has more general versions, which apply to nonsmooth curves, and give information relating flexes and bitangents, nodes and cusps. See [**G-H78**] for a thorough discussion.

Rational normal curves have fascinated geometers for centuries, beginning with conics; they have a multitude of minimizing properties both in the algebraic and the differential categories. See [**G-H78**] or [**Harris92**] for lots more. Elliptic normal curves is the subject of the monograph [**Hulek86**].

Chapter VI. Algebraic Curves and the Riemann-Roch Theorem

1. Algebraic Curves

Throughout this section X will be a compact Riemann surface of genus g.

Separating Points and Tangents. It is a basic and highly nontrivial result that a compact Riemann surface has nonconstant meromorphic functions on it. However, almost all Riemann surfaces which one stumbles across in nature have plenty of them. The theory involved in producing meromorphic functions for an unknown compact Riemann surface X is rather technical analysis and functional analysis. After one has access to meromorphic functions, however, the theory is completely algebraic, or at least can be made so. Therefore to get an overview of the subject we may be excused if we simply assume that the Riemann surfaces under discussion have a decent supply of meromorphic functions.

Let us say that a meromorphic function f on a Riemann surface X has *multiplicity one* at a point $p \in X$ if either f is holomorphic at p and $\text{ord}_p(f - f(p)) = 1$, or f has a simple pole at p. This is exactly equivalent to the associated map F from X to the Riemann Sphere having multiplicity one at p.

With this terminology, let us make a specific definition for having "lots" of meromorphic functions.

DEFINITION 1.1. Let S be a set of meromorphic functions on a compact Riemann surface X. We say that S *separates points* of X if for every pair of distinct points p and q in X there is a meromorphic function $f \in S$ such that $f(p) \neq f(q)$. We say that S *separates tangents* of X if for every point $p \in X$ there is a meromorphic function $f \in S$ which has multiplicity one at p. A compact Riemann surface X is an *algebraic curve* if the field $\mathcal{M}(X)$ of global meromorphic functions separates the points and tangents of X.

We want to explicitly allow poles in the functions considered above. This means that if f has a pole at p and not at q, then $f(p) \neq f(q)$.

The terminology of "separating points" needs no explanation; that of "separating tangents" is a bit more obscure. The idea comes from calculus of several variables, and is more suggestive when working with higher-dimensional varieties than simply curves. Recall that given a map F between manifolds, the derivative F' is naturally, at each point p, a map from the tangent space at p to the tangent space at $F(p)$. To say that F "separates tangents" should mean that the derivative F' is a 1-1 map on the tangent vectors. In our case of a meromorphic function f on X defined at a point p, we may consider it as a holomorphic map F from X to the Riemann Sphere, which in local coordinates can be written as a holomorphic function g from $\mathbb{C}$ to $\mathbb{C}$, sending 0 to 0. Here the tangent space is 1-dimensional, and the derivative map at 0 is simply multiplication by the number $g'(0)$. Hence we should say that f separates the tangents at p if $g'(0) \neq 0$; and this exactly says that $\operatorname{mult}_p(F) = 1$.

Also note that when $\operatorname{ord}_p(f - f(p)) = 1$, f may be used as a local coordinate near p. Therefore the separating tangents condition is equivalent to saying that, at every point $p \in X$, there is a local coordinate which extends to a meromorphic function on all of X.

We have actually seen these conditions in a slightly different guise in Section 4 of Chapter V. The conditions that a map ϕ_D to projective space be a holomorphic embedding imply that the space $L(D)$ generates a field of functions which separates the points and tangents of X.

Finally note that if X is an algebraic curve, then for every $p \in X$ we can find a global meromorphic function g on X such that $\operatorname{ord}_p(g) = 1$: take a function f exhibiting the separation of tangents at p, and use either $g = f - f(p)$ if f is holomorphic at p or $g = 1/f$ if f has a simple pole at p.

The reader should check the following examples. The first uses the rational functions on $\mathbb{C}_\infty$.

EXAMPLE 1.2. The Riemann Sphere $\mathbb{C}_\infty$ is an algebraic curve.

Using ratios of theta-functions, we have the following.

EXAMPLE 1.3. Any complex torus $\mathbb{C}/L$ is an algebraic curve.

Using ratios of homogeneous polynomials of the same degree, one can check the next three examples.

EXAMPLE 1.4. Any smooth projective plane curve is an algebraic curve.

EXAMPLE 1.5. Any smooth projective curve in $\mathbb{P}^n$ is an algebraic curve.

EXAMPLE 1.6. The Riemann surface obtained by resolving the nodes of a projective plane curve with nodes is an algebraic curve.

Given a hyperelliptic surface, or more generally any cyclic covering of the line, which is given by an equation of the form $y^d = h(x)$, one has available as a global meromorphic function any rational function of x and y. These are enough to prove that these surfaces are algebraic.

EXAMPLE 1.7. Any hyperelliptic Riemann surface is an algebraic curve.

EXAMPLE 1.8. Any cyclic covering of the line is an algebraic curve.

The basic analytic result from which we will proceed is the following.

THEOREM 1.9. *Every compact Riemann surface is an algebraic curve.*

As noted above, this is a rather deep theorem, using tools of analysis and functional analysis. Indeed, it is not easy to see that a compact Riemann surface has any nonconstant meromorphic functions at all! We will say a little more about this later. For now, our basic point of view will be to take the algebraicity as an assumption and proceed from there. We will head towards a proof of the Riemann-Roch Theorem in this chapter, which gives some quantitative information about meromorphic functions (it describes the dimension of the space of functions $L(D)$). We will see that, given the Riemann-Roch Theorem for a compact Riemann surface, it is easy to prove that it is an algebraic curve. Thus the "qualitative" information that a Riemann surface is an algebraic curve is seen to be equivalent to the more quantitative statement of Riemann-Roch.

As a consequence of the Riemann-Roch Theorem, we will be able to show that any algebraic curve can be holomorphically embedded into projective space. Moreover, we will see that any global meromorphic function on a smooth projective curve is a rational function. Therefore the entire field $\mathcal{M}(X)$ of global meromorphic functions on X consist entirely of rational functions, given some projective embedding of X. For this reason the field $\mathcal{M}(X)$ is sometimes called the *rational function field*, or simply the *function field*, of X.

Constructing Functions with Specified Laurent Tails. Our first job is to parlay the existence of meromorphic functions for an algebraic curve X into slightly more specific statements, which will be useful later. This we now do in a series of lemmas.

LEMMA 1.10. *Let X be an algebraic curve, and let $p \in X$. Then for any integer N there is a global meromorphic function f on X with $\mathrm{ord}_p(f) = N$.*

PROOF. We have already remarked that we can produce a global meromorphic function g on X such that $\mathrm{ord}_p(g) = 1$, using the hypothesis that $\mathcal{M}(X)$ separates tangents. The function $f = g^N$ then has order N at p. □

A Laurent polynomial $r(z) = \sum_{i=n}^{m} c_i z^i$ is called a *Laurent tail* of a Laurent series $h(z)$ if the Laurent series starts with $r(z)$, i.e., if the series $h - r$ has all of its terms higher than the top degree term of r.

We next note that on an algebraic curve we can produce a global meromorphic function f with any given Laurent tail at a point p.

LEMMA 1.11. *Let X be an algebraic curve. Fix a point p on X and a local coordinate z centered at p. Fix any Laurent polynomial $r(z)$ in z. Then there exists a global meromorphic function f on X whose Laurent series at p has $r(z)$ as a Laurent tail.*

PROOF. Write $r(z) = \sum_{i=n}^{m} c_i z^i$ with c_n and c_m nonzero. Therefore r has $m - n + 1$ terms. We will proceed by induction on the number of terms. If r has a single term cz^m, then all that is being required (up to a constant factor) is that the function f have order m at p. This is possible by Lemma 1.10.

Now suppose that $r = \sum_{i=n}^{m} c_i z^i$ has at least two terms, with lowest term $c_n z^n$. Again, by Lemma 1.10 (or by the previous single term case), we can find a global meromorphic function h with $c_n z^n$ as a Laurent tail. Let $s(z)$ be the Laurent polynomial which is the tail of the Laurent series of $h - r$ at p, up through the z^m term; note that $s(z)$ has fewer terms than $r(z)$. Therefore by induction there is a global meromorphic function g whose Laurent series at p has s as a tail. Then the function $f = h - g$ has r as a Laurent tail. □

The above lemma is a kind of approximation result: any Laurent series at a single point can be approximated (up to arbitrary order) by a global meromorphic function. We now begin to analyze what is possible simultaneously at several points.

LEMMA 1.12. *Let X be an algebraic curve. Then for any two points p and q in X, there is a global meromorphic function f on X with a zero at p and a pole at q.*

PROOF. Since $\mathcal{M}(X)$ separates points of X, we see that there is a global meromorphic function g on X such that $g(p) \neq g(q)$. By replacing g by $1/g$ if necessary, we may assume that p is not a pole of g; by replacing g by $g - g(p)$, we may assume in fact that $g(p) = 0$. If q is a pole of g, we are done; if not, then $f = g/(g(q) - g)$ has the required properties. □

This can be extended to any number of points by a simple induction.

LEMMA 1.13. *Let X be an algebraic curve. Then for any finite number of points $p, q_1, \ldots, q_n$ in X, there is a global meromorphic function f on X with a zero at p and a pole at each q_i.*

PROOF. This goes by induction on the number n of q's. The $n = 1$ case is the previous lemma. Suppose $n \geq 2$. Let g be a global meromorphic function on X with a zero at p and a pole at $q_1, \ldots, q_{n-1}$, which exists by the induction assumption. Let h be a global meromorphic function on X with a zero at p and a pole at q_n. Then for large m, the function $f = g + h^m$ has the required zeroes and poles.

Indeed, f certainly has a zero at p. Fix one of the q_i's with $i \leq n - 1$, so that g has a pole there; then if h is holomorphic at q_i, then f has a pole at q_i for every m. If h also has a pole at q_i, then for large m the pole of h^m will be of larger order than the pole of g, and so the sum will have a pole. Finally, consider q_n, where h has a pole. Then no matter what behaviour g has at q_n, for large m, f will have a pole there. □

In fact, we can be more specific about achieving various values (with orders) at collections of points.

LEMMA 1.14. *Let X be an algebraic curve. Then for any finite number of points $p, q_1, \ldots, q_n$ in X, and any $N \geq 1$, there is a global meromorphic function f on X with $\mathrm{ord}_p(f-1) \geq N$ and $\mathrm{ord}_{q_i}(f) \geq N$ for each i.*

PROOF. Let g be a global meromorphic function with a zero at p and a pole at each q_i. Then $f = 1/(1 + g^N)$ has the required properties. □

This in turn can be generalized, to produce a single function which approximates any given behaviour at a finite set of points. By this we essentially mean that we can find a single function which agrees with a collection of given functions up to arbitrary order at a finite set of points. This generalizes Lemma 1.11 to any finite set.

LEMMA 1.15 (LAURENT SERIES APPROXIMATION). *Suppose that X is an algebraic curve. Fix a finite number of points $p_1, \ldots, p_n$ in X, choose a local coordinate z_i at each p_i, and finally choose Laurent polynomials $r_i(z_i)$ for each i. Then there is a global meromorphic function f on X such that for every i, f has r_i as a Laurent tail at p_i.*

PROOF. Fix an integer N larger than every exponent of every term of every r_i. Extend each r_i by adding zero terms, and consider each r_i as a Laurent polynomial with terms of degree less than N. We will find an f that agrees with r_i up through the terms of order less than N. For f to have r_i as a Laurent tail at p_i is equivalent to the inequality $\mathrm{ord}_{p_i}(f - r_i) \geq N$. By Lemma 1.11, there are global meromorphic functions g_i on X such that g_i has r_i as a Laurent tail at p_i. Let M be the minimum of the orders $\mathrm{ord}_{p_i}(r_i)$, which is the same as the minimum of the orders $\mathrm{ord}_{p_i}(g_i)$. By Lemma 1.14, there are global meromorphic functions h_i on X such that for each i, $\mathrm{ord}_{p_i}(h_i - 1) \geq N - M$ and $\mathrm{ord}_{p_j}(h_i) \geq N - M$ for $j \neq i$.

Consider then the function $f = \sum_i h_i g_i$. At a point p_i, the term $h_i g_i$ has r_i as its Laurent tail; all terms at the other points are zero up through order $N - 1$. Therefore at p_i, f has r_i as its Laurent tail, and is the desired function. □

We actually will only require the following, which is a much simpler statement: that we can achieve any given set of orders at a finite set of points.

COROLLARY 1.16. *Let X be an algebraic curve. Fix a finite number of points $p_1, \ldots, p_n$ in X, and a finite number of integers m_i. Then there exists a global meromorphic function f on X such that $\mathrm{ord}_{p_i} = m_i$ for each i.*

The Laurent Series Approximation Lemma can be formulated in a purely algebraic setting, and as such forms the basis for the algebraic study of function fields. There are analogues for number fields (that is, finite extensions of

$\mathbb{Q}$) as well. The reader may wish to consult the first sections of [**Lang82**] or [**Deuring73**] for a purely algebraic treatment.

The Transcendence Degree of the Function Field $\mathcal{M}(X)$. The simple bound on the dimension of the spaces $L(D)$ given above leads directly to a bound on the transcendence degree for the function field $\mathcal{M}(X)$ on an algebraic curve X.

PROPOSITION 1.17. *Let X be an algebraic curve. Then the function field $\mathcal{M}(X)$ is a finitely generated extension field of $\mathbb{C}$ of transcendence degree exactly one.*

PROOF (TRANSCENDENCE DEGREE). The statement that the transcendence degree is one is the more elementary one. Since X is an algebraic curve, there is a nonconstant element of $\mathcal{M}(X)$; so the transcendence degree of $\mathcal{M}(X)$ must be at least one. Suppose that it is at least two; let f and g be algebraically independent elements of $\mathcal{M}(X)$. Let D be a nonnegative divisor on X such that f and g are both in $L(D)$. (Choosing D to be greater than the divisors of poles $\mathrm{div}_\infty(f)$ of f and $\mathrm{div}_\infty(g)$ suffices.) Note then that for any i and j at least zero, we have

$$f^i g^j \in L(nD) \text{ if } i+j \leq n.$$

Hence $L(nD)$ has every monomial of degree at most n in f and g in it, and these monomials are linearly independent since f and g are algebraically independent. Therefore

$$\dim L(nD) \geq (n^2 + 3n + 2)/2,$$

which is the number of these monomials. On the other hand, since D (and nD) are positive divisors, we have the crude bound

$$\dim L(nD) \leq 1 + \deg(nD) = 1 + \deg(D)n.$$

For large n we obtain a contradiction: the dimension is $L(nD)$ is not growing fast enough. □

The finite generation is slightly more involved. First choose a nonconstant meromorphic function f on X. Denote by $\mathbb{C}(f)$ the field of all rational expressions in the function f; $\mathbb{C}(f)$ is isomorphic to the field $\mathbb{C}(t)$ of rational functions in one variable t, where a rational function $r(t)$ corresponds to the rational expression $r(f)$. We then have a chain of fields

$$\mathbb{C} \subset \mathbb{C}(f) \subseteq \mathcal{M}(X);$$

in order to show that $\mathcal{M}(X)$ is finitely generated over $\mathbb{C}$ it suffices to show that $\mathcal{M}(X)$ is a finite algebraic extension of $\mathbb{C}(f)$.

We require a lemma.

1. ALGEBRAIC CURVES

LEMMA 1.18. *Let A be a divisor on a compact Riemann surface X, and let $D = \mathrm{div}_\infty(f)$ be the divisor of poles of some nonconstant meromorphic function f on X. Then there is an integer $m > 0$ and a meromorphic function g on X such that $A - \mathrm{div}(g) \leq mD$. Moreover, g can be taken to be a polynomial in f: $g = r(f)$ for some polynomial $r(t) \in \mathbb{C}[t]$.*

PROOF (OF THE LEMMA). Let $p_1, \ldots, p_k$ be the points in the support of A which are not poles of f, and which have $A(p_i) \geq 1$. Then $f(p_i)$ is a number in $\mathbb{C}$, and so $f - f(p_i)$ has a zero at p_i, to at least order one; moreover, its poles are the same as the poles of f. Hence $(f - f(p_i))^{A(p_i)}$ has a zero at p_i to at least order $A(p_i)$, and $(f - f(p))^{A(p_i)}$ has no poles other than the poles of f. Taking the product over all these points p_i, which are not poles of f, of these factors gives a meromorphic function g which is a polynomial in f such that $A - \mathrm{div}(g)$ is positive only at the poles of f.

Therefore for some integer m, $A - \mathrm{div}(g) \leq mD$, where D is the divisor of poles of f. □

By applying this lemma with $A = -\mathrm{div}(h)$ for meromorphic h on X, we immediately obtain the following:

COROLLARY 1.19. *Let X be a compact Riemann surface, and let f and h be nonconstant meromorphic functions on X. Then there is a polynomial $r(t) \in \mathbb{C}[t]$ such that the function $r(f)h$ has no poles outside of the poles of f. In this case there is an integer m such that $r(f)h \in L(mD)$, where $D = \mathrm{div}_\infty(f)$ is the divisor of poles of f.*

This in turn gives the following lower bound on the dimension of the space $L(mD)$ for large m:

LEMMA 1.20. *Fix a meromorphic function f on a compact Riemann surface, and let $D = \mathrm{div}_\infty(f)$. Suppose that $[\mathcal{M}(X) : \mathbb{C}(f)] \geq k$. Then there is a constant m_0 such that for all $m \geq m_0$,*

$$\dim L(mD) \geq (m - m_0 + 1)k$$

PROOF. Suppose that $g_1, \ldots, g_k$ are elements of $\mathcal{M}(X)$ which are linearly independent over $\mathbb{C}(f)$. By the previous corollary, for each i there is a nonzero polynomial $r_i(t)$ such that the poles of $h_i = r_i(f)g_i$ occur only at the poles of f. Note that the functions $h_1, \ldots, h_k$ are also linearly independent over $\mathbb{C}(f)$, and there is an integer m_0 such that $h_i \in L(m_0 D)$.

Now for any integer $m \geq m_0$, the functions $f^i h_j$ are in $L(mD)$ as long as $i \leq m - m_0$, since $f \in L(D)$. These are all linearly independent over $\mathbb{C}$, so

$$\dim L(mD) \geq (m - m_0 + 1)k$$

for $m \geq m_0$ as claimed. □

We can now finish the proof of Proposition 1.17.

PROOF (FINITE GENERATION OF $\mathcal{M}(X)$ OVER $\mathbb{C}$). In fact it is the case that $[\mathcal{M}(X) : \mathbb{C}(f)] \leq \deg(D)$, where $D = \operatorname{div}_\infty(f)$. Suppose on the contrary that $[\mathcal{M}(X) : \mathbb{C}(f)] \geq 1 + \deg(D)$. By the above lemma, we have that there is an integer m_0 such that for all $m \geq m_0$,

$$\dim L(mD) \geq (m - m_0 + 1)(1 + \deg(D)).$$

However the crude bound for $L(mD)$ gives that

$$\dim L(mD) \leq 1 + \deg(mD) = 1 + m\deg(D),$$

so that

$$1 + m\deg(D) \geq (m - m_0 + 1)(1 + \deg(D))$$

which is silly for large m. □

In fact we can be more precise concerning the index of $\mathbb{C}(f)$ in $\mathcal{M}(X)$. This uses the Laurent Series Approximation Lemma (it actually just uses its Corollary 1.16).

PROPOSITION 1.21. *Let f be a nonconstant meromorphic function on an algebraic curve X. Then*

$$[\mathcal{M}(X) : \mathbb{C}(f)] = \deg(D),$$

where D is the divisor $\operatorname{div}_\infty(f)$ of poles of f.

PROOF. We have just seen above that the index is at most $\deg(D)$, so we must only show the other inequality. Write the polar divisor as $D = \sum_i n_i p_i$, with each $n_i \geq 1$, and consider functions g_{ij}, where g_{ij} has a pole at p_i to order j, and no zero or pole at any of the other p_k's. This is possible using Corollary 1.16.

We claim that $\{g_{ij} \mid 1 \leq j \leq n_i\}$ are linearly independent over $\mathbb{C}(f)$. This will suffice to complete the proof.

Suppose to the contrary that there is a linear relation

$$\sum c_{ij}(f) g_{ij} = 0$$

with coefficients rational functions of f. By clearing denominators we may assume that in fact the c_{ij}'s are polynomials in f. Therefore the only poles of $c_{ij}(f)$ are at the points p_k, and if c_{ij} has degree d, then $c_{ij}(f)$ has a pole of order exactly dn_k at p_k.

Choose i_0 and j_0 such that the degree d_0 of $c_{i_0 j_0}$ is maximal (and in case of a tie choose the highest j). Renumber so that $i_0 = 1$, and divide through by $c_{i_0 j_0}$ to produce a relation

$$\sum d_{ij}(f) g_{ij} = 0$$

with $d_{1j_0} = 1$ and all other d_{ij}'s being rational functions of f whose denominators have degree at least as large as the numerators. In particular, all of these functions have nonnegative order at the poles $\{p_k\}$ of f, and these orders are all multiples of the order of f, that is, $\mathrm{ord}_{p_k}(d_{ij}(f))$ is a nonnegative multiple of n_k for each i and j.

Look at the order of this expression at the point p_1. First consider the terms with $i \neq 1$. Then g_{ij} has order 0 at p_1, and d_{ij} as nonnegative order; so all of these terms have nonnegative order.

Next consider the terms with $i = 1$. Since the orders of the d_{1j}'s are nonnegative multiples of n_1 at p_1, and the order of g_{1j} is $-j$, which is between $-n_1$ and -1, the only way a term can have negative order is if d_{1j} has order 0, and in this case the term will have order $-j$. Such a term does exist, namely the $1j_0$ term: the term $d_{1j_0}g_{1j_0} = g_{1j_0}$ has order exactly $-j_0$ at p_1. The g_{ij}'s have distinct orders, so if we consider the maximum j among all those with $\mathrm{ord}_{p_1}(d_{ij}) = 0$, we see that we have a term with a pole of order j, which cannot be cancelled by any other term in the entire sum. Since the sum is supposed to be identically zero, this is a contradiction. □

Computing the Function Field $\mathcal{M}(X)$. This precise result concerning the function field $\mathcal{M}(X)$ for an algebraic curve allows us to compute $\mathcal{M}(X)$ in most cases of interest. To see how this goes, we need to remark that all of the statements made in this section about $\mathcal{M}(X)$ in fact hold for *any* field of global meromorphic functions which separates points and tangents of X; this is all that was ever used in the proofs. In particular, this holds for the index statement above, and we can conclude the following:

If $\mathcal{F}$ is any field of global meromorphic functions on X containing the constant functions, which separates the points and tangents of X, then for every nonconstant $f \in \mathcal{F}$, we have
$$[\mathcal{F} : \mathbb{C}(f)] = \deg(\mathrm{div}_\infty(f)).$$

On the other hand, this is true for $\mathcal{M}(X)$ also, and $\mathcal{F} \subseteq \mathcal{M}(X)$. Therefore we must have $\mathcal{F} = \mathcal{M}(X)$, and we have proved the following.

PROPOSITION 1.22. *Suppose that X is an algebraic curve, and $\mathcal{F}$ is any field of global meromorphic functions on X containing the constant functions, which separates the points and tangents of X. Then $\mathcal{F} = \mathcal{M}(X)$. In particular, if S is any set of global meromorphic functions on X which separates the points and tangents of X, then S generates the function field $\mathcal{M}(X)$ as a field extension of $\mathbb{C}$.*

We have the following corollary, which we leave to the reader to check.

COROLLARY 1.23.
  (i) *The function field of the Riemann Sphere $\mathbb{C}_\infty$ is the field generated by the affine coordinate z.*
  (ii) *The function field of a complex torus $\mathbb{C}/L$ is generated by ratios of theta functions.*

(iii) *The function field of a smooth projective curve or of the resolution of a projective plane curve with nodes is the field of rational functions. That is, if X is embedded in projective space $\mathbb{P}^n$, with coordinates $[x_0 : x_1 : \cdots : x_n]$, then the ratios x_i/x_j generate $\mathcal{M}(X)$.*
(iv) *The function field of a hyperelliptic Riemann surface, and more generally of a cyclic covering of the line defined by $y^d = h(x)$, is generated by x and y.*

Problems VI.1

A. Let D be a very ample divisor on a compact Riemann surface X. Show that the field generated by the functions in $L(D)$ separate the points and tangents of X.
B. In Problem A, is it true that the functions in $L(D)$ separate the points and tangents of X?
C. Show that the Riemann Sphere $\mathbb{C}_\infty$ is an algebraic curve.
D. Show that a complex torus $\mathbb{C}/L$ is an algebraic curve.
E. Show that a smooth projective curve is an algebraic curve.
F. Show that a hyperelliptic Riemann surface is an algebraic curve.
G. Show that the Riemann surface obtained by resolving the nodes of a projective plane curve with nodes is an algebraic curve.
H. Show that a cyclic covering of the line is an algebraic curve.
I. Let X be an algebraic curve. Show, using the compactness of X, that there are a finite number of global meromorphic functions on X which separate the points and tangents of X.
J. Show that if the conclusion of Corollary 1.16 holds for a compact Riemann surface X, then X is an algebraic curve.
K. Prove Corollary 1.23.
L. Let G be a finite group acting effectively on an algebraic curve X.
   i. Show that G acts on the function field $\mathcal{M}(X)$.
   ii. Show that the function field of the quotient Riemann surface X/G is the field of invariants $\mathcal{M}(X)^G$.
   iii. Show that X/G is an algebraic curve.

2. Laurent Tail Divisors

In the Laurent Series Approximation Lemma, we saw the need for having a collection of Laurent tails defined at a finite set of points of X. It is useful to make a group out of this kind of data.

Definition of Laurent Tail Divisors. Let X be a compact Riemann surface. For each point p in X, choose once and for all a local coordinate z_p centered at p. This allows us to associate, to each meromorphic function defined near p, a Laurent series in the coordinate z_p.

2. LAURENT TAIL DIVISORS

DEFINITION 2.1. A *Laurent tail divisor* on X is a finite formal sum

$$\sum_p r_p(z_p) \cdot p,$$

where $r_p(z_p)$ is a Laurent polynomial in the coordinate z_p, that is, a Laurent series with a finite number of terms. The set of Laurent tail divisors on X forms a group under formal addition, and will be denoted by $\mathcal{T}(X)$.

Note the similarity with the concept of ordinary divisors, where a simple integer is associated to finitely many points of X.

We use ordinary divisors to "filter" the group of Laurent tail divisors in the following way. Given an ordinary divisor D on X, we may consider the subgroup

$$\mathcal{T}[D](X) = \{\sum_p r_p \cdot p \quad | \quad \text{for all } p \text{ with } r_p \neq 0, \text{the top term of } r_p$$
$$\text{has degree strictly less than } -D(p)\}.$$

As an example, take the trivial ordinary divisor $D = 0$. Then $\mathcal{T}[0](X)$ is the group of Laurent tail divisors $\sum_p r_p \cdot p$ such that every term of each r_p has strictly negative degree.

There is a natural truncation map from $\mathcal{T}(X)$ to $\mathcal{T}[D](X)$ which simply takes each Laurent polynomial r_p and removes all terms of degree $-D(p)$ and higher. Similarly, if D_1 and D_2 are two divisors with $D_1 \leq D_2$, then there is a natural truncation map

$$t = t_{D_2}^{D_1} : \mathcal{T}[D_1](X) \to \mathcal{T}[D_2](X)$$

defined by removing from each r_p all terms of degree $-D_2(p)$ and higher.

Given a meromorphic function f, and any divisor D, we can define a multiplication operator

$$\mu_f = \mu_f^D : \mathcal{T}[D](X) \to \mathcal{T}[D - \operatorname{div}(f)](X)$$

defined by sending $\sum r_p \cdot p$ to the suitable truncation of $\sum (fr_p) \cdot p$. Note that μ_f^D is an isomorphism, with inverse equal to $\mu_{1/f}^{D-\operatorname{div}(f)}$. This isomorphism is related to the isomorphism between $L(D)$ and $L(D - \operatorname{div}(f))$ discussed in Section 3 of Chapter V.

There is a Laurent tail version of the divisor function. This requires fixing a given ordinary divisor D on X; then we have the map

$$\alpha_D : \mathcal{M}(X) \to \mathcal{T}[D](X)$$

defined by sending the meromorphic function f to the sum $\sum_p r_p \cdot p$, where r_p is the truncation of the Laurent series $f(z_p)$ of f in terms of z_p, removing all terms of order $-D(p)$ and higher.

Note that α_D commutes with the truncation maps, in the following sense. If $D_1 \leq D_2$, then the composition of α_{D_1} with $t_{D_2}^{D_1}$ is just α_{D_2}:

$$\alpha_{D_2} : \mathcal{M}(X) \xrightarrow{\alpha_{D_1}} \mathcal{T}[D_1](X) \xrightarrow{t_{D_2}^{D_1}} \mathcal{T}[D_2](X).$$

This divisor map α_D is compatible with the multiplication operators μ in the following sense: if f and g are meromorphic functions on X then

(2.2) $$\mu_f(\alpha_D(g)) = \alpha_{D-\text{div}(f)}(fg)$$

for any divisor D.

If $D(p) = 0$, then $\alpha_D(f)$ is a Laurent tail divisor whose Laurent polynomial at p has only strictly negative degree terms; it is exactly the strictly negative terms of the Laurent series for f at p. In particular, this term of $\alpha_D(f)$ is zero if and only if f is holomorphic at p.

The space $L(D)$ of global meromorphic functions on X with poles bounded by D has a nice interpretation using this circle of ideas. For a function f to get into $L(D)$, it may have no terms of order less than $-D(p)$ at each p. This exactly says that upon truncating at the $-D(p)$ level and higher, we get zero at every point. Thus

$$L(D) = \ker(\alpha_D).$$

Mittag-Leffler Problems and $H^1(D)$. If we take a Laurent tail divisor $Z \in \mathcal{T}[D](X)$, we may ask whether it is in the image of α_D, i.e., whether there is a global meromorphic function on X with precisely those tails. We have addressed a similar question before, in the Laurent Series Approximation Lemma; there we saw that we can have any combination of Laurent tails at a finite set of points, but we had no control over what happens at all other points of X.

Note that if a point p has $r_p = 0$ in Z and $D(p) = 0$ (which happens for all but finitely many points p), then a preimage of Z under α_D must be holomorphic at p. Therefore we see that the construction of a preimage under α_D to a Laurent tail divisor Z is much harder than the Laurent Series Approximation Lemma might suggest; we are requiring that a global function f not only have specified Laurent tails at a finite number of points, but also that it be holomorphic everywhere else.

This problem of constructing functions with specified Laurent tails at a finite number of points, and no other poles, is called the *Mittag-Leffler Problem* for the Riemann surface X. It is clearly of fundamental importance in the function theory of X.

As an example, again consider the case $D = 0$; a Laurent tail divisor in $\mathcal{T}[0](X)$ is determined by giving a Laurent polynomial r_i at a finite number of points $\{p_i\}$, all of whose terms have strictly negative degree. A preimage of this Laurent tail divisor under α_0 is a global meromorphic function f such that $f - r_i$ is holomorphic at each p_i, and f is holomorphic everywhere else. The strictly

2. LAURENT TAIL DIVISORS

negative part of a Laurent series is sometimes called the *principal part* of the series; thus this Mittag-Leffler problem is often referred to as "specifying a finite number of principal parts".

Algebraically, the Mittag-Leffler problem of constructing preimages to α_D is measured by the cokernel. We give this cokernel a special notation:

$$H^1(D) := \operatorname{coker}(\alpha_D) = \mathcal{T}[D](X)/\operatorname{image}(\alpha_D).$$

By definition, a Laurent tail divisor $Z \in \mathcal{T}[D](X)$ is in the image of α_D if and only if its class in $H^1(D)$ is zero. Hence this space measures the failure of being able to solve Mittag-Leffler problems on X.

It would be nice to know that this space is finite-dimensional; this we will see shortly.

The definition and further analysis of the space $H^1(D)$ is facilitated by the use of exact sequences. Recall that a sequence of $\mathbb{C}$-linear maps

$$U \xrightarrow{a} V \xrightarrow{b} W$$

between complex vector spaces is said to be *exact at V* if $\ker(b) = \operatorname{image}(a)$. A longer sequence of maps is exact if it is exact at each interior space. A *short exact sequence* is an exact sequence of five spaces and four maps such that the first and last spaces are both $\{0\}$.

We see immediately from the definitions that for any ordinary divisor D on X, we have an exact sequence

$$0 \to L(D) \to \mathcal{M}(X) \xrightarrow{\alpha_D} \mathcal{T}[D](X) \to H^1(D) \to 0,$$

which we can write as a short exact sequence

$$0 \to \mathcal{M}(X)/L(D) \xrightarrow{\alpha_D} \mathcal{T}[D](X) \to H^1(D) \to 0.$$

Comparing H^1 Spaces. Suppose that $D_1 \leq D_2$, so that the truncation map $t : \mathcal{T}[D_1](X) \to \mathcal{T}[D_2](X)$ is defined. In this case we also have $L(D_1) \subseteq L(D_2)$. Since the truncation commutes with the α maps, we obtain an induced map between the short exact sequences

$$
\begin{array}{ccccccc}
0 \to & \mathcal{M}(X)/L(D_1) & \xrightarrow{\alpha_{D_1}} & \mathcal{T}[D_1](X) & \to & H^1(D_1) & \to 0 \\
& \downarrow & & t \downarrow & & \downarrow & \\
0 \to & \mathcal{M}(X)/L(D_2) & \xrightarrow{\alpha_{D_2}} & \mathcal{T}[D_2](X) & \to & H^1(D_2) & \to 0
\end{array}
$$

where the two squares in the diagram commute. The vertical maps in this diagram are all onto, so by the snake lemma we obtain a short exact sequence for the kernels of these vertical maps. Let us analyze these kernels.

Firstly, the kernel of the map on the left from $\mathcal{M}(X)/L(D_1)$ to $\mathcal{M}(X)/L(D_2)$ is simply $L(D_2)/L(D_1)$; therefore

$$\dim \ker(\mathcal{M}(X)/L(D_1) \to \mathcal{M}(X)/L(D_2)) = \dim L(D_2) - \dim L(D_1).$$

Secondly, the kernel of the truncation map $t : \mathcal{T}[D_1](X) \to \mathcal{T}[D_2](X)$ is the space of those Laurent tail divisors $\sum_p r_p \cdot p$ such that the top term of r_p has order less than $-D_1(p)$ and the bottom term has order at least $-D_2(p)$ for each p. Hence we obtain exactly $D_2(p) - D_1(p)$ possible monomials in z_p which are allowed to appear in an element of the kernel, namely those monomials z_p^k with $-D_2(p) \leq k < -D_1(p)$. The conditions at each p are independent, and so the total dimension of the kernel of t is $\sum_p (D_2(p) - D_1(p))$, which is exactly the difference of the degrees of the two divisors:

$$\dim \ker(t : \mathcal{T}[D_1](X) \to \mathcal{T}[D_2](X)) = \deg(D_2) - \deg(D_1).$$

Thirdly, let us denote by $H^1(D_1/D_2)$ the kernel of the induced map on the right from $H^1(D_1)$ to $H^1(D_2)$. Since the snake lemma gives us a short exact sequence of kernels

$$0 \to L(D_2)/L(D_1) \to \ker(t) \to H^1(D_1/D_2) \to 0,$$

and we have seen that $\ker(t)$ is finite-dimensional, we have immediately that $H^1(D_1/D_2)$ is also finite-dimensional. Moreover, by the computations of the dimensions above we have proved the following.

LEMMA 2.3. *Suppose that D_1 and D_2 are ordinary divisors on a compact Riemann surface X, with $D_1 \leq D_2$. Then*

$$\dim H^1(D_1/D_2) = [\deg(D_2) - \dim L(D_2)] - [\deg(D_1) - \dim L(D_1)].$$

Note that we can expand the sequence of kernels above and obtain a "long exact sequence" of spaces

$$0 \to L(D_1) \to L(D_2) \to \ker(t) \cong \mathbb{C}^{\deg(D_2 - D_1)} \to H^1(D_1) \to H^1(D_2) \to 0.$$

The Finite-Dimensionality of $H^1(D)$. Our next order of business is to prove that the spaces $H^1(D)$ for an algebraic curve X are finite-dimensional. Thus there are only finitely many linear conditions on a given Laurent tail divisor Z to be able to solve the Mittag-Leffler problem for Z.

We have already seen that when $D_1 \leq D_2$, the space $H^1(D_1/D_2)$ is finite-dimensional; this space measures the "difference" between $H^1(D_1)$ and $H^1(D_2)$. Therefore if either one of these spaces is finite-dimensional, so is the other. The approach is then to prove that for some single divisor D the space $H^1(D)$ is finite-dimensional, and then deduce that all such spaces are finite-dimensional using this trick.

We begin with a lemma.

LEMMA 2.4. *Fix a nonconstant global meromorphic function f on an algebraic curve X, and let D be its divisor of poles: $D = \mathrm{div}_\infty(f)$. Then for large m the dimension of $H^1(0/mD)$ is constant, independent of m.*

PROOF. Applying Lemma 2.3 with $D_1 = 0$ and $D_2 = mD$, we obtain

$$\dim H^1(0/mD) = m\deg(D) - \dim L(mD) + 1,$$

using the fact that X is compact so $\dim L(0) = 1$. Recall that by Proposition 1.21, $[\mathcal{M}(X) : \mathbb{C}(f)] = \deg(D)$ as field extensions; then by Lemma 1.20, we see that there is an integer m_0 such that

$$\dim L(mD) \geq (m - m_0 + 1)\deg(D)$$

for $m \geq m_0$. Using this in the above formula gives

$$\dim H^1(0/mD) \leq m\deg(D) - (m - m_0 + 1)\deg(D) + 1 = 1 + \deg(D)(m_0 - 1),$$

which is independent of m and gives a uniform bound for $\dim H^1(0/mD)$ for large m.

Now if $0 < m_1 < m_2$, we have $0 < m_1 D < m_2 D$, so that $H^1(0)$ maps onto $H^1(m_1 D)$, which maps onto $H^1(m_2 D)$. The composition of these maps is of course the natural one from $H^1(0)$ to $H^1(m_2 D)$; hence the kernel $H^1(0/m_1 D)$ of the map from $H^1(0)$ to $H^1(m_1 D)$ is contained in the kernel $H^1(0/m_2 D)$ of the map to $H^1(m_2 D)$:

$$H^1(0/m_1 D) \subseteq H^1(0/m_2 D).$$

Thus we see that $\dim H^1(0/mD)$ is nondecreasing as m increases. Since we have just seen that it is uniformly bounded, it must stabilize eventually. □

Fixing a meromorphic function f and its divisor of poles D, we see that there is a constant M (depending only on D) such that

$$\deg(mD) - \dim L(mD) \leq M$$

for all $m \geq 0$, since this is simply $\dim H^1(0/mD) - 1$. This statement generalizes to any divisor.

LEMMA 2.5. *For any algebraic curve X, there is an integer M such that*

$$\deg(A) - \dim L(A) \leq M$$

for every divisor A on X.

PROOF. Fix a meromorphic function f on X and let $D = \text{div}_\infty(f)$ be its divisor of poles. Let M be such that

$$\deg(mD) - \dim L(mD) \leq M$$

for all $m \geq 0$; such an M exists as noted above.

Now let A be any divisor on X. By Lemma 1.18, there is a meromorphic function g on X and an integer m such that $B = A - \text{div}(g) \leq mD$. Note that

$\deg(B) = \deg(A)$ and $L(B) \cong L(A)$, so that $\deg(A) - \dim L(A) = \deg(B) - \dim L(B)$. Therefore

$$\begin{aligned}\deg(A) - \dim L(A) &= \deg(B) - \dim L(B) \\ &= [\deg(mD) - \dim L(mD)] - \dim H^1(B/mD) \\ &\leq \deg(mD) - \dim L(mD) \\ &\leq M\end{aligned}$$

as desired. □

Hence there is a divisor A_0 on X such that $\deg(A_0) - \dim L(A_0)$ is maximal.

LEMMA 2.6. *For this divisor A_0, we have $H^1(A_0) = 0$.*

PROOF. Suppose that $H^1(A_0) \neq 0$. Then there is a Laurent tail divisor Z in $\mathcal{T}[A_0](X)$ which is not $\alpha_{A_0}(f)$ for any meromorphic function f on X. By increasing A_0 to a divisor B we may truncate Z to 0, i.e., $t(Z) = 0$ in $\mathcal{T}[B](X)$. Therefore the class of $t(Z)$ in $H^1(B)$ is certainly zero. Hence the class of Z in $H^1(A_0)$ is in fact in the kernel $H^1(A_0/B)$; thus this kernel is nonzero. But by Lemma 2.3,

$$1 \leq \dim H^1(A_0/B) = [\deg(B) - \dim L(B)] - [\deg(A_0) - \dim L(A_0)],$$

which is nonpositive by the maximality of $\deg(A_0) - \dim L(A_0)$. This is a contradiction, proving the lemma. □

This is the critical step in finite-dimensionality.

PROPOSITION 2.7. *For any divisor D on an algebraic curve X, $H^1(D)$ is a finite-dimensional vector space over $\mathbb{C}$.*

PROOF. Let A_0 be as above, and write $D - A_0 = P - N$, where P and N are nonnegative divisors. Then $H^1(A_0)$ surjects onto $H^1(A_0 + P)$, so that $H^1(A_0 + P) = 0$ also. Therefore $H^1(A_0 + P - N) \cong H^1(A_0 + P - N/A_0 + P)$, which is finite-dimensional. Since $D = A_0 + P - N$, we are done. □

Problems VI.2

A. Given a meromorphic function f, and any divisor D, show that the multiplication operator

$$\mu_f = \mu_f^D : \mathcal{T}[D](X) \to \mathcal{T}[D - \text{div}(f)](X),$$

defined by sending $\sum r_p \cdot p$ to the suitable truncation of $\sum (fr_p) \cdot p$, is an isomorphism. Show that its inverse is $\mu_{1/f}^{D-\text{div}(f)}$.

B. Show that if f and g are global meromorphic functions on X, and D is an ordinary divisor on X, then $\mu_f(\alpha_D(g)) = \alpha_{D-\text{div}(f)}(fg)$ as elements of $\mathcal{T}[D - \text{div}(f)](X)$.

C. Show that if $D_1 \leq D_2$, then
$$\alpha_{D_2} = t_{D_2}^{D_1} \circ \alpha_{D_1}.$$

D. If you have never heard of the snake lemma, prove from first principles that if one has two short exact sequences, and onto maps from the spaces of one to the spaces of the other, such that the two obvious squares commute, then the kernels of the maps form a short exact sequence.

E. Let X be the Riemann Sphere $\mathbb{C}_\infty$. Show that $H^1(0) = 0$ by explicitly finding a preimage under α_0 for any Laurent tail divisor Z in $T[0](X)$. (Hint: use partial fractions.)

F. Let X be the Riemann Sphere $\mathbb{C}_\infty$, and let p be the point $z = 0$. Considering p as an ordinary divisor on X, show that $H^1(-p) = 0$, again by explicitly finding preimages under α_{-p}.

G. With the same notation as in the previous problem, consider the Laurent tail divisor $Z = z \cdot p$; note that $Z \in T[-2p](X)$. Show that Z is not in the image of α_{-2p}, and conclude that $H^1(-2p) \neq 0$.

H. Let $X = \mathbb{C}/L$ be a complex torus, and let p be the zero of the group law on X. Let z be the local coordinate on X at p, and consider the Laurent tail divisor $Z = z^{-1} \cdot p$; note that $Z \in T[0](X)$. Show that Z is not in the image of α_0, and conclude that $H^1(0) \neq 0$ for a complex torus.

I. Let X be a complex torus. Fix a finite number of points p_i on X, with local coordinate z_i at p_i. Consider the Laurent tail divisor $Z = \sum_i c_i z_i^{-1} \cdot p_i$; note that it is in $T[0](X)$. Show that $Z = \alpha_0(f)$ for some global meromorphic function f on X if and only if $\sum_i c_i = 0$.

J. Show that if $D_1 \sim D_2$, then $H^1(D_1) \cong H^1(D_2)$, by showing that an isomorphism is induced from an appropriate multiplication operator on the corresponding Laurent tail spaces.

3. The Riemann-Roch Theorem and Serre Duality

In this section we will prove the Riemann-Roch Theorem, which is the foundation of the theory of algebraic curves, giving a precise answer for the dimension of the space $L(D)$.

The Riemann-Roch Theorem I. The finite-dimensionality of the $H^1(D)$'s allows us to split the dimension of the kernels $H^1(D_1/D_2)$ as

$$\dim H^1(D_1/D_2) = \dim H^1(D_1) - \dim H^1(D_2)$$

if $D_1 \leq D_2$. By plugging this into the formula given by Lemma 2.3 and rearranging a bit we see that

$$\dim L(D_1) - \deg(D_1) - \dim H^1(D_1) = \dim L(D_2) - \deg(D_2) - \dim H^1(D_2)$$

if $D_1 \leq D_2$. Noting that any two divisors have a common maximum, we conclude that this quantity

$$\dim L(D) - \deg(D) - \dim H^1(D)$$

is constant over all divisors D. When $D = 0$, this quantity is simply $1 - \dim H^1(0)$. Thus we obtain the following.

THEOREM 3.1 (THE RIEMANN-ROCH THEOREM: FIRST FORM). *Let D be a divisor on an algebraic curve X. Then*

$$\dim L(D) - \dim H^1(D) = \deg(D) + 1 - \dim H^1(0).$$

The problem of computing the dimension of the space $L(D)$ of meromorphic functions on X with poles bounded by D is called the *Riemann-Roch Problem*. It is of fundamental importance in the theory of Riemann surfaces. It was solved by Riemann and Roch in the last century.

The above description of $\dim L(D)$ is somewhat unsatisfying, since we have simply traded our problem to the computation of the dimension of $H^1(D)$ (and of $H^1(0)$). Since as we have noted above H^1 spaces are also directly related to the existence of meromorphic functions, it is fair to say that we have no surprises yet. The real power of the Riemann-Roch Theorem comes after adequately identifying the H^1's. This is afforded by the Serre Duality Theorem, which we discuss next.

The Residue Map. The space $H^1(D)$ measures whether a Laurent tail divisor Z can be the truncation of a meromorphic function f on X. There is another measure we can make concerning this question, which is based on the Residue Theorem. For illustration suppose we have a Laurent tail divisor $Z = \sum r_p \cdot p$ in $T[0](X)$, so that the terms r_p of Z have only negative exponents appearing. To ask whether $Z = \alpha_0(f)$ is exactly to ask whether there is a meromorphic function f on X such that at each point $p \in X$, the negative terms in the Laurent series for f form exactly the Laurent tail r_p. Thus we are trying to specify the Laurent tails at all poles of f.

Suppose that such an f exists, and we are given a holomorphic 1-form ω on X. Then $f\omega$ can have poles only at the poles of f, and the negative terms in the Laurent series for the meromorphic 1-form $f\omega$ are determined just by the negative terms in the Laurent series for f (and ω of course). In other words, the negative terms of the Laurent series for $f\omega$ at p are just the negative terms of the Laurent series for $r_p\omega$, if $\alpha_0(f) = \sum r_p \cdot p$.

Now the Residue Theorem comes into play; it states that

$$\sum_p \mathrm{Res}_p(f\omega) = 0,$$

which reduces by the above analysis to

(3.2) $$\sum_p \mathrm{Res}_p(r_p\omega) = 0$$

if $\alpha_0(f) = \sum r_p \cdot p$ as desired. Therefore (3.2) is a necessary condition on the Laurent tail divisor $Z = \sum r_p \cdot p$ for Z to be $\alpha_0(f)$ for some f.

3. THE RIEMANN-ROCH THEOREM AND SERRE DUALITY

The Serre Duality Theorem says that these conditions, suitably generalized for any divisor D, are necessary and sufficient for the existence of the function f; moreover linearly independent ω's give independent conditions. Therefore the space $H^1(D)$ can be identified with a space of 1-forms (or, more precisely, the dual of such a space).

To get started, suppose that D is a divisor on X and ω is a meromorphic 1-form on X in the space $L^{(1)}(-D)$. Therefore by definition $\operatorname{div}(\omega) \geq D$, i.e., $\operatorname{ord}_p(\omega) \geq D(p)$ for all $p \in X$. Therefore we may write

$$\omega = \left(\sum_{n=D(p)}^{\infty} c_n z_p^n\right) dz_p$$

in the local coordinate z_p at p, for every p.

Next suppose that f is a meromorphic function on X. Write $f = \sum_k a_k z_p^k$ near p. Computing the residue of $f\omega$ at p, we find that

$$\operatorname{Res}_p(f\omega) = \text{coefficient of } (1/z_p)dz_p \text{ in } (\sum_k a_k z_p^k \cdot \sum_{n=D(p)}^{\infty} c_n z_p^n) dz_p$$

$$= \sum_{n=D(p)}^{\infty} c_n a_{-1-n}$$

so that this residue depends only on those coefficients a_i for f with $i < -D(p)$.

Saying this another way, we see that the residue depends only on the Laurent tail divisor $\alpha_D(f)$, which simply encodes the truncations of the Laurent series for f at every point, at exactly this exponent.

Therefore we may define a *residue* map

$$\operatorname{Res}_\omega : \mathcal{T}[D](X) \to \mathbb{C} \quad \text{for } \omega \in L^{(1)}(-D)$$

by setting

$$\operatorname{Res}_\omega\left(\sum r_p \cdot p\right) = \sum_p \operatorname{Res}_p(r_p \omega).$$

What we have just seen above is that

$$\sum_p \operatorname{Res}_p(f\omega) = \operatorname{Res}_\omega(\alpha_D(f)) \text{ when } \omega \in L^{(1)}(-D).$$

Since $\sum_p \operatorname{Res}_p(f\omega) = 0$ by the Residue Theorem, we have that

$$\operatorname{Res}_\omega(\alpha_D(f)) = 0 \quad \text{when } \omega \in L^{(1)}(-D).$$

In other words, $\operatorname{Res}_\omega$ vanishes on the image of α_D when $\omega \in L^{(1)}(-D)$. Hence $\operatorname{Res}_\omega$ descends to a linear functional

$$\operatorname{Res}_\omega : H^1(D) \to \mathbb{C}$$

which we then think of as an element of the dual space $H^1(D)^*$. Thus we obtain a linear map, also called the *residue map*,

$$\text{Res}: L^{(1)}(-D) \to H^1(D)^*$$

sending $\omega \in L^{(1)}(-D)$ to the linear functional Res_ω on $H^1(D)$.

Serre Duality. The Duality Theorem of Serre states that this is an isomorphism.

THEOREM 3.3 (SERRE DUALITY). *For any divisor D on an algebraic curve X, the map*

$$\text{Res}: L^{(1)}(-D) \to H^1(D)^*$$

is an isomorphism of complex vector spaces. In particular, for any canonical divisor K on X,

$$\dim H^1(D) = \dim L^{(1)}(-D) = \dim L(K - D).$$

The proof naturally breaks into showing that Res is injective, and then surjective. The injectivity is the easier part.

PROOF (INJECTIVITY OF Res). Suppose that $\omega \in L^{(1)}(-D)$, $\omega \neq 0$, such that $\text{Res}(\omega)$ is the identically zero map on $H^1(D)$. This means that

$$\sum_p \text{Res}_p(r_p \omega) = 0$$

for every $\sum r_p \cdot p \in T[D]$. Fix a point p with local coordinate $z = z_p$; since $\omega \in L^{(1)}(-D)$, we must have $\text{ord}_p(\omega) \geq D(p)$. Write $k = \text{ord}_p(\omega)$; hence $-1 - k < -D(p)$ and so the Laurent tail divisor $z^{-1-k} \cdot p$ is in $T[D](X)$. But if we write

$$\omega = \left(\sum_{n=k}^{\infty} c_n z^n \right) dz$$

where the lowest coefficient $c_k \neq 0$, then

$$\begin{aligned}\text{Res}_\omega(z^{-1-k} \cdot p) &= \text{Res}_p(z^{-1-k} \sum_{n=k}^{\infty} c_n z^n dz) \\ &= c_k,\end{aligned}$$

which is not zero. This contradiction shows that $\text{Res}(\omega)$ cannot be the identically zero map on $H^1(D)$ unless $\omega = 0$. □

The surjectivity follows from two lemmas. Before proceeding to them, let us note that the dual space $H^1(D)^*$ to $H^1(D)$ can be identified with the space of linear functionals on $T[D](X)$ which vanish on the image $\alpha_D(\mathcal{M}(X))$ of the meromorphic functions. This allows us to pull our computations back to the $T[D](X)$ level, which is convenient.

Note that if $\phi : T[D](X) \to \mathbb{C}$ is linear and vanishes on $\alpha_D(\mathcal{M}(X))$, and f is any meromorphic function, then $\phi \circ \mu_f : T[D + \operatorname{div}(f)](X) \to \mathbb{C}$ is also linear and vanishes on $\alpha_{D+\operatorname{div}(f)}(\mathcal{M}(X))$, since

$$\phi(\mu_f(\alpha_{D+\operatorname{div}(f)}(g))) = \phi(\alpha_D(fg)) = 0$$

using (2.2).

LEMMA 3.4. *Suppose that ϕ_1 and ϕ_2 are two linear functionals on $H^1(A)$ for some divisor A. Then there is a positive divisor C and nonzero meromorphic functions f_1, f_2 in $L(C)$ such that*

$$\phi_1 \circ t_A^{A-C-\operatorname{div}(f_1)} \circ \mu_{f_1} = \phi_2 \circ t_A^{A-C-\operatorname{div}(f_2)} \circ \mu_{f_2}$$

as functionals on $H^1(A-C)$. In other words, the two maps on $T[A-C](X)$ in the diagram

$$\begin{array}{ccc}
 & T[A - C - \operatorname{div}(f_1)](X) \xrightarrow{t} T[A](X) & \\
\nearrow \mu_{f_1} & & \searrow \phi_1 \\
T[A - C](X) & & \mathbb{C} \\
\searrow \mu_{f_2} & & \nearrow \phi_2 \\
 & T[A - C - \operatorname{div}(f_2)](X) \xrightarrow{t} T[A](X) &
\end{array}$$

are equal for some C and some $f_1, f_2 \in L(C) - \{0\}$.

PROOF. Suppose no such divisor C and functions f_i exist. Then for every divisor C, the $\mathbb{C}$-linear map

$$L(C) \times L(C) \to H^1(A-C)^*$$

defined by sending a pair (f_1, f_2) to $\phi_1 \circ t_A^{A-C-\operatorname{div}(f_1)} \circ \mu_{f_1} - \phi_2 \circ t_A^{A-C-\operatorname{div}(f_2)} \circ \mu_{f_2}$ is injective. Therefore for every C, we must have

(3.5) $$\dim H^1(A-C) \geq 2 \dim L(C).$$

Now for C large and positive, the Riemann-Roch Theorem applied to the divisor $A - C$ gives

$$\begin{aligned}
\dim H^1(A-C) &= \dim L(A-C) - \deg(A-C) - 1 + \dim H^1(0) \\
&\leq \dim L(A) - \deg(A) - 1 + \dim H^1(0) + \deg(C),
\end{aligned}$$

which for fixed A grows at most like $a + \deg(C)$ for some constant a. On the other hand, Riemann-Roch for the divisor C implies that

$$\dim L(C) \geq \deg(C) + 1 - \dim H^1(0),$$

so $2 \dim L(C)$ grows at least like $b + 2 \deg(C)$ for a constant b. These two growth rates are incompatible with (3.5), giving a contradiction. $\square$

At first blush the above lemma seems miraculous: one can make any two functionals look the same after pulling them back far enough. But the key lies in the use of the multiplication operators μ_{f_1} and μ_{f_2} above. We are trying to prove that the $H^1(D)$ spaces are isomorphic (after dualizing) to spaces of 1-forms. If we recall that given any two meromorphic 1-forms ω_1 and ω_2, there is a meromorphic function f such that $\omega_2 = f\omega_1$, then it is much less surprising that given any two functionals on $H^1(D)$, they "differ" by appropriate multiplication operators.

The second lemma is more elementary.

LEMMA 3.6. *Suppose that D_1 is a divisor on X with $\omega \in L^{(1)}(-D_1)$, so that $\mathrm{Res}_\omega : T[D_1](X) \to \mathbb{C}$ is well defined. Suppose that $D_2 \geq D_1$ and that Res_ω vanishes on the kernel of $t : T[D_1](X) \to T[D_2](X)$. Then in fact $\omega \in L^{(1)}(-D_2)$.*

PROOF. Suppose on the contrary that ω is not in $L^{(1)}(-D_2)$; this means that there is a point $p \in X$ with $k = \mathrm{ord}_p(\omega) < D_2(p)$. Consider the Laurent tail divisor $Z = z_p^{-k-1} \cdot p$. Then $Z \in \ker(t)$, but $\mathrm{Res}_\omega(Z) \neq 0$. This contradiction proves the lemma. □

Finally we note that the map Res_ω is compatible with the multiplication map μ_f in the following sense. Suppose that f is a meromorphic function on X and $\omega \in L^{(1)}(-D)$. Then $f\omega \in L^{(1)}(-D - \mathrm{div}(f))$ and

$$\mathrm{Res}_\omega \circ \mu_f = \mathrm{Res}_{f\omega}$$

as functionals on $T[D + \mathrm{div}(f)]$.

We can now finish the proof of the Serre Duality Theorem.

PROOF (SURJECTIVITY OF Res). Fix a divisor D on X and a nonlinear functional $\phi : H^1(D) \to \mathbb{C}$, which we consider as a functional on $T[D](X)$, zero on $\alpha_D(\mathcal{M}(X))$. Choose any nonzero meromorphic differential form ω, and let $K = \mathrm{div}(\omega)$. Find a divisor A such that $A \leq D$ and $A \leq K$. Note then that $\omega \in L^{(1)}(-A)$, so Res_ω is well defined on $T[A](X)$. Denote by $\phi_A = \phi \circ t : T[A](X) \to \mathbb{C}$ the composition of ϕ with the truncation map from $T[A](X)$ to $T[D](X)$. Thus ϕ_A and Res_ω are both linear functionals on $T[A](X)$. Hence by Lemma 3.4, there is a positive divisor C and meromorphic functions $f_1, f_2 \in L(C)$ such that

$$\phi_A \circ t_A^{A-C-\mathrm{div}(f_1)} \circ \mu_{f_1} = \mathrm{Res}_\omega \circ t_A^{A-C-\mathrm{div}(f_2)} \circ \mu_{f_2}$$

as functionals on $H^1(A - C)$. Now $\mathrm{Res}_\omega \circ t_A^{A-C-\mathrm{div}(f_2)}$ is simply the functional Res_ω, acting on $T[A - C - \mathrm{div}(f_2)](X)$; and $\mathrm{Res}_\omega \circ \mu_{f_2}$ is exactly $\mathrm{Res}_{f_2\omega}$ on $T[A - C](X)$. Therefore we have that

$$\phi_A \circ t_A^{A-C-\mathrm{div}(f_1)} \circ \mu_{f_1} = \mathrm{Res}_{f_2\omega}$$

as functionals on $\mathcal{T}[A - C](X)$. Composing with μ_{1/f_1}, which is the inverse of μ_{f_1}, we find that

$$\phi_A \circ t_A^{A-C-\mathrm{div}(f_1)} = \mathrm{Res}_{f_2\omega} \circ \mu_{1/f_1} = \mathrm{Res}_{(f_2/f_1)\omega}$$

as functionals on $\mathcal{T}[A - C - \mathrm{div}(f_1)](X)$. Note that $(f_2/f_1)\omega \in L^{(1)}(-A + C + \mathrm{div}(f_1))$, and the above shows that $\mathrm{Res}_{(f_2/f_1)\omega}$ vanishes on the kernel of $t = t_A^{A-C-\mathrm{div}(f_1)}$. Therefore by Lemma 3.6, we see that $(f_2/f_1)\omega \in L^{(1)}(-A)$, and so $\phi_A = \mathrm{Res}_{(f_2/f_1)\omega}$. Noting that $\phi_A = \phi \circ t_D^A$, we see that $\mathrm{Res}_{(f_2/f_1)\omega}$ vanishes on the kernel of t_D^A, so that in fact $(f_2/f_1)\omega \in L^{(1)}(-D)$ and $\phi = \mathrm{Res}_{(f_2/f_1)\omega} = \mathrm{Res}((f_2/f_1)\omega)$. □

If we fix a canonical divisor $K = \mathrm{div}(\omega)$ for some meromorphic 1-form ω on X, then using the isomorphism of Lemma 3.11 of Chapter V, we see that

$$\dim L^{(1)}(-D) = \dim L(K - D),$$

which, combined with the isomorphism of the Residue map between $L^{(1)}(-D)$ and the dual of $H^1(D)$, proves that $\dim H^1(D)$ is equal to the above dimensions; this checks the final statement of the Serre Duality Theorem.

The Equality of the Three Genera. The first application of Serre Duality is to identify the term $\dim H^1(0)$ appearing in the first form of the Riemann-Roch Theorem.

First we note that for any canonical divisor K on an algebraic curve of genus g, we have

(3.7) $$\deg(K) = 2g - 2,$$

by Proposition 1.14 of Chapter V.

Secondly, applying Serre Duality to a canonical divisor K, we see that

(3.8) $$\dim H^1(K) = 1,$$

since $H^1(K)$ is Serre dual to $L(K - K) = L(0)$, which has dimension one. Moreover we also have that

(3.9) $$\dim H^1(0) = \dim L(K),$$

since these spaces are also Serre dual to each other.

Finally we apply Riemann-Roch to a canonical divisor K, obtaining

$$\dim L(K) - \dim H^1(K) = \deg(K) + 1 - \dim H^1(0),$$

which using (3.9), (3.7), and (3.8) we write as

$$2\dim H^1(0) = \deg(K) + 1 + \dim H^1(K) = (2g - 2) + 1 + 1 = 2g.$$

Therefore

(3.10) $$\dim H^1(0) = \dim L^{(1)}(0) = \dim L(K) = g.$$

Note that (at least) two statements are being made here. Firstly, the topological genus g is exactly the mystery term $\dim H^1(0)$ in the Riemann-Roch Theorem. This mystery term is sometimes referred to as the *arithmetic genus* of X; such a term always appears in generalizations of the Riemann-Roch Theorem to higher dimensions.

Secondly, since the space $L^{(1)}(0)$ is exactly the space $\Omega^1(X)$ of global holomorphic 1-forms on X, we see that the dimension of this space is also exactly g. The dimension of this space is a priori an analytic invariant, depending very much on the complex structure. Some authors call $\dim \Omega^1(X)$ the *analytic genus* of X.

Thus we see that all three genera, namely
- the topological genus g,
- the arithmetic genus $\dim H^1(0)$, and
- the analytic genus $\dim \Omega^1(X) = \dim L^{(1)}(0) = \dim L(K)$

are all equal; this generalizes in higher dimensions to a theorem called the Hirzebruch-Riemann-Roch Theorem.

The Riemann-Roch Theorem II. Rolling all of these things together we obtain the most useful form of the Riemann-Roch Theorem.

THEOREM 3.11 (THE RIEMANN-ROCH THEOREM: SECOND FORM). *Let X be an algebraic curve of genus g. Then for any divisor D and any canonical divisor K, we have*

$$\dim L(D) - \dim L(K - D) = \deg(D) + 1 - g.$$

It was Riemann's theorem that

$$\dim L(D) \geq \deg(D) + 1 - g,$$

and then Roch provided the error term.

The Riemann-Roch Theorem is often expressed in terms of the dimension of the complete linear system $|D|$ rather than the dimension of the space $L(D)$. In this form it becomes

$$\dim |D| - \dim |K - D| = \deg(D) + 1 - g,$$

where here we must adopt the convention that if a linear system is empty, its dimension is -1.

When the degree of D is large enough, then the degree of $K - D$ is small enough so that $L(K - D)$ is automatically zero. This gives the following computation for such "big" divisors:

COROLLARY 3.12. *Let D be a divisor of degree at least $2g - 1$ on an algebraic curve X. Then $H^1(D) = 0$ and*

$$\dim L(D) = \deg(D) + 1 - g.$$

Problems VI.3

A. Check that the map Res_ω is compatible with the multiplication map μ_f in the following sense. Suppose that f is a meromorphic function on X and $\omega \in L^{(1)}(-D)$. Then $f\omega \in L^{(1)}(-D - \text{div}(f))$ and

$$\text{Res}_\omega \circ \mu_f = \text{Res}_{f\omega}$$

as functionals on $\mathcal{T}[D + \text{div}(f)](X)$.

B. Show that if D is a positive divisor of degree at least $g+1$, then there is a nonconstant function in $L(D)$.

C. Let D be a divisor on an algebraic curve X of genus g, such that $\deg(D) = 2g - 2$ and $\dim L(D) = g$. Show that D is a canonical divisor.

D. Let X be an algebraic curve and D a divisor on X with $\deg(D) > 0$. Show that $H^1(K + D) = 0$.

E. Check the Riemann-Roch Theorem in the case when X is the Riemann Sphere, and when X is a complex torus.

F. Show that if the Riemann-Roch Theorem is true for a divisor D, then it is true for the divisor $K - D$.

G. Show that if $g \geq 2$ and $m \geq 2$ then

$$\dim L(mK) = (g-1)(2m-1).$$

H. Show that if X is a hyperelliptic curve of genus g defined by $y^2 = h(x)$, then the space $\Omega^1(X)$ of holomorphic 1-forms on X is

$$\Omega^1(X) = \{p(x)\frac{dx}{y} \mid \deg(p) \leq g - 1\}.$$

(See Problem IV.1G.)

I. Show that if X is a smooth projective plane curve of degree d, defined by an affine equation $f(u,v) = 0$, then

$$\Omega^1(X) = \{p(u,v)\frac{du}{\partial f/\partial v} \mid \deg(p) \leq d - 3\}.$$

(See Problem IV.1E.)

J. Show that on an algebraic curve of genus one, there is a nowhere zero holomorphic 1-form ω, which is unique up to a constant factor.

K. Show that if $D \geq 0$ then $\deg(D) - g \leq \dim |D| \leq \deg(D)$.

Further Reading

The definition of an algebraic curve given here is somewhat nonstandard; most introductory texts take an "embedded" point of view and define an algebraic curve as a projective curve. This has the immediate advantage of applying directly to any ground field, not just the complex numbers. Our definition generalizes well to higher dimensions; the fundamental paper discussing these ideas is [**Moishezon67**].

The Laurent series approximation results follow [**Lang82**] very closely ; see also [**Deuring73**].

The construct of Laurent tail divisors is a poorly disguised version of adeles, first introduced for these purposes in [**Weil38**]. This method of approaching the Riemann-Roch theorem is taken in [**Lang82**] and [**Serre59**], where the adeles are called *repartitions*. The reader may also want to compare this approach with that taken in [Mumford76].

We have seen in this chapter that the function field $\mathcal{M}(X)$ has taken center stage, with the curve X itself retreating somewhat. Taken to its logical conclusion, in fact the curve X may be dispensed with altogether: it can be recovered from the field $\mathcal{M}(X)$. Moreover the reader eager for generalizations sees quickly that this approach permits an arbitrary ground field. There are good sections in [**Hartshorne77**] and [**Lang82**] with this more general point of view.

Chapter VII. Applications of Riemann-Roch

1. First Applications of Riemann-Roch

In this section we collect some of the more elementary, yet basic, applications of the Riemann-Roch Theorem.

How Riemann-Roch implies Algebraicity. Note that on a very basic level, the Riemann-Roch Theorem can be seen to simply guarantee the existence of nonzero meromorphic functions on X; if $\deg(D) > g$, then $\dim L(D) > 1$ so there is a nonconstant function in $L(D)$. In fact, Riemann-Roch implies algebraicity in the following sense:

PROPOSITION 1.1. *If X is a compact Riemann surface which satisfies the Riemann-Roch Theorem for every divisor D, then X is an algebraic curve.*

PROOF. First we show that $\mathcal{M}(X)$ separates the points of X. Fix two points p and q on X, and consider the divisor $D = (g+1) \cdot p$. By Riemann's inequality, we see that $\dim L(D) \geq \deg(D) + 1 - g = 2$; hence there is a nonconstant function $f \in L(D)$. This function f must have a pole, and the only poles allowed are at p, so f has a pole at p and no other poles. In particular f does not have a pole at q, and f then separates p and q.

Secondly we show that $\mathcal{M}(X)$ separates the tangents of X. Fix a point $p \in X$, and consider the divisors $D_n = n \cdot p$. For large n, $\dim L(D_n) = n + 1 - g$ by Corollary 3.12 of Chapter VI; hence there are functions in $L(D_{n+1})$ which are not in $L(D_n)$ for large n. This implies that for large n, there are functions f_n with a pole of order exactly n at p and no other poles. The ratio f_n/f_{n+1} then has a simple zero at p. □

Criterion for a Divisor to be Very Ample. Using the Riemann-Roch Theorem, we can give a cheap criterion in terms of the degree of a divisor D for the map ϕ_D to projective space to be an embedding.

PROPOSITION 1.2. *Let X be an algebraic curve of genus g. Then any divisor D with $\deg(D) \geq 2g + 1$ is very ample, that is, the complete linear system $|D|$ has no base points and the associated holomorphic map ϕ_D to projective space* *is a holomorphic embedding onto a smooth projective curve of degree equal to* $\deg(D)$.

PROOF. We need to check that $\dim L(D - p - q) = \dim L(D) - 2$ for any points p and q on X. Since both D and $D - p - q$ have degree at least $2g - 1$, we have that $H^1(D) = H^1(D - p - q) = 0$ by Corollary 3.12 of Chapter VI, and

$$\dim L(D) = \deg(D) + 1 - g \text{ and } \dim L(D - p - q) = \deg(D - p - q) + 1 - g.$$

Since the degrees differ by two, the result follows. □

Every Algebraic Curve is Projective. Here we just want to remark that the basic example of an algebraic curve, namely that of a smooth projective curve, is in fact the "only" example:

PROPOSITION 1.3. *Every algebraic curve X can be holomorphically embedded into projective space.*

The proof simply requires constructing a very ample divisor D on X, and by Proposition 1.2, we only need to find a divisor of degree at least $2g + 1$, where g is the genus of X. Pick any point p, and use $D = (2g + 1) \cdot p$.

Actually, using this divisor proves somewhat more. Recall that if a complete linear system $|D|$ is used to define a holomorphic map ϕ_D to projective space $\mathbb{P}^n$, then the set of hyperplane divisors is exactly the complete linear system (Corollary 4.14 of Chapter V). In particular, if $D = (2g + 1) \cdot p$ for some point p, then there is a hyperplane H such that $\phi_D^*(H) = (2g+1) \cdot p$. Set-theoretically, this implies that the inverse image of H is just the point p; therefore the complement $X - p$ is embedded via ϕ_D into $\mathbb{P}^n - H$, which is isomorphic to $\mathbb{C}^n$. Hence:

PROPOSITION 1.4. *If X is an algebraic curve and $p \in X$, then $X - p$ can be embedded into affine space $\mathbb{C}^n$.*

Curves of Genus Zero are Isomorphic to the Riemann Sphere. We are now in a position to answer some basic questions concerning possible complex structures on 2-manifolds. In particular, we can now show that there is only one complex structure on the 2-sphere. The basis for this is the following, which is a reworking of Proposition 4.11 of Chapter II.

LEMMA 1.5. *Let X be a compact Riemann surface. Suppose that for some point $p \in X$, $L(p)$ has dimension greater than one. Then X is isomorphic to the Riemann Sphere.*

PROOF. The hypothesis implies that there is a nonconstant meromorphic function f in $L(p)$. This function must have poles, but the only pole which is allowed is a simple pole at p. Therefore f has a simple pole at p and no other poles. In this case the associated holomorphic map $F : X \to \mathbb{C}_\infty$ has degree one, and is therefore an isomorphism. □

The contrapositive statement is the form in which this is most used:

COROLLARY 1.6. *If X is a compact Riemann surface of genus at least one, then the space of functions $L(p)$ consists only of constant functions, for any point $p \in X$.*

Lemma 1.5 is the basis for the "classification" of curves of genus zero.

PROPOSITION 1.7. *Let X be an algebraic curve of genus 0. Then X is isomorphic to the Riemann Sphere $\mathbb{C}_\infty$.*

PROOF. Fix any point $p \in X$. Since the canonical divisor K on X has degree $2g - 2 = -2$, then the divisor $K - p$ has degree -3. This is strictly negative, so $L(K - p) = 0$. Applying Riemann-Roch to the divisor p, we find that

$$\dim L(p) = \deg(p) + 1 - g + \dim L(K - p) = 2.$$

We conclude using Lemma 1.5 that X is isomorphic to the Riemann Sphere. □

Curves of Genus One are Cubic Plane Curves. If we apply the criterion of Proposition 1.2 to an algebraic curve X of genus one, we see that any divisor of degree 3 is very ample. Since by Riemann-Roch, $\dim L(D) = 3$ if $\deg(D) = 3$, we see that the holomorphic map ϕ_D would map X to the plane $\mathbb{P}^2$. Since $\deg(D) = 3$, the hyperplane divisor is of degree 3, and so the image is a smooth cubic curve. Therefore:

PROPOSITION 1.8. *Every algebraic curve of genus one is isomorphic to a smooth projective plane cubic curve.*

Curves of Genus One are Complex Tori. Our other example of a curve of genus one is given by a complex torus, and we are now in a position to prove that every curve of genus one is a complex torus. Since we will return to this in Chapter VIII, we will only sketch the construction.

Let X be a curve of genus one, and let $\pi : Y \to X$ be its universal covering space. We know from topology that as a topological space $Y = \mathbb{R}^2$ and the fundamental group $\mathbb{Z} \times \mathbb{Z}$ of X acts on Y by two independent translations. We need only show that $Y \cong \mathbb{C}$, as a Riemann surface.

Using Riemann-Roch, specifically (3.7) and (3.10), we see that if $K_0 = \text{div}(\omega_0)$ is a canonical divisor on X, then $\deg(K_0) = 0$ and $\dim L(K_0) = 1$. If $f \in L(K_0)$, then $\omega = f\omega_0$ is a *holomorphic* 1-form, and still $\text{div}(\omega)$ has degree 0; therefore $\text{div}(\omega) = 0$ and ω has no zeroes or poles.

We use ω to define an isomorphism $\phi : Y \to \mathbb{C}$. Consider the pullback $\pi^*\omega$; note that this is a holomorphic 1-form on Y with no zeroes. Fix a point $p_0 \in Y$, and for $p \in Y$, choose a path γ_p from p_0 to p. Set

$$\phi(p) = \int_{\gamma_p} \pi^*\omega$$

Since Y is simply connected and $\pi^*\omega$ is holomorphic, this integral depends only on the endpoints, and so is well defined. Moreover it is clear that ϕ is holomorphic: integrals depend holomorphically on their endpoints. The map ϕ is the desired isomorphism between Y and $\mathbb{C}$:

PROPOSITION 1.9. *Every algebraic curve of genus one is isomorphic to a complex torus.*

Curves of Genus Two are Hyperelliptic. Let X be an algebraic curve of genus $g = 2$. Consider a canonical divisor K, which has degree $2g - 2 = 2$. Since $\dim L(K) = g = 2$, we may assume that $K > 0$; moreover we conclude that there is a nonconstant function $f \in L(K)$. The associated holomorphic map $F : X \to \mathbb{C}_\infty$ has degree 2. Therefore:

PROPOSITION 1.10. *Every algebraic curve of genus two is hyperelliptic.*

Clifford's Theorem. Most of the easy applications of the Riemann-Roch Theorem give some criterion for $H^1(D)$ to be zero, and therefore we get a formula for $\dim L(D)$. When both of these spaces are nonzero, the problem of computing them becomes harder. A divisor D such that $D \geq 0$ (so that $\dim L(D) \geq 1$) *and* $H^1(D) \neq 0$ is called a *special divisor*; the dimension of $H^1(D)$ was called the *index of speciality* when the subject was younger.

Note that D is a special divisor if and only if both $\dim L(D) \geq 1$ and $\dim L(K - D) \geq 1$, where K is a canonical divisor.

Riemann-Roch says that the difference of these numbers is $\deg(D) + 1 - g$. Given that they are both strictly positive, we can develop an inequality for their sum, and deduce an inequality for $\dim L(D)$ itself. We require a lemma.

LEMMA 1.11. *Let D_1 and D_2 be two divisors on a compact Riemann surface X. Then*

$$\dim L(D_1) + \dim L(D_2) \leq \dim L(\min\{D_1, D_2\}) + \dim L(\max\{D_1, D_2\}).$$

PROOF. By the definition of these spaces, we see immediately that

$$L(D_1) \cap L(D_2) = L(\min\{D_1, D_2\}).$$

Since $D_i \leq \max\{D_1, D_2\}$, we have that $L(D_i) \subseteq L(\max\{D_1, D_2\})$ for both i, so that

$$L(D_1) + L(D_2) \subseteq L(\max\{D_1, D_2\}).$$

If W_1 and W_2 are two subspaces of a vector space, we know that

$$\dim(W_1) + \dim(W_2) = \dim(W_1 + W_2) + \dim(W_1 \cap W_2);$$

applying this in our situation gives

$$\begin{aligned}\dim L(D_1) + \dim L(D_2) &= \dim(L(D_1) + L(D_2)) + \dim(L(D_1) \cap L(D_2)) \\ &\leq \dim L(\max\{D_1, D_2\}) + \dim L(\min\{D_1, D_2\})\end{aligned}$$

which is the desired inequality. □

This simple remark is what we will apply to D and $K - D$.

LEMMA 1.12. *Let D be a divisor on an algebraic curve X of genus g. Suppose that $\dim L(D) \geq 1$ and $\dim L(K - D) \geq 1$. Then $\dim L(D) + \dim L(K - D) \leq 1 + g$.*

PROOF. To say that $\dim L(D) \neq 0$ is equivalent to having the complete linear system $|D|$ nonempty; we may therefore choose a positive divisor $D_1 \sim D$. Similarly choose a positive divisor $D_2 \sim K - D$. Since both D_1 and D_2 are positive, so is their minimum; i.e., $0 \leq \min\{D_1, D_2\}$. Moreover their maximum is bounded by their sum, i.e., $\max\{D_1, D_2\} \leq D_1 + D_2$.

If the maximum was exactly equal to the sum, and the minimum equal to zero, then we would have

$$\begin{aligned}
\dim L(D) + \dim L(K - D) &= \dim L(D_1) + \dim L(D_2) \\
&\leq \dim L(\max\{D_1, D_2\}) + \dim L(\min\{D_1, D_2\}) \\
&= \dim L(D_1 + D_2) + \dim L(0) \\
&= \dim L(K) + \dim L(0) = g + 1
\end{aligned}$$

and this would finish the proof.

Now the maximum is the sum and the minimum is zero exactly when D_1 and D_2 have disjoint support. This we will arrange with a slightly more careful construction.

Choose D_2 arbitrarily in $|K - D|$ as above. Write $|D| = F + |M|$, where F is the fixed part of $|D|$ and the linear system $|M|$ has no base points. Since $|M|$ has no base points, there is a divisor $D_3 \in |M|$ whose support is disjoint from D_2. Moreover $\dim L(D_3) = \dim L(M) = \dim L(D)$. Furthermore $D_3 + D_2 \leq F + D_3 + D_2 \sim D + (K - D) = K$, so that $\dim L(D_3 + D_2) \leq \dim L(K) = g$. Therefore we have, arguing as above, that

$$\begin{aligned}
\dim L(D) + \dim L(K - D) &= \dim L(D_3) + \dim L(D_2) \\
&\leq \dim L(\max\{D_3, D_2\}) + \dim L(\min\{D_3, D_2\}) \\
&= \dim L(D_3 + D_2) + \dim L(0) \\
&\leq g + 1.
\end{aligned}$$

□

Adding this inequality for the sum $\dim L(D) + \dim L(K - D)$ to the Riemann-Roch equality for the difference $\dim L(D) - \dim L(K - D)$ we obtain Clifford's Theorem:

THEOREM 1.13 (CLIFFORD'S THEOREM). *Let D be a special divisor on an algebraic curve X, that is, both $L(D)$ and $L(K - D)$ are nonzero. Then*

$$2 \dim L(D) \leq \deg(D) + 2.$$

Clifford's Theorem is often expressed in terms of the dimension of the complete linear system $|D|$; recall that $|D|$ is naturally a projective space of dimension one less than $\dim L(D)$. Therefore if D is a special divisor we have

$$\dim |D| \leq \frac{1}{2} \deg(D).$$

The Canonical System is Base-Point-Free. Let $|K|$ be the canonical linear system on an algebraic curve X, that is, $|K|$ consists of the divisors of the holomorphic 1-forms on X. If X has genus 0, this system is empty. For higher genera, we can show that the canonical system has no base points.

LEMMA 1.14. *The canonical linear system $|K|$ on an algebraic curve X of genus $g \geq 1$ is base-point-free.*

PROOF. Fix a point $p \in X$. We must show that $L(K - p) \neq L(K)$, and for this it suffices to show that $\dim L(K - p) = \dim L(K) - 1 = g - 1$.

Now since $\dim L(p) = 1$ by Corollary 1.6, we have using Riemann-Roch that

$$1 = \dim L(p) = \dim L(K - p) + \deg(p) + 1 - g,$$

which gives $\dim L(K - p) = g - 1$ as desired. □

The Existence of Meromorphic 1-Forms. The Riemann-Roch and Serre Duality Theorems are directed towards answering questions concerning the existence of meromorphic functions. What about meromorphic 1-forms? It should come as no surprise, since 1-forms are used as the basic tool in the proof of these theorems, that with a bit of thought the existence of meromorphic 1-forms can be analyzed.

Suppose one wants to study 1-forms in the same spirit as functions: prescribe certain zeroes and poles, in the data of a divisor D, and study the meromorphic 1-forms with poles bounded by D. This is exactly the space $L^{(1)}(D)$, and can be directly related to a space of functions via a canonical divisor K: $L^{(1)}(D) \cong L(D + K)$. So this type of question is a bit too easy at this point.

More interesting is to specify, in addition to prescribed poles, also prescribed residues. Let us consider the case when we seek a meromorphic 1-form with prescribed simple poles, with prescribed residues, and no other poles.

To fix notation, choose a finite set of points $\{p_i\}$ and a corresponding set of complex numbers $\{r_i\}$. Does there exist a meromorphic 1-form with simple poles at the p_i's, with residue r_i at p_i, and no other poles? Clearly by the Residue Theorem we must have $\sum_i r_i = 0$. In fact this is a sufficient condition.

PROPOSITION 1.15. *Given an algebraic curve X, a finite set of points $\{p_i\}$ on X, and a corresponding set of complex numbers $\{r_i\}$, there is a meromorphic 1-form ω on X with simple poles at the p_i's, no other poles, and $\mathrm{Res}_{p_i}(\omega) = r_i$ for each i, if and only if $\sum_i r_i = 0$.*

PROOF. We will assume that the genus $g \geq 1$; if $g = 0$, then X is the Riemann Sphere, and one can write down the 1-form ω explicitly in this case.

Let D be the divisor $D = \sum_i p_i$ of the desired poles. Since $g \geq 1$, the canonical linear system is base-point-free by Lemma 1.14; hence we may choose a nonnegative canonical divisor K which has none of the p_i's in its support. Let ω_0 be a meromorphic 1-form on X whose divisor is K; since $K \geq 0$, ω_0 is in fact holomorphic. We will find our desired 1-form ω as $f\omega_0$, for a suitable meromorphic function f.

Choose a local coordinate z_i at each p_i. For each i, we may write $\omega_0 = (c_i + z_i g_i) \mathrm{d}z_i$ where $g_i(z_i)$ is holomorphic in z_i, and moreover $c_i \neq 0$ (else p_i would be a zero of ω_0, and hence would appear in K). Consider the Laurent tail divisor Z which is supported at the p_i's, and whose value at p_i is the Laurent tail $(r_i/c_i)z_i^{-1}$. We may consider this Laurent tail divisor Z as being in the truncated space of Laurent tail divisors $\mathcal{T}[K](X)$.

We claim that a solution to our problem is exactly given by a global meromorphic function f such that $\alpha_K(f) = Z$. Such an f clearly will have no poles except at the p_i's and the points in the support of K. However at any point q in the support of K, the order of pole allowed in f is no more than the order of zero of ω_0; hence the 1-form $f\omega_0$ will not have any pole at such a q. Moreover, the pole of f at p_i will be simple, as will the pole of $f\omega_0$; and the residue of $f\omega_0$ will be exactly r_i as required.

Now we may solve this Mittag-Leffler problem and find the desired function f if and only if the class of Z is zero in $H^1(K)$. We have seen in Equation (3.8) of Chapter VI that this is a 1-dimensional space. Therefore there is exactly one linear condition on such Z's for the function f to exist. We already know one such linear condition: the sum of the residues r_i must be zero. Hence this is the only linear condition, and it is sufficient for the existence of f (and also of ω). □

One could take a more theoretical point of view with this proof at the expense of introducing a new space of Laurent tails. This is the space of *Laurent tail 1-form divisors*, which are defined exactly as the space of Laurent tail divisors, but with additional appropriate dz's everywhere. Let us denote this space by $\mathcal{T}^1[D](X)$. There is a natural divisor map from the space $\mathcal{M}^1(X)$ of meromorphic 1-forms on X; its kernel is the space $L^{(1)}(D)$ of 1-forms with poles bounded by D.

Choosing a meromorphic 1-form ω_0 whose canonical divisor is K, we have a natural map from $\mathcal{T}[K](X)$ to $\mathcal{T}^1[0](X)$, induced by multiplication by ω_0. Indeed, the diagram below

$$\begin{array}{ccccccc} 0 & \to & L(K) & \to & \mathcal{M}(X) & \to & \mathcal{T}[K](X) \\ & & \downarrow & & \downarrow & & \downarrow \\ 0 & \to & \Omega^1(X) & \to & \mathcal{M}^1(X) & \to & \mathcal{T}^1[0](X) \end{array}$$

has exact rows, and the vertical maps are all isomorphisms, given by multiplication by ω_0.

Since the right map on the top row has a 1-dimensional cokernel $H^1(K)$, so does the right map on the bottom row. But this map sends a meromorphic 1-form to the negative parts of its Laurent series at every point. Therefore we see that there is exactly one linear condition on a Laurent tail 1-form divisor in $\mathcal{T}^1[0](X)$ to be the set of Laurent tails of a global meromorphic 1-form; this condition is of course that the sum of the residues is zero.

Problems VII.1

A. Let X be an algebraic curve and D a divisor on X with $\deg(D) > 0$. Recall that $\dim L(D) \leq 1 + \deg(D)$. Show that equality holds if and only if X has genus zero. (This has been given before as Problem V.3H; if you didn't do it then, do it now.)

B. Let X be an algebraic curve of genus $g \geq 2$ and D a divisor on X with $\deg(D) > 0$. Show that if $\deg(D) \leq 2g - 3$ then $\dim L(D) \leq g - 1$. Show that if $\deg(D) = 2g - 2$, then $\dim L(D) \leq g$. Therefore we see that among divisors of degree $2g - 2$, the canonical divisors have the most sections.

C. Let X be an algebraic curve of genus g. Show that if $g \geq 3$, then mK is very ample for every $m \geq 2$. Show that if $g = 2$, then mK is very ample for every $m \geq 3$. Show that if $g = 2$, then ϕ_{2K} maps X to a smooth projective plane conic, and that this map has degree 2.

D. Let X be an algebraic curve of genus $g \geq 1$. Prove that equality holds in Clifford's Theorem if D is either a principal divisor or a canonical divisor.

E. Show that if X is hyperelliptic of genus g, $\pi : X \to \mathbb{P}^1$ is the double covering map, and $D = \pi^*(E)$ for any positive divisor E of degree at most $g - 1$ on $\mathbb{P}^1$, then D is special and equality holds in Clifford's Theorem for D. Does equality hold when E has larger degree?

F. Show that the previous two problems encompass all cases of equality in Clifford's Theorem. (Hint: use induction on the degree of D.)

G. Show that given an algebraic curve X, and a point $p \in X$, there is a global meromorphic 1-form ω on X with a double pole at p and no other poles. Conclude that if z is a local coordinate on X centered at p, there is a meromorphic 1-form ω on X whose Laurent series at p has the form dz/z^2 with no other terms of negative degree, and no other poles other than at p.

H. Generalize the previous problem: given an algebraic curve X, an integer $n \geq 2$, a point $p \in X$, and a local coordinate z at p, show that there is a meromorphic 1-form ω on X with a pole of order n at p, no other poles, and whose Laurent series at p has the form dz/z^n with no other terms of negative degree.

I. Using the previous problem, and Proposition 1.15, show that given an algebraic curve X, a finite set of points p_i, (with local coordinates z_i), and a finite set of Laurent tails $r_i(z_i)$ all of whose terms are negative, show that

2. The Canonical Map

there is a meromorphic 1-form ω on X whose Laurent series has $r_i(z_i)dz_i$ as its terms of negative degree for each i if and only if the sum of the coefficients of the $z_i^{-1}dz_i$ terms are zero.

2. The Canonical Map

The Canonical Map for a Curve of Genus at Least Three. Let X be an algebraic curve of genus 3 or more. Let K be a canonical divisor on X. Note that the complete linear system $|K|$ is exactly the set of divisors of holomorphic 1-forms on X.

We have seen in Lemma 1.14 that $|K|$ is base-point-free; hence the associated holomorphic map

$$\phi_K : X \to \mathbb{P}^{g-1}$$

is defined. This map, undoubtedly the most important map for the theory of algebraic curves, is called the *canonical map* for X. A basic question arises: when is the canonical map an embedding? There is no chance unless the genus is at least 3, so we assume this.

We have seen in Proposition 4.20 of Chapter V that ϕ_K will fail to be an embedding if and only if there are points p and q on X ($q = p$ is possible) such that $\dim L(K-p-q) \neq \dim L(K)-2$. This can only happen if $\dim L(K-p-q) = \dim L(K) - 1 = g - 1$ since $|K|$ has no base points. Using Riemann-Roch we have that

$$\dim L(K-p-q) = \deg(K-p-q) + 1 - g + \dim L(p+q) = g - 3 + \dim L(p+q).$$

Hence the canonical map will fail to be an embedding if and only if there exist two points p and q such that $\dim L(p+q) = 2$.

If this happens, then any nonconstant function $f \in L(p+q)$ gives a degree 2 map to the Riemann Sphere, and so X is hyperelliptic by Proposition 4.11 of Chapter III. Conversely, if X is hyperelliptic and $\pi : X \to \mathbb{C}_\infty$ is the degree 2 mapping, then the inverse image divisor $p+q$ of ∞ has degree 2 and $\dim L(p+q) = 2$. Therefore we have proved the following.

PROPOSITION 2.1. *Let X be an algebraic curve of genus $g \geq 3$. Then the canonical map is an embedding if and only if X is not hyperelliptic. If X is not hyperelliptic, the canonical map embeds X into $\mathbb{P}^{g-1}$ as a smooth projective curve of degree $2g - 2$.*

The only additional remark necessary is to note that since $\deg(K) = 2g - 2$ by (3.7) of Chapter VI, then the degree of the image of the canonical map is also $2g - 2$, by Proposition 4.23 of Chapter V.

The Canonical Map for a Hyperelliptic Curve. For a hyperelliptic curve X of genus $g \geq 2$, the canonical linear system $|K|$ still has no base points, and so the canonical map ϕ_K is still defined. However, as we saw above, ϕ_K is not an embedding. We can easily see what ϕ_K is in this case.

Suppose that X is defined by $y^2 = h(x)$, where $h(x)$ is a polynomial of degree $2g+1$ or $2g+2$ with distinct roots. We have seen (Problem VI.3H) that the space of holomorphic 1-forms on X is

$$\Omega^1(X) = \{p(x)\frac{dx}{y} \mid \deg(p) \leq g-1\}.$$

In particular, if we use the canonical divisor $K = \mathrm{div}(dx/y)$, then a basis for the space $L(K)$ is $\{1, x, x^2, \ldots, x^{g-1}\}$. Hence the canonical map is given by

$$\phi_K = [1 : x : x^2 : \cdots : x^{g-1}].$$

If we denote by $\pi : X \to \mathbb{P}^1$ the double covering map, sending (x,y) to x, we see that the canonical map ϕ_K is the composition of the double covering map π with the holomorphic Veronese map $\nu_{g-1} : \mathbb{P}^1 \to \mathbb{P}^{g-1}$ given by a basis for the functions in $L((g-1)\cdot\infty)$. The image of this Veronese map ν_{g-1} is a *rational normal curve* (of degree $g-1$ in $\mathbb{P}^{g-1}$). We have shown the following:

PROPOSITION 2.2. *For a hyperelliptic curve X of genus $g \geq 2$, the canonical map ϕ_K is the composition of the double covering map and a Veronese map. In particular, the image of ϕ_K is a rational normal curve Y of degree $g-1$ in $\mathbb{P}^{g-1}$, and the map $\phi_K : X \to Y$ has degree 2.*

The above also shows that

the double covering map for a hyperelliptic curve of genus $g \geq 2$ is unique

since is it the canonical map after all.

Finding Equations for Smooth Projective Curves. Suppose that D is a very ample divisor on an algebraic curve, inducing the holomorphic embedding $\phi_D : X \to \mathbb{P}^n$. The image (which we will also call X) is a smooth projective curve. We are now in a position to be precise about the hypersurfaces (defined by the vanishing of homogeneous polynomials $F = 0$) on which X might lie.

It is convenient to introduce the notation $\mathcal{P}(n,k)$ for the vector space of homogeneous polynomials of degree k in the $n+1$ homogeneous variables of $\mathbb{P}^n$. Note that

(2.3) $$\dim \mathcal{P}(n,k) = \binom{n+k}{k}.$$

Fix a degree k, and a homogeneous polynomial F_0 of degree k such that F_0 is *not* identically zero on X. Consider the intersection divisor $\mathrm{div}(F_0)$ on X. Since the hyperplane divisors on X are exactly the divisors in the linear system $|D|$ (by Corollary 4.14 of Chapter V), we see that $\mathrm{div}(F_0) \sim kD$ since F_0 has degree k.

Suppose then that F is another homogeneous polynomial of degree k, so that the ratio $f = F/F_0$ is a meromorphic function on X. It is clear that f has poles bounded by $\operatorname{div}(F_0)$; hence we obtain a $\mathbb{C}$-linear map

$$R_k : \mathcal{P}(n, k) \to L(kD)$$

defined by sending a homogeneous polynomial F to the ratio F/F_0. (Here we are suppressing the use of the isomorphism between $L(\operatorname{div}(F_0))$ and $L(kD)$ for notational convenience.) The map R_k should be thought of as a restriction map; we are taking a ratio F/F_0 and restricting it to a function on X when we map F to $L(kD)$.

In any case we see that a polynomial F is in the kernel of R_k if and only if F vanishes identically on X. Therefore the kernel of R_k is formed by the equations of the hypersurfaces in $\mathbb{P}^n$ containing X.

Now as the degree k grows, by (2.3), the dimension of the space $\mathcal{P}(n, k)$ grows like $k^n/n!$. By Riemann-Roch, the dimension of $L(kD)$ equals $\deg(D)k + 1 - g$. Clearly for k large we will have lots of equations in the kernel!

For a local complete intersection curve, which is *defined* by the vanishing of certain polynomials $F_1, \ldots, F_k$, one obtains for free many equations which vanish on X: any linear combination $\sum_i G_i F_i$ for example, where the degree of G_i is chosen so that all the degrees of the products $G_i F_i$ are the same. What we are seeing above is the first step in showing that any projective curve is a local complete intersection: in any case we have lots of equations which vanish.

Note that when $k = 1$, the restriction map R_1 is always 1-1. This is because X is a nondegenerate curve in $\mathbb{P}^n$, and lies inside no hyperplane.

For a specific divisor D about which we know the dimensions of the spaces $L(kD)$, we can be quite precise, since the dimension of the kernel of R_k is at least the difference between the dimension of $\mathcal{P}(n, k)$ and the dimension of $L(kD)$. For example, if $\deg(D) \geq g$, then $\deg(kD) \geq kg \geq 2g - 1$ if $k \geq 2$, so that $H^1(kD) = 0$ for $k \geq 2$ by Corollary 3.12 of Chapter VI. Therefore:

LEMMA 2.4. *Suppose that D is a very ample divisor on an algebraic curve X of degree at least g. Then for every $k \geq 2$,*

$$\dim \ker(R_k) \geq \binom{n+k}{k} - \deg(D)k - 1 + g$$

Classification of Curves of Genus Three. Let us apply the above analysis of equations to classify, via the canonical embedding, curves of low genus. We begin with a nonhyperelliptic curve X of genus 3, where the canonical map ϕ_K maps X into $\mathbb{P}^2$ as a smooth curve of degree 4.

We should fully expect X to be the zeroes of a polynomial F of degree 4, and we can discover F by the method outlined above. Indeed, by Lemma 2.4, we see that

$$\dim \ker(R_4) \geq \binom{2+4}{4} - 4\deg(K) - 1 + 3 = 15 - 16 + 2 = 1$$

so that there is indeed a quartic polynomial F vanishing on X. This polynomial must be irreducible, since no polynomial of degree less than four can vanish on X, for degree reasons.

There cannot be two independent polynomials of degree 4 vanishing on X; the zeroes of two independent polynomials would form a finite set in $\mathbb{P}^2$. So we conclude that $\ker(R_4)$ is one-dimensional, generated by the quartic equation F vanishing on X.

If F vanishes on X, then so does any multiple GF for any homogeneous polynomial of degree $k - 4$. This gives, for $k \geq 5$, a subspace of $\ker(R_k)$ of dimension $\dim \mathcal{P}(2, k-4) = (k-2)(k-3)/2$. Moreover this is also the lower bound given by Lemma 2.4:

$$\dim \ker(R_k) \geq (k+2)(k+1)/2 - 4k + 2 = (k-2)(k-3)/2.$$

Finally we note that this must be all of $\ker(R_k)$: if a polynomial H vanished on X, and was not a multiple of F, then either H and F have a common factor (of degree less than four) which must vanish on X, or their common zero locus is a finite set. In either case this is a contradiction.

Therefore we conclude that *every polynomial vanishing on X is a multiple of the quartic polynomial F.*

The analysis of precisely which polynomials vanish on a subset of projective space is practically the defining problem of algebraic geometry.

We have shown:

PROPOSITION 2.5. *Let X be an algebraic curve of genus 3. Then either X is hyperelliptic (defined by an equation of the form $y^2 = h(x)$ where h has degree 7 or 8) or the canonical map ϕ_K for X embeds X into the projective plane $\mathbb{P}^2$ as a smooth plane quartic curve defined by the vanishing of a quartic polynomial.*

Actually, we obtain slightly more from the computations above. We have shown that $\dim \ker(R_k)$ is exactly the lower bound given by Lemma 2.4. This means that the map $R_k : \mathcal{P}(2, k) \to L(kK)$ must be surjective. Passing to the linear systems, we see that any pluricanonical divisor in $|kK|$ is an intersection divisor $\mathrm{div}(G)$ for some homogeneous polynomial G of degree k. Therefore the system of intersection divisors for any degree k is complete.

Classification of Curves of Genus Four. Let X be a nonhyperelliptic curve of genus 4, whose canonical map ϕ_K embeds X into $\mathbb{P}^3$ as a curve of degree 6.

From Lemma 2.4, we see that

$$\dim \ker(R_2) \geq \binom{5}{2} - 2 \cdot 6 - 1 + 4 = 1$$

and

$$\dim \ker(R_3) \geq \binom{6}{3} - 3 \cdot 6 - 1 + 4 = 5.$$

Hence our first conclusion is that there is a quadratic polynomial F vanishing on X, since $\ker(R_2) \neq 0$. In fact F is unique, up to a constant multiple. Suppose on the contrary that there were two independent quadratics F and F_1 vanishing on X. Choose a general hyperplane $H \cong \mathbb{P}^2$, and restrict to H. The polynomials F and F_1 restrict to quadratics in the variables of H, and by Bezout's theorem (which is elementary in the case of two conics, even if they are singular) the common zeroes of F and F_1 in H consists of at most 4 points. However the intersection of X with H consists of 6 points; so X cannot be contained in the zeroes of both F and F_1.

Since F vanishes on X, so do the cubic polynomials xF, yF, zF, and wF, where $[x : y : z : w]$ are the variables of $\mathbb{P}^3$. In other words, for any linear polynomial L, the cubic polynomial LF will vanish on X and hence lie in $\ker(R_3)$. This gives an obvious 4-dimensional subspace of $\ker(R_3)$, but we have seen above that $\dim \ker(R_3) \geq 5$. Hence there must be a cubic polynomial G, not a multiple of F, which vanishes on X.

Again, intersecting with a general hyperplane H, it is easy to see that F and G have 6 zeroes on H; since X intersects H in 6 points, and X is contained in the common zeroes of F and G, we see that X must equal the common zeroes of F and G, at least on the general hyperplane H. Making a more delicate analysis in case there are fewer than 6 points in the intersection of H with X one can see that in fact X must equal the common zeroes of F and G in $\mathbb{P}^3$. Therefore we have proved:

PROPOSITION 2.6. *Let X be an algebraic curve of genus 4. Then either X is hyperelliptic (defined by an equation of the form $y^2 = h(x)$ where h has degree 9 or 10) or the canonical map ϕ_K for X embeds X into $\mathbb{P}^3$ as a smooth curve of degree 6 defined by the vanishing of a quadratic and a cubic polynomial.*

A more careful analysis shows that the canonical curve X is the complete intersection of the quadratic F and the cubic G, and that every polynomial vanishing on X is a linear combination of F and G.

The Geometric Form of Riemann-Roch. The Riemann-Roch Theorem has a beautiful expression when interpreted in terms of points on the canonical curve. Let X be a nonhyperelliptic curve of genus g, canonically embedded in $\mathbb{P}^{g-1}$. Fix a positive divisor $D = p_1 + \cdots + p_d$ of degree d (there may be repetitions among the points p_i). If the points p_i are distinct, their span is a linear subspace of $\mathbb{P}^{g-1}$, which may be expressed, if it is not the whole space, as the intersection of the hyperplanes containing all of the points.

It is this way of thinking of the span which generalizes to the case when the points are not distinct. We say that a hyperplane H *contains the divisor D* if $\mathrm{div}(H) \geq D$. Hence if $D(p) = 0$, there is no condition on H at p; if $D(p) = 1$, H must pass through p; and if $D(p) \geq 2$, not only must H pass through p, but the order of the function h defining $\mathrm{div}(H)$ must be at least $D(p)$. Note that this definition depends on the curve X, not just the points of D in canonical space.

For an arbitrary divisor D on the canonical curve X, we therefore define the *span of D*, denoted $\mathrm{span}(D)$, to be the intersection of the hyperplanes $H \subset \mathbb{P}^{g-1}$ such that H contains D, i.e., $\mathrm{div}(H) \geq D$; if there are no such hyperplanes, the span of D is taken to be the whole space $\mathbb{P}^{g-1}$.

For the canonical embedding, the hyperplane divisors are exactly the divisors in the canonical linear system $|K|$. For a hyperplane divisor to contain a given divisor D is a set of linear conditions on the hyperplane, and so for fixed D the hyperplane divisors containing D form a linear subsystem of $|K|$. Since every one of these divisors is at least D, we see that D is in the fixed part of this linear system. So after removing D, we have a linear system of divisors in $|K - D|$.

Conversely, given any divisor E in $|K - D|$, we see that $D + E$ is a canonical divisor, and hence corresponds to a hyperplane H which obviously contains D. Therefore we have a 1-1 correspondence between hyperplanes containing D and the linear system $|K - D|$.

The set of hyperplanes containing D is the same as the set of hyperplanes containing the span of D, and this set forms a linear space inside the dual projective space $(\mathbb{P}^{g-1})^*$ parametrizing all the hyperplanes. The dimension of this linear space (of hyperplanes containing $\mathrm{span}(D)$) is complementary to the dimension of the span of D; since this linear space of hyperplanes is isomorphic to the linear system $|K - D|$, we see that

$$\dim |K - D| + \dim \mathrm{span}(D) = g - 2.$$

(In this formula the dimension of the empty set must be taken to be -1.) By Serre Duality this also has an interpretation in terms of $H^1(D)$:

$$\dim H^1(D) = g - 1 - \dim \mathrm{span}(D).$$

These remarks are the basis for the "geometric" version of the Riemann-Roch Theorem:

THEOREM 2.7 (THE RIEMANN-ROCH THEOREM: GEOMETRIC FORM). *Let X be a nonhyperelliptic algebraic curve of genus g, canonically embedded in $\mathbb{P}^{g-1}$. Fix a positive divisor D on X, considered as a divisor on the canonical curve. Then*

$$\dim |D| = \deg(D) - 1 - \dim \mathrm{span}(D),$$

or, equivalently,
$$\dim L(D) = \deg(D) - \dim \operatorname{span}(D).$$

Classification of Curves of Genus Five. Let X be a nonhyperelliptic curve of genus 5, whose canonical map ϕ_K embeds X into $\mathbb{P}^4$ as a curve of degree 8.

From Lemma 2.4, we see that
$$\dim \ker(R_2) \geq \binom{6}{2} - 2 \cdot 8 - 1 + 5 = 3$$

so that there are at least 3 linearly independent quadratics F_1, F_2, and F_3 vanishing on X. Now the precise analysis is starting to get too advanced to give a proper proof, but the bottom line is that most of the time X is the complete intersection of these three quadrics.

The case when X is not a complete intersection of the three quadrics is now easy to understand, given the Geometric Form of the Riemann-Roch Theorem. Suppose that there is a linear system Q of dimension one and degree 3 on X; recall that such a linear system Q is called a g_3^1. Suppose that $D = p + q + r$ is a divisor in the linear system. By the Geometric Form of Riemann-Roch, we see that
$$\dim \operatorname{span}(p + q + r) = \deg(D) - 1 - \dim |D| = 3 - 1 - 1 = 1,$$

so that the three points p, q, and r are *collinear*, and so lie on a line ℓ in canonical space $\mathbb{P}^4$.

Hence any quadratic polynomial vanishing on all of X restricts to a quadratic polynomial on the line ℓ, which vanishes at the three points p, q, and r; since the quadratic on the line can have at most two roots if it is nonzero, it must be identically zero on the line. We conclude that any polynomial vanishing on all of X also vanishes on the line. Moreover there is a 1-dimensional family of such lines, since the linear system Q has dimension one. The union of these lines is called a *scroll*.

A nonhyperelliptic algebraic curve which has a g_3^1 is called a *trigonal* curve. The precise statement for curves of genus 5 is as follows:

PROPOSITION 2.8. *Let X be an algebraic curve of genus 5. Then either X is hyperelliptic (defined by an equation of the form $y^2 = h(x)$ where h has degree 11 or 12) or the canonical map ϕ_K for X embeds X into $\mathbb{P}^4$ as a smooth curve of degree 8. If X is not trigonal, then X is a complete intersection defined by the vanishing of three independent quadratic polynomials. If X is trigonal, then X lies on a scroll which is the intersection of the quadrics containing X.*

The Space $L(D)$ for a General Divisor. The Geometric Form of the Riemann-Roch Theorem is quite useful in understanding the dimension of $L(D)$ for a "general" divisor D. What do we mean here by the word "general"? Unfortunately, there is no strict definition: the precise use of the word changes from one lemma to the next. However there is a principle at work in using the word; let us try to explain it.

In the Geometric Form of the Riemann-Roch Theorem, the dimension of $L(D)$ is expressed in terms of the dimension of the span of D. If D is a positive divisor of degree d, consisting of d distinct points, then its span can have dimension at most $d-1$ (or $g-1$, if $d \geq g$): two points span a line, three at most a 2-plane, etc. Moreover one sees immediately that if one chooses the points at random, so to speak, they will span the maximum dimension possible: for points to be linearly dependent, some equations need to be satisfied among their coordinates.

To be more precise, we may consider a positive divisor D of degree d as being associated to a point in X^d (if we order the points appearing in D). The space X^d becomes a parameter space for all divisors of degree d, up to a choice of the ordering of the points. We say a statement about divisors of degree d is true for a "general" divisor of degree d if it is true for all divisors parametrized by a dense open set in X^d.

It is an exercise to show that, since the canonical curve spans $\mathbb{P}^{g-1}$, the property that a divisor of degree d on the canonical curve has a span of the maximum dimension $d-1$ (or $g-1$, if $d \geq g$) is true for a general divisor.

The language is most often used in the following way: one says that "the general divisor of degree d satisfies Property P", and by this it is meant that the subset of X^d parametrizing those divisors which have the Property P (whatever it is) is a dense open subset of X^d.

With all of this said, we have the following immediate corollary of the Geometric Form of Riemann-Roch.

COROLLARY 2.9. *Let X be a nonhyperelliptic curve of genus g.*
  a. *For a general positive divisor D of degree $d \leq g$, $\dim L(D) = 1$ and $\dim H^1(D) = g - d$, so that $\dim |D| = 0$ and $|D| = \{D\}$.*
  b. *For a general positive divisor D of degree $d \geq g$, $H^1(D) = 0$ and $\dim L(D) = d + 1 - g$, so that $\dim |D| = d - g$.*

Therefore we see that "special" positive divisors, i.e., those divisors $D > 0$ with $H^1(D) \neq 0$, are actually general if $d < g$, so beware! On the other hand, the above corollary states that special divisors of degree $d \geq g$ are *not* general. The word "special" has a definite precise meaning for divisors (while the word "general", as we have seen does not); the opposite of "special" is *nonspecial*.

An application to automorphisms is the

COROLLARY 2.10. *Suppose that X is a nonhyperelliptic algebraic curve of genus g. Then any nontrivial automorphism of X has at most $2g + 2$ fixed* *points.*

PROOF. By Corollary 2.9, we may choose $g+1$ general points $p_1, \ldots, p_{g+1}$ on X, and find a meromorphic function f on X with simple poles at each p_i and no other poles. If $\sigma \in \text{Aut}(X)$ and is not the identity, then $g = f - f \circ \sigma$ has at most $2g + 2$ poles, namely the p_i's and the $\sigma^{-1}(p_i)$'s. Therefore g has at most $2g + 2$ zeroes also. But any point fixed by σ is a zero of g. □

A Few Words on Counting Parameters. Given a complex manifold M, one of the first questions one asks is: what is the dimension of M? If M is a parameter space whose points correspond to some given objects, the dimension of M is said to be the *number of parameters* for the given objects. Another use of the terminology is to say that the given objects *depend on k parameters*, if the dimension of the parameter space is k.

For example, if one wants to classify divisors of degree d on an algebraic curve X, then we have seen above that a reasonable parameter space to choose is the d-fold product X^d. This actually parametrizes divisors whose points are ordered, so we would say that divisors on X of degree d whose points are ordered depend on d parameters.

Finding the number of parameters on which a set of objects or a given set of constructions depend is called "counting parameters", and it can be a pleasant hobby; moreover the skill of counting parameters is occasionally quite useful.

There is really one basic tool for counting parameters, namely a dimension formula relating the dimension of two manifolds if there is a map between them. It is the following.

THEOREM 2.11 (DIMENSION THEOREM). *Suppose M and N are connected complex manifolds and $F : M \to N$ is a C^∞ onto map such that the fibers $F^{-1}(n)$ for $n \in N$ are also complex manifolds, all of the same dimension. Then*
$$\dim(M) = \dim(N) + \dim F^{-1}(n)$$
for $n \in N$.

Probably the simplest proof, conceptually, relies on a bit of differential geometry (or differential topology). There are some technicalities, but the sketch is as follows. Pick a general point $p \in M$; then the differential DF of the map F sends the tangent space $T_p(M)$ to M at p linearly to the tangent space $T_q(N)$ to N at $q = F(p)$, and DF is onto (since F is) with kernel equal to the tangent space to the fiber $F^{-1}(q)$. Since the tangent spaces are vector spaces with the same dimension as the manifolds, the dimension theorem follows from linear algebra.

The hypotheses may be relaxed significantly if the map is holomorphic; then one needs only to know the dimension of the *general* fiber of F. The reader may see [**Shafarevich77**] for a discussion in the algebraic category.

As an example, let us compute the number of parameters for straight lines in $\mathbb{P}^n$. Part of the fun of counting parameters is that you can do it without constructing the parameter space at all! The approach is to assume a space exists, and then find what its dimension must be.

The space parametrizing lines in $\mathbb{P}^n$ is called a *Grassmann variety*, and is denoted by $\mathbb{G}(1,n)$. The points of $\mathbb{G}(1,n)$ are in 1-1 correspondence with the lines of $\mathbb{P}^n$. If ℓ is a line of $\mathbb{P}^n$, let us denote by $[\ell]$ the point of the Grassmann corresponding to ℓ.

Consider the space of triples
$$\mathcal{I} = \{([\ell],p,q) \mid p,q \in \ell \text{ and } p \neq q\} \subset \mathbb{G}(1,n) \times \mathbb{P}^n \times \mathbb{P}^n.$$

If we denote by Δ the diagonal inside $\mathbb{P}^n \times \mathbb{P}^n$, we have the two obvious projections

$$\begin{array}{ccc} & \mathcal{I} & \\ \pi_1 \swarrow & & \searrow \pi_2 \\ \mathbb{G}(1,n) & & \mathbb{P}^n \times \mathbb{P}^n - \Delta \end{array}$$

and since any line contains two points, and any pair of distinct points lies on a line, both projections are onto.

Now $\dim(\mathbb{P}^n \times \mathbb{P}^n - \Delta) = 2n$; moreover the fibers of π_2 are all singletons, since through two distinct points there passes a unique line. Hence by the Dimension Theorem, $\dim(\mathcal{I}) = 2n$.

The fiber of π_1 over a "point" $[\ell]$ are those triples $([\ell],p,q)$ with p and q coming from ℓ; this is isomorphic to $\ell \times \ell -$ (the diagonal of ℓ), and so since ℓ is a line, this fiber has dimension 2.

Hence by the Dimension Theorem one more time, we see that

(2.12) $$\dim(\mathbb{G}(1,n)) = 2n - 2,$$

and so we would say that *lines in $\mathbb{P}^n$ depend on $2n - 2$ parameters*.

The space $\mathcal{I}$ used in the count of parameters above is called an *incidence space*; there is no strict definition of an incidence space, but it always is a space of pairs, triples, or in general n-tuples of objects, some of which meet or lie in the others.

Riemann's Count of $3g - 3$ Parameters for Curves of Genus g. Without further ado let us proceed directly to the mother of all parameter counts: on how many parameters do algebraic curves of genus g depend? For low genus we can count parameters without much technique. For example, for $g = 0$ we have seen that the only algebraic curve is the Riemann Sphere, up to isomorphism; hence there are no parameters.

For genus one, we have seen that complex tori depend on one parameter, namely τ in the upper half-plane $\mathbb{H}$, up to the action of the discrete group $SL(2,\mathbb{Z})$. Since the action of the group is discrete, the orbit space which actually parametrizes complex tori has dimension equal to the dimension of the upper half-plane, which is one.

2. THE CANONICAL MAP

Before proceeding to the count of parameters for general curves of genus g, let us count the parameters for unordered sets of k distinct points on the Riemann Sphere, up to automorphisms. Since the automorphism group acts triply transitively on the sphere (i.e., any three points may be taken to any other by an automorphism), there are no parameters if $k \leq 3$. If we denote by P_k the space parametrizing unordered sets of k distinct points in $\mathbb{C}_\infty$ up to automorphisms, we see that there is a natural map

$$\phi : (\mathbb{C}_\infty - \{0, 1, \infty\})^{k-3} - \Delta \to P_k$$

(where Δ is the set of $(k-3)$-tuples for which there is some duplication) which sends a $(k-3)$-tuple $(p_1, \ldots, p_{k-3})$ of distinct points to the unordered set $\{0, 1, \infty, p_1, \ldots, p_{k-3}\}$, up to automorphisms. Since any three points may be taken to 0, 1, and ∞ by an automorphism, we see that ϕ is an onto map. Moreover the fibers of ϕ are finite: given an unordered set of k points, there are only finitely many ways to order the set, and for each ordering there is a unique automorphism sending the first three points to 0, 1, and ∞. We conclude that:

(2.13)   the number of parameters for $k \geq 3$ unordered points in $\mathbb{C}_\infty$ (or $\mathbb{P}^1$), up to automorphisms, is $k - 3$.

We may immediately apply the above count of parameters in the case of curves of genus two. Every such curve is hyperelliptic, in a unique way (the double covering map is the canonical map). The double covering π is determined, by the monodromy argument, simply by the six branch points in $\mathbb{P}^1$. Moreover these branch points are unordered, and are only defined up to a choice of coordinate in $\mathbb{P}^1$, or equivalently, up to automorphism. Therefore curves of genus 2 depend on 3 parameters.

In general, the same argument shows that hyperelliptic curves of genus $g \geq 2$ depend on $2g - 1$ parameters; the canonical double covering has $2g + 2$ branch points.

We can use the monodromy construction now to count the number of parameters for branched coverings of $\mathbb{P}^1$. Fix an integer g, and consider holomorphic maps $F : X \to \mathbb{P}^1$ of degree $2g - 1$, branched at $6g - 4$ points, over each of which there is a single ramification point of multiplicity 2. Let us denote by $\{F : X \to \mathbb{P}^1\}_g$ the space of such maps. By Hurwitz's formula, the curve X will have genus g. Moreover such maps $F : X \to \mathbb{P}^1$ are determined, up to the choice of permutations, by the branch points. In other words, since the branch points are unordered, and since we have not chosen a coordinate on $\mathbb{P}^1$, there is a map

$$\alpha : \{F : X \to \mathbb{P}^1\}_g \to P_{6g-4}$$

sending a map to its set of branch points. By the monodromy construction, this is onto, with finite fibers: if one fixes the branch points, then there are only

finitely many ways to choose the permutations. Therefore these spaces have the same dimension:
$$\dim\{F : X \to \mathbb{P}^1\}_g = 6g - 7.$$

There are maps of degree $2g - 1$ which are branched over fewer points, either with a higher multiplicity ramification point or with a branch point having more than one ramification point lying above it. It is easy to see that these maps depend on fewer parameters, however, and in fact they are in some sense "limits" of the more general maps described above. We will ignore these in what follows.

Now there is another description of the space $\{F : X \to \mathbb{P}^1\}_g$, given by the alternate way of constructing a holomorphic map to $\mathbb{P}^1$, namely by using linear systems. A map from a curve X to $\mathbb{P}^1$ of degree $2g - 1$ is given by a base-point-free pencil Q of degree $2g - 1$, that is, a base-point-free linear system Q of dimension 1 and degree $2g - 1$. (This linear system Q would be a g^1_{2g-1}.) Let us denote by $\{(X, g^1_{2g-1})\}$ the space of pairs whose first coordinate is an algebraic curve X of genus g and whose second is a g^1_{2g-1} on X. Since the g^1_{2g-1} pencil Q gives a map $F : X \to \mathbb{P}^1$, and vice versa, we see that
$$\dim\{(X, g^1_{2g-1})\} = \dim\{F : X \to \mathbb{P}^1\}_g = 6g - 7$$
also.

Recall that a pencil Q is a linear subspace inside a complete linear system $|D|$ of dimension one. If Q has degree $2g - 1$, then so does D of course, and so by Riemann-Roch, we have that $\dim |D| = \deg(D) - g = g - 1$. Let us denote by $\{(X, g^{g-1}_{2g-1})\}$ the space parametrizing pairs $(X, |D|)$, where X is an algebraic curve of genus g and $|D|$ is a complete linear system of degree $2g - 1$ on X. We have a natural map
$$\beta : \{(X, g^1_{2g-1})\} \to \{(X, g^{g-1}_{2g-1})\}$$
sending a pencil to its complete linear system. This is an onto map, and the fiber of β over a pair $(X, |D|)$ is the set of pairs (X, Q) where $Q \subset |D|$ is a pencil. Therefore the fiber is a Grassmann variety, parametrizing the lines in the projective space $|D|$. Since $\dim |D| = g - 1$, by (2.12) we see that the fibers of β have dimension $2g - 4$. Hence by the dimension theorem, we have that
$$\dim\{(X, g^{g-1}_{2g-1})\} = (6g - 7) - (2g - 4) = 4g - 3.$$

Next consider the space $\{(X, D_{2g-1})\}$ of pairs (X, D) where X is an algebraic curve of genus g and D is a divisor of degree $2g - 1$ on X. The natural map
$$\gamma : \{(X, D_{2g-1})\} \to \{(X, g^{g-1}_{2g-1})\}$$
sending a divisor to its complete linear system is of course onto, and the fiber of γ over a point $(X, |D|)$ is the set of pairs (X, E) with $E \in |D|$; since $|D|$ is a projective space of dimension $g - 1$, this fiber has dimension $g - 1$. Therefore
$$\dim\{(X, D_{2g-1})\} = (4g - 3) + (g - 1) = 5g - 4.$$

Finally consider the space $\{X_g\}$ of curves of genus g. We have the projection map
$$p_1 : \{(X, D_{2g-1})\} \to \{X_g\}$$
sending a pair (X, D) to X, which is of course onto. The dimension of the fiber over a "point" X the space of divisors on X of degree $2g - 1$, and this space has dimension $2g - 1$. Therefore
$$\dim\{X_g\} = (5g - 4) - (2g - 1) = 3g - 3,$$
or, using the other language,

(2.14)   the number of parameters for curves of genus $g \geq 2$ is $3g - 3$.

Pictorially, we have arrived at the count of $3g - 3$ parameters by analyzing the spaces and maps

$$\{(X, D_{2g-1})\} \qquad \{(X, g^1_{2g-1})\} \cong \{F : X \to \mathbb{P}^1\}_g \quad (\mathbb{C}_\infty - \{0, 1, \infty\})^{6g-7}$$
$$2g-1 \downarrow \quad \searrow g-1 \quad \swarrow 2g-4 \qquad \searrow 0 \quad \swarrow 0$$
$$\{X_g\} \qquad \{(X, g^{g-1}_{2g-1})\} \qquad\qquad P_{6g-4}$$

where we have written the dimensions of the fibers next to the maps.

We have ignored in the above all questions concerning the constructions of the appropriate spaces; this is in fact nontrivial. The space $\{X_g\}$ parametrizing curves of genus g is usually denoted by $\mathcal{M}_g$, and is called the *moduli space for curves of genus g*.

Problems VII.2

A. Check that for a smooth plane curve of degree 4, the bound given by Lemma 2.4 for the dimension of the space of polynomials of degree k vanishing on X is trivial for $k \leq 3$.

B. Prove that the common zeroes of two polynomials F and G in $\mathbb{P}^2$ forms a finite set, unless F and G have a common factor. (You may have to consult a book on Algebra for this.)

C. Let $X \subset \mathbb{P}^n$ be a smooth nondegenerate projective curve, i.e., X does not lie in a hyperplane. Show that the general divisor D of degree d on X has $\dim \mathrm{span}(D) = \min\{d - 1, n\}$.

D. Let $\mathbb{G}(k, n)$ be the Grassmann variety parametrizing k-planes in $\mathbb{P}^n$. Compute the dimension of $\mathbb{G}(k, n)$ in the same spirit as was done above for $\mathbb{G}(1, n)$; the answer is $(k + 1)(n - k)$.

E. Compute the dimension of the space of trigonal curves. For which genera does one expect all curves to be trigonal?

F. Count parameters for curves of degree 4 in $\mathbb{P}^2$ and show that, up to linear automorphisms of $\mathbb{P}^2$, they depend on 6 parameters as they should.

G. Count parameters for the intersection of a quadric and a cubic surface in $\mathbb{P}^3$, and show that up to linear automorphisms of $\mathbb{P}^3$ they depend on 9 parameters.

H. Count parameters for the intersection of three general quadrics in $\mathbb{P}^4$ and show that up to linear automorphisms these curves depend on 12 parameters.

I. Fix the genus g. What is the lowest integer k such that for every curve X of genus g there is a holomorphic map $F : X \to \mathbb{P}^1$ of degree k?

3. The Degree of Projective Curves

Given a smooth projective curve $X \subset \mathbb{P}^n$, we have available three discrete invariants: the genus g of X (which is intrinsic to the curve X), the degree d of X (which depends on the embedding of X into projective space), and the dimension n of the ambient projective space. In this section we will develop the first ideas needed to understand how these invariants are related.

The Minimal Degree. We assume that X is a nondegenerate subset of $\mathbb{P}^n$, i.e., X lies in no hyperplane. If we let H be a hyperplane of $\mathbb{P}^n$, and write $D = \text{div}(H)$ as its hyperplane divisor, we have that the set of hyperplane divisors on X forms a linear system Q, which is a subsystem of its complete linear system $|D|$. The dimension of the linear system Q is then the ambient dimension n; hence we have that $\dim |D| \geq n$. Expressed in terms of the dimension of the vector space $L(D)$, we have that $\dim L(D) \geq n + 1$.

Moreover the degree d of X is the degree of any hyperplane divisor, and so $d = \deg(D)$. Since D is a positive divisor, we have the basic bound that $\dim L(D) \leq d + 1$. Therefore:

PROPOSITION 3.1. *If $X \subset \mathbb{P}^n$ is a nondegenerate smooth projective curve, then*

$$\deg(X) \geq n.$$

This fits well with what we know about plane curves, where $n = 2$: if the degree of X is less than two, X must be a line, and therefore be degenerate!

In $\mathbb{P}^3$, we see that the minimum degree is 3, and this is achieved by the twisted cubic curve. Any curve of degree 2 in $\mathbb{P}^3$ must lie in a plane: conics are fundamentally planar objects.

Rational Normal Curves. Recall that a rational normal curve is, up to changes of coordinates in $\mathbb{P}^n$, the image of the holomorphic map $\nu_n : \mathbb{P}^1 \to \mathbb{P}^n$ sending $[s : t]$ to $[s^n : s^{n-1}t : \cdots : st^{n-1} : t^n]$. This is a nondegenerate smooth projective curve of degree n in $\mathbb{P}^n$, and therefore has minimal possible degree.

Suppose that $X \subset \mathbb{P}^n$ is nondegenerate and has degree $d = n$. Then, analyzing the argument above, we see that the inequalities in

$$n = \dim(Q) \leq \dim |D| = \dim L(D) - 1 \leq d$$

must be equalities. Therefore the linear system Q of hyperplane divisors is the complete linear system $|D|$ (so that $n = \dim(Q) = \dim|D|$) and also $\dim L(D) = 1 + \deg(D)$.

This latter equality implies that X has genus zero, and is therefore isomorphic to $\mathbb{P}^1$. The first equality says that any divisor of degree n is a hyperplane divisor, and in particular that we may choose the hyperplane H so that its divisor is $D = n \cdot \infty$. In this case a basis for $L(D)$ is $\{1, z, z^2, \ldots, z^n\}$ using an affine coordinate z; and then the associated map to $\mathbb{P}^n$ is exactly the map giving the rational normal curve. We have proved the following.

PROPOSITION 3.2. *Suppose X is a nondegenerate smooth projective curve in $\mathbb{P}^n$ of minimal degree n. Then X is a rational normal curve.*

This is just one of the many extremal properties of rational normal curves.

The genus of the curve X does not enter into the bound on the degree, but as we have seen, curves achieving the bound have genus zero. Therefore the bound is better for curves of positive genus: if $g \geq 1$ then $d \geq n + 1$. This simple line of reasoning is exhausted rather quickly, but it suggests that we can do better by somehow incorporating the genus into the analysis. This turns out to be a bit involved, but leads to a wonderful bound, discovered by Castelnuovo, which is sharp for all genera.

In order to get to this, we must understand a bit about tangent hyperplanes.

Tangent Hyperplanes. Suppose that $X \subset \mathbb{P}^n$ is a nondegenerate smooth curve, and H is a hyperplane in $\mathbb{P}^n$. We know that if p is a point on the curve X, then H passes through p if and only if $\mathrm{div}(H) \geq p$, i.e., if and only if the point p is in the support of the hyperplane divisor.

DEFINITION 3.3. We say that the hyperplane H is *tangent to X at p* if $\mathrm{div}(H) \geq 2 \cdot p$.

The following justifies the use of this terminology.

LEMMA 3.4. *A hyperplane H is tangent to X at p if and only if H contains the tangent line to X at p.*

PROOF. We may choose coordinates in $\mathbb{P}^n$ so that $p = [1 : 0 : 0 : \cdots : 0]$ and the tangent line is spanned by p and $q = [0 : 1 : 0 : 0 : \cdots : 0]$. In this case there is a local coordinate z centered at p such that near p, X has a parametrization of the form $[1 : z : g_2(z) : \cdots : g_n(z)]$ where each $g_i(z)$ is holomorphic in z, and $g_i(0) = g_i'(0) = 0$.

Suppose now that H is a hyperplane through p, and is defined by the equation $\sum_{i=1}^n c_i x_i = 0$ (the x_0 term is zero since we assume that H contains p). To compute the value of the hyperplane divisor $\mathrm{div}(H)$ at p, we choose the auxiliary polynomial x_0 (which does not vanish at p), form the ratio $h = (\sum_{i=1}^n c_i x_i)/x_0$, and take $\mathrm{ord}_p(h)$.

In the local coordinate z, we see that

$$h(z) = c_1 z + \sum_{i=1}^n c_i g_i(z);$$

since each g_i vanishes at 0 to order at least two, h will have order at least two at p if and only if $c_1 = 0$.

Therefore $\operatorname{div}(H)(p) \geq 2$ if and only if $c_1 = 0$, and this in turn is equivalent to having the point $q \in H$. Since p is already in H, then $q \in H$ if and only if the tangent line $\overline{pq} \subset H$. □

A hyperplane which is *not* tangent to X at any point is called a *transverse* hyperplane to X. Most hyperplanes are transverse:

LEMMA 3.5. *Suppose that $X \subset \mathbb{P}^n$ is a nondegenerate smooth curve. Then the general hyperplane H is transverse to X (in the sense that the set of transverse hyperplanes is an open dense subset of the dual projective space of all hyperplanes).*

PROOF. Clearly the property of being tangent to X is a closed condition. We will show that the set of transverse hyperplanes forms an open dense set in the dual projective space of hyperplanes of $\mathbb{P}^n$. The first job is to show that there is a transverse hyperplane.

Choose two hyperplanes H_1 and H_2 such that $H_1 \cap X$ is disjoint from $H_2 \cap X$. Let Q be the pencil generated by H_1 and H_2, which therefore has no base points. The pencil Q corresponds to a holomorphic map $F : X \to \mathbb{P}^1$, and the inverse image divisors of F are exactly the divisors in the pencil Q. If we choose a point $p \in \mathbb{P}^1$ which is not one of the finitely many branch points of F, then the inverse image divisor $D = F^{-1}(p)$ will consist of d distinct points, each with multiplicity one. Since D is a member of the pencil Q, D is also a hyperplane divisor, and the hyperplane H such that $\operatorname{div}(H) = D$ is transverse to X.

Therefore the open set of transverse hyperplanes is nonempty. To see that it is dense, let H_1 be a tangent hyperplane. Choose another hyperplane H_2 such that $H_1 \cap X$ is disjoint from $H_2 \cap X$. Make the same construction as above: we saw that all but finitely many hyperplane divisors in the pencil generated by $\operatorname{div}(H_1)$ and $\operatorname{div}(H_2)$ come from transverse hyperplanes. Therefore H_1 is a limit of transverse hyperplanes, and the set of transverse hyperplanes is dense. □

We can be a bit more precise:

LEMMA 3.6. *Suppose that $X \subset \mathbb{P}^n$ is a nondegenerate smooth curve. Then through every point p of X there is a transverse hyperplane. Hence the general hyperplane of $\mathbb{P}^n$ through p is transverse to X. Moreover for fixed p there are only finitely many points $q \in X$ such that no hyperplane through p and q is transverse to X.*

PROOF. The argument is very similar to that given above, although we apply it to two hyperplanes H_1 and H_2 which only have p in their common support. Then the pencil Q generated by H_1 and H_2 has only p as a base point, so removing this base point gives a linear system $Q - p$ without base points. By the argument given above, using the associated map to $\mathbb{P}^1$, we may find a divisor D of $Q - p$ consisting of $d - 1$ distinct points, none of which are p. Then $D + p$ is a hyperplane divisor, and the associated hyperplane is transverse to X and contains p.

This proves all but the final statement, which follows since the divisor D above may be chosen to be any inverse image divisor of a nonbranch point of the map to $\mathbb{P}^1$. $\square$

Flexes and Bitangents. We need to dig a bit deeper into these sorts of issues, and show that the general hyperplane which *is* tangent to X is not more tangent than it needs to be. To be more precise, we will say that a hyperplane H meets X at p with multiplicity k if $\operatorname{div}(H)(p) = k$. The hyperplane H is said to be *transverse* to X at p if H meets X at p with multiplicity one, and is said to be *simply tangent* to X at p if it meets X at p with multiplicity two. A point $p \in X$ is called an *inflection point*, or a *flex* point, if *every* tangent hyperplane to X at p meets X at p with multiplicity at least three: there are no simply tangent hyperplanes at p.

LEMMA 3.7. *Suppose that $X \subset \mathbb{P}^n$ is a nondegenerate smooth curve (with $n \geq 2$). Then X has only finitely many flex points.*

PROOF. The assumption that $n \geq 2$ is meant only to exclude the straight line, which is its own tangent at all points, with infinite multiplicity!

We will show that the set of flex points on X is discrete; since X is compact, the finiteness follows.

Suppose that p_0 is a flex point of X. Choose coordinates so that $p_0 = [1 : 0 : \cdots : 0]$ and the tangent line to X at p_0 is spanned by p_0 and $q = [0 : 1 : 0 : 0 : \cdots : 0]$. Then there is a local coordinate z on X centered at p_0 such that X is parametrized near p_0 by $[1 : z : g_2(z) : \cdots : g_n(z)]$, where each $g_i(z)$ is holomorphic. Moreover we have that for each $i \geq 2$, $g_i(0) = 0$ (since $z = 0$ corresponds to p_0), and we may assume that $g_i'(0) = 0$ (by the choice of the tangent line) and that $g_i''(0) = 0$ (since we assume that p_0 is a flex point).

If the homogeneous coordinates of $\mathbb{P}^n$ are $[x_0 : \cdots : x_n]$, and if H is a hyperplane defined by $\sum c_i x_i = 0$, then for H to be tangent to X at the point p_a with coordinate $z = a$ we must have

$$c_0 + c_1 a + \sum_{i=2}^{n} g_i(a) c_i = 0 \text{ and}$$

$$c_1 + \sum_{i=2}^{n} g_i'(a) c_i = 0,$$

the first equation says that H passes through the point p_a, and the second that H is tangent to X at p_a. We may solve these equations for c_0 and c_1, and deduce that the hyperplanes tangent to X at p_a are parametrized by the homogeneous coordinates $[c_2 : \cdots : c_n]$.

The multiplicity of the tangency is determined by the order of vanishing of the holomorphic function

$$\begin{aligned} h(z) &= c_0 + c_1 z + \sum_{i=2}^{n} g_i(z) c_i \\ &= (a \sum_{i=2}^{n} g_i'(a) c_i - \sum_{i=2}^{n} g_i(a) c_i) - (\sum_{i=2}^{n} g_i'(a) c_i) z + \sum_{i=2}^{n} g_i(z) c_i \\ &= \sum_{i=2}^{n} [g_i(z) - g_i(a) - g_i'(a)(z-a)] c_i \end{aligned}$$

at $z = a$. By construction, h vanishes at $z = a$ to order at least 2: H is tangent to X at the point p_a.

If we set $f(z) = \sum_{i=2}^{n} g_i(z) c_i$, we see, using the Taylor series expansion of f, that h vanishes at $z = a$ to order three or more if and only if $f''(a) = 0$.

Hence if there are flex points arbitrarily close to p_0, we conclude that for every choice of c_i's, the function f'' has zeroes arbitrarily close to 0; since it is holomorphic, we must have that f'' is identically zero near 0, and hence f is linear. This being true for all choices of the c_i's, we then conclude that $g_i(z)$ is linear for each i. But then X is a straight line! This contradiction proves the lemma. □

A hyperplane H is said to be *bitangent* to X if it is tangent to X at two (or more) distinct points. For a hyperplane to be bitangent at two points p and q on X, it must be the case that p and q have the same tangent line. As with flex points, this is a rare occurrence:

LEMMA 3.8. *Suppose that $X \subset \mathbb{P}^n$ is a nondegenerate smooth curve (with $n \geq 2$). Then there are only finitely many pairs of distinct points p and q with the same tangent line.*

PROOF. Again it suffices to show the discreteness of the set of points p whose tangent line is also tangent at another point of X. For this we begin as with the flex points. Suppose that p_0 and q_0 share a tangent line to X. Choose coordinates so that $p_0 = [1 : 0 : \cdots : 0]$ and $q_0 = [0 : 1 : 0 : 0 : \cdots : 0]$, so that the tangent line to X at p_0 is spanned by p_0 and q_0. Then there is a local coordinate z on X centered at p_0 such that X is parametrized near p_0 by $[1 : z : g_2(z) : \cdots : g_n(z)]$, where each $g_i(z)$ is holomorphic. Moreover we have that for each $i \geq 2$, $g_i(0) = 0$ (since $z = 0$ corresponds to p_0), and we may assume that $g_i'(0) = 0$ (by the choice of the tangent line).

For a point p_a near p, parametrized by $z = a$, its homogeneous coordinates are $p_a = [1 : a : g_2(a) : \cdots : g_n(a)]$. We assume that there are arbitrarily small values of a for which p_a is a bitangent point. Let $t_a = [0 : 1 : g_2'(a) : \cdots : g_n'(a)]$ be the derivative point, so that p_a and t_a span the tangent line to X at p_a. We are assuming that there is another point $q_a \in X$ having this tangent line; if so, we must have $q_a = t_a + w(a) p_a$ for some (holomorphic) function w, and so we may write $q_a = [w(a) : 1 + w(a)a : g_2'(a) + w(a)g_2(a) : \cdots : g_n'(a) + w(a)g_n(a)]$.

The tangent to X at q_a is spanned by q_a and its derivative point

$$\begin{aligned} r_a &= [w'(a) : w(a) + w'(a)a : g_2''(a) + w'(a)g_2(a) + w(a)g_2'(a) : \\ &\quad \cdots : g_n''(a) + w'(a)g_n(a) + w(a)g_n'(a)]. \end{aligned}$$

That this tangent line is the same as that for p_a means that this point is a linear combination of p_a and t_a, and so we may write $r_a = \lambda(a) p_a + \mu(a) t_a$ for constants λ and μ, which depend on a.

Equating the first two coefficients gives, up to scaling, that $\lambda = w'(a)$ and $\mu = w(a)$. Then equating further coefficients we see that $g_i''(a) + w'(a)g_i(a) + w(a)g_i'(a) = w'(a)g_i(a) + w(a)g_i'(a)$, which forces $g_i''(a)$ to be identically zero. Therefore, as in the previous lemma, we conclude that each function g_i is linear, so that X is a straight line, contradicting the nondegeneracy. □

The above lemmas combine to give the following.

COROLLARY 3.9. *Suppose that $X \subset \mathbb{P}^n$ is a nondegenerate smooth curve of degree d (with $n \geq 2$).*
  a. *The general hyperplane H in $\mathbb{P}^n$ is such that its divisor $\mathrm{div}(H)$ consists of d distinct points $\{p_i\}$, each having $\mathrm{div}(H)(p_i) = 1$ (i.e., the general hyperplane is transverse to X).*
  b. *For all but finitely many points p of X, the general tangent hyperplane H to X at p is such that $\mathrm{div}(H) = 2 \cdot p + q_3 + \cdots + q_d$ with all q_i distinct and unequal to p (i.e., H is neither a flexed tangent nor a bitangent hyperplane).*

Monodromy of the Hyperplane Divisors. Given a smooth projective curve X of degree d, and a transverse hyperplane H, its divisor $\mathrm{div}(H) = p_1 + \cdots + p_d$ consists of d points, each having multiplicity one in the hyperplane divisor. We may consider these points as an unordered set of d points on X, and a priori they have no particular structure to them, in the sense that there is no obvious way to partition this set into any meaningful subsets.

If fact our goal is now to prove that there is *no* extra structure to be found among the points of a general hyperplane divisor on X. For this we make a monodromy construction similar in spirit to that discussed in Section 4 of Chapter III.

Let $(\mathbb{P}^n)^*$ be the dual space parametrizing all the hyperplanes of $\mathbb{P}^n$. By Lemma 3.5, inside $(\mathbb{P}^n)^*$ is the open dense subset $\mathcal{T}$ of transverse hyperplanes. Fix a transverse hyperplane $H_0 \in \mathcal{T}$, and let $\operatorname{div}(H_0) = p_1 + p_2 + \cdots + p_d$.

As one varies the hyperplane H_0, the points of intersection p_i also vary; if one has a path γ in $\mathcal{T}$ starting at H_0, then in fact each point p_i may be followed as the hyperplanes vary along the path. To be explicit, define the space

$$\mathcal{T}_d = \{(H, p_1, \ldots, p_d) \mid H \in \mathcal{T} \text{ and } \operatorname{div}(H) = \sum_i p_i\}$$

whose points parametrize transverse hyperplanes together with an ordering of the points in their divisors.

It is straightforward to check that the projection

$$\pi : \mathcal{T}_d \to \mathcal{T}$$

sending $(H, p_1, \ldots, p_d)$ to H is a covering space, in the sense of topology, of degree $d!$. Note that although it is easy to see that it is a covering space, it is not obvious at all that $\mathcal{T}_d$ is connected. This is one of our main goals in fact.

By the path-lifting property of covering spaces, any loop in $\mathcal{T}$ based at H_0 may be lifted to a path in $\mathcal{T}_d$. If γ is such a loop, then the lift of γ which starts at the point $(H_0, p_1, \ldots, p_d)$ will end at a point $(H_0, p_{\sigma(1)}, \ldots, p_{\sigma(d)})$ for some permutation σ of the indices on the points.

This permutation depends only on the homotopy class of the loop γ, and so we obtain a group homomorphism

$$\rho : \pi_1(\mathcal{T}, H_0) \to S_d$$

from the fundamental group of $\mathcal{T}$ to the symmetric group S_d, called the *hyperplane monodromy representation* of the curve $X \subset \mathbb{P}^n$.

The Surjectivity of the Monodromy. The image of the hyperplane monodromy representation is the group of permutations of the points $p_1, \ldots, p_d$ which can be achieved by varying the hyperplane H_0 around a loop in the space $\mathcal{T}$. Suppose that some subset $R \subset \{p_1, \ldots, p_d\}$ of points enjoyed some special property, which distinguished R from all other subsets. Suppose further that as we varied the hyperplane, that special property was preserved as we follow the points of the subset. Then, given any loop in $\mathcal{T}$, the permutation corresponding to that loop would have to preserve the subset R. Hence the image of the hyperplane monodromy representation would be contained in the subgroup consisting of those permutations which preserved the indices of R.

Similar considerations would apply if there were some set of subsets, all of which enjoyed some distinguishing property, and that property was preserved under variation: the image of the hyperplane monodromy representation would fail to be the full symmetric group, but would be contained in a subgroup consisting of those permutations which preserved the set of subsets.

Thus the fact that there can be no meaningful way to consistently partition the set of points of a general hyperplane divisor may be expressed by the following mathematical statement.

PROPOSITION 3.10. *Let $X \subset \mathbb{P}^n$ be a smooth nondegenerate curve of degree d. Then the hyperplane monodromy representation*

$$\rho : \pi_1(\mathcal{T}, H_0) \to S_d$$

is surjective.

The proof of the proposition proceeds in several steps. The first is to choose a simply tangent hyperplane H, whose divisor is of the form $\operatorname{div}(H) = 2p + q_3 + \cdots + q_d$, with p and the q_i's distinct points of X. We may further assume that neither p nor any of the q_i's are equal to any point p_i appearing in the divisor of H_0, i.e., that $\operatorname{div}(H)$ and $\operatorname{div}(H_0)$ have disjoint supports.

Consider the line V joining H and H_0 in the dual projective space; this is a pencil of hyperplanes, without base points, and therefore gives a map $F : X \to \mathbb{P}^1$. The point of $\mathbb{P}^1$ corresponding to H_0 is not a branch point of F, while the point b corresponding to H *is* a branch point. By Lemma 4.6 of Chapter III, the cycle structure of the permutation representing a small loop around b is $(2, 1, 1, \ldots, 1)$, i.e., this permutation is a simple transposition. Therefore:

LEMMA 3.11. *There is a simple transposition in the image of the hyperplane monodromy representation ρ.*

This is a good start, but of course we need more. We mentioned above that the space $\mathcal{T}_d$ is connected, but that this was not obvious. In fact the connectedness of $\mathcal{T}_d$ is equivalent to the surjectivity of the hyperplane monodromy representation ρ: if $\mathcal{T}_d$ is connected, then for any permutation σ, we could find a path $\tilde{\gamma}$ in $\mathcal{T}_d$ joining $(H_0, p_1, \ldots, p_d)$ to $(H_0, p_{\sigma(1)}, \ldots, p_{\sigma(d)})$, and the image γ in $\mathcal{T}$ of this path would be a loop for which $\rho([\gamma]) = \sigma$.

We will settle for a bit less for now, which will in the end also suffice to prove the surjectivity of ρ. Let

$$\mathcal{T}_2 = \{(H, p_1, p_2) \mid H \in \mathcal{T} \text{ and } p_1 + p_2 \leq \operatorname{div}(H)\}$$

be the space whose points parametrize transverse hyperplanes together with two ordered points in their divisors. We have the second projection $\pi_2 : \mathcal{T}_2 \to X \times X$ sending (H, p_1, p_2) to (p_1, p_2).

LEMMA 3.12. *The image of the second projection π_2 is connected.*

PROOF. Suppose we fix (p_1, p_2) in the image of π_2. Consider the points (p_1, p), as p varies in X; for this pair to be in the image of π_2, there must be a transverse hyperplane H through p_1 and p. By Lemma 3.6, we conclude that for all but finitely many p's, the pair (p_1, p) is in the image of π_2.

If we consider the map $\alpha : \text{image}(\pi_2) \to X$ sending (p_1, p_2) to p_1, we see that α is onto (since through every point there is a transverse hyperplane) and that the fibers of α are connected. Therefore the image of π_2 is connected. □

After dealing with the image of π_2, we now turn to its fibers.

LEMMA 3.13. *The fibers of π_2 are connected.*

PROOF. Fix a pair (p_1, p_2) in the image of π_2; then the fiber $\pi_2^{-1}(p_1, p_2)$ consists of the set of transverse hyperplanes H which contain p_1 and p_2. This is a nonempty set by assumption, and the property of being transverse is general, so this fiber is an open dense set in the linear space of all hyperplanes through p_1 and p_2.

To show connectedness, assume that H_1 and H_2 are both transverse hyperplanes containing p_1 and p_2. We argue as in the proof of Lemma 3.6. Consider the pencil Q generated by H_1 and H_2. Let F be the fixed part of Q, which since H_1 is transverse, is a divisor of the form $p_1 + p_2 + q_1 + \cdots + q_k$ with all points distinct. Consider the pencil $Q - F$, and the associated map to $\mathbb{P}^1$; for all but finitely many points $\lambda \in \mathbb{P}^1$, the inverse image divisor D_λ will be such that the divisor $D_\lambda + F$ is the divisor of a transverse hyperplane. Therefore there are only finitely many hyperplanes in the pencil Q which are not transverse to X, and so H_1 and H_2 may be connected by a path which in fact lies inside this pencil. □

We now have all the tools at hand to address the surjectivity of the hyperplane monodromy representation.

PROOF (OF PROPOSITION 3.10). By the previous two lemmas, we see that $\pi_2 : \mathcal{T}_2 \to X \times X$ has connected image and connected fibers; therefore the domain $\mathcal{T}_2$ is connected. Therefore for any four indices i, j, k, ℓ, there is a path $\tilde{\gamma}$ in $\mathcal{T}_2$ joining (H_0, p_i, p_j) to (H_0, p_k, p_ℓ). The image γ in $\mathcal{T}$ is a loop based at H_0, such that $\rho([\gamma])$ sends i to k and j to ℓ. We conclude that the image of ρ is *doubly transitive*: any two indices may be sent to any other two.

By Lemma 3.11, there is a transposition (ij) in the image of ρ. If k and ℓ are any two indices, by the double transitivity there is a permutation σ in the image of ρ such that $\sigma(i) = k$ and $\sigma(j) = \ell$. Then the permutation $\sigma(ij)\sigma^{-1}$ is the transposition $(k\ell)$ switching k and ℓ, and is in the image of ρ. We conclude that the image of ρ contains every such transposition $(k\ell)$.

Since every permutation in S_d may be written as a product of transpositions, ρ is surjective. □

The General Position Lemma. Our primary application of the surjectivity of the hyperplane monodromy representation is to prove that there can be no nontrivial dependencies among the points of a general hyperplane divisor.

We say that a set of points $\{p_1, \ldots, p_d\}$ in $\mathbb{P}^{n-1}$ are *in general position* if any subset of n or fewer points are linearly independent.

LEMMA 3.14 (THE GENERAL POSITION LEMMA). *Let X be a nondegenerate smooth projective curve in $\mathbb{P}^n$. Then for a general hyperplane H of $\mathbb{P}^n$, the points of $H \cap X$ are in general position in the projective space H.*

PROOF. Let $p_1, \ldots, p_d$ be the points of $H \cap X$, and write $p_i = [a_{i0} : a_{i1} : \cdots : a_{in}]$ for each i. If we form the matrix $A = (a_{ij})$, we see (since all the points lie in the hyperplane H) that every determinant of a minor of size $n+1$ must vanish.

In order that the points $p_{i_1}, \ldots, p_{i_n}$ be linearly independent, we must have that there is a determinant of a minor of size n, using the rows indexed by $i_1, \ldots, i_n$, which is not zero.

To be explicit, if $\underline{i} = \{i_1, \ldots, i_n\}$ is a subset of the indices for the points and $\underline{j} = \{j_1, \ldots, j_n\}$ is a subset of the indices for the coordinates, let us denote by $d(\underline{i}, \underline{j})$ the determinant of the corresponding minor of size n. Hence the points $p_{i_1}, \ldots, p_{i_n}$ are dependent in H if and only if for all $\underline{j}$, we have $d(\underline{i}, \underline{j}) = 0$.

Denote by $Z_{\underline{i}}$ the set of elements $(H, p_1, \ldots, p_d) \in \mathcal{T}_d$ such that $\{p_{i_1}, \ldots, p_{i_n}\}$ are dependent. This subset $Z_{\underline{i}}$ is defined by the vanishing of all the determinants $d(\underline{i}, \underline{j})$ as $\underline{j}$ varies, and therefore is a closed subset of $\mathcal{T}_d$. Since the coordinates of the points p_i vary holomorphically with the coefficients of the hyperplane H, these determinants are also holomorphic functions of these coefficients; hence if $Z_{\underline{i}}$ is not the whole space, its complement is open and dense in $\mathcal{T}_d$.

Hyperplanes H for which the points of $H \cap X$ are in general position are those parametrized by the intersection of the complements of the subsets $Z_{\underline{i}}$. To prove the General Position Lemma, we must show that this intersection is open and dense. By the above, this would follow if the intersection is nonempty.

Suppose on the contrary that the intersection is empty. Since the complements are open and dense if nonempty, for the intersection to be empty it must be the case that one or more of the subsets $Z_{\underline{i}}$ is the whole space $\mathcal{T}_d$.

On the other hand it must be the case that at least one of the $Z_{\underline{i}}$'s is *not* the whole space: since X is nondegenerate, we can find points $p_{i_1}, \ldots, p_{i_n}$ on X which are linearly independent, and the hyperplane H they lie in will not have $(H, p_1, \ldots, p_n)$ in $Z_{\underline{i}}$.

Therefore we may choose a hyperplane H_0 to use as a base point for $\mathcal{T}$ with the following property for its points $p_1, \ldots, p_d$ of intersection with X: there is a subset $A = \{p_{k_1}, \ldots, p_{k_n}\}$ of points which are linearly independent in H, and another subset $B = \{p_{i_1}, \ldots, p_{i_n}\}$ of points such that $Z_{\underline{i}}$ is the whole space $\mathcal{T}_d$. Then no loop in $\mathcal{T}$ based at H_0 can take the set B to the set A: following the points of B, one always obtains dependent points.

This violates the surjectivity of the hyperplane monodromy representation, and proves the General Position Lemma. □

Points Imposing Conditions on Hypersurfaces. Suppose that $p_1, \ldots, p_d$ are distinct points in projective space $\mathbb{P}^{n-1}$, which has homogeneous coordinates $[x_1 : \cdots : x_n]$. (We are using projective space of dimension $n-1$ here simply to apply the notation to a hypersurface in $\mathbb{P}^n$.) Consider the vector space $\mathcal{P}(n-1,k)$ of homogeneous polynomials of degree k in the n variables $x_1, \ldots, x_n$.

For a homogeneous polynomial F to vanish at a point p is a single linear condition on the coefficients of F: if $p = [a_1 : \cdots : a_n]$ and $F = \sum c_{i_1,\ldots,i_n} x_1^{i_1} \cdots x_n^{i_n}$, then $F(p) = 0$ if and only if the coefficients $c_{i_1,\ldots,i_n}$ satisfy the linear equation $\sum c_{i_1,\ldots,i_n} a_0^{i_1} \cdots a_n^{i_n} = 0$.

The vanishing of F at a point is of course equivalent to having the hypersurface defined by $F = 0$ pass through that point. It is more common to use the hypersurface language than the polynomial language.

More generally, the vanishing of F at the set of points $p_1, \ldots, p_d$ is d linear equations on the coefficients of F. These linear equations may or may not be independent. To determine the amount of independence precisely, let us define $\mathcal{P}(n-1,k)(-p_1-p_2-\cdots-p_d)$ to be the subspace of $\mathcal{P}(n-1,k)$ consisting of those homogeneous polynomials of degree k which vanish at all of the points $p_1, \ldots, p_d$.

DEFINITION 3.15. *We say that the set of points $\{p_1, \ldots, p_d\}$ imposes r conditions on hypersurfaces of degree k if the codimension of the linear subspace $\mathcal{P}(n-1,k)(-p_1-p_2-\cdots-p_d)$ of polynomials of degree k which vanish at the points, in the space $\mathcal{P}(n-1,k)$, is r. We say that the set of points $\{p_1, \ldots, p_d\}$ imposes independent conditions on hypersurfaces of degree k if they impose the maximum of d conditions.*

Of course the number of conditions imposed by a set of points in projective space is one of the first pieces of information one would like to know about the homogeneous ideal of that set of points.

We introduce these ideas now in order to apply them to obtain Castelnuovo's inequality relating the degree and genus of a smooth projective curve $X \subset \mathbb{P}^n$. Fix a hyperplane H, with hyperplane divisor $D = \operatorname{div}(H)$. The idea is to estimate the number $\dim L(mD)$ in two ways: first by analyzing the number of conditions imposed by the points of D, and secondly by applying the Riemann-Roch theorem.

By the General Position Lemma, we may assume that the points of $D = p_1 + \cdots + p_d$ are in general position in the hyperplane H. This allows us to apply the following estimate.

LEMMA 3.16. *Suppose that $p_1, \ldots, p_d$ are points in general position in $\mathbb{P}^{n-1}$. Then the number r of conditions imposed by $p_1, \ldots, p_d$ on hypersurfaces of degree k is at least*

$$\begin{array}{ll} 1 + k(n-1) & \text{if } 1 + k(n-1) \leq d, \text{ or} \\ d & \text{if } 1 + k(n-1) \geq d. \end{array}$$

PROOF. Let $s = \min\{d, 1 + k(n-1)\}$; we are claiming that the number of conditions imposed is at least s. Choose s of the points, and renumber them so that they are $p_1, \ldots, p_s$. Fix homogeneous coordinates for these p_i's so that evaluation of homogeneous polynomials is possible. Define the evaluation map

$$E : \mathcal{P}(n-1, k) \to \mathbb{C}^s$$

by sending a homogeneous polynomial F to $(F(p_1), \ldots, F(p_s))$.

This is a linear map, and clearly the subspace $\mathcal{P}(n-1,k)(-p_1 - \cdots - p_d)$, consisting of those polynomials vanishing at all of the points, is contained in the kernel of the map E. Therefore the number of conditions imposed is at least the codimension of the kernel of E, which is exactly the dimension of the image of E. Therefore it suffices to show that E is surjective.

There are at most $k(n-1)$ points in the set $p_1, \ldots, p_{s-1}$; therefore we may partition them into k subsets $A_1, \ldots, A_k$, each of which has at most $n-1$ points. Since the points are in general position, for each i we may find a hyperplane H_i passing through the points of the subset A_i and not through p_s. If H_i is defined by the linear equation F_i, then the product $F = F_1 \cdots F_k$ is a polynomial of degree k vanishing at all the points $p_1, \ldots, p_{s-1}$, and not at p_s. Therefore after appropriately scaling F, we see that $E(F)$ is the standard basis vector in $\mathbb{C}^s$ which has a 1 in the sth spot and zeroes elsewhere.

But the choice of the point p_s among the s points was arbitrary; by symmetry we see each standard basis vector of $\mathbb{C}^s$ is in the image of E, and so E is surjective as desired. □

We apply the above lemma to the points of a hyperplane divisor to obtain the following result.

PROPOSITION 3.17. *Let X be a nondegenerate smooth projective curve in $\mathbb{P}^n$ of degree d. Let D be a hyperplane divisor on X. Then*

$$\dim L(kD) - \dim L((k-1)D) \geq \begin{cases} 1 + k(n-1) & \text{if } 1 + k(n-1) \leq d \\ d & \text{if } 1 + k(n-1) \geq d. \end{cases}$$

PROOF. Since the dimension of the spaces $L(\ell D)$ are the same for all hyperplane divisors D, we may assume that D is the divisor of a transverse hyperplane H_0 and the d points of D are in general position in H_0. We may choose coordinates in $\mathbb{P}^n$ so that H_0 is defined by the linear equation $x_0 = 0$.

Given a homogeneous polynomial $F \in \mathcal{P}(n, k)$, consider the ratio F/x_0^k. This is a ratio of homogeneous polynomials of the same degree, and so is a meromorphic function on X; moreover it has poles bounded by D. Therefore we obtain a linear map

$$\beta : \mathcal{P}(n, k) \to L(kD)$$

defined by $\beta(F) = F/x_0^k$.

At a point p_i of the divisor D, we see that $\operatorname{div}(\beta(F))(p_i) = \operatorname{div}(F)(p_i) - k$; hence $\beta(F)$ has a pole of order exactly k at p_i if (and only if) F does not vanish at p_i. Therefore if we consider the subspace $L((k-1)D) \subset L(D)$, its preimage under β is exactly the subspace $\mathcal{P}(n,k)(-D)$ of homogeneous polynomials which vanish at all the points of D.

Hence by the First and Second Isomorphism Theorems for vector spaces, we have

(3.18)
$$\begin{aligned}
\mathcal{P}(n,k)/\mathcal{P}(n,k)(-D) &\cong \text{image}(\beta)/(\text{image}(\beta) \cap L((k-1)D)) \\
&\cong (\text{image}(\beta) + L((k-1)D))/L((k-1)D) \\
&\subseteq L(kD)/L((k-1)D)
\end{aligned}$$

and therefore

$$\dim L(kD) - \dim L((k-1)D) \geq \dim \mathcal{P}(n,k) - \dim \mathcal{P}(n,k)(-D).$$

To finish, we use the map

$$\zeta : \mathcal{P}(n,k) \to \mathcal{P}(n-1,k),$$

which simply sets x_0 equal to zero in a polynomial: $\zeta(F(x_0, x_1, \ldots, x_n)) = F(0, x_1, \ldots, x_n)$. This is an onto linear map, and using the same notation as above, the preimage $\zeta^{-1}(\mathcal{P}(n-1,k)(-D))$ of the subspace of polynomials in the variables of H which vanish at the points of D contains the subspace $\mathcal{P}(n,k)(-D)$. Therefore

$$\begin{aligned}
\dim \mathcal{P}(n,k) - \dim \mathcal{P}(n,k)(-D) &\geq \dim \mathcal{P}(n,k) - \dim \zeta^{-1}(\mathcal{P}(n-1,k)(-D)) \\
&= \dim \mathcal{P}(n-1,k) - \dim \mathcal{P}(n-1,k)(-D).
\end{aligned}$$

Now the result follows from the previous lemma. $\square$

Castelnuovo's Bound. It is more convenient to express the bound of the previous Proposition in terms of the multiple k directly. To this end define

$$m = \left[\frac{d-1}{n-1}\right]$$

and note that with this definition, we have that

$$\dim L(kD) - \dim L((k-1)D) \geq \begin{cases} 1 + k(n-1) & \text{if } k \leq m \\ d & \text{if } k \geq m. \end{cases}$$

Now for large ℓ we form the telescoping sum

$$\dim L(\ell D) - \dim L(0) = \sum_{k=1}^{\ell} [\dim L(kD) - \dim L((k-1)D)]$$

and the inequality on the differences, combined with knowing that $\dim L(0) = 1$, gives

$$\begin{aligned}\dim L(\ell D) &\geq 1 + \sum_{k=1}^{m}[1 + k(n-1)] + \sum_{k=m+1}^{\ell} d \\ &= 1 + m + (n-1)m(m+1)/2 + d(\ell - m).\end{aligned}$$

Now (finally!) we can apply Riemann-Roch: for large ℓ, $H^1(\ell D) = 0$ so that

$$\dim L(\ell D) = d\ell + 1 - g$$

where g is the genus of X. If we write

$$d - 1 = m(n-1) + \varepsilon, \text{ with } 0 \leq \varepsilon < n - 1,$$

then combining this with the above gives

$$\begin{aligned}g &= d\ell + 1 - \dim L(\ell D) \\ &\leq d\ell + 1 - [1 + m + (n-1)m(m+1)/2 + d(\ell - m)] \\ &= -m - (n-1)m(m+1)/2 + [1 + m(n-1) + \varepsilon]m \\ &= (n-1)m(m-1)/2 + m\varepsilon,\end{aligned}$$

which is the classical bound of Castelnuovo:

THEOREM 3.19. *Let X be a nondegenerate smooth projective curve in $\mathbb{P}^n$ of degree d and genus g. Write*

$$d - 1 = m(n-1) + \varepsilon \text{ with } m \geq 1 \text{ and } 0 \leq \varepsilon < n - 1.$$

Then

$$g \leq (n-1)\frac{m(m-1)}{2} + m\varepsilon.$$

A surprisingly large number of curves which we have encountered up to now actually achieve the Castelnuovo bound for the genus and degree. We leave the following computations to the reader to check.

EXAMPLE 3.20. Suppose that X is a smooth plane curve of degree d. Then $g = (d-1)(d-2)/2$ by Plücker's formula, which achieves the Castelnuovo bound (with $n = 2$, $m = d - 1$, and $\varepsilon = 0$).

EXAMPLE 3.21. Any rational normal curve of degree n in $\mathbb{P}^n$ achieves the Castelnuovo bound (with $g = 0$, $m = 1$, and $\varepsilon = 0$). Any elliptic normal curve of degree $n + 1$ in $\mathbb{P}^n$ achieves the Castelnuovo bound (with $g = 1$, $m = 1$, and $\varepsilon = 1$ if $n \geq 3$; if $n = 3$ then $m = 2$ and $\varepsilon = 0$).

EXAMPLE 3.22. Suppose that X is an algebraic curve of genus g, and D is a divisor on X of degree $d \geq 2g+1$. Then D is very ample and the associated map ϕ_D maps X isomorphically onto a smooth projective curve in $\mathbb{P}^{d-g}$ of degree d. The image is a curve achieving the Castelnuovo bound (with $n = d - g$, and $m = 1$, $\varepsilon = g$ if $d > 2g+1$, while $m = 2$, $\varepsilon = 0$ if $d = 2g+1$).

EXAMPLE 3.23. Suppose that X is a nonhyperelliptic curve of genus $g \geq 3$. Then the canonical image of X in $\mathbb{P}^{g-1}$ achieves the Castelnuovo bound (with $d = 2g - 2$, $n = g - 1$, $m = 2$, and $\varepsilon = 1$).

Curves of Maximal Genus. Smooth projective curves achieving the Castelnuovo bound are sometimes called *Castelnuovo curves*, or *Castelnuovo extremal curves*; they are the curves with maximum possible genus for their degree, or minimal possible degree for their genus.

The examples above show that there are several important classes of curves, including canonical curves, which do have maximal genus. For such curves one ought to be able to say something specific, by analyzing the inequalities in the proof of the Castelnuovo bound: all these inequalities must be equalities for curves of maximal genus.

Let us focus on the inequalities of Proposition 3.17: that these are equalities leads to the following conclusion.

PROPOSITION 3.24. *Let $X \subset \mathbb{P}^n$ be a smooth projective curve achieving the Castelnuovo bound for the maximum genus. Fix a hyperplane divisor $D = \operatorname{div}(H)$ on X, where we choose coordinates so that the hyperplane H is defined by $x_0 = 0$. Then for every $k \geq 0$, the natural map*

$$\beta_k : \mathcal{P}(n, k) \to L(kD)$$

defined by $\beta_k(F) = F/x_0^k$ is onto.

PROOF. The proof proceeds by induction on k, and for $k = 0$ the result is trivial: both $\mathcal{P}(n, k)$ and $L(0)$ are the constant functions, and β_0 is the identity.

Assume then that β_{k-1} is onto. Reading the proof of Proposition 3.17, we see that the inclusion

$$(\operatorname{image}(\beta_k) + L((k-1)D))/L((k-1)D) \subseteq L(kD)/L((k-1)D)$$

of (3.18) must be an equality. Hence every function $f \in L(kD)$ is equal to $\beta_k(F)$ (for some $F \in \mathcal{P}(n, k)$) modulo an element of $L((k-1)D)$; this element, by induction, may be written as $\beta_{k-1}(G)$ for some $G \in \mathcal{P}(n, k-1)$. Hence $f = \beta_k(F) + \beta_{k-1}(G) = \beta_k(F + x_0 G)$ is in the image of β_k. □

Applying this with $k = 1$, we see that every function in $L(D)$ is of the form L/x_0 for some linear homogeneous polynomial L. In terms of the corresponding hyperplane divisors, this gives that these curves are embedded by complete linear systems:

COROLLARY 3.25. *Let $X \subset \mathbb{P}^n$ be a smooth projective curve achieving the Castelnuovo bound for the maximum genus. If D is a hyperplane divisor for X, then the hyperplane linear system is the complete linear system $|D|$: every divisor in $|D|$ is a hyperplane divisor.*

That this is true for the canonical curves, and curves already embedded by complete linear systems, is obvious. It is not for plane curves however.

Moving to higher degree k, recall that if $f \in L(k_1 D)$ and $g \in L(k_2 D)$ then $fg \in L((k_1 + k_2)D)$. In particular, by induction, if $f_1, \ldots, f_n$ are all in $L(D)$, then any homogeneous polynomial in the f_i's of degree k is in $L(kD)$.

The usual notation for the vector space of homogeneous polynomial expressions of degree k in arbitrary vectors from a vector space V is $\text{Symm}^k(V)$. We then have the following:

COROLLARY 3.26. *Let $X \subset \mathbb{P}^n$ be a smooth projective curve achieving the Castelnuovo bound for the maximum genus. If D is a hyperplane divisor for X, then for all $k \geq 1$ the natural map*

$$\text{Symm}^k(L(D)) \to L(kD)$$

(sending a polynomial expression in functions in $L(D)$ to the actual function it represents in $L(kD)$) is onto.

In general, a curve with this property is called *projectively normal*. That this is true for a canonical curve is a theorem of Max Noether:

COROLLARY 3.27. *Let $X \subset \mathbb{P}^{g-1}$ be a canonical curve. Then for all $k \geq 1$ the natural map*

$$\text{Symm}^k(L(K)) \to L(kK)$$

is onto.

The kernel of this map is of course those polynomial expressions which vanish on X: this is the kth homogeneous piece of the homogeneous ideal of X, because the canonical curve is embedded by the functions in $L(K)$. Since the dimension of $L(K)$ is g, then

$$\dim \text{Symm}^k(L(K)) = \binom{g-1+k}{k},$$

by (2.3), while for $k \geq 2$, $\dim L(kK) = (2k-1)(g-1)$ by Riemann-Roch. Therefore we see that, in the $k = 2$ case, for example, a canonical curve of genus g lies on exactly $g(g+1)/2 - 3(g-1) = (g-2)(g-3)/2$ linearly independent quadric hypersurfaces. This sharpens the inequality we had in Lemma 2.4.

A deeper analysis, in fact using these quadrics through the curve, leads ultimately to a complete classification of curves of maximal genus. Even to state the classification requires some knowledge of the theory of algebraic surfaces, and we will forego this. The reader may consult [**ACGH85**] for a complete treatment.

Problems VII.3

A. Let X be a nondegenerate smooth projective curve in $\mathbb{P}^n$ of degree d. Show that if the hyperplane divisor D is a special divisor, then $d \geq 2n$.

B. Let X be a nondegenerate smooth projective curve in $\mathbb{P}^n$. Show that the property of being tangent to X is a closed condition on the dual space of hyperplanes.

C. Suppose that X is a smooth projective plane curve of degree d defined by $F(x, y, z) = 0$. Define the *Hessian* H_F of F to be the polynomial

$$H_F(x,y,z) = \det \begin{pmatrix} \partial^2 F/\partial x^2 & \partial^2 F/\partial x \partial y & \partial^2 F/\partial x \partial z \\ \partial^2 F/\partial y \partial x & \partial^2 F/\partial y^2 & \partial^2 F/\partial y \partial z \\ \partial^2 F/\partial z \partial x & \partial^2 F/\partial z \partial y & \partial^2 F/\partial z^2 \end{pmatrix}.$$

Show that $\mathrm{div}(H_F)$ is supported at the flex points of X; moreover if $p \in X$ and $\mathrm{div}(H_F)(p) = k$, then the tangent line ℓ to X at p meets X exactly $k+2$ times at p, in the sense that $\mathrm{div}(\ell)(p) = k + 2$.

D. Check that the projection $\pi_1 : \mathcal{T}_d \to \mathcal{T}$ is a covering space in the sense of topology of degree $d!$.

E. For $k \leq d$, define the space

$$\mathcal{T}_k = \{(H, p_1, \ldots, p_k) \mid H \in \mathcal{T} \text{ and } p_1 + \cdots + p_k \leq \mathrm{div}(H)\}$$

be the space whose points parametrize transverse hyperplanes together with k ordered points in their divisors. Show that $\mathcal{T}_k$ is connected.

F. Show that the coordinates of the points of intersection of a hyperplane H with a smooth projective curve $X \subset \mathbb{P}^n$ vary holomorphically with the coefficients of H. (That is, if we fix all but one coefficient of H, then the coordinates depend holomorphically on that last coefficient.)

G. 1. Show that two points always impose two conditions on hypersurfaces of any degree.

2. Show that three points impose three conditions on hypersurfaces of degree $k \geq 2$, and impose three conditions on hyperplanes if and only if the three points are not collinear.

3. Show that four points in the plane impose independent conditions on conics if and only if they are not collinear.

4. Show that five points in the plane impose independent conditions on conics if and only if no four are collinear.

5. Show that d points in $\mathbb{P}^n$ impose independent conditions on hypersurfaces of large enough degree.

H. Check that the curves of Examples 3.20–3.23 achieve the Castelnuovo bound as claimed.

I. Show that for every $d \geq n \geq 2$, there are smooth projective curves of degree d in $\mathbb{P}^n$ achieving the Castelnuovo bound for the maximum genus.

J. Assume that $X \subset \mathbb{P}^n$ is a smooth projective curve of degree d achieving the Castelnuovo bound for the maximum genus. Fix a positive integer k such that $s = 1 + k(n-1) < d$. Suppose that $D = \sum_i p_i$ is a general hyperplane divisor on X, and that F is a polynomial of degree k in the $n-1$ variables of the hyperplane vanishing on s of the points of D. Show that F must vanish at all d points of D.

K. Assume that $X \subset \mathbb{P}^3$ is a smooth projective curve of degree d achieving the Castelnuovo bound for the maximum genus. Show that its genus g is $(d/2-1)^2$ if d is even, and is $(d-1)(d-3)/2$ if d is odd. Show that if $d \geq 7$ then X lies on a unique quadric surface.

4. Inflection Points and Weierstrass Points

Recall that, given a smooth projective curve $X \subset \mathbb{P}^n$ and a point $p \in X$, a hyperplane H through p is said to be tangent to X at p if $\text{div}(H) \geq 2 \cdot p$. We call H a flexed tangent hyperplane if in fact $\text{div}(H) \geq 3 \cdot p$, and p a flex point of X if every tangent hyperplane is in fact flexed.

These ideas can be considerably refined, and in this section we indicate how this goes.

Gap Numbers and Inflection Points of a Linear System. For a smooth projective curve X embedded in projective space by a complete linear system $|D|$, the hyperplanes passing through a point $p \in X$ correspond to the subspace $L(D-p) \subset L(D)$. Those tangent to X at p correspond to $L(D-2p)$, and those flexed to X at p correspond to $L(D-3p)$.

Therefore the study of tangent and flexed hyperplanes may be reduced to the study of the subspaces $L(D-kp) \subset L(D)$. Subspaces of $L(D)$ for a fixed divisor D are linear systems, and with this point of view, we may formulate the problem for *any* linear system $Q \subset |D|$ on an algebraic curve X. We assume that Q is nonempty, so that the vector subspace $V \subset L(D)$ corresponding to Q is nonzero.

In this case, for any point $p \in X$, we have the nested sequence of subspaces
$$V(-np) = V \cap L(D-np) = \{f \in V \mid \text{ord}_p(f) \geq -D(p) + n\}.$$
Clearly
$$V(-np) \subseteq V(-(n-1)p)$$
and the sequence of subspaces eventually arrives at $\{0\}$: if $k > \deg(D)$ then $L(D-kp) = 0$. Moreover the subspaces $V(-np) \subseteq V(-(n-1)p)$ are either equal or differ in dimension by exactly one. The linear system corresponding to $V(-np)$ will be denoted by $Q(-np)$; these are those elements D of Q which satisfy $D \geq np$.

DEFINITION 4.1. An integer $n \geq 1$ is said to be a *gap number* for Q at p if $V(-np) \neq V(-(n-1)p)$, or, equivalently, if $\dim V(-np) = \dim V(-(n-1)p) - 1$. The set of gap numbers for Q at p is denoted by $G_p(Q)$.

Of course the dimension criteria can be expressed in terms of the linear systems too: $n \in G_p(Q)$ if and only if $\dim Q(-np) = \dim Q(-(n-1)p) - 1$.

Recall that a linear system Q is called a g_d^r if $\dim Q = r$ and $\deg(Q) = d$. We leave the following elementary remarks to the reader.

LEMMA 4.2. *Let Q be a nonempty g_d^r on an algebraic curve X, and fix a point $p \in X$. Then:*
   a. *The set of gap numbers $G_p(Q)$ is a finite set, and $\#G_p(Q) = 1 + r$.*
   b. *$G_p(Q) \subset \{1, 2, \ldots, 1+d\}$.*
   c. *The point p is a base point of Q if and only if $1 \notin G_p(Q)$.*
   d. *$1 + d \in G_p(Q)$ if and only if $d \cdot p \in Q$.*
   e. *$G_p(|0|) = \{1\}$ for every point p.*

Imposing an extra order of zero on the functions of V is a single linear condition; therefore one might expect that these conditions are in general independent, and that each subspace $V(-kp)$ is indeed smaller than the previous subspace, until of course we reach the zero space. This phenomenon happens exactly when the set of gap numbers $G_p(Q)$ is equal to the set $\{1, 2, 3, \ldots, r+1\}$. We will say that p is an *inflection point* for the linear system Q if $G_p(Q)$ is *not* equal to this set of the first $r+1$ integers.

The filtration of the space V by the subspaces $V(-np)$ can be used to give a basis for V. Suppose that $r = \dim Q$, so that $r + 1 = \dim(V)$ and we write the gap numbers for Q at p in increasing order as $G_p(Q) = \{n_1 < n_2 < \cdots < n_{r+1}\}$. Then for each $i = 1, \ldots, r+1$ we may choose a function $f_i \in V(-(n_i - 1)p) - V(-n_i p)$, which must therefore satisfy $\mathrm{ord}_p(f_i) = n_i - 1 - D(p)$. The set of f_i's then give a basis for V, and any basis for V with these orders at p is called an *inflectionary basis* for V with respect to p.

If z is a local coordinate centered at p, and we set $g_i = z^{D(p)} f_i$, then a local formula for the holomorphic map $\phi_Q : X \to \mathbb{P}^r$ associated to the linear system Q is given by $\phi_Q(z) = [g_1 : g_2 : \cdots : g_{r+1}]$, which, after multiplying by appropriate constants, has the form

$$\phi_Q(z) = [z^{n_1 - 1} + \cdots : z^{n_2 - 1} + \cdots : \cdots : z^{n_{r+1} - 1} + \ldots].$$

Moreover if p is not a base point of Q, then the first gap number $n_1 = 1$ and the above formula is defined at p, not just in a punctured neighborhood.

In any case we see that p is *not* an inflection point for Q if there is a basis $\{f_1, \ldots, f_{r+1}\}$ for V such that $\mathrm{ord}_p(f_i) = i - 1 - D(p)$ for each i. (This will necessarily be an inflectionary basis for V with respect to p.)

The Wronskian Criterion. Fix a local coordinate z centered at p, and any basis $\{h_k\}$ for V. Set $g_k = z^{D(p)} h_k$ for each k, so that every g_k is holomorphic at p. In order that p not be an inflection point for Q, it is necessary and sufficient that we be able to find a linear combination $\sum_k c_k g_k$ having order $i - 1$ at p for each $i = 1, \ldots, r+1$: then the corresponding linear combinations of the h_k's will be a basis $\{f_i\}$ for V, and will satisfy $\mathrm{ord}_p(f_i) = i - 1 - D(p)$ for every i.

By Taylor's Theorem, the Laurent series for g_k at $z = 0$ (which is a Taylor series) has the form

$$g_k = g_k(0) + g_k'(0)z + g_k^{(2)}(0)\frac{z^2}{2!} + \cdots + g_k^{(r)}(0)\frac{z^r}{r!} + \cdots.$$

In order that we be able to form a linear combination of the g_k's to achieve any order between 0 and r at $z = 0$, it is necessary and sufficient that the square matrix

$$\begin{pmatrix} g_1(0) & g_1'(0) & g_1^{(2)}(0) & \cdots & g_1^{(r)}(0) \\ g_2(0) & g_2'(0) & g_2^{(2)}(0) & \cdots & g_2^{(r)}(0) \\ \vdots & \vdots & \vdots & \cdots & \vdots \\ g_{r+1}(0) & g_{r+1}'(0) & g_{r+1}^{(2)}(0) & \cdots & g_{r+1}^{(r1)}(0) \end{pmatrix}$$

be invertible.

The reader familiar with the elementary theory of ordinary differential equations will recognize the above matrix as the *Wronskian* of the set of functions $g_1, \ldots, g_{r+1}$, evaluated at $z = 0$. To be specific, if we have $r + 1$ functions $g_1, \ldots, g_{r+1}$, we define their *Wronskian* to be the function

$$W_z(g_1, \ldots, g_{r+1})(z) = \det \begin{pmatrix} g_1(z) & g_1'(z) & g_1^{(2)}(z) & \cdots & g_1^{(r)}(z) \\ g_2(z) & g_2'(z) & g_2^{(2)}(z) & \cdots & g_2^{(r)}(z) \\ \vdots & \vdots & \vdots & \cdots & \vdots \\ g_{r+1}(z) & g_{r+1}'(z) & g_{r+1}^{(2)}(z) & \cdots & g_{r+1}^{(r)}(z) \end{pmatrix},$$

which is holomorphic if every $g_i(z)$ is holomorphic. The subscript on the W is meant to indicate what variable one is taking all the derivatives with respect to.

The above discussion allows us to conclude the following.

LEMMA 4.3. *Let X be an algebraic curve and Q a linear system on X, corresponding to a subspace $V \subset L(D)$. Then a point $p \in X$ with local coordinate z is an inflection point for Q if and only if for any basis $\{f_1, \ldots, f_{r+1}\}$ for V, the Wronskian $W_z(z^{D(p)}f_1, \ldots, z^{D(p)}f_{r+1})$ is zero at p.*

Just as in the theory of O.D.E's, the Wronskian may be used to determine if a set of functions are linearly dependent.

LEMMA 4.4. *If $g_1, \ldots, g_{r+1}$ are linearly independent holomorphic functions defined in a neighborhood of $z = 0$, then the Wronskian $W_z(g_1, \ldots, g_{r+1})(z)$ is not identically zero near $z = 0$.*

We indicate a proof in the Problems.

COROLLARY 4.5. *For a fixed linear system Q on an algebraic curve X, there are only a finite number of inflection points.*

PROOF. As usual, we prove finiteness by proving discreteness. Fix a point $p \in X$; then there is a neighborhood U of p such that for all $q \in U$, we have $D(q) = 0$ if $q \neq p$. Fix a basis $\{g_1, \ldots, g_{r+1}\}$ for V. By the above analysis, we have that q is an inflection point for Q if and only if the Wronskian $W_z(g_1, \ldots, g_{r+1})$ is zero at q. Since the g_i's are linearly independent, this Wronskian is not identically zero; and since it is holomorphic, it has discrete zeroes. Hence after shrinking U there will be no inflection points in $U - \{p\}$. □

Higher-order Differentials. The natural question immediately arises: how many inflection points does a linear system Q have? This is a global matter, and to get to a satisfactory answer we must understand how the Wronskian globalizes.

Now the Wronskian is built up from the derivatives of the functions involved, and so one might expect forms to enter into the picture. However the derivatives in the Wronskian are of higher order than one, and so we should be suspicious that 1-forms are the right tool.

In fact the correct approach is to define *higher-order forms* on a Riemann surface. This we now briefly do, leaving all of the details to the reader.

DEFINITION 4.6. A *meromorphic n-fold differential* on an open set $V \subset \mathbb{C}$ is an expression μ of the form

$$\mu = f(z)(\mathrm{d}z)^n,$$

where f is a meromorphic function on V. We say that μ is a meromorphic n-fold differential *in the coordinate z*.

The language of "differentials" is used instead of "forms" to avoid confusion between the idea of n-fold differentials and n-forms.

The compatibility condition for meromorphic n-fold differentials is as follows:

DEFINITION 4.7. Suppose that $\mu_1 = f(z)(\mathrm{d}z)^n$ is a meromorphic n-fold differential in the coordinate z, defined on an open set V_1. Also suppose that $\mu_2 = g(w)(\mathrm{d}w)^n$ is a meromorphic n-fold differential in the coordinate w, defined on an open set V_2. Let $z = T(w)$ define a holomorphic mapping from the open set V_2 to V_1. We say that μ_1 *transforms to* μ_2 *under* T if $g(w) = f(T(w))T'(w)^n$.

Transporting the notion of meromorphic n-fold differentials from the complex plane to a Riemann surface is now done in the usual way:

DEFINITION 4.8. Let X be a Riemann surface. A *meromorphic n-fold differential* on X is a collection of meromorphic n-fold differentials $\{\mu_\phi\}$, one for each chart $\phi : U \to V$ in the variable of the target V, such that if two charts $\phi_i : U_i \to V_i$ (for $i = 1, 2$) have overlapping domains, then the associated meromorphic n-fold differential μ_1 transforms to μ_2 under the change of coordinate mapping $T = \phi_1 \circ \phi_2^{-1}$.

4. INFLECTION POINTS AND WEIERSTRASS POINTS

Of course as always it is enough to give a collection of meromorphic n-fold differentials on the charts of a single atlas for X.

Note immediately that a 1-fold differential is just a 1-form. More generally, if $\omega_1, \ldots, \omega_n$ are meromorphic 1-forms, we may define their *product* $\mu = \omega_1 \cdots \omega_n$ locally by the formula

$$\mu = f_1 \cdots f_n (\mathrm{d}z)^n,$$

where $\omega_i = f_i(z)\mathrm{d}z$ locally. In particular if ω is a meromorphic 1-form, its power ω^n is a meromorphic n-fold differential.

As promised, the Wronskian provides the main example of these objects.

LEMMA 4.9. *Let X be an algebraic curve, and let $g_1, \ldots, g_\ell$ be meromorphic functions on X. Then*

$$W_z(g_1(z), \ldots, g_\ell(z))(\mathrm{d}z)^{\ell(\ell-1)/2}$$

defines a meromorphic $\ell(\ell-1)/2$-fold differential on X.

PROOF. Since each g_i is meromorphic, so is the Wronskian $W_z(g_1, \ldots, g_\ell)$; hence we certainly have a meromorphic $\ell(\ell-1)/2$-fold differential locally. We must only check that these local formulas transform to each other as meromorphic n-fold differentials under changes of coordinates. Suppose that the change of coordinates mapping is $z = T(w)$. One can show easily by induction that for any k,

$$\frac{\mathrm{d}^k g_i(T(w))}{\mathrm{d}w^k} = T'(w)^k \frac{\mathrm{d}^k g_i(z)}{\mathrm{d}z^k} + \sum_{j=0}^{k-1} \alpha_{jk}(z) \frac{\mathrm{d}^j g_i(z)}{\mathrm{d}z^j},$$

where the functions $\alpha_{jk}(z)$ are holomorphic functions. Therefore the matrix $(\mathrm{d}^k g_i(T(w))/\mathrm{d}w^k)$, whose determinant is the Wronskian W_w in the w coordinate, can be brought via column operations to the matrix $(T'(w)^k \mathrm{d}^k g_i(z)/\mathrm{d}z^k)$. Factoring out from each column the factor $T'(w)^k$, we are left with the matrix whose determinant is the Wronskian W_z in the z coordinate. The total number of $T'(w)$ factors is $0 + 1 + \cdots + (\ell-1) = \ell(\ell-1)/2$, and so

$$W_w = \det\left(\frac{\mathrm{d}^k g_i(T(w))}{\mathrm{d}w^k}\right) = T'(w)^{\ell(\ell-1)/2} \det\left(\frac{\mathrm{d}^k g_i(z)}{\mathrm{d}z^k}\right) = T'(w)^{\ell(\ell-1)/2} W_z$$

proving the result. □

This meromorphic differential defined locally by the Wronskian will be denoted by $W(g_1, \ldots, g_\ell)$, without the subscript.

As is the case for meromorphic functions and meromorphic 1-forms, at any point p the *order* of a meromorphic n-fold differential is well defined, by

$$\mathrm{ord}_p(f(z)(\mathrm{d}z)^n) = \mathrm{ord}_p(f(z)).$$

Moreover the same language of zeroes and poles is used as usual. This allows us to define the *divisor* $\mathrm{div}(\mu)$ of a meromorphic n-fold differential μ to be

$$\mathrm{div}(\mu) = \sum_p \mathrm{ord}_p(\mu) \cdot p.$$

For any divisor D on X, we denote by $L^{(n)}(D)$ the vector space of meromorphic n-fold differentials whose poles are bounded by D:

$$L^{(n)}(D) = \{\text{meromorphic } n\text{-fold differentials } \mu \mid \mathrm{div}(\mu) \geq -D\}.$$

We will need to know in which of these spaces the Wronskian lies.

LEMMA 4.10. *Let X be an algebraic curve, D a divisor on X, and let $f_1, \ldots, f_\ell$ be meromorphic functions in $L(D)$. Then the meromorphic n-fold differential $W(f_1, \ldots, f_\ell)$ has poles bounded by ℓD:*

$$W(f_1, \ldots, f_\ell) \in L^{(\ell(\ell-1)/2)}(\ell D).$$

PROOF. This lemma is simply based on the remark that if all the g_i's are holomorphic, so is the Wronskian. Fix a point $p \in X$ with local coordinate z. Then for each i, $\mathrm{ord}_p(g_i) \geq -D(p)$, so that $z^{D(p)} g_i$ is holomorphic at p. Hence the Wronskian $W_z(z^{D(p)} g_1, \ldots, z^{D(p)} g_\ell)$ is holomorphic at p. But the Wronskian is multilinear, so that $W_z(z^{D(p)} g_1, \ldots, z^{D(p)} g_\ell) = z^{\ell D(p)} W_z(g_1(z), \ldots, g_\ell(z))$. Since this is holomorphic at p, we have that $\mathrm{ord}_p(W(g_1(z), \ldots, g_\ell(z))) \geq -\ell D(p)$ as claimed. □

The Number of Inflection Points. Returning to the problem of computing inflection points of a linear system $Q \subseteq |D|$, assume that $\dim Q = r$ and that $\{f_1, \ldots, f_{r+1}\}$ is a basis for the corresponding subspace $V \subseteq L(D)$.

Consider the Wronskian $W(f_1, \ldots, f_{r+1})$. If one changes the basis of V, then the Wronskian changes by the determinant of the change of basis matrix, and so the Wronskian is well defined (up to scalar constant) by the linear system Q itself, and not by the choice of basis. We will therefore denote it by $W(Q)$ when convenient; Lemma 4.10 then implies that

$$W(Q) \in L^{(r(r+1)/2)}((r+1)D)$$

if $r = \dim Q$.

Counting the zeroes and poles of $W(Q)$ is based on the following remark, which is the analogue of Lemma 3.11 of Chapter V.

LEMMA 4.11. *Let X be an algebraic curve, D a divisor on X, and $K = \mathrm{div}(\omega)$ a canonical divisor. Then the multiplication map*

$$\zeta : L(D + nK) \to L^{(n)}(D)$$

defined by $\zeta(f) = f\omega^n$ is an isomorphism of vector spaces.

4. INFLECTION POINTS AND WEIERSTRASS POINTS

PROOF. Of course since ω is a meromorphic 1-form, and f is a meromorphic function, then $f\omega^n$ is a meromorphic n-fold differential. Moreover the multiplication map is clearly linear in f, and is 1-1.

To show that $f\omega^n$ has poles bounded by D, fix a point $p \in X$ and a local coordinate z at p, and write $\omega = g(z)dz$. Then $\omega^n = g(z)^n(dz)^n$, so that

$$\operatorname{ord}_p(f\omega^n) = \operatorname{ord}_p(f) + n\operatorname{ord}_p(g) = \operatorname{ord}_p(f) + nK(p) \geq -D(p)$$

if $f \in L(D + nK)$; hence we see that ζ does map $L(D + nK)$ to $L^{(n)}(D)$.

Finally to see that ζ is onto, we note that if $\mu = h(z)(dz)^n \in L^{(n)}(D)$, and $\omega = g(z)dz$, then $f = h/g^n$ is a meromorphic function in $L(D + nK)$, which is defined *globally*. □

This allows us to compute the sum of the orders of the Wronskian:

COROLLARY 4.12. *Let X be an algebraic curve and Q a linear system on X with $r = \dim Q$. Then*

$$\deg(\operatorname{div}(W(Q))) = \sum_p \operatorname{ord}_p(W(Q)) = r(r+1)(g-1).$$

PROOF. Let $n = r(r+1)/2$, so that by Lemma 4.10 we have that the Wronskian differential $W(Q)$ is an element of the space $L^{(n)}((r+1)D)$. Then by Lemma 4.11 there is a meromorphic 1-form ω and a meromorphic function f such that $W(Q) = f\omega^n$. Then

$$\begin{aligned}
\sum_p \operatorname{ord}_p(W(Q) &= \sum_p \operatorname{ord}_p(f\omega^n) \\
&= \sum_p [\operatorname{ord}_p(f) + n\operatorname{ord}_p(\omega)] \\
&= n\sum_p \operatorname{ord}_p(\omega) \quad (\text{since } \sum_p \operatorname{ord}_p(f) = 0) \\
&= n(2g-2) = r(r+1)(g-1),
\end{aligned}$$

using the fact that $\deg(\operatorname{div}(\omega)) = 2g - 2$. □

In order to obtain a good formula, we must have a computation of the order $\operatorname{ord}_p(W(Q))$, and relate this to the gap numbers for an inflection point. Fix a point $p \in X$, with a local coordinate z. If $\{f_1, \ldots, f_{r+1}\}$ is a basis for V, recall that by Lemma 4.3, p is an inflection point for $|D|$ if and only if the Wronskian $W_z(z^{D(p)}f_1, \ldots, z^{D(p)}f_{r+1})$ is zero at p. Since

$$\begin{aligned}
\operatorname{ord}_p(W_z(z^{D(p)}f_1, \ldots, z^{D(p)}f_{r+1})) &= \operatorname{ord}_p(z^{(r+1)D(p)}W_z(f_1, \ldots, f_{r+1})) \\
&= (r+1)D(p) + \operatorname{ord}_p(W(Q)),
\end{aligned}$$
(4.13)

we have a clean link between the order of vanishing of $W_z(z^{D(p)}f_1, \ldots, z^{D(p)}f_{r+1})$ (which measures inflectionary behaviour) and $\operatorname{ord}_p(W(Q))$, for which we have a global formula.

The final ingredient is provided by the

LEMMA 4.14. *If $G_p(Q) = \{n_1 < n_2 < \cdots < n_{r+1}\}$, and $\{f_1, \ldots, f_{r+1}\}$ is a basis for V, then*

$$\operatorname{ord}_p(W_z(z^{D(p)}f_1, \ldots, z^{D(p)}f_{r+1})) = \sum_{i=1}^{r+1}(n_i - i).$$

PROOF. To compute the order of the Wronskian, we may choose any basis for V, and it is convenient to choose an inflectionary basis; hence we may assume that for each i, $f_i = z^{n_i - 1 - D(p)} + \ldots$; setting $g_i = z^{D(p)}f_i$, then it is the order of

$$W(z) = W_z(g_1, \ldots, g_{r+1}) = W_z(z^{n_1-1} + \ldots, \ldots, z^{n_{r+1}-1} + \ldots)$$

which we must compute.

The lowest term of the Taylor series for the determinant which computes this Wronskian may be obtained by considering the determinant of the lowest terms of the entries only, if this determinant is not zero. In other words, if

$$Y(z) = W_z(z^{n_1-1}, \ldots, z^{n_{r+1}-1})$$

has its lowest possible term nonzero, then this lowest term is also the lowest term of $W(z)$.

Expanding the determinant for $Y(z)$ into the usual sum of products over permutations, one sees readily that that *every* term of the sum is a monomial whose exponent is $\sum_{i=1}^{r+1}(n_i - i)$. Hence the determinant for $Y(z)$ is a single monomial, and moreover the coefficient of this monomial is itself a determinant, of the matrix N whose j^{th} column N_j is

$$N_j = \begin{pmatrix} 1 \\ n_j - 1 \\ (n_j - 1)(n_j - 2) \\ \vdots \\ (n_j - 1)(n_j - 2) \cdots (n_j - r - 1) \end{pmatrix}.$$

It is a standard exercise to show that N is invertible when all of the integers n_i are distinct; the matrix N is a relative of a Vandermonde matrix, and we leave the details to the reader. Therefore the determinant of N is nonzero, proving that the lowest term of $W(z)$ is the term whose exponent is $\sum_{i=1}^{r+1}(n_i - i)$, which proves the result. □

With this in hand, it is natural to define the *inflectionary weight* of a point p with respect to a linear system Q to be the sum

$$w_p(Q) = \sum_{i=1}^{r+1}(n_i - i),$$

where the gap numbers are $G_p(Q) = \{n_1 < n_2 < \cdots < n_{r+1}\}$. We note that the inflectionary weight is positive exactly when the point is an inflection point.

Putting everything together yields the following.

THEOREM 4.15. *Let X be an algebraic curve of genus g, and let Q be a g_d^r on X, that is, a linear system on X of degree d, with $r = \dim Q$. Then*

$$\sum_{p \in X} w_p(Q) = (r+1)(d + rg - r).$$

PROOF. Choose a basis $f_1, \ldots, f_{r+1}$ for the subspace V of $L(D)$ corresponding to Q. We compute

$$\begin{aligned}
\sum_p w_p(Q) &= \sum_p \mathrm{ord}_p(W_z(z^{D(p)}f_1, \ldots, z^{D(p)}f_{r+1})) \quad (\text{by Lemma 4.14}) \\
&= \sum_p [(r+1)D(p) + \mathrm{ord}_p(W(Q))] \quad (\text{using (4.13)}) \\
&= (r+1)d + r(r+1)(g-1) = (r+1)(d + rg - r)
\end{aligned}$$

using Corollary 4.12. □

Flex Points of Smooth Plane Curves. We can use the inflectionary weight formula of Theorem 4.15 to compute the number of flex points to a smooth projective plane curve X of degree d. Recall that a line H in $\mathbb{P}^2$ is said to meet X at p with multiplicity k if $\mathrm{div}(H)(p) = k$.

This multiplicity is related to the inflectionary weight as follows. Let Q be the hyperplane linear system, which is a g_d^2 on X. For any point $p \in X$, since p is not a base point of Q, we have $n_1 = 1$. Moreover since the general line through p is not tangent to X at p, we have $n_2 = 2$. The linear subsystem $Q(-p)$ consists of the divisors of the lines through p, and the system $Q(-2p)$ consists of the single divisor of the tangent line to X at p. If that tangent line H meets X at p with multiplicity k, then $\mathrm{div}(H) \in Q(-kp)$ but $\mathrm{div}(H) \notin Q(-(k+1)p)$: $Q(-(k+1)p)$ is the empty linear system. We conclude that the 3 gap numbers of $G_p(Q)$ must be $n_1 = 1$, $n_2 = 2$, and $n_3 = k + 1$.

Therefore the inflectionary weight $w_p(Q) = k - 2$. Recalling that Plücker's formula says that $g = (d-1)(d-2)/2$, Theorem 4.15 now gives the following.

COROLLARY 4.16. *Let X be a smooth projective plane curve of degree d. Then X has exactly $3d(d-2)$ flex points, where a flex point p whose tangent line meets X at p with multiplicity k is counted $k - 2$ times.*

Weierstrass Points. There is of course one linear system on any algebraic curve which enjoys a special position, and that is the canonical linear system. To study the inflectionary behaviour of points for the canonical system is to study the tangent hyperplanes to the canonical curve, at least when the curve is not hyperelliptic.

Inflection points for the canonical linear system have a special name: they are called the *Weierstrass points* for the curve. The *weight* of a Weierstrass point is its inflectionary weight for the canonical linear system. Since $\dim |K| = g - 1$, each point $p \in X$ has exactly g gap numbers for the canonical linear system.

Since the canonical linear system has dimension $g - 1$ and degree $2g - 2$, Theorem 4.15 gives immediately a count for the number of Weierstrass points:

COROLLARY 4.17. *Let X be an algebraic curve of genus $g \geq 1$. Then there are*

$$g^3 - g = g(g-1)(g+1)$$

Weierstrass points on X, each counted according to their weight.

The Weierstrass points are a set of marked points on X; they have properties which the other points of X do not. This is because no choices are necessary to construct the canonical linear system on X. The matter perhaps becomes clearer after applying Riemann-Roch, which we now explain.

Fix a point $p \in X$. Then an integer n is a gap number for $|K|$ at p if and only if $L(K - np) \neq L(K - (n-1)p)$. By Riemann-Roch, we have

$$\dim L(K - \ell p) = (2g - 2 - \ell) + 1 - g + \dim L(\ell p) = g - 1 - \ell + \dim L(\ell p)$$

for every ℓ. Therefore

$$\begin{aligned}
\dim L(K - (n-1)p) - \dim L(K - np) &= [g - 1 - (n-1) + \dim L((n-1)p)] \\
&\quad - [g - 1 - n + \dim L(np)] \\
&= 1 + \dim L((n-1)p) - \dim L(np),
\end{aligned}$$

so that n is a gap number for $|K|$ at p if and only if $L((n-1)p) = \dim L(np)$. Therefore the gap numbers for $|K|$ at p may be computed by looking at the nested sequence of vector spaces of meromorphic functions

$$\{0\} \subset L(0) \subseteq L(p) \subseteq L(2p) \subseteq \cdots \subseteq L((n-1)p) \subseteq L(np) \subseteq \cdots.$$

At each stage, passing from $L((n-1)p)$ to $L(np)$, the dimension either stays the same, or increases by 1. When it stays the same, n is a gap number for $|K|$ at p. Therefore we see that n is a gap number for $|K|$ at p if and only if there is no meromorphic function f with a pole of order exactly n at p and having no other poles.

Eventually, indeed for $n \geq 2g - 1$, $H^1(np) = 0$ and $\dim L(np) = n + 1 - g$, so that the dimensions increase by one from then on. They start with $\dim L(0 \cdot p) = 1$, so that there are exactly g integers where the dimension does *not* increase, and these are the set of gap numbers $G_p(|K|)$ for $|K|$ at p.

In any case we see that p is a Weierstrass point on X if and only if the set of gap numbers $G_p(|K|)$ is *not* equal to the set $\{1, 2, \ldots, g\}$ of the first g integers. This exactly means that there is no increase all the way from $L(0)$ to

$L(gp)$. Hence a point p is a Weierstrass point on X if and only if $L(gp)$ has a nonconstant function in it, or, equivalently, if and only if $\dim L(gp) \geq 2$.

The set of gap numbers for the canonical system enjoys the following special property: their complement (inside the natural numbers) forms a semigroup under addition. That is, if n and m are *not* gap numbers in $G_p(|K|)$, then neither is their sum $n+m$. This is a result of the interpretation in terms of the existence of meromorphic functions: if n and m are not gap numbers, then there are functions f and g with poles of order exactly n and m respectively and having no other poles. Then the product fg has a pole of order $n+m$ at p and no other poles, so that $n+m$ is not a gap number.

Weierstrass points, since they are intrinsic to the curve, must be permuted by any automorphism. This remark is the basis for the

THEOREM 4.18. *An algebraic curve X of genus at least two has only finitely many automorphisms.*

PROOF. We will only sketch the proof, leaving the details to the interested reader. First one shows separately that if X is hyperelliptic, then any automorphism must commute with the canonical double covering $\pi : X \to \mathbb{P}^1$, and so descend to an automorphism of $\mathbb{P}^1$ which permutes the $2g+2$ branch points. Therefore we obtain a homomorphism $\rho : \text{Aut}(X) \to S_{2g+2}$, it is easy to see that the kernel has order two, generated by the canonical involution for π.

If X is not hyperelliptic, then it is an exercise to check that there must be $k \geq 2g+6$ Weierstrass points. Any automorphism must permute these, and so again there is a homomorphism $\rho : \text{Aut}(X) \to S_k$. Any automorphism in the kernel fixes the k Weierstrass points. Hence by Corollary 2.10, the kernel is trivial, and so $\text{Aut}(X)$ is finite. □

Problems VII.4

A. Prove Lemma 4.2.

B. Show that if D is very ample, then $2 \in G_p(|D|)$.

C. If D is very ample, inducing the holomorphic embedding $\phi_D : X \to \mathbb{P}^n$, show that p is a flex point of the image curve if and only if $3 \notin G_p(|D|)$.

D. If X has genus zero, and $\deg(D) = d$, show that $G_p(|D|) = \{1, 2, \ldots, d+1\}$.

E. If X has genus one, and $\deg(D) = d$, show that $G_p(|D|) = \{1, 2, \ldots, d\}$ unless $d \cdot p \sim D$; in this case, show that $G_p(|D|) = \{1, 2, \ldots, d-1, d+1\}$. Show that there are exactly d^2 such points on X.

F. Let X be a hyperelliptic curve, with the canonical double covering map $F : X \to \mathbb{P}^1$. Let D be an inverse image divisor of F, of degree 2. Show that $G_p(|D|) = \{1, 2\}$ if $\text{mult}_p(F) = 1$, while $G_p(|D|) = \{1, 3\}$ if $\text{mult}_p(F) = 2$. Note that there are exactly $2g+2$ such points on X.

G. Show that $W_z(hg_1, \ldots, hg_\ell) = h^\ell W_z(g_1, \ldots, g_\ell)$.

H. Let $g_1, \ldots, g_\ell$ be holomorphic, and for integers $n_1, \ldots, n_\ell$, define the function

$D(n_1, \ldots, n_\ell)$ to be the determinant

$$D(n_1, \ldots, n_\ell) = \det \begin{pmatrix} g_1^{(n_1)} & g_1^{(n_2)} & \cdots & g_1^{(n_\ell)} \\ g_2^{(n_1)} & g_2^{(n_2)} & \cdots & g_2^{(n_\ell)} \\ \vdots & \vdots & \cdots & \vdots \\ g_\ell^{(n_1)} & g_\ell^{(n_2)} & \cdots & g_\ell^{(n_\ell)} \end{pmatrix}.$$

Note then that the Wronskian is $W_z(g_1, \ldots, g_\ell) = D(0, 1, 2, \ldots, \ell - 1)$. Show the following "product rule": the derivative of the function $D(n_1, \ldots, n_\ell)$ is the sum

$$D(n_1, \ldots, n_\ell)' = \sum_{i=1}^\ell D(n_1, \ldots, n_{i-1}, n_i + 1, n_{i+1}, \ldots, n_\ell).$$

I. Assume that the functions g_i are all holomorphic and that the Wronskian $W_z(g_1, \ldots, g_\ell)$ is identically zero in a neighborhood of $z = 0$. Then all derivatives of W_z must be zero at $z = 0$. Use the product rule above to conclude that all of the vectors

$$v_n = \begin{pmatrix} g_1^{(n)}(0) \\ g_2^{(n)}(0) \\ \vdots \\ g_\ell^{(n)}(0) \end{pmatrix}$$

must lie in a $(\ell - 1)$-dimensional subspace of $\mathbb{C}^\ell$.

J. Continuing with the previous problem, conclude that there are constants c_i such that $\sum_i c_i g_i^{(n)}(0) = 0$ for every $n \geq 0$. Conclude that $g(z) = \sum_i c_i g_i(z)$ is identically zero, so that the functions $\{g_i\}$ are linearly dependent.

K. Check that if $\omega_1, \ldots, \omega_n$ are meromorphic 1-forms, defined locally by $\omega_i = f_i(z)dz$, then their product $\mu = \omega_1 \cdots \omega_n$ defined locally by the formula

$$\mu = f_1 \cdots f_n (dz)^n$$

is a meromorphic n-fold differential.

L. Check that if A is an invertible matrix of constants and $(f_1, \ldots, f_\ell) = (g_1, \ldots, g_\ell)A$, then $W(f_1, \ldots, f_\ell) = \det(A) W(g_1, \ldots, g_\ell)$.

L. Check the details of the proof of Lemma 4.14.

M. Show that for a base-point-free pencil Q (that is, a g_d^1), defining a holomorphic map $\phi_Q : X \to \mathbb{P}^1$, we have $w_p(Q) = \text{mult}_p(\phi_Q) - 1$. Show that Theorem 4.15 gives exactly Hurwitz's formula in this case.

N. Show that if Q is a g_d^r on an algebraic curve X, such that $w_p(Q) = 0$ for all points p, then X has genus zero, $r = d + 1$, and Q is the complete linear system of divisors of degree d. Hence the only curve $X \subset \mathbb{P}^d$ with no inflection points (for the hyperplane linear system) is the rational normal curve of degree d.

O. Show that a smooth plane cubic curve has exactly 9 inflection points, each of whose tangent line meets it with multiplicity 3. Find these inflection points for the curve given by $y^2 z = x^3 - xz^2$.

P. Show that if a linear system Q has the divisor F as its fixed part, and if $p \in X$, then the first gap number for Q at p is $n_1 = F(p) + 1$.

Q. Consider the projective line $X = \mathbb{P}^1$, and the complete linear system of divisors of degree 3 on X. Suppose that Q is a linear subsystem of dimension r, that is, Q is a g_3^r on $\mathbb{P}^1$.

1. If $r = 3$ show that there are no inflection points for Q.
2. If $r = 2$ and Q is base-point-free show that there are either three inflection points, each with gap numbers $\{1, 2, 4\}$, or two inflection points, one with gap numbers $\{1, 2, 4\}$ and one with gap numbers $\{1, 3, 4\}$. Show that in the first case the associated holomorphic map ϕ_Q maps $\mathbb{P}^1$ onto a cubic plane curve with a single node, and in the second case onto a cubic plane curve with a single cusp.
3. If $r = 1$ and Q is base-point-free show that there can be either 2, 3, or 4 inflection points, with inflectionary weights 1 or 2, summing to 4. Give examples of all the numerical possibilities.
4. What can happen if Q has base points?

R. Let X be an algebraic curve of genus $g \geq 2$.

1. Show that $n \in G_p(|K|)$ if and only if there is a holomorphic 1-form ω on X with $\mathrm{ord}_p(\omega) = n - 1$.
2. Show that $2 \notin G_p(|K|)$ if and only if X is hyperelliptic and p is a ramification map for the double covering $\pi : X \to \mathbb{P}^1$. If so, show that $G_p(|K|) = \{1, 3, 5, \ldots, 2g - 1\}$.
3. Conversely, show that if X is hyperelliptic, then there are exactly $2g + 2$ Weierstrass points on X, each having this set (of the first g odd numbers) as the set of gap numbers.

S. Let X be a nonhyperelliptic curve of genus $g \geq 3$. Write $G_p(|K|) = \{n_1 < n_2 < \cdots < n_g\}$.

1. Show that $n_1 = 1$.
2. Show that $n_i \leq 2i - 2$ for every $i \geq 2$.
3. Conclude that the canonical inflectionary weight $w(p) = \sum_{i=1}^{g}(n_i - i)$ is at most $(g-1)(g-2)/2$.
4. Show that there are at least $2g + 6$ Weierstrass points on X.

Further Reading

All of the "elementary" applications of Riemann-Roch given in the first section are completely standard and are discussed in most texts. There are several proofs of Clifford's Theorem; the reader may want to compare the one here with those in [**Hartshorne77**], [**G-H78**], and [**ACGH85**]. For more on counting parameters, leading to Riemann's count of $3g - 3$ parameters for curves of genus g, see [**G-H78**]. The dimension theorem for complex manifolds is proved there; in the algebraic category it is proved for example in [**Hartshorne77**] and [**Shafarevich77**], and in the real category one may consult [**Warner71**].

One's education in the subject of curves is not complete without having spent some time with Mumford's excellent monograph [**Mumford78**]; in the first two lectures are discussed curves and moduli spaces.

We have approached the Castelnuovo theory from the same point of view as in [**ACGH85**]; it is also discussed in [**G-H78**], [**Harris80**], [**Harris82**], and [**Narasimhan92**]. The "primary" reference is [**Castelnuovo1889**]. This has become a very active area of current research, and the literature is beginning to be quite large. It is not an overstatement to say that more has been done in Castelnuovo theory in the last 15 years than in the previous 85.

Most of the discussion of inflection points does not require the Riemann-Roch Theorem; only in the application to Weierstrass points is it used. A proof of the finiteness of automorphisms can be read in [**G-H78**] or [**ACGH85**], for example.

Chapter VIII. Abel's Theorem

1. Homology, Periods, and the Jacobian

The First Homology Group. Recall from Chapter III the definition of the first homology group of a compact Riemann surface X, as the quotient of the group of closed chains on X modulo the subgroup of boundary chains:

$$H_1(X, \mathbb{Z}) = \operatorname{CLCH}(X)/\operatorname{BCH}(X).$$

For a compact Riemann surface X of genus g, this is a free abelian group of rank $2g$. A standard set of generators for this group can be obtained using a certain representation of X as a polygon with $4g$ sides, appropriately identified in pairs.

The Standard Identified Polygon. Let $\mathcal{P} = \mathcal{P}_g$ be a polygon with $4g$ sides, labeled (in counterclockwise order) as a_i, b_i, a_i', b_i' as i runs from 1 to g. Direct these sides so that a_i and b_i are directed counterclockwise, and a_i' and b_i' clockwise, for each i. A compact orientable 2-manifold of genus g is obtained by identifying the sides a_i with a_i' and b_i with b_i' for every i, in the given directions. We call this representation of a surface of genus g the *standard identified polygon* representation.

Note that all $4g$ vertices on the boundary of $\mathcal{P}$ are identified to one point of X. Therefore the curves a_i and b_i considered on X are closed paths. These closed paths, considered as closed chains on X, freely generate the first homology group $H_1(X, \mathbb{Z})$.

Periods of 1-Forms. Let ω be a closed $\mathcal{C}^\infty$ 1-form on X. If D is any triangulated subset of X, then using Stoke's theorem we have

$$\int_{\partial D} \omega = \iint_D d\omega = \iint_D 0 = 0.$$

Therefore the integral of ω around any boundary chain is 0. Hence the integrals of ω around any closed chain only depends on the homology class of the chain, and so for any homology class $[c] \in H_1(X, \mathbb{Z})$, the integral

$$\int_{[c]} \omega = \int_c \omega$$

is well defined.

In particular, since every holomorphic 1-form is closed, we see that integrals of holomorphic 1-forms around homology classes in $H_1(X,\mathbb{Z})$ are well defined. Hence for every homology class $[c]$, we obtain a well defined functional on the space $\Omega^1(X)$ of holomorphic 1-forms, given by integration around c:

$$\int_{[c]} : \Omega^1(X) \to \mathbb{C}.$$

DEFINITION 1.1. A linear functional $\lambda : \Omega^1(X) \to \mathbb{C}$ is a *period* if it is $\int_{[c]}$ for some homology class $[c] \in H_1(X,\mathbb{Z})$.

We note that the set Λ of periods forms a subgroup of the dual space $\Omega^1(X)^*$.

The Jacobian of a Compact Riemann Surface. The quotient space of functionals on $\Omega^1(X)$ modulo the period subgroup is of fundamental importance in the study of compact Riemann surfaces.

DEFINITION 1.2. Let X be a compact Riemann surface. The *Jacobian* of X, denoted by $\mathrm{Jac}(X)$, is the quotient group

$$\mathrm{Jac}(X) = \frac{\Omega^1(X)^*}{\Lambda}$$

of functionals on the space of holomorphic 1-forms modulo periods.

We can be more down-to-earth and describe the Jacobian using bases. Let $\omega_1, \ldots, \omega_g$ be a basis for $\Omega^1(X)$. This allows us to identify the dual space $\Omega^1(X)^*$ with the space of column vectors $\mathbb{C}^g$, by associating a functional λ to the column vector $(\lambda(\omega_1), \ldots, \lambda(\omega_g))^\top$. In particular, every period is associated to such a vector: the period corresponding to the homology class of a closed chain c is the vector $(\int_c \omega_1, \ldots, \int_c \omega_g)^\top$. In this way we have

$$\mathrm{Jac}(X) \cong \frac{\mathbb{C}^g}{\Lambda},$$

where here Λ is the subgroup of $\mathbb{C}^g$ associated to the periods.

In any case, we note that $\mathrm{Jac}(X)$ is an abelian group.

EXAMPLE 1.3. The Jacobian of the Riemann Sphere is trivial: $\mathrm{Jac}(\mathbb{C}_\infty) = \{0\}$. Indeed, the space of holomorphic 1-forms $\Omega^1(\mathbb{C}_\infty)$ is trivial.

The next example is more serious; we leave it to the reader to check the details.

EXAMPLE 1.4. Let X be a complex torus $\mathbb{C}/L$. Then $\mathrm{Jac}(X) \cong X$.

Problems VIII.1

A. Show that all $4g$ vertices on the boundary of the standard identified polygon $\mathcal{P}$ are identified to a single point.

B. Show that the Jacobian of a complex torus $X = \mathbb{C}/L$ is isomorphic to X by explicitly showing that the subgroup of periods $\Lambda \subset \mathbb{C}$ is a lattice which is homothetic to the defining lattice L for X, i.e., there is a nonzero complex number μ such that $\mu\Lambda = L$.

C. Let X be a hyperelliptic curve given with a degree 2 holomorphic map $F : X \to \mathbb{P}^1$. Let b_1 and b_2 be two of the branch points of F in $\mathbb{P}^1$, and let γ be a path in $\mathbb{P}^1$ starting at b_1 and ending at b_2, not passing through any other branch points of F. If r_1 and r_2 are the points of X lying above b_1 and b_2 respectively, then the path γ lifts to two paths γ_1 and γ_2 from r_1 to r_2. Hence $\gamma_1 - \gamma_2$ is a closed chain on X. Show that such closed chains generate the first homology group $H_1(X, \mathbb{Z})$.

D. Using the ideas of the previous problem, compute the subgroup of periods of the genus two hyperelliptic curve defined by $y^2 = x^6 - 1$.

2. The Abel-Jacobi Map

In order to fully exploit the construction of the Jacobian of an algebraic curve X, we need to more explicitly relate the Jacobian to X itself. This is provided by the Abel-Jacobi map.

The Abel-Jacobi Map A on X. Choose a base point p_0 on the compact Riemann surface X. For each point $p \in X$, choose a path γ_p on X from p_0 to p. Define the map

$$A : X \to \Omega^1(X)^*$$

by sending p to the functional given by integration along γ_p:

$$A(p)(\omega) = \int_{\gamma_p} \omega.$$

This function is not well defined: if one chooses a different path γ'_p from p_0 to p, then the value of $A(p)$ changes by the functional which is integration around the closed chain $\gamma_p - \gamma'_p$. In other words, $A(p)$ is well defined modulo the subgroup of periods. Therefore we do obtain a well defined map

$$A : X \to \text{Jac}(X).$$

This map is called the *Abel-Jacobi map* for X. It depends on the base point p_0.

If we choose a basis $\{\omega_i\}$ for $\Omega^1(X)$, we may consider the Abel-Jacobi map A as mapping to $\mathbb{C}^g/\Lambda$, by the formula

$$A(p) = \Big(\int_{p_0}^p \omega_1, \ldots, \int_{p_0}^p \omega_g\Big)^\top \mod \Lambda.$$

In this setting recall that Λ is the subgroup of periods in $\mathbb{C}^g$.

The Extension of A to Divisors. We may extend the Abel-Jacobi map from X to the group $\text{Div}(X)$ of divisors on X, by defining

$$A(\sum n_p p) = \sum n_p A(p).$$

This gives us a group homomorphism, also called the Abel-Jacobi map, and also denoted by A:

$$A : \text{Div}(X) \to \text{Jac}(X).$$

Of more immediate importance is the restriction of this map to the subgroup of divisors of degree 0 on X:

$$A_0 : \text{Div}_0(X) \to \text{Jac}(X).$$

Independence of the Base Point. The following lemma shows that the choice of base point in the definition of the Abel-Jacobi map is irrelevant for A_0.

LEMMA 2.1. *The Abel-Jacobi map A_0 defined on divisors of degree 0 is independent of the chosen base point p_0 on X.*

PROOF. Suppose a new base point p_0' is chosen. Let γ be a path from p_0 to p_0' on X. Then in the formula for $A(p)$, if we change the base point from p_0 to p_0', we see that the image vector changes exactly by $j = (\int_\gamma \omega_1, \ldots, \int_\gamma \omega_g)^\top$ mod Λ. This element $j \in \text{Jac}(X)$ is independent of p; hence if $\sum n_p = 0$, then $A(\sum n_p p)$ changes by $\sum n_p j = j \sum n_p = 0$. □

Statement of Abel's Theorem. We are now in a position to state the main theorem of this chapter. Recall that if f is a meromorphic function on a compact Riemann surface X, then its divisor $\text{div}(f)$ (a principal divisor) has degree 0. We have seen that in general this is not a sufficient condition however for a divisor to be a principal divisor. Abel's Theorem gives us the sharp criterion.

THEOREM 2.2 (ABEL'S THEOREM). *Let X be a compact Riemann surface of genus g. Let D be a divisor of degree 0 on X. Then D is the divisor of a meromorphic function on X if and only if $A_0(D) = 0$ in $\text{Jac}(X)$.*

The proof of Abel's theorem will occupy the rest of this chapter.

Problems VIII.2
A. Show that the Abel-Jacobi map $A : X \to \text{Jac}(X)$ is an isomorphism of groups when X is a complex torus $\mathbb{C}/L$, and when the base point p_0 is chosen to be 0 mod L, by explicitly computing the integrals.
B. Reconcile the statement given above for Abel's Theorem in general with the versions of Abel's Theorem previously stated for the Riemann Sphere and a complex torus.
C. Show that the chosen base point p_0 is sent by the Abel-Jacobi map A to the origin of $\text{Jac}(X)$.

D. Suppose that $A : X \to \mathrm{Jac}(X)$ is defined with a base point p_0, and $A' : X \to \mathrm{Jac}(X)$ is defined with a base point p'_0. Show that $A' = \tau \circ A$, where τ is the translation (in the group law of the Jacobian $\mathrm{Jac}(X)$) by the element $A(p'_0)$: $\tau(j) = j - A(p'_0)$ for $j \in \mathrm{Jac}(X)$.

3. Trace Operations

Let $F : X \to Y$ be a nonconstant holomorphic map between compact Riemann surfaces. Given any kind of function or form on Y, say σ, there is a very natural notion of "pullback" which constructs a corresponding function or form $F^*\sigma$ on X.

In this section we will develop a converse operation, associating to a function or form on X its *trace* on Y.

The Trace of a Function. Let h be a meromorphic function on X. Suppose that $q \in Y$, which is not a branch point of F. If F has degree d, then there are exactly d preimages of q in X; call them $p_1, \ldots, p_d$. Moreover, since the multiplicity of F at each p_i is one, there is a chart domain U containing q such that $F^{-1}(U)$ is the disjoint union of chart domains $V_1, \ldots, V_d$, with $F|_{V_i}$ biholomorphic for each i.

Let $\phi_i : U \to V_i$ be the inverse of $F|_{V_i}$. Define a function $\mathrm{Tr}(h)$ on U, called the *trace* of h, by

$$\mathrm{Tr}(h) = \sum_{i=1}^{d} h \circ \phi_i = \sum_{i=1}^{d} \phi_i^*(h|_{V_i}).$$

In other words, the trace function evaluated at a point $q \in Y$ is simply the sum over the preimages of the original function on X:

$$\mathrm{Tr}(h)(q) = \sum_{p \in F^{-1}(q)} h(p);$$

this at least is true if h has no poles at the preimages of q. We note that if h is holomorphic at each p_i, then $\mathrm{Tr}(h)$ is holomorphic at q.

We leave it as an exercise to check that this gives a well defined meromorphic function $\mathrm{Tr}(h)$ on the complement of the branch locus in Y.

What happens at the branch points of F? By the classification of singularities, we have that $\mathrm{Tr}(h)$ has either an essential singularity, a pole, or a removable singularity at the branch points. The claim is that the branch points are at worst poles of $\mathrm{Tr}(h)$, and if h is holomorphic at each preimage point, then again $\mathrm{Tr}(h)$ is holomorphic.

The critical computation to make is to assume that the branch point q has a single preimage p, with multiplicity m. By the local normal form for F, we may choose a local coordinate z centered at q and w centered at p such that the map F has the form $z = w^m$.

Assume then that h has a Laurent series $\sum_n c_n w^n$ at p. Let $\zeta = \exp(2\pi i/m)$. Then for nonzero z's, the value of $\mathrm{Tr}(h)(z)$ is obtained by summing the values of the preimages, as noted above. The preimages of any nonzero $z = w^m$ are the points $\zeta^i w$, for $i = 0, \ldots, m-1$. Hence

$$\begin{aligned}
\text{Tr}(h)(z) = \text{Tr}(h)(w^m) &= \sum_{i=0}^{m-1} h(\zeta^i w) \\
&= \sum_{i=0}^{m-1} \sum_n c_n (\zeta^i w)^n \\
&= \sum_n c_n [\sum_{i=0}^{m-1} \zeta^{in}] w^n.
\end{aligned}$$

Now if m divides n, every ζ^{in} is 1. However if m does not divide n, the sum of these roots of unity is zero. Therefore only the terms with $n = km$ survive; we obtain

$$\text{Tr}(h)(z) = \sum_k m c_{km} w^{km} = \sum_k m c_{km} z^k.$$

This formula shows that if h is meromorphic at p, then $\text{Tr}(h)$ is meromorphic at q; and if h is holomorphic at p, then $\text{Tr}(h)$ is holomorphic at q.

Now in the general case that q has more than one preimage point, the function $\text{Tr}(h)$ is simply the sum of the traces obtained from neighborhoods of each preimage point. Hence again we see that $\text{Tr}(h)$ is holomorphic (respectively meromorphic) if h is holomorphic (respectively meromorphic) at each preimage point.

The Trace of a 1-Form. The same principles used above in the construction of the trace of a meromorphic function can be applied to construct the trace of a meromorphic 1-form also. Let ω be a meromorphic 1-form on X. For a point $q \in Y$ which is not a branch point of F, use the same notation as above and get a chart domain U containing q whose inverse image is a disjoint union of chart domains V_i, on which F is an isomorphism. Again let ϕ_i be the inverse of $F|_{V_i}$. Then define $\text{Tr}(\omega)$ on U by

$$\text{Tr}(\omega) = \sum_{i=1}^d \phi_i^*(\omega|_{V_i}).$$

This defines a meromorphic 1-form at all nonbranch points of Y.

To see that $\text{Tr}(\omega)$ extends nicely across the branch points, we make a computation similar to what we did above for the functions. Again we may assume that q has a single preimage p, with multiplicity m, and we may choose centered coordinates z and w such that F has the form $z = w^m$. We then write $\omega = h(w)dw$, where h has a Laurent series $\sum_n c_n w^n$. Note that $dz = mw^{m-1}dw$, so we may write $\omega = [h(w)/mw^{m-1}]dz$. We have the formula (valid for nonzero $z = w^m$)

$$\begin{aligned}
\mathrm{Tr}(\omega) &= \sum_{i=0}^{m-1} h(\zeta^i w)/m(\zeta^i w)^{m-1} dz \\
&= \sum_{i=0}^{m-1} (1/m) \sum_n c_n (\zeta^i w)^{n-m+1} dz \\
&= (1/m) \sum_n c_n [\sum_{i=0}^{m-1} \zeta^{i(n-m+1)}] w^{n-m+1} dz.
\end{aligned}$$

This time the only terms which survive are those with m dividing $n - m + 1$, i.e., those with $n = km - 1$ for some k. Therefore we have

(3.1) $$\mathrm{Tr}(\omega) = \sum_k c_{km-1} z^{k-1} dz,$$

which shows that $\mathrm{Tr}(\omega)$ is also meromorphic at q, and indeed is holomorphic if ω is holomorphic at the preimages of q.

If there is more than one preimage of q, a sum is taken over those preimages. This gives a local definition of the 1-form $\mathrm{Tr}(\omega)$ in all situations; we leave to the reader to check that these local definitions all transform to one another, and therefore define a global meromorphic 1-form.

This 1-form $\mathrm{Tr}(\omega)$ is called the *trace* of ω.

The Residue of a Trace. The formula above for the trace of a meromorphic 1-form allows us to conclude the following.

LEMMA 3.2. *Let $F : X \to Y$ be a nonconstant holomorphic map between compact Riemann surfaces. Let ω be a meromorphic 1-form on X. Then for any point $q \in Y$,*

$$\mathrm{Res}_q(\mathrm{Tr}(\omega)) = \sum_{p \in F^{-1}(q)} \mathrm{Res}_p(\omega).$$

PROOF. It suffices to check that the residues are equal if the preimage of q is a single point p, with multiplicity m. Using the notation above, we have that $\mathrm{Res}_p(\omega) = c_{-1}$, the coefficient of w^{-1} in the Laurent series for ω. By (3.1), this is also the coefficient of z^{-1} in the Laurent series for $\mathrm{Tr}(\omega)$. □

An Algebraic Proof of the Residue Theorem. The notion of the trace operation for meromorphic 1-forms can be used to give a different proof of the Residue Theorem for algebraic curves. The proof is more algebraic, in the sense that it does not utilize the theory of integration. Indeed, it can be adapted to give a proof of the Residue Theorem in more general settings (i.e., for algebraic curves over arbitrary fields).

We want to show that for any meromorphic 1-form ω on an algebraic curve X, the sum of the residues of ω is zero. The idea is very simple: find a nonconstant holomorphic map F from X to the Riemann Sphere, verify the Residue Theorem for $\mathrm{Tr}(\omega)$, and finally show that the Residue Theorem for $\mathrm{Tr}(\omega)$ implies it for ω.

The first part is immediate when X is an algebraic curve; simply choose any nonconstant meromorphic function f on X and let F be the associated holomorphic map to $\mathbb{C}_\infty$. The assumption that X is algebraic is used simply to guarantee the existence of a single such f.

Now consider the meromorphic 1-form $\mathrm{Tr}(\omega)$ on $\mathbb{C}_\infty$. As we have seen, we may choose an affine coordinate z and write

$$\mathrm{Tr}(\omega) = r(z)\mathrm{d}z,$$

where $r(z)$ is a *rational* function of z. Any rational function of z may be expanded into partial fractions, and so we may write $r(z)$ as a sum of terms, each of the form $c(z-a)^n$ (where n may be positive or negative, and $a, c \in \mathbb{C}$). To show that the sum of the residues of $\mathrm{Tr}(\omega)$ is zero, it suffices to show this for each of these terms.

So consider the meromorphic 1-form $c(z-a)^n \mathrm{d}z$ on $\mathbb{C}_\infty$. If $n \leq -2$, then this has a unique pole at the point a, with residue zero. If $n = -1$, then this has a simple pole at a with residue c and a simple pole at ∞ with residue $-c$, and no other poles. If $n \geq 0$, then this has a unique pole at ∞ (of order at least 2) with residue zero. Therefore in every case the sum of the residues of $c(z-a)^n \mathrm{d}z$ is zero, and this proves the Residue Theorem for $\mathrm{Tr}(\omega)$.

Finally Lemma 3.2 is all that is required to lift the Residue Theorem from $\mathrm{Tr}(\omega)$ to ω. This simply involves organizing the points of X into the fibers of F:

$$\begin{aligned}\sum_{p \in X} \mathrm{Res}_p(\omega) &= \sum_{q \in \mathbb{C}_\infty} \sum_{p \in F^{-1}(q)} \mathrm{Res}_p(\omega) \\ &= \sum_{q \in \mathbb{C}_\infty} \mathrm{Res}_q(\mathrm{Tr}(\omega)) \text{ using Lemma 3.2} \\ &= 0 \text{ by the Residue Theorem for } \mathrm{Tr}(\omega).\end{aligned}$$

This finishes the "algebraic proof" of the Residue Theorem.

Integration of a Trace. Let γ be a path on Y, and let ω be a meromorphic 1-form on X whose poles do not lie on the preimage of the points of γ. In this case the integral of $\mathrm{Tr}(\omega)$ along γ is well defined. One can relate this to an integral of ω as follows.

Away from the branch points of F, one has exactly d preimages of every point on Y, and F is a local homeomorphism. Therefore, at all but finitely many points of γ, we may lift γ to exactly d paths $\gamma_1, \ldots, \gamma_d$. These paths come together at ramification points of F which lie above any branch points lying on γ; but in any case we may take the closure of these lifted paths and obtain d lifts which we also denote by γ_i.

DEFINITION 3.3. The *pullback* of a path γ on Y is the chain $F^*\gamma = \gamma_1 + \cdots + \gamma_d$.

3. TRACE OPERATIONS

Extending by linearity, we may define the pullback of any chain on Y. With this construction, we have the following formula for the integral of a trace:

LEMMA 3.4. *Let $F : X \to Y$ be a nonconstant holomorphic map between compact Riemann surfaces. Let ω be a holomorphic 1-form on X. Let γ be a chain on Y. Then*

$$\int_{F^*\gamma} \omega = \int_\gamma \mathrm{Tr}(\omega).$$

PROOF. The integrals in question do not "see" the branch points of F, and we may assume that γ is a path not through any branch point. In this case the left side of the equation is a sum of integrals of ω, adding up the integrals along the lifted curves γ_i. The right side of the equation is an integral of the sum of the appropriate ω's. These are clearly the same. □

Proof of Necessity in Abel's Theorem. We are now in a good position to prove one half of Abel's Theorem.

PROOF (NECESSITY). Suppose that $D = \mathrm{div}(f)$ is the divisor of a nonconstant meromorphic function on the compact Riemann surface X. Let $F : X \to \mathbb{C}_\infty$ be the corresponding holomorphic map of degree d to the Riemann Sphere. On the sphere, choose a path γ from ∞ to 0, not passing through any branch points of F (other than possibly ∞ and 0). In this case the pullback chain $F^*\gamma = \sum_{i=1}^d \gamma_i$ is a sum of paths, each joining a pole of f to a zero of f. Write $q_i = \gamma_i(0)$ and $p_i = \gamma_i(1)$, so that $\sum_i p_i$ is the divisor of zeroes of f and $\sum_i q_i$ is the divisor of poles of f; in particular, we may write the divisor D of f as

$$D = \sum_{i=1}^d (p_i - q_i).$$

Choose a basis $\omega_1, \ldots, \omega_g$ for $\Omega^1(X)$, and choose a base point $x \in X$. For each i, choose a path α_i from x to p_i and a path β_i from x to q_i. Then the Abel-Jacobi mapping A_0 applied to D is exactly

$$A_0(D) = \sum_{i=1}^d \left(\int_{\alpha_i} \omega_1, \cdots \int_{\alpha_i} \omega_g\right)^\top - \left(\int_{\beta_i} \omega_1, \cdots \int_{\beta_i} \omega_g\right)^\top \quad \mathrm{mod}\ \Lambda.$$

For each i, let η_i be the closed path $\alpha_i - \gamma_i - \beta_i$. Since this is a closed path, the vector

$$\left(\int_{\eta_i} \omega_1, \cdots \int_{\eta_i} \omega_g\right)^\top$$

is a period in Λ for each i. Therefore, we may subtract it from the above formula for $A_0(D)$, and obtain

$$\begin{aligned}A_0(D) &= \sum_{i=1}^{d}(\int_{\gamma_i}\omega_1,\cdots\int_{\gamma_i}\omega_g)^\top \mod \Lambda \\ &= (\int_{F^*\gamma}\omega_1,\cdots\int_{F^*\gamma}\omega_g)^\top \mod \Lambda.\end{aligned}$$

Now consider the jth coordinate of this vector, which is the integral $\int_{F^*\gamma}\omega_j$. Using Lemma 3.4, we see that this is equal to $\int_\gamma \text{Tr}(\omega_j)$. Since ω_j is holomorphic on X, $\text{Tr}(\omega_j)$ is holomorphic on the Riemann Sphere $\mathbb{C}_\infty$. But since $\mathbb{C}_\infty$ has genus 0, there are no nonzero holomorphic 1-forms on $\mathbb{C}_\infty$. Hence $\text{Tr}(\omega_j) = 0$ for each j, and the integral in question is then also 0. Therefore $A_0(D) = 0$ in $\text{Jac}(X)$, and we have proved the necessity of this condition for D to be a principal divisor. □

Problems VIII.3

A. Check that if $F : X \to Y$ is a nonconstant map between compact Riemann surfaces, and h is a meromorphic function on X, then the prescription given above for defining $\text{Tr}(h)$ makes $\text{Tr}(h)$ into a meromorphic function on the complement of the branch locus in Y. (The only serious part is to check that $\text{Tr}(h)$ is well defined.)

B. Suppose that $F : X \to Y$ is a nonconstant holomorphic map between compact Riemann surfaces. Show that $\text{Tr} : \mathcal{M}(X) \to \mathcal{M}(Y)$ is a group homomorphism.

C. Let X be a hyperelliptic curve defined by the equation $y^2 = h(x)$, and let $\pi : X \to \mathbb{C}_\infty$ be the hyperelliptic double covering sending $(x,y) \in X$ to x. Consider the meromorphic function $r(x) + s(x)y$ on X, where r and s are rational functions of x. Compute its trace on $\mathbb{C}_\infty$.

D. Suppose that $F : X \to Y$ is a nonconstant holomorphic map of degree d between compact Riemann surfaces. Show that if g is a meromorphic function on Y, then $\text{Tr}(F^*(g)) = dg$. More generally show that if g is meromorphic on Y and h is meromorphic on X then $\text{Tr}(F^*(g)h) = g\,\text{Tr}(h)$. Hence if we make $\mathcal{M}(X)$ into a $\mathcal{M}(Y)$-vector space via F^*, then Tr is $\mathcal{M}(Y)$-linear.

E. Show that the definition of $\text{Tr}(\omega)$ at a nonbranch point of Y is well defined.

F. Suppose that $F : X \to Y$ is a nonconstant holomorphic map of degree d between compact Riemann surfaces. Show that if ω is a meromorphic 1-form on Y, then $\text{Tr}(F^*(\omega)) = d\omega$. More generally show that if ω is a meromorphic 1-form on Y and h is a meromorphic function on X then $\text{Tr}(hF^*(\omega)) = \text{Tr}(h)\omega$.

G. Let X be a hyperelliptic curve defined by the equation $y^2 = h(x)$, and let $\pi : X \to \mathbb{C}_\infty$ be the hyperelliptic double covering sending $(x,y) \in X$ to x.

Consider the meromorphic 1-form $(r(x) + s(x)y)dx$ on X, where r and s are rational functions of x. Compute its trace on $\mathbb{C}_\infty$.

4. Proof of Sufficiency in Abel's Theorem

Lemmas Concerning Periods. Recall the description of the compact Riemann surface X of genus g as the standard identified polygon $\mathcal{P}$ with its $4g$ sides $\{a_i, b_i, a_i', b_i'\}_{i=1}^g$. Recall that a_i and a_i' are identified in X to a closed path a_i, and b_i and b_i' are also identified in X to a closed path b_i. Around the polygon $\mathcal{P}$ appear the directed sides in the order: a_i, b_i, a_i' backwards, then b_i' backwards.

For any 1-form σ on X, we set

$$A_i(\sigma) = \int_{a_i} \sigma \quad \text{and} \quad B_i(\sigma) = \int_{b_i} \sigma$$

to be the values of the integrals of σ along these closed paths on X. For a fixed 1-form σ, these $2g$ numbers are called the *periods* of σ. Note the dual use of the word. The numbers A_i are called the *a-periods*, and the B_i are called the *b-periods*.

Let σ be any closed $\mathcal{C}^\infty$ 1-form on X. This form may be considered as a form on $\mathcal{P}$, also. Choose a base point x in the interior of $\mathcal{P}$. For any point $p \in \mathcal{P}$, define

$$f_\sigma(p) = \int_x^p \sigma,$$

where the integral is taken on any path from x to p which lies entirely inside $\mathcal{P}$. This function is well defined on $\mathcal{P}$ since σ is closed, and so integrals are independent of path since $\mathcal{P}$ is simply connected.

We note that the function f_σ is $\mathcal{C}^\infty$ on $\mathcal{P}$, and indeed $df_\sigma = \sigma$ by the Fundamental Theorem of Calculus. In particular, if σ is holomorphic, then so is f_σ.

LEMMA 4.1. *Let σ and τ be closed $\mathcal{C}^\infty$ 1-forms on X. Then*

$$\int_{\partial \mathcal{P}} f_\sigma \tau = \sum_{i=1}^g A_i(\sigma) B_i(\tau) - A_i(\tau) B_i(\sigma).$$

Here $\partial \mathcal{P} = \sum_{i=1}^g (a_i + b_i - a_i' - b_i')$ as a chain. The equality holds even if τ is a meromorphic 1-form on X with no poles along any of the curves a_i or b_i.

PROOF. For any point p on the side a_i, denote by p' the corresponding point on a_i' which is to be identified with p in X. Let α_p be a path inside $\mathcal{P}$ from p to p'. Then

$$f_\sigma(p) - f_\sigma(p') = \int_x^p \sigma - \int_x^{p'} \sigma = -\int_{\alpha_p} \sigma,$$

since integrals of σ around closed paths are zero. Now on X, the path α_p is a closed path, which is homotopic to the path b_i. Therefore, since σ is closed, we

have
$$f_\sigma(p) - f_\sigma(p') = -B_i(\sigma) \text{ for every } p \in a_i.$$

Make a similar analysis for a point q on b_i. Let q' be the corresponding point on b'_i, and let β_q be a path inside $\mathcal{P}$ from q to q'. Then
$$f_\sigma(q) - f_\sigma(q') = \int_x^q \sigma - \int_x^{q'} \sigma = -\int_{\beta_q} \sigma$$
exactly as above. Now this time the curve β_q, considered on X, is homotopic to a'_i, which is $-a_i$; hence
$$f_\sigma(q) - f_\sigma(q') = A_i(\sigma) \text{ for every } q \in b_i.$$

Finally note that since τ is a 1-form on X, its values along a_i and a'_i are equal, and similarly for b_i and b'_i. Hence

$$\begin{aligned}
\int_{\partial \mathcal{P}} f_\sigma \tau &= \sum_{i=1}^g \left(\int_{a_i} - \int_{a'_i} + \int_{b_i} - \int_{b'_i} \right) f_\sigma \tau \\
&= \sum_{i=1}^g \left(\int_{p \in a_i} [f_\sigma(p) - f_\sigma(p')] \tau + \int_{q \in b_i} [f_\sigma(q) - f_\sigma(q')] \tau \right) \\
&= \sum_{i=1}^g \left(\int_{p \in a_i} [-B_i(\sigma)] \tau + \int_{q \in b_i} [A_i(\sigma)] \tau \right) \\
&= \sum_{i=1}^g [-B_i(\sigma) A_i(\tau) + A_i(\sigma) B_i(\tau)]
\end{aligned}$$

as promised. $\square$

The above lemma is used in the following.

LEMMA 4.2. *Suppose that ω is a nonzero holomorphic 1-form on X. Then*
$$\operatorname{Im} \sum_{i=1}^g A_i(\omega) \overline{B_i(\omega)} < 0.$$

PROOF. Locally, write $\omega = f(z) dz$, so that $\overline{\omega} = \overline{f(z)} d\overline{z}$. Then $\omega \wedge \overline{\omega} = |f|^2 dz \wedge d\overline{z} = -2i |f|^2 dx \wedge dy$. Hence
$$\operatorname{Im} \iint_X \omega \wedge \overline{\omega} < 0.$$

Now use Lemma 4.1 with $\sigma = \omega$ and $\tau = \overline{\omega}$. We note that
$$\begin{aligned}
\int_{\partial \mathcal{P}} f_\omega \overline{\omega} &= \iint_\mathcal{P} d(f_\omega \overline{\omega}) \quad \text{by Stoke's Theorem} \\
&= \iint_\mathcal{P} (df_\omega \wedge \overline{\omega} + f_\omega d\overline{\omega}) \\
&= \iint_\mathcal{P} \omega \wedge \overline{\omega}
\end{aligned}$$

since $df_\omega = \omega$ and $d\overline{\omega} = 0$ (ω is holomorphic). We conclude using Lemma 4.1 that
$$\operatorname{Im} \sum_{i=1}^{g} [A_i(\omega) B_i(\overline{\omega}) - A_i(\overline{\omega}) B_i(\omega)] < 0.$$

Let $z_0 = x_0 + iy_0$ be the sum $\sum_i A_i(\omega) B_i(\overline{\omega})$. Since $A_i(\overline{\omega}) = \overline{A_i(\omega)}$ and the same for the B_i's, the sum above is simply $z_0 - \overline{z_0} = 2iy_0$. Therefore $y_0 < 0$ as desired. □

COROLLARY 4.3. *Suppose that ω is a holomorphic 1-form on X with $A_i(\omega) = 0$ for every i. Then $\omega = 0$. The same conclusion holds if $B_i(\omega) = 0$ for every i.*

PROOF. If $A_i(\omega) = 0$ for every i, then the sum above is clearly 0; this is a contradiction if $\omega \neq 0$. The same argument is used if $B_i(\omega) = 0$ for each i. □

Several properties of the a- and b-periods of holomorphic 1-forms can be nicely expressed by introducing the following matrices. Let $\omega_1, \ldots, \omega_g$ be a basis for the space $\Omega^1(X)$ of holomorphic 1-forms on X. Let $\mathbf{A}$ be the $g \times g$ matrix whose ij^{th} entry is the a-period $A_i(\omega_j)$. Similarly let $\mathbf{B}$ be the $g \times g$ matrix whose ij^{th} entry is the b-period $B_i(\omega_j)$. These matrices are called the *period matrices* of X (relative to this choice of basis for $\Omega^1(X)$ and this choice of paths $\{a_i, b_i\}$ generating $H_1(X, \mathbb{Z})$).

LEMMA 4.4. *Both $\mathbf{A}$ and $\mathbf{B}$ are nonsingular matrices.*

PROOF. Suppose that $\mathbf{A}$ is singular, and let $\mathbf{c} = (c_1, \ldots, c_g)^\top$ be a nonzero vector such that $\mathbf{Ac} = 0$. Let $\omega = \sum_j c_j \omega_j$; this is a nonzero holomorphic 1-form on X. But then
$$A_i(\omega) = \sum_j c_j A_i(\omega_j) = 0 \text{ for each } i,$$
since it is the ith entry of the product $\mathbf{Ac}$. This is a contradiction by Corollary 4.3, since $\omega \neq 0$.

The proof for the $\mathbf{B}$ matrix is identical. □

These period matrices satisfy a symmetry property also.

LEMMA 4.5. *The period matrices satisfy the identity $\mathbf{A}^\top \mathbf{B} = \mathbf{B}^\top \mathbf{A}$.*

PROOF. Fix indices j and k, and apply Lemma 4.1 with $\sigma = \omega_j$ and $\tau = \omega_k$. Note that
$$\int_{\partial \mathcal{P}} f_{\omega_j} \omega_k = \iint_{\mathcal{P}} d(f_{\omega_j} \omega_k) = \iint_{\mathcal{P}} \omega_j \wedge \omega_k + f_{\omega_j} d\omega_k = 0$$
since $\omega_j \wedge \omega_k = 0$ (both are of type $(1,0)$) and $d\omega_k = 0$ (ω_k is holomorphic, hence closed). Therefore we conclude using Lemma 4.1 that
$$\sum_{i=1}^{g} A_i(\omega_j) B_i(\omega_k) = \sum_{i=1}^{g} A_i(\omega_k) B_i(\omega_j).$$

The left side is the jk^{th} entry of $\mathbf{A}^\top \mathbf{B}$, and the right side is the jk^{th} entry of $\mathbf{B}^\top \mathbf{A}$. □

The Proof of Sufficiency. We are now in a position to prove the other half of Abel's Theorem. One more lemma is required.

LEMMA 4.6. *Let D be a divisor of degree 0 on a compact Riemann surface X, such that $A_0(D) = 0$ in $\mathrm{Jac}(X)$. Then there is a meromorphic 1-form ω on X such that*
- *ω has simple poles at the points where $D \neq 0$, and has no other poles.*
- *$\mathrm{Res}_p(\omega) = D(p)$ for each point $p \in X$.*
- *The a- and b-periods of ω are integral multiples of $2\pi i$.*

PROOF. Since the only condition that a meromorphic 1-form exist with simple poles and prescribed residues is that the sum of the residues is zero (Proposition 1.15 of Chapter VII), we see that there is a meromorphic 1-form satisfying the first two conditions, since D has degree 0. Let τ be such a form. Note that for any constants c_i, the form $\omega = \tau - \sum_{i=1}^{g} c_i \omega_i$ will also satisfy the first two conditions. We therefore seek constants c_i such that ω also satisfies the third condition.

We may assume that no point where $D(p) \neq 0$ lies on any of the curves a_i or b_i.

For each $k = 1, \ldots, g$ define
$$\rho_k = \frac{1}{2\pi i} \sum_{i=1}^{g} (A_i(\omega_k) B_i(\tau) - A_i(\tau) B_i(\omega_k)).$$

Using Lemma 4.1, we see that
$$\rho_k = \frac{1}{2\pi i} \int_{\partial \mathcal{P}} f_{\omega_k} \tau = \sum_{p \in \mathcal{P}} \mathrm{Res}_p(f_{\omega_k} \tau) = \sum_{p \in X} \mathrm{Res}_p(f_{\omega_k} \tau)$$

using the ordinary Residue Theorem for integrals of meromorphic functions on complex domains. Since f_{ω_k} is holomorphic, and τ has simple poles exactly at the points where $D(p) \neq 0$, with residue $D(p)$, we see that $\mathrm{Res}_p(f_{\omega_k} \tau) = f_{\omega_k}(p) \cdot D(p)$. Hence
$$\rho_k = \sum_p D(p) f_{\omega_k}(p) = \sum_p D(p) \int_{p_0}^{p} \omega_k$$

where p_0 is the chosen base point on X. By definition, this is exactly the kth coordinate of the Abel-Jacobi map applied to the divisor D. By assumption, $A_0(D) = 0$; we therefore conclude that the vector $(\rho_1, \ldots, \rho_g)^\top$ is a period vector in Λ.

Hence we may write
$$(\rho_1, \ldots, \rho_g) = \sum_{i=1}^{g} m_i (A_i(\omega_1), \ldots, A_i(\omega_g)) - \sum_{i=1}^{g} n_i (B_i(\omega_1), \ldots, B_i(\omega_g))$$

4. PROOF OF SUFFICIENCY IN ABEL'S THEOREM

for some integers n_i and m_i, i.e., there are integers n_i, m_i such that

$$\rho_k = \sum_i m_i A_i(\omega_k) - \sum_i n_i B_i(\omega_k)$$

for every k. On the other hand by definition we have

$$\rho_k = \sum_i \frac{B_i(\tau)}{2\pi i} A_i(\omega_k) - \sum_i \frac{A_i(\tau)}{2\pi i} B_i(\omega_k).$$

Comparing these two we see that

$$\sum_{i=1}^g (B_i(\tau) - 2\pi i m_i) A_i(\omega_k) = \sum_{i=1}^g (A_i(\tau) - 2\pi i n_i) B_i(\omega_k)$$

for each k. Let $\mathbf{a}$ be the vector with i^{th} coordinate $A_i(\tau) - 2\pi i n_i$, and let $\mathbf{b}$ be the vector with i^{th} coordinate $B_i(\tau) - 2\pi i m_i$. The above equation can be expressed succinctly as

$$\mathbf{A}^\top \mathbf{b} = \mathbf{B}^\top \mathbf{a}.$$

Now consider the sequence of linear transformations

$$\mathbb{C}^g \xrightarrow{\alpha} \mathbb{C}^{2g} \xrightarrow{\beta} \mathbb{C}^g$$

where in terms of matrices

$$\alpha = \begin{pmatrix} \mathbf{A} \\ \mathbf{B} \end{pmatrix} \text{ and } \beta = \begin{pmatrix} \mathbf{B}^\top & -\mathbf{A}^\top \end{pmatrix}.$$

Since both $\mathbf{A}$ and $\mathbf{B}$ are nonsingular, both α and β have maximal rank g. Moreover, by the symmetry property of Lemma 4.5, we have that $\beta \circ \alpha = 0$. We conclude by dimension reasons that the sequence is exact, i.e. that $\ker(\beta) = \operatorname{image}(\alpha)$.

What we have seen above is that the $2g$-dimensional vector $\begin{pmatrix} \mathbf{a} \\ \mathbf{b} \end{pmatrix}$ is in the kernel of β. Therefore it is in the image of α, and there is a vector $\mathbf{c}$ such that

$$\mathbf{Ac} = \mathbf{a} \text{ and } \mathbf{Bc} = \mathbf{b}.$$

This is the desired vector to use to alter the given 1-form τ. Let c_j be the jth coordinate of $\mathbf{c}$, and let $\omega = \tau - \sum_j c_j \omega_j$. Compute the periods of ω:

$$\begin{aligned} A_i(\omega) &= A_i(\tau) - \sum_j c_j A_i(\omega_j) = A_i(\tau) - [\, i^{th} \text{ coordinate of } \mathbf{Ac} = \mathbf{a}] \\ &= A_i(\tau) - [A_i(\tau) - 2\pi i n_i] = 2\pi i n_i, \end{aligned}$$

and

$$\begin{aligned} B_i(\omega) &= B_i(\tau) - \sum_j c_j B_i(\omega_j) = B_i(\tau) - [\, i^{th} \text{ coordinate of } \mathbf{Bc} = \mathbf{b}] \\ &= B_i(\tau) - [B_i(\tau) - 2\pi i m_i] = 2\pi i m_i. \end{aligned}$$

Thus ω satisfies the required conditions. □

The above lemma is really the heart of the matter; from here on the proof is rather formal. Suppose that D is a divisor of degree 0 with $A_0(D) = 0$ in $\text{Jac}(X)$. Let ω be the meromorphic 1-form satisfying the conditions of Lemma 4.6. Fixing a base point $p_0 \in X$, define the function

$$f(p) = \exp(\int_{p_0}^{p} \omega).$$

Note that since the periods of ω are integral multiples of $2\pi i$, and the residues of ω are integers, the function f is well defined, independent of the path chosen from p_0 to p. Moreover f is clearly holomorphic wherever ω is, that is, away from the support of the divisor D.

We will show that f is meromorphic, and $D = \text{div}(f)$. Suppose that p is in the support of D, with $D(p) = n \neq 0$. Then near p, in terms of a local coordinate z centered at p, ω may be written as

$$\omega = \frac{n}{z} + g(z)$$

where $g(z)$ is holomorphic. Hence near p, the integral $\int_{p_0}^{z} \omega$ has the form $n \ln(z) + h(z)$ for a holomorphic function h. Therefore f has the form

$$f(z) = z^n e^{h(z)}$$

which is clearly meromorphic. Moreover $\text{ord}_p(f) = n$, and so $\text{div}(f) = D$.

This proves the sufficiency of the condition stated in Abel's theorem, and completes the proof.

Riemann's Bilinear Relations. Lemma 4.5 is the first of two relations on the period matrices $\mathbf{A}$ and $\mathbf{B}$ which together are known as the *Riemann bilinear relations*. The first was a consequence of Lemma 4.1, and the second follows from Lemma 4.2.

We note that since the a-period matrix $\mathbf{A}$ is nonsingular, we may change the basis $\omega_1, \ldots, \omega_g$ for $\Omega^1(X)$ and achieve $\mathbf{A} = \mathbf{I}$. Such a basis for $\Omega^1(X)$ is said to be *normalized* (with respect to the generators $\{a_i, b_i\}$ for $H_1(X, \mathbb{Z})$). If such a basis is used, the b-period matrix is said to be *normalized*.

LEMMA 4.7. *Any normalized b-period matrix $\mathbf{B}$ is symmetric and has positive definite imaginary part.*

PROOF. If $\mathbf{A} = \mathbf{I}$, the symmetry of $\mathbf{B}$ follows from Lemma 4.5. Fix real numbers $c_1, \ldots, c_g$, not all zero, and let $\omega = \sum_j c_j \omega_j$. By Lemma 4.2, we have

$$\text{Im} \sum_{i=1}^{g} A_i(\omega) \overline{B_i(\omega)} < 0.$$

Expanding this, and noting that $A_i(\omega) = c_i$, we obtain

$$\text{Im} \sum_{i,j} c_i c_j \overline{B_i(\omega_j)} < 0$$

since each c_j is real. Writing $\mathbf{c} = (c_1, \ldots, c_g)^\top$, and dispensing with the conjugation, we have that
$$\operatorname{Im}(\mathbf{c}^\top \mathbf{B} \mathbf{c}) > 0,$$
which exactly says that $\operatorname{Im} \mathbf{B}$ is a positive definite real matrix. $\square$

COROLLARY 4.8. *The $2g$ columns of the period matrices $\mathbf{A}$ and $\mathbf{B}$ are linearly independent over $\mathbb{R}$.*

PROOF. It suffices to show this after changing to a normalized basis, in which case $\mathbf{A} = \mathbf{I}$ and $\mathbf{B}$ is a normalized matrix of b-periods. Suppose that the columns are not linearly independent over $\mathbb{R}$. Then there are real g-dimensional vectors $\mathbf{a}$ and $\mathbf{b}$ such that
$$\begin{pmatrix} \mathbf{I} & \mathbf{B} \end{pmatrix} \begin{pmatrix} \mathbf{a} \\ \mathbf{b} \end{pmatrix} = 0,$$
i.e, $\mathbf{a} + \mathbf{B}\mathbf{b} = 0$. Taking imaginary parts, we see that $\operatorname{Im} \mathbf{B}\mathbf{b} = 0$, forcing $\mathbf{b} = 0$ by Lemma 4.7. Then $\mathbf{a} = 0$ is immediate. $\square$

The Jacobian and the Picard Group. Note that the columns of the period matrices form a basis for the subgroup of periods $\Lambda \subset \mathbb{C}^g$. Since they are linearly independent over $\mathbb{R}$, they form a *lattice* in $\mathbb{C}^g$. We may take the columns of the period matrices as a real basis for $\mathbb{C}^g$ over $\mathbb{R}$, and then conclude that, as an abelian group, $\operatorname{Jac}(X)$ is isomorphic to $2g$ copies of the circle group $\mathbb{R}/\mathbb{Z}$. In particular $\operatorname{Jac}(X)$ is compact.

We see then that $\operatorname{Jac}(X)$ is a g-dimensional generalization of a complex torus $\mathbb{C}/(\mathbb{Z} + \mathbb{Z}\tau)$.

DEFINITION 4.9. A group of the form $\mathbb{C}^g/\Lambda$, where Λ is a subgroup of $\mathbb{C}^g$ generated by $2g$ vectors which are independent over $\mathbb{R}$, is called a *g-dimensional complex torus*.

Thus the Jacobian of a compact Riemann surface of genus g is a g-dimensional complex torus.

It is a theorem of Jacobi, which we will not prove here, that the Abel-Jacobi mapping $A_0 : \operatorname{Div}_0(X) \to \operatorname{Jac}(X)$ is surjective. Abel's Theorem states that the kernel of A_0 is exactly the subgroup $\operatorname{PDiv}(X)$ of principal divisors. Therefore we obtain the isomorphism
$$\frac{\operatorname{Div}_0(X)}{\operatorname{PDiv}(X)} \cong \operatorname{Jac}(X).$$

The group of divisors modulo principal divisors is called the *Picard group* of X, and is denoted by $\operatorname{Pic}(X)$:
$$\operatorname{Pic}(X) = \operatorname{Div}(X)/\operatorname{PDiv}(X).$$

If we denote by $\text{Pic}^0(X)$ the subgroup of $\text{Pic}(X)$ given by classes of divisors of degree 0, then the above isomorphism is exactly that

$$\text{Pic}^0(X) \cong \text{Jac}(X).$$

This is often how Abel's Theorem is stated.

Problems VIII.4

A. Let σ be any closed C^∞ 1-form on X, considered as a 1-form on the standard identified polygon $\mathcal{P}$. Fix a base point x in the interior of $\mathcal{P}$, and for any point $p \in \mathcal{P}$, define

$$f_\sigma(p) = \int_x^p \sigma,$$

where the integral is taken on any path from x to p which lies entirely inside $\mathcal{P}$. Show that f_σ is well defined and C^∞ on $\mathcal{P}$. Show that $df_\sigma = \sigma$. Show that if σ is holomorphic, then so is f_σ.

B. Let ω be a meromorphic 1-form on X whose periods are all integral multiples of $2\pi i$, and whose residues are all integers. Fixing a base point $p_0 \in X$, define the function

$$f(p) = \exp(\int_{p_0}^p \omega).$$

Show that f is well defined, independent of the path chosen from p_0 to p. Show that f is holomorphic wherever ω is.

C. Let X be a compact Riemann surface of genus at least one. Show that the Abel-Jacobi map $A: X \to \text{Jac}(X)$ is 1-1.

D. For any divisor D on X, show that $A^{-1}(A(D))$ is the linear equivalence class of D.

E. Let $\text{Div}_d^+(X)$ denote the set of nonnegative divisors of degree d on X. Note that $\text{Div}_0^+(X) = \{0\}$ and that $\text{Div}_1^+(X) = X$. Consider the restriction A_d of the Abel-Jacobi map to $\text{Div}_d^+(X)$:

$$A_d : \text{Div}_d^+(X) \to \text{Jac}(X).$$

Show that for any nonnegative divisor D of degree d, the fiber $A_d^{-1}(A_d(D))$ is the complete linear system $|D|$. (Hence the fiber is a projective space, of dimension equal to $\dim |D|$.)

F. Let X be a curve of genus 2 with a positive canonical divisor K. Show that the fiber of $A_2 : \text{Div}_2^+(X) \to \text{Jac}(X)$ over $A_2(K)$ is isomorphic to a projective line, but that all other nonempty fibers are singletons. Generalize this to other hyperelliptic curves of higher genus.

G. Show that the derivative of the Abel-Jacobi map $A: X \to \text{Jac}(X)$ is the canonical map for X, in the following sense. Fix a point $p \in X$, and consider A as a map between two complex manifolds. Let $T_p(X)$ denote the 1-dimensional tangent space to X at p. Let $\mathbb{C}^g$ be the g-dimensional tangent space to $\text{Jac}(X)$. Show that the derivative dA of A, at the point p, maps $T_p(X)$ to $\mathbb{C}^g$ and is 1-1. This then gives a 1-dimensional subspace of $\mathbb{C}^g$ for every point $p \in X$; and hence a map from X to $\mathbb{P}^{g-1}$. This map is the canonical map for X.

5. Abel's Theorem for Curves of Genus One

The Abel-Jacobi Map is an Embedding. An immediate consequence of Abel's Theorem is that the Abel-Jacobi mapping embeds a compact Riemann surface X into its Jacobian:

PROPOSITION 5.1. *Let X be a compact Riemann surface of genus $g \geq 1$. Then the Abel-Jacobi mapping $A : X \to \text{Jac}(X)$ is 1-1.*

PROOF. Suppose that $A(p) = A(q)$, with $p \neq q$. Then on the divisor level, $A(p - q) = 0$ in $\text{Jac}(X)$, so that $p - q$ is a principal divisor. Hence there is a meromorphic function f on X with a simple zero at p, a simple pole at q, and no other poles. The associated holomorphic map $F : X \to \mathbb{C}_\infty$ would then be a nonconstant map of degree one, and hence an isomorphism. Since $g \geq 1$, this is a contradiction. $\square$

Every Curve of Genus One is a Complex Torus. Now suppose X has genus one. Then $\text{Jac}(X)$ is itself a complex torus of dimension one, and is therefore a Riemann surface of genus one also. Moreover the Abel-Jacobi map $A : X \to \text{Jac}(X)$ is a holomorphic map. This follows from the local definition of the map as integration: locally, A sends p to $\int_{p_0}^p \omega$, where ω is a holomorphic 1-form on X, and this is a holomorphic function of p.

Therefore the Abel-Jacobi map for a curve of genus one is a 1-1 holomorphic map between compact Riemann surfaces. Hence it is an isomorphism; and the choice of a base point p_0 is the only choice made in the definition of A. This proves the following, which we sketched in Section VII.1.

PROPOSITION 5.2. *Every compact Riemann surface of genus one is isomorphic to a complex torus. Moreover given any point $p_0 \in X$ there is an isomorphism A of X with the complex torus $\text{Jac}(X)$ such that $A(p_0) = 0$.*

In particular, we see that every curve of genus one is an abelian group, the group law being induced by the Abel-Jacobi isomorphism. Any point on the curve may serve as the origin of the group law.

Note that the isomorphism $A : X \to \text{Jac}(X)$ induces the group homomorphism $\text{Div}(X) \to \text{Jac}(X)$ which sends a divisor D to the sum in $\text{Jac}(X)$ of its points. When $\text{Jac}(X)$ is identified with X, we simply obtain the map sending a formal sum $\sum n_p \cdot p$ to the actual sum $\sum n_p p$ in the group law of X. This is the most common formulation of the Abel-Jacobi mapping for a curve of genus one: the map $A : \text{Div}(X) \to X$ sending a formal sum to the actual sum.

The Group Law on a Smooth Projective Plane Cubic. Consider a smooth projective plane cubic curve X, which necessarily has genus one by Plücker's formula. By the above Proposition, X is isomorphic to a complex torus, and so is an abelian group.

Let us denote the origin of the group law by p_0; this can be an arbitrary point of X, and the choice of p_0 essentially determines the group law, as we shall now see.

Consider a linear homogeneous polynomial $H(X, Y, Z) = aX + bY + cZ$ in the variables of the projective plane. Since X is a cubic curve, the hyperplane divisor $\text{div}(H)$ has degree 3, and we may write it as a formal sum $\text{div}(H) = p_1 + p_2 + p_3$ for three points p_1, p_2, and p_3 on X. If the p_i's are distinct, then we say the line given by $H = 0$ is *transverse* to X; otherwise we say that the line is *tangent* to X. If all three p_i are equal, then $H = 0$ is a *flexed tangent* to X, and the point is a *flex point* of X.

Define $\ell(H) = p_1 + p_2 + p_3$ to be the sum of the three points, in the group law of X; $\ell(H)$ is a point of X.

We claim that this point is the same for all lines H. Indeed, if H' is another homogeneous linear polynomial, then the two divisors $\text{div}(H)$ and $\text{div}(H')$ are linearly equivalent by Lemma 2.3 of Chapter V. Therefore the divisor $D = \text{div}(H) - \text{div}(H')$ is a principal divisor, and so by Abel's Theorem $A(D) = p_0$ in group law of X. But $A(D) = \ell(H) - \ell(H')$, so that $\ell(H) = \ell(H')$ as points of X.

Let us denote this point by ℓ; we will call this special point the *collinearity point* of the plane cubic. The collinearity point is the sum, in the group law of X, of any three collinear points on X.

Suppose that $p \in X$ and $3p = \ell$ in the group law. (By the description of X as a complex torus, we know that there are exactly 9 such points on X.) Consider the tangent line L to X at p: the intersection divisor for this tangent line must have the form $p + p + q$ for some point $q \in X$. But then $2p + q = \ell = 3p$ in the group law, forcing $p = q$; we conclude that p is a flex point of X.

Conversely, any flex point p clearly satisfies $3p = \ell$ in the group law. Hence we have shown:

LEMMA 5.3. *There are exactly 9 flex points on a smooth projective cubic curve X, given by the nine solutions p to the equation $3p = \ell$ in the group law of X.*

We knew the number of flexes already, from Corollary 4.16 of Chapter VII.

What is this mystery point ℓ on X? Take the tangent line to the cubic X at the origin point p_0. This line has intersection divisor $p_0 + p_0 + q$ for some point q, since it is tangent to X at p_0. Therefore $2p_0 + q = \ell$; since p_0 is the origin of the group law, we see that $q = \ell$. Thus

> *The collinearity point ℓ is the third point of intersection of the tangent line at the origin point p_0 with the cubic X.*

Now suppose we are lucky enough to have chosen one of the 9 flex points as the origin point. Then the point ℓ is also p_0, and the entire analysis simplifies; the description of the group law is particularly nice.

PROPOSITION 5.4. *Suppose that X is a smooth projective plane cubic curve whose origin p_0 is a flex point. Then three points on X are collinear if and only if they sum to zero in the group law of X. The inverse law in the group of X has the form: $-p$ is the third point of intersection of the line joining p_0 and p with X. The group law itself has the form: $p + q$ is the inverse of the third point of intersection of the line joining p and q with X. When $p = q$, this law becomes: $2p$ is the inverse of the third point of intersection of the tangent line to X at p with X.*

PROOF. Since the origin p_0 is a flex point, the collinearity point ℓ is p_0; the collinear condition in general is that three points are collinear (that is, form the intersection divisor of X with a line) if and only if they sum to ℓ in the group law. This proves the first statement.

To check the inverse law, note that $p + q = 0$ in the group if and only if $p + q + p_0 = 0$ in the group, since $p_0 = 0$. This happens if and only if p, q, and p_0 are collinear.

To check the sum law, we note that $p + q + r = 0$ in the group if and only if r is the inverse of $p + q$. Since this is also the condition that the points be collinear, we are done.

Finally the doubling law is similar to the above, with $p = q$. □

Problems VIII.5

A. Show that if X has genus one, then the Abel-Jacobi map $A : X \to \text{Jac}(X)$ is holomorphic.

B. Suppose a smooth cubic curve X is given by the equation $Y^2 Z = X^3 + aXZ^2 + bZ^3$, for constants a and b, with $4a^3 + 27b^2 \neq 0$ (to insure smoothness). Use the flex point $p_0 = [0:1:0]$ as the origin of the group law on X. Find the formula for $p + q$ in the group law of X, given the homogeneous coordinates of p and q. Find the formula for $2p$, also. (Hint: use this exercise to get friendly with a symbolic algebra package.)

C. The hyperelliptic curve X defined by $y^2 = x^4 + 1$ has genus one. Choose a convenient point and write down the group law on X in terms of the affine coordinates x and y. Read Fermat's proof that this curve has no points (x, y) with x and y both rational numbers; show that his infinite descent argument is essentially "dividing by 2" in the group law.

Further Reading

Again, the monograph [**Mumford78**] cannot be recommended too much at this point, especially lecture III. There is a good guide to the existing literature there also. Most other recent texts cover this material, and some go somewhat deeper, e.g. [**Gunning72**], [**R-F74**], [**Gunning76**], [**Reyssat89**] and [**Narasimhan92**].

There are good sections for this material also in [**G-H78**] and [**ACGH85**]; the latter especially is an excellent source for further material.

The reader will want to also look at the article [**Smith89**] for a nice continuation of the theory.

Chapter IX. Sheaves and Čech Cohomology

The ideas of sheaves have become pervasive in modern geometry. Although the language of sheaves takes some getting used to, one can get quite far by thinking of a sheaf as a way of organizing functions and forms which satisfy local properties. With this point of view, we will see that virtually all of the sheaves which will be introduced in this section have been introduced before without using the sheaf language.

1. Presheaves and Sheaves

A sheaf is most naturally defined as something called a *presheaf*, with an extra condition to be explained later.

Presheaves. If one wants to organize together the functions or forms which satisfy some local property, the first step is to define, for each open set U, those objects which satisfy the property. This is most conveniently done by employing the language of a *presheaf*, which is the first ingredient in the definition of a sheaf.

DEFINITION 1.1. Let X be a topological space. A *presheaf of groups* $\mathcal{F}$ on X is a collection of groups $\mathcal{F}(U)$, one for every open set U, and a collection of group homomorphisms $\rho_V^U : \mathcal{F}(U) \to \mathcal{F}(V)$ whenever $V \subseteq U$, such that
- $\mathcal{F}(\emptyset)$ is the trivial group with one element,
- $\rho_U^U = \text{id}$ on $\mathcal{F}(U)$, and
- if $W \subseteq V \subseteq U$, then $\rho_W^U = \rho_W^V \circ \rho_V^U$.

The homomorphisms ρ_V^U are called the *restriction maps* for the presheaf.

One has the notion of a presheaf of rings also: each group $\mathcal{F}(U)$ is in fact a ring, and the restriction maps are ring homomorphisms.

The elements of $\mathcal{F}(U)$ are commonly called the *sections of $\mathcal{F}$ over U*.

The elements of $\mathcal{F}(X)$, which are the sections of $\mathcal{F}$ over the entire space X, are called the *global sections* of $\mathcal{F}$.

Examples of Presheaves. The basic example of a presheaf on a space X is to take a group G and define a presheaf G^X on X by setting $G^X(U)$ for any nonempty open set U to be the set of all functions from U to G:

$$G^X(U) = G^U = \{f : U \to G\} \text{ for } U \neq \emptyset.$$

There are no conditions on the functions here at all. The set $G^X(U)$ is given the group structure of pointwise multiplication: $(f \cdot g)(x) = f(x) \cdot g(x)$. The restriction maps are given by the usual restriction of functions. If we set $G^X(\emptyset) = \{0\}$ by fiat, the conditions of a presheaf are clearly satisfied.

If we take a ring R instead of a group G, then the same construction gives a presheaf of rings R^X on X.

Now arbitrary functions into a group are usually not very interesting geometrically; no attempt is made to respect the topology of X, after all. One usually incorporates the topology by requiring the functions to be at least continuous; this of course requires putting a topology on the group G (or the ring R).

Our favorite example of a ring with a topology is the complex numbers $\mathbb{C}$, so let us start from there. In order to define a presheaf of complex-valued functions on a space X, one needs only to specify the additional properties you want the functions to satisfy, and check that under restriction these properties are preserved. Then the presheaf axioms will be automatic, and a presheaf will be defined.

What additional properties (besides continuity) will be interesting to us? As you may have suspected, differentiability and analyticity are of primary concern. All of these properties can be defined *at a point*; then they hold for an open set U if and only if they hold at each point of U.

Suppose that $\mathcal{P}$ is a property of functions, which is defined initially at points; i.e., a function f defined in a neighborhood of a point p has property $\mathcal{P}$ at p if and only if some condition holds at the point p. We usually extend the definition to having the property on open sets by declaring that a function f has property $\mathcal{P}$ on U if and only if f has property $\mathcal{P}$ at every point of U.

If this is the case, then clearly such a property will be preserved under restriction to a smaller open set: if the property holds at all points of U, and $V \subset U$, then obviously the property will hold at all points of V. Hence in this situation we will *always* get a presheaf of functions $\mathcal{F}_\mathcal{P}$, defined by setting $\mathcal{F}_\mathcal{P}(U)$ to be those functions defined on U and satisfying property $\mathcal{P}$ on U.

The reader can easily check that many of the examples below are of this type.

EXAMPLE 1.2. Let $\mathcal{C}_X^\infty(U)$ be the set of all $\mathcal{C}^\infty$ functions $f : U \to \mathbb{C}$. This is a presheaf of rings $\mathcal{C}_X^\infty$ (or simply $\mathcal{C}^\infty$) on X.

EXAMPLE 1.3. Let X be a Riemann surface. Let $\mathcal{O}_X(U)$ be the set of all holomorphic functions $f : U \to \mathbb{C}$. This is a presheaf of rings $\mathcal{O}_X$ (or simply $\mathcal{O}$) on X.

EXAMPLE 1.4. Let X be a Riemann surface. Let $\mathcal{O}_X^*(U)$ be the set of all nowhere zero holomorphic functions $f : U \to \mathbb{C}^*$. This is a presheaf of groups

$\mathcal{O}_X^*$ (or simply $\mathcal{O}^*$) on X; the group action is pointwise multiplication.

EXAMPLE 1.5. Let X be a Riemann surface. Let $\mathcal{M}_X(U)$ be the set of all meromorphic functions on U. This is a presheaf of rings $\mathcal{M}_X$ (or simply $\mathcal{M}$) on X. Note that if U is connected, then $\mathcal{M}(U)$ is actually a field.

EXAMPLE 1.6. Let X be a Riemann surface. Let $\mathcal{M}_X^*(U)$ be the set of all meromorphic functions on U which are not identically zero on any connected component of U. This is a presheaf of groups $\mathcal{M}_X^*$ (or simply $\mathcal{M}^*$) on X, under multiplication. If U is connected, then $\mathcal{M}^*(U)$ is simply the multiplicative group of the field $\mathcal{M}(U)$.

EXAMPLE 1.7. Let X be a Riemann surface. Let $\mathcal{H}_X(U)$ be the set of all harmonic functions on U. This is a presheaf of groups $\mathcal{H}_X$ (or simply $\mathcal{H}$) on X.

EXAMPLE 1.8. Let X be a Riemann surface, and let D be a divisor on X. Let $\mathcal{O}_X[D](U)$ be the set of all meromorphic functions on U which satisfy the condition that
$$\mathrm{ord}_p(f) \geq -D(p) \text{ for all } p \in U.$$
This is a presheaf of groups $\mathcal{O}_X[D]$ (or simply $\mathcal{O}[D]$) on X. Note that $\mathcal{O}_X[0]$ is exactly the presheaf $\mathcal{O}_X$ of holomorphic functions on X.

EXAMPLE 1.9. Let X be a topological space, and let G be a group. Let $G_X(U)$ be the group G, for every nonempty open set $U \subseteq X$. Continuing with the function terminology, we can consider $G_X(U)$ as the set of all constant functions from U to G. Every restriction map is the identity, unless the subset is the empty set, in which the restriction map is zero. This is a presheaf of groups on X, called a *constant* presheaf.

We can generalize all of this to forms instead of functions immediately. Essentially all properties of forms which have been introduced survive after restriction, so we immediately obtain presheaves. These include:

- $\mathcal{E}_X^1 : \mathcal{E}^1(U) = \{\mathcal{C}^\infty \text{ 1-forms on } U\}$.
- $\mathcal{E}_X^{1,0} : \mathcal{E}^{1,0}(U) = \{\mathcal{C}^\infty \text{ (1,0)-forms on } U\}$.
- $\mathcal{E}_X^{0,1} : \mathcal{E}^{0,1}(U) = \{\mathcal{C}^\infty \text{ (0,1)-forms on } U\}$.
- $\Omega_X^1 : \Omega^1(U) = \{\text{holomorphic 1-forms on } U\}$.
- $\overline{\Omega}_X^1 : \overline{\Omega}^1(U) = \{\text{complex conjugates of holomorphic 1-forms on } U\}$.
  (These are locally of the form $\overline{f(z)}d\overline{z}$ where f is holomorphic in z.)
- $\mathcal{M}_X^{(1)} : \mathcal{M}^{(1)}(U) = \{\text{meromorphic 1-forms on } U\}$.
- $\Omega_X^1[D]$:
  $\Omega^1[D](U) = \{\text{meromorphic 1-forms on } U \text{ with poles bounded by } D\}$.
  That is, we require $\mathrm{ord}_p(\omega) \geq -D(p)$ for all $p \in U$.
- $\mathcal{E}_X^2 : \mathcal{E}^2(U) = \{\mathcal{C}^\infty \text{ 2-forms on } U\}$.

As we have done above, it is common to drop the subscript denoting the surface if there is no possibility of confusion.

The Sheaf Axiom. We have seen above that one obtains a presheaf anytime one has a property of a function or form defined on an open set which survives under restriction to any open subset. In other words, if we have that whenever the property holds for an open set U, then it also holds for any open subset $V \subset U$, then we have a presheaf of functions or forms.

The sheaf axiom essentially states the converse: a property should hold if and only if it holds on the subsets. To be precise, let us say that a property P is *local* if whenever $\{U_i\}$ is an open cover of an open set U, then the property holds on U if and only if it holds on each U_i. The following axiom captures this idea for general presheaves.

DEFINITION 1.10 (THE SHEAF AXIOM). Suppose $\mathcal{F}$ is a presheaf on X, U is an open subset of X, and $\{U_i\}$ is an open covering of U. We say that $\mathcal{F}$ *satisfies the sheaf axiom* for U and $\{U_i\}$ if whenever one has elements $s_i \in \mathcal{F}(U_i)$ which "agree on the intersections" in the sense that the restrictions to intersections are equal, i.e.,

$$\rho^{U_i}_{U_i \cap U_j}(s_i) = \rho^{U_j}_{U_i \cap U_j}(s_j) \text{ for every } i \text{ and } j,$$

then these sections s_i patch together uniquely to give a section on U; that is, there is a unique $s \in \mathcal{F}(U)$ such that

$$\rho^{U}_{U_i}(s) = s_i \text{ for each } i.$$

We say that $\mathcal{F}$ is a *sheaf* if it satisfies the sheaf axiom for every open set U and every open covering $\{U_i\}$ of U.

The uniqueness of the section $s \in \mathcal{F}(U)$ guaranteed by the sheaf axiom can be phrased quite usefully as follows. Suppose that s and t are two sections in $\mathcal{F}(U)$ which agree on the open sets U_i for every i, that is,

$$\rho^{U}_{U_i}(s) = \rho^{U}_{U_i}(t) \text{ for every } i.$$

Then we conclude using the sheaf axiom that $s = t$.

This is often used simply to conclude that a section s of a sheaf is zero: it is zero if its restriction to an open cover gives zero on each subset.

If one has a presheaf of functions or forms on X which is defined by some property P which is a local property in the sense given above, then the presheaf is automatically a sheaf. This is because the agreement of the functions or forms on the overlap intersections automatically gives a well defined unique function or form on the open set U, and one must only check that it satisfies the property P. This is exactly what the "local" aspect of the property P insures.

All of the presheaves defined above are sheaves, except for the "constant" presheaf. This is because all of the properties defining the functions or forms in these presheaves are local in the sense above. Specifically, the property of a function or form being $\mathcal{C}^\infty$, harmonic, holomorphic, or meromorphic are clearly local. The type of a 1-form is a local property. Finally, having poles bounded by a given divisor D is a local property for meromorphic functions or 1-forms.

Locally Constant Sheaves. Note however that the property of being constant is not a local property for a function. Specifically, if an open set is disconnected into two open disjoint subsets, then a function may be constant on each of the subsets, but with different values; it is then not constant on the whole set. Thus a constant presheaf is never a sheaf, unless the group is trivial or the space enjoys the property that any two open sets intersect.

It should not be a surprise that we can remedy this by considering the functions which are *locally constant*. If $U \subseteq X$ is an open subset of the space X, and G is an arbitrary group, then a function $f : U \to G$ is *locally constant* if for every $p \in U$ there is a neighborhood V of p with $V \subset U$ such that f is constant on V. Note that this is equivalent to asking for an open covering $\{U_i\}$ of U such that f is constant on each U_i.

It is immediate that the property of being locally constant is local in the sense above. Therefore the locally constant functions into a group G forms a sheaf, which is usually denoted by $\underline{G}$. The most commonly used groups are actually all rings: the integers, the reals, and the complexes. The associated sheaves of locally constant functions are denoted by

$$\underline{\mathbb{Z}}, \quad \underline{\mathbb{R}}, \quad \text{and} \quad \underline{\mathbb{C}}$$

respectively.

It is common to call these sheaves simply *constant sheaves*; the reader is expected to recall that a constant sheaf actually consists of locally constant functions.

Skyscraper Sheaves. Suppose that one gives, for each point $p \in X$, a different group G_p. Then to any open set U of X, one may assign the direct product of all groups G_p for $p \in U$ to U:

$$\mathcal{S}(U) = \prod_{p \in U} G_p.$$

One has the restriction maps given by the natural projections. It is elementary to check that this gives a sheaf on X. Since there is absolutely no requirement that the different groups at different points be related in any way, we might call such a sheaf a *totally discontinuous* sheaf.

An example of this is to use the same group G at every point of X. Then one simply obtains the sheaf of all functions from X to G, which we have seen before and denoted by G^X.

One of the most common types of totally discontinuous sheaves is obtained by giving a single group G at a single point p, and assigning the trivial group $\{0\}$ to all other points of X. This kind of sheaf is also called a *skyscraper sheaf* (out of the drab skyline the group G rises majestically at the chosen point p). Such a skyscraper sheaf would be denoted by G_p. Note that this type of skyscraper sheaf has

$$G_p(U) = \begin{cases} \{0\} & \text{if } p \notin U \\ G & \text{if } p \in U. \end{cases}$$

For example, the skyscraper sheaf $\mathbb{C}_p$ is the complex numbers at the point p, and is $\{0\}$ at all other points.

Given a totally discontinuous sheaf $\mathcal{S}$, a section $s \in \mathcal{S}(U)$ may be "evaluated" at a point p in U by setting $s(p)$ equal to the pth coordinate of s (which is an element of the group G_p associated to the point p). The *support* of a section $s \in \mathcal{S}(U)$ is the set of points $p \in U$ such that $s(p) \neq 0$. In general, a totally discontinuous sheaf can have sections with arbitrary support sets.

Notice that the sections of the skyscraper sheaf G_p can only have the point p (or the empty set) for support.

A slight variation of the G_p construction is actually more useful in many instances, and is also referred to as a skyscraper sheaf. Again one gives a group G_p for every point $p \in X$, but to each open set U one assigns only the subgroup of the above direct product consisting of sections with discrete support. That is, the sections are those U-tuples $(g_p \in G_p : p \in U)$ such that the set of $p \in U$ with $g_p \neq 0$ forms a discrete set. This also forms a sheaf, and this case arises more often in practice.

Here it is not so much that the sheaf itself resembles a skyscraper on the landscape, but each section of the sheaf does!

Note that a totally discontinuous sheaf is a skyscraper sheaf if and only if one uses the trivial group at all but a discrete set of points of X. In particular, the skyscraper sheaf G_p is a skyscraper sheaf in this second sense.

A more complicated example is given by assigning the group of integers to every point of a Riemann surface. Then to every open set U one obtains the group of functions from U to $\mathbb{Z}$, which are discretely supported. Such a function is exactly a divisor on U! Therefore we obtain the (skyscraper) sheaf of divisors $\mathcal{D}iv_X$.

Another particular case of a skyscraper sheaf is given by the Laurent tail divisor construction. Fix an ordinary divisor D on a Riemann surface X, and choose a local coordinate z_p at every point $p \in X$. To each p associate the group of Laurent tails truncated at $-D(p)$, that is, the group of Laurent polynomials in z_p whose top term has degree strictly less than $-D(p)$. We will denote by $\mathcal{T}_X[D]$ the skyscraper sheaf with these groups at each point.

A relative version of this can be made in case one has two ordinary divisors D_1 and D_2 on X with $D_1 \leq D_2$. Then for each p we associate the group of Laurent polynomials in z_p whose top term has degree strictly less than $-D_1(p)$ and whose lowest term has degree at least $-D_2(p)$. The skyscraper sheaf associated to this choice at each point is denoted by $\mathcal{T}_X[D_1/D_2]$.

Global Sections on Compact Riemann Surfaces. Let X be a compact Riemann surface. In this case many of the sheaves introduced above have well-understood groups of global sections.

If G is any group, then $G^X(X)$ is the group of all functions from X to G. Similarly, $\mathcal{C}^\infty(X)$ is the ring of all $\mathcal{C}^\infty$ complex-valued functions on X. Similarly for the other $\mathcal{C}^\infty$ sheaves of forms. (These examples do not use that X is compact.)

Since every holomorphic or harmonic function on a compact Riemann surface is constant, we have that

$$\mathcal{O}_X(X) = \mathcal{H}_X(X) = \mathbb{C} \text{ and } \mathcal{O}_X^*(X) = \mathbb{C}^*.$$

Since X is connected, the global sections of the sheaf of meromorphic functions $\mathcal{M}_X$ forms a field: this is exactly the field of meromorphic functions on X, which we have been denoting by $\mathcal{M}(X)$. The global sections of $\mathcal{M}_X^*$ is the multiplicative group of this field.

If we fix a divisor D on X, then the global sections of the sheaf $\mathcal{O}_X[D]$ of meromorphic functions with poles bounded by D is exactly the space $L(D)$:

$$\mathcal{O}_X[D](X) = L(D).$$

The connectivity of X insures that the global sections of any of the locally constant sheaves is simply the group itself. Thus

$$\underline{\mathbb{Z}}(X) = \mathbb{Z}, \quad \underline{\mathbb{R}}(X) = \mathbb{R}, \quad \text{and} \quad \underline{\mathbb{C}}(X) = \mathbb{C}.$$

Turning to 1-forms, we see that the global sections of Ω_X^1 of the sheaf of holomorphic 1-forms is the space of global holomorphic 1-forms, which we have been denoting by $\Omega^1(X)$. Similarly, if D is a divisor on X, then

$$\Omega_X^1[D](X) = L^{(1)}(D)$$

is the space of global meromorphic 1-forms with poles bounded by D.

A totally discontinuous sheaf $\mathcal{S}$ with groups G_p at each point p has the direct product of all of the groups as the group of global sections:

$$\mathcal{S}(X) = \prod_{p \in X} G_p.$$

In particular, the skyscraper sheaf G_p constructed with a single group G at the point p and the group $\{0\}$ at all other points has global sections equal to the group G. For example,

$$\mathbb{C}_p(X) = \mathbb{C}.$$

For a skyscraper sheaf $\mathcal{T}$ formed by taking the sections of a totally discontinuous sheaf which have discrete support, the group of global sections are those X-tuples with discrete support. If X is compact, then we must have only finite

support; therefore the group of global sections in this case is the direct sum of the point groups:
$$T(X) = \bigoplus_{p \in X} G_p \quad \text{if } X \text{ is compact.}$$

The global sections of the sheaf $\mathcal{D}iv_X$ of divisors is simply the group of divisors $\text{Div}(X)$ on X.

In the case of the discrete skyscraper sheaf of Laurent tail divisors $\mathcal{T}_X[D]$, the space of global sections on a compact Riemann surface is simply the space $T[D](X)$ introduced in Chapter VI, Section 2. The relative version $\mathcal{T}_X[D_1/D_2]$ has as its space of global sections the space of Laurent polynomials which at every point p has terms of degree at least $-D_2(p)$ and strictly less than $-D_1(p)$. This leaves only $D_2(p) - D_1(p)$ possible monomials; therefore on a compact Riemann surface the space of global sections of this sheaf is a vector space of dimension $\deg(D_2) - \deg(D_1)$.

Restriction to an Open Subset. Let $\mathcal{F}$ be a sheaf on a space X, and let $Y \subset X$ be an open set. Then it is easy to check that by only taking open subsets inside of Y, we obtain a sheaf $\mathcal{F}|_Y$ on Y; if $U \subset Y$ is open, then U is also open in X, and we set
$$\mathcal{F}|_Y(U) = \mathcal{F}(U).$$

We only note that if we restrict the sheaf of holomorphic functions $\mathcal{O}_X$ on X to the open subset Y, we obtain Y's sheaf of holomorphic functions; that is,
$$\mathcal{O}_X|_Y = \mathcal{O}_Y.$$

Problems IX.1

A. Check that the presheaves of functions $\mathcal{C}_X^\infty$, $\mathcal{O}_X$, $\mathcal{O}_X^*$, $\mathcal{M}_X$, $\mathcal{M}_X^*$, $\mathcal{H}_X$, and $\mathcal{O}_X[D]$ are all indeed sheaves on a Riemann surface X.

B. Check that the presheaves of forms $\mathcal{E}_X^1$, $\mathcal{E}_X^{1,0}$, $\mathcal{E}_X^{0,1}$, Ω_X^1, $\overline{\Omega}_X^1$, $\mathcal{M}_X^1$, $\Omega_X^1[D]$, and $\mathcal{E}_X^2$ are all indeed sheaves on a Riemann surface X.

C. Let S and T be two sets, and suppose for each index i there is given a function $f_i : S \to T$. Also suppose that a set R is given, with a function $g : R \to S$. We say that the diagram

$$R \xrightarrow{g} S \begin{array}{c} \xrightarrow{f_1} \\ \xrightarrow{f_2} \\ \vdots \end{array} T$$

is *exact* if $f_i \circ g = f_j \circ g$ for every pair of indices i and j, and if $s \in S$ with $f_i(s) = f_j(s)$ for every pair of indices i and j, then $s = g(r)$ for some $r \in R$. Express the sheaf axiom in terms of exactness of such a diagram of sets.

D. Show that if G is a group, and X is a topological space in which every pair of nonempty open sets intersect, then the constant presheaf G_X (defined by setting $G_X(U) = G$ for every nonempty open subset $U \subseteq X$) is indeed a sheaf on X.
E. Check that the general skyscraper construction produces a sheaf.
F. Prove that a totally discontinuous sheaf is a skyscraper sheaf if and only if the trivial group is used at all but a discrete set of points of X.
G. Let $\mathcal{F}$ and $\mathcal{G}$ be two sheaves of abelian groups on X. Define the direct sum sheaf $\mathcal{F} \oplus \mathcal{G}$ by setting

$$\mathcal{F} \oplus \mathcal{G}(U) = \mathcal{F}(U) \oplus \mathcal{G}(U)$$

for an open set $U \subseteq X$. Define the restriction maps for $\mathcal{F} \oplus \mathcal{G}$ using the restriction maps of $\mathcal{F}$ and of $\mathcal{G}$, and show that $\mathcal{F} \oplus \mathcal{G}$ is a sheaf on X.
H. Prove that if X is a Riemann surface and Y is an open subset, then

$$\mathcal{O}_X|_Y = \mathcal{O}_Y.$$

In general, show that if Z and Y are both open subsets, with $Z \subset Y \subset X$, and $\mathcal{F}$ is a sheaf on X, then

$$(\mathcal{F}|_Y)|_Z = \mathcal{F}|_Z.$$

I. Let $\mathcal{F}$ be a sheaf on X, and fix a point $p \in X$. Consider the disjoint union set $D = \sqcup_{U \ni p} \mathcal{F}(U)$ (note that the union is only over the open neighborhoods of p). Define an equivalence relation on D as follows: if $f \in \mathcal{F}(U)$ and $g \in \mathcal{F}(V)$ then we declare $f \sim g$ if there is an open neighborhood W of p contained in $U \cap V$ such that $\rho_W^U(f) = \rho_W^V(g)$. Show that this is an equivalence relation on D. The set of equivalence classes is called the *stalk* of $\mathcal{F}$ at p, and is denoted by $\mathcal{F}_p$. Note that there is a natural map $\pi : \mathcal{F}(U) \to \mathcal{F}_p$ for any open neighborhood U of p, sending a section of $\mathcal{F}$ over U to its equivalence class in the stalk.
J. Show that given any two elements s_1 and s_2 in the stalk $\mathcal{F}_p$, there is an open neighborhood U of p such that both s_i are represented by sections of $\mathcal{F}$ over U.
K. Show that if $\mathcal{F}$ is a sheaf of groups, then the stalk $\mathcal{F}_p$ inherits the group structure in the following way: if s_1 and s_2 are two elements of $\mathcal{F}_p$, then find an open neighborhood U of p and sections f_1 and f_2 in $\mathcal{F}(U)$ representing s_1 and s_2 respectively; then set $s_1 + s_2$ to be the equivalence class of $f_1 + f_2$ in $\mathcal{F}(U)$. Show this is well defined and gives a group structure on the stalk. Show that for any U the map $\pi : \mathcal{F}(U) \to \mathcal{F}_p$ is a homomorphism.
L. Show that the stalks of a locally constant sheaf $\underline{G}$ are all isomorphic to the group G.
M. If X is a Riemann surface, show that the stalks of the sheaf $\mathcal{O}_X$ of holomorphic functions on X are all isomorphic (after a choice of local coordinate) to the ring of convergent power series $\mathbb{C}\{z\}$ in one variable over $\mathbb{C}$.

N. What are the stalks of a totally discontinuous sheaf? What are the stalks of a skyscraper sheaf?

O. Define the sheaf of meromorphic n-fold differentials on a Riemann surface X.

2. Sheaf Maps

The concept of a sheaf assigns algebraic objects to geometric ones. No algebraic construction is complete until the maps between the algebraic objects are described. In the case of sheaves the maps are called sheaf maps, or sheaf homomorphisms.

Definition of a Map between Sheaves. Let $\mathcal{F}$ and $\mathcal{G}$ be two sheaves on a space X.

DEFINITION 2.1. A *sheaf map*, or *sheaf homomorphism*, from $\mathcal{F}$ to $\mathcal{G}$ is a collection of homomorphisms

$$\phi_U : \mathcal{F}(U) \to \mathcal{G}(U),$$

for every open set $U \subset X$, which commute with the restriction maps ρ for the two sheaves, that is,

$$\begin{array}{ccc} \mathcal{F}(U) & \stackrel{\phi_U}{\to} & \mathcal{G}(U) \\ \rho_V^U \downarrow & & \downarrow \rho_V^U \\ \mathcal{F}(V) & \stackrel{\phi_V}{\to} & \mathcal{G}(V) \end{array}$$

commutes whenever $V \subset U$ is an open subset. (Note that the restriction maps on the left are those for the sheaf $\mathcal{F}$, and those on the right are for the sheaf $\mathcal{G}$.) The sheaf map is denoted by ϕ.

If $\mathcal{F}$ and $\mathcal{G}$ are sheaves of groups, then the maps ϕ_U must be group homomorphisms. If they are sheaves of rings, then the maps must be ring homomorphisms.

We remark that the identity map $\mathrm{id} : \mathcal{F} \to \mathcal{F}$ is always a sheaf map, and that the composition of sheaf maps is a sheaf map.

Inclusion Maps. The first kind of sheaf maps one encounters are inclusions. These arise simply when, for every open set U, the group $\mathcal{F}(U)$ is a subgroup of the group $\mathcal{G}(U)$. The requirement of commutativity with the restriction maps is usually immediate in these situations.

Let us simply list the inclusions among the sheaves we have introduced so far.

- Constant sheaves: $\underline{\mathbb{Z}} \subset \underline{\mathbb{R}} \subset \underline{\mathbb{C}}$.
- Holomorphic/Meromorphic sheaves: $\underline{\mathbb{C}} \subset \mathcal{O}_X \subset \mathcal{M}_X$.
- Nonzero Holomorphic/Meromorphic sheaves: $\mathcal{O}_X^* \subset \mathcal{M}_X^*$.

- Sheaves of functions with bounded poles: if $D_1 \leq D_2$ are divisors on X, then $\mathcal{O}_X[D_1] \subset \mathcal{O}_X[D_2]$.
- Sheaves of $\mathcal{C}^\infty$ functions: $\mathcal{O}_X \subset \mathcal{H}_X \subset \mathcal{C}_X^\infty$.
- Sheaves of $\mathcal{C}^\infty$ 1-forms: $\Omega_X^1 \subset \mathcal{E}_X^{1,0} \subset \mathcal{E}_X^1$ and $\overline{\Omega}_X^1 \subset \mathcal{E}_X^{0,1} \subset \mathcal{E}_X^1$.
- Sheaves of meromorphic 1-forms: $\Omega_X^1 \subset \mathcal{M}_X^{(1)}$, $\Omega_X^1[D] \subset \mathcal{M}_X^{(1)}$ for any divisor D, and $\Omega_X^1[D_1] \subset \Omega_X^1[D_2]$ whenever $D_1 \leq D_2$.
- Sheaves of Laurent Tail Divisors: if $D_1 \leq D_2$ then $\mathcal{T}[D_2] \subset \mathcal{T}[D_1]$ and $\mathcal{T}[D_1/D_2] \subset \mathcal{T}[D_1]$.

Differentiation Maps. The second type of map which arises immediately is induced by differentiation, in all of its varied forms. Such maps always commute with restriction, and give sheaf maps. We have encountered all of these maps before, but without the sheaf terminology. We have:

- Differentiation of functions:

$$\begin{aligned}
d &: \mathcal{C}_X^\infty \to \mathcal{E}_X^1, \\
\partial &: \mathcal{C}_X^\infty \to \mathcal{E}_X^{1,0}, \\
\overline{\partial} &: \mathcal{C}_X^\infty \to \mathcal{E}_X^{0,1}, \\
\partial\overline{\partial} &: \mathcal{C}_X^\infty \to \mathcal{E}_X^2, \\
d(=\partial) &: \mathcal{O}_X \to \Omega_X^1, \\
d &: \mathcal{H}_X \to \Omega_X^1 \oplus \overline{\Omega}_X^1, \text{ and} \\
d(=\partial) &: \mathcal{M}_X \to \mathcal{M}_X^{(1)}.
\end{aligned}$$

- Differentiation of 1-forms:

$$\begin{aligned}
d &: \mathcal{E}_X^1 \to \mathcal{E}_X^2, \\
\partial(=d) &: \mathcal{E}_X^{0,1} \to \mathcal{E}_X^2, \text{ and} \\
\overline{\partial}(=d) &: \mathcal{E}_X^{1,0} \to \mathcal{E}_X^2.
\end{aligned}$$

Restriction or Evaluation Maps. The most general type of restriction map takes a function or form on X and suitably restricts it to a closed subset. In the case of Riemann surfaces, the most natural closed subset is simply a single point. In this case restriction maps tend to be simply evaluation maps of some form.

The simplest type of restriction is to take a nonzero meromorphic function on a Riemann surface X and attach at every point its order. This gives an integer at every point, which is zero outside of a discrete set; therefore we have a sheaf map div from the sheaf of nonzero meromorphic functions to the sheaf of divisors:

$$\text{div} : \mathcal{M}_X^* \to \mathcal{D}iv_X.$$

Fix a point p on a Riemann surface X, and denote by $\mathbb{C}_p$ the skyscraper sheaf supported at p, with ring the complex numbers $\mathbb{C}$ at p, and $\{0\}$ at all other points.

The first form of evaluation simply takes a function f and evaluates it at p. This gives a sheaf map
$$\mathrm{eval}_p : \mathcal{C}_X^\infty \to \mathbb{C}_p,$$
which on any open set U containing p sends the $\mathcal{C}^\infty$ function f defined on U to the constant $f(p)$. On an open set not containing p the sheaf map is identically zero.

More generally, fix a divisor D on X. Locally near the point p, the sections of $\mathcal{O}_X[D]$ are meromorphic functions whose order is at least $-D(p)$ at p. Therefore the Laurent series for such a function in terms of a local coordinate z at p can have terms no lower that the $z^{-D(p)}$ term. Therefore we can assign to such a function the coefficient of its lowest possible term, giving a sheaf map
$$\mathrm{eval}_p : \mathcal{O}_X[D] \to \mathbb{C}_p$$
sending $f = \sum_{n \geq -D(p)} c_n z^n$ to the coefficient $c_{-D(p)}$. Note that this map depends on the choice of local coordinate, unless $D(p) = 0$, in which case it is simply the evaluation map given above.

Finally, taking the residue of a meromorphic 1-form at a point p can be phrased as a sheaf map. Specifically, we have
$$\mathrm{Res} : \mathcal{M}_X^{(1)} \to \mathbb{C}_p$$
defined on an open set U containing p by sending a meromorphic 1-form ω on U to the residue $\mathrm{Res}_p(\omega)$. This residue map can also be defined on any of the sheaves $\Omega^1[D]$.

Multiplication Maps. There are several maps between sheaves which are defined by fixing a single global function or form, and multiplying by that function or form. These by their very nature always commute with restriction and give a map of sheaves.

For example, fix a global meromorphic function f on a Riemann surface X. Then for any open set $U \subset X$, multiplication by f gives a group homomorphism from $\mathcal{M}(U)$ to itself, which commutes with restriction; we obtain a sheaf map
$$\mu_f : \mathcal{M}_X \to \mathcal{M}_X.$$

This multiplication map can be applied to various subsheaves of $\mathcal{M}$. Fix a divisor D; then if $\mathrm{div}(g) \geq -D$ on an open set U, we will have $\mathrm{div}(fg) = \mathrm{div}(f) + \mathrm{div}(g) \geq \mathrm{div}(f) - D$ on U. Hence multiplication by f gives a natural map
$$\mu_f : \mathcal{O}_X[D] \to \mathcal{O}_X[D - \mathrm{div}(f)].$$

2. SHEAF MAPS

We can also multiply forms by f; if we do this with meromorphic 1-forms we obtain a sheaf map

$$\mu_f : \mathcal{M}_X^{(1)} \to \mathcal{M}_X^{(1)},$$

and if D is any divisor on X, we similarly have the sheaf map

$$\mu_f : \Omega_X^1[D] \to \Omega_X^1[D - \text{div}(f)].$$

Now fix a global meromorphic 1-form ω on X. Clearly sending a meromorphic function f to $f\omega$ gives a sheaf map

$$\mu_\omega : \mathcal{M}_X \to \mathcal{M}_X^{(1)}.$$

Again we may look at subsheaves; fixing a divisor D, we see that we have

$$\mu_\omega : \mathcal{O}_X[D] \to \Omega_X^1[D - \text{div}(\omega)]$$

using a computation identical to that given above.

Truncation Maps. Suppose that D is an ordinary divisor on a Riemann surface X. Choosing a local coordinate z_p at each point p, we have the discrete skyscraper sheaf of Laurent Tail divisors $\mathcal{T}_X[D]$ on X constructed by taking at each p the group of those Laurent polynomials in z_p whose top term has degree strictly less than $-D(p)$. If f is meromorphic on an open set U, then at each point p we may write the Laurent series for f in terms of z_p and truncate this series at the $-D(p)$ terms and higher. This gives a Laurent tail divisor, and this assignment gives a sheaf map

$$\alpha_D : \mathcal{M}_X \to \mathcal{T}_X[D]$$

for each D.

If $D_1 \leq D_2$, we may truncate the Laurent tail divisors in $\mathcal{T}_X[D_1]$ at the $-D_2(p)$ level for each p; this gives a sheaf map

$$t_{D_2}^{D_1} : \mathcal{T}_X[D_1] \to \mathcal{T}_X[D_2].$$

Finally, if $D_1 \leq D_2$, then we may take any meromorphic function whose poles are bounded by D_2 and truncate its Laurent series at the $-D_1$ level and higher; this gives a sheaf map

$$\alpha_{D_1/D_2} : \mathcal{O}_X[D_2] \to \mathcal{T}_X[D_1/D_2].$$

The Exponential Map. Given a holomorphic function f on an open set U of a Riemann surface X, the function $\exp(2\pi i f)$ is also holomorphic, and nowhere zero on U. This exponentiation commutes with restriction, and is a homomorphism from the additive group $\mathcal{O}_X(U)$ to the multiplicative group $\mathcal{O}_X^*(U)$; therefore we obtain the *exponential map* of sheaves

$$\exp(2\pi i -) : \mathcal{O}_X \to \mathcal{O}_X^*.$$

The Kernel of a Sheaf Map. Suppose that $\phi : \mathcal{F} \to \mathcal{G}$ is a sheaf map between two sheaves of groups on X. Define a subsheaf $\mathcal{K} \subset \mathcal{F}$, called the *kernel* of ϕ, by declaring that
$$\mathcal{K}(U) = \ker(\phi_U),$$
that is, the group associated to the open set U is exactly the kernel of the group homomorphism $\phi_U : \mathcal{F}(U) \to \mathcal{G}(U)$. Note that if ρ denotes the restriction map for $\mathcal{F}$, and if $f \in \mathcal{K}(U)$, then for any open subset $V \subset U$, the restriction $\rho_V^U(f)$ is in the group $\mathcal{K}(V)$, since
$$\phi_V(\rho_V^U(f)) = \rho_V^U(\phi_U(f)) = \rho_V^U(0) = 0.$$
It is immediate that, using the restriction maps from $\mathcal{F}$, the kernel $\mathcal{K}$ is a presheaf on X.

LEMMA 2.2. *The kernel presheaf is a sheaf.*

PROOF. We must check that the sheaf axiom holds. Let $\{U_i\}$ be a collection of open sets of X, and let $U = \cup_i U_i$ be their union. Suppose that $s_i \in \mathcal{K}(U_i)$ are given, such that for each i and j, the restrictions agree on the overlaps:
$$\rho_{U_i \cap U_j}^{U_i}(s_i) = \rho_{U_i \cap U_j}^{U_j}(s_j).$$
Since $\mathcal{F}$ is a sheaf, there exists a unique element s of $\mathcal{F}(U)$ such that $\rho_{U_i}^U(s) = s_i$ for each i. We must show that this element s is actually in the kernel $\mathcal{K}(U)$. For this we must check that $\phi_U(s) = 0$ in $\mathcal{G}(U)$.

Let $t_i = \rho_{U_i}^U(\phi_U(s))$ be the restriction to U_i of this section $\phi_U(s)$ of $\mathcal{G}(U)$. By the commutativity of ϕ with the restriction maps, we have that
$$t_i = \phi_{U_i}(\rho_{U_i}^U(s)) = \phi_{U_i}(s_i) = 0 \text{ for each } i,$$
since s_i are sections of the kernel sheaf. Now we use the sheaf axiom for $\mathcal{G}$ to conclude that, since t_i is zero for each i, the original section $\phi_U(s)$ is zero too. □

1-1 and Onto Sheaf Maps. Since sheaves are built to encode local properties of functions and forms, various properties of sheaf maps are also defined using local ideas only. The properties of being injective and surjective are of this type. Let $\phi : \mathcal{F} \to \mathcal{G}$ be a sheaf map.

DEFINITION 2.3. We say that ϕ is 1-1, or *injective*, if for every point p and every open set U containing p, there is an open subset $V \subset U$ containing p such that ϕ_V is 1-1. We say that ϕ is *onto*, or *surjective*, if for every point p every open set U containing p, and every section $f \in \mathcal{G}(U)$, there is an open subset $V \subset U$ containing p such that ϕ_V hits the restriction of f to V.

Note that we don't require all the maps ϕ_U to be either 1-1 or onto, but only "eventually" 1-1 or onto, in the sense above.

Actually, the definition for 1-1 can be considerably shortened:

LEMMA 2.4. *The following are equivalent for a sheaf map $\phi : \mathcal{F} \to \mathcal{G}$:*

(i) ϕ is 1-1.
(ii) ϕ_U is 1-1 for every open subset $U \subset X$.
(iii) The kernel sheaf for ϕ is the identically zero sheaf.

PROOF. Clearly (ii) and (iii) are equivalent and (ii) implies (i); hence we need only show that (i) implies (ii). Let U be an arbitrary open set, and let $s \in \mathcal{F}(U)$ with $\phi_U(s) = 0$ in $\mathcal{G}(U)$. We must show that $s = 0$.

For this it suffices to show that $\rho_V^U(s)$ is zero in $\mathcal{F}(V)$ for each subset V of an open covering of U, by the sheaf axiom.

For every point $p \in U$, since ϕ is 1-1, by definition there is an open subset $V_p \subset U$ containing p such that ϕ_{V_p} is 1-1. Let $s_p = \rho_{V_p}^U(s)$ be the restriction of s to V_p. Since the V_p's cover U it suffices to show that each s_p is zero. Since ϕ_{V_p} is 1-1, it then suffices to show that $\phi_{V_p}(s_p) = 0$. But

$$\phi_{V_p}(s_p) = \phi_{V_p}(\rho_{V_p}^U(s)) = \rho_{V_p}^U(\phi_U(s)) = \rho_{V_p}^U(0) = 0,$$

which proves the result. □

The analogous lemma is *not* true for onto maps of sheaves. Think of a sheaf map $\phi : \mathcal{F} \to \mathcal{G}$ as defining some operator on functions or forms (e.g., differentiation). To say that ϕ_U is onto for a particular open set U is to say that one can always solve the equation $\phi_U(f) = g$ for any $g \in \mathcal{G}(U)$, with some $f \in \mathcal{F}(U)$ (depending of course on g).

To say that the sheaf map ϕ is onto however means that one may not be able to solve all such equations on the open set U; it means exactly that one can solve the equation *on a possibly smaller* neighborhood of any point $p \in U$.

The prototype for this concept is the exponential mapping

$$\exp(2\pi i -) : \mathcal{O}_X \to \mathcal{O}_X^*$$

on a Riemann surface X. Suppose for example that X is the complex plane with the origin removed, i.e., $X = \mathbb{C}^*$. Consider the function $g(z) = 1/z$, which is nowhere zero on X and is therefore an element of $\mathcal{O}^*(X)$. There is no holomorphic function $f(z)$ such that $\exp(2\pi i f) = 1/z$; hence the exponential mapping is not onto for the open set X. However, at any point $p \in \mathbb{C}^*$, there is a branch of the logarithm $\ln(z)$ defined near p, and $f(z) = (-1/2\pi i)\ln(z)$ is a solution to the desired equation near p.

What we have just shown is that the exponential sheaf map is onto in this case. Indeed, since a branch of the logarithm exists at any nonzero point, the exponential sheaf map is onto for any Riemann surface.

Other examples of onto sheaf maps are given below.

EXAMPLE 2.5. Locally, any holomorphic 1-form may be written as df, for a holomorphic function f. Therefore the sheaf map $d : \mathcal{O} \to \Omega^1$ is onto.

EXAMPLE 2.6. Let $\mathcal{K}$ be the kernel of the sheaf map $d : \mathcal{E}^1 \to \mathcal{E}^2$, that is, $\mathcal{K}$ is the sheaf of d-closed 1-forms. Then since $ddf = 0$ for any $\mathcal{C}^\infty$ function f, we have that d maps the sheaf of $\mathcal{C}^\infty$ functions into $\mathcal{K}$. This map is onto by Poincaré's Lemma.

EXAMPLE 2.7. Locally, any $\mathcal{C}^\infty$ function g can be written as $\overline{\partial}f$; this is Dolbeault's Lemma. Therefore the sheaf map $\overline{\partial} : \mathcal{C}^\infty \to \mathcal{E}^{0,1}$ is onto. Applying conjugation gives that the sheaf map $\partial : \mathcal{C}^\infty \to \mathcal{E}^{1,0}$ is onto. At the 1-form level, this implies that the maps $\overline{\partial} : \mathcal{E}^{1,0} \to \mathcal{E}^2$ and $\partial : \mathcal{E}^{0,1} \to \mathcal{E}^2$ are both onto. Hence also $d : \mathcal{E}^1 \to \mathcal{E}^2$ is onto, and $\partial\overline{\partial} : \mathcal{C}^\infty \to \mathcal{E}^2$ is onto.

EXAMPLE 2.8. One can always write down a convergent Laurent series with any given order and any given lowest coefficient. Therefore the sheaf map $\text{eval}_p : \mathcal{O}[D] \to \mathbb{C}_p$ is onto for any divisor D and any point p.

EXAMPLE 2.9. Similarly, locally one can write down a meromorphic 1-form with an arbitrary residue at a given point. Therefore the sheaf map $\text{Res} : \mathcal{M}^{(1)} \to \mathbb{C}_p$ is onto for any point p. If D is any divisor such that $D(p) \geq 1$, then the sheaf map $\text{Res} : \Omega^1[D] \to \mathbb{C}_p$ is onto. It is the identically zero map if $D(p) \leq 0$: one is not allowing any pole of the 1-form at p, and so certainly the residue at p will be zero.

EXAMPLE 2.10. Since any Laurent polynomial can be realized by a meromorphic function locally, the sheaf map $\alpha_D : \mathcal{M}_X \to \mathcal{T}_X[D]$ is onto for a Riemann surface X. For the same reasons, if $D_1 \leq D_2$, the sheaf map $\alpha_{D_1/D_2} : \mathcal{O}_X[D_2] \to \mathcal{T}_X[D_1/D_2]$ is onto. In particular, any possible order for a meromorphic function can be locally achieved; therefore the divisor map div from the sheaf of nonzero meromorphic functions $\mathcal{M}_X^*$ to the sheaf of divisors $\mathcal{D}iv_X$ is onto.

EXAMPLE 2.11. If $D_1 \leq D_2$ are two ordinary divisors on a Riemann surface X, then the truncation map $t_{D_2}^{D_1} : \mathcal{T}_X[D_1] \to \mathcal{T}_X[D_2]$ is an onto map of sheaves.

Short Exact Sequences of Sheaves. As is the case with the theory of vector spaces and modules in general, there is a notion of a short exact sequence of sheaves which is quite useful. Recall that a short exact sequence of abelian groups is a sequence

$$0 \to A \to B \xrightarrow{\phi} C \to 0,$$

where the map ϕ from B to C is onto, and the map from A to B is an isomorphism onto the kernel of ϕ. We could in fact take A to simply be the kernel of ϕ, and then the map from A to B is an inclusion. This formulation is good enough for the sheafification of this concept.

DEFINITION 2.12. We say that a sequence of sheaf maps

$$0 \to \mathcal{K} \to \mathcal{F} \xrightarrow{\phi} \mathcal{G} \to 0$$

is a *short exact sequence of sheaves* if the sheaf map ϕ is onto, and the sheaf $\mathcal{K}$ is the kernel sheaf of ϕ.

Thus one obtains a short exact sequence of sheaves anytime one has an onto sheaf map for which you can identify the kernel.

EXAMPLE 2.13. On a Riemann surface, the sequence
$$0 \to \underline{\mathbb{C}} \to \mathcal{O} \stackrel{d=\partial}{\to} \Omega^1 \to 0$$
is exact, since the kernel of d in this setting is exactly the constant functions.

EXAMPLE 2.14. The sequence
$$0 \to \underline{\mathbb{Z}} \to \mathcal{O} \stackrel{\exp(2\pi i-)}{\to} \mathcal{O}^* \to 0$$
is exact, since the kernel of the exponential map is exactly the integer-valued functions.

EXAMPLE 2.15. The sequence
$$0 \to \mathcal{O} \to \mathcal{C}^\infty \stackrel{\overline{\partial}}{\to} \mathcal{E}^{0,1} \to 0$$
is exact, since a function f is holomorphic if and only if the Cauchy-Riemann equations are satisfied, and this is exactly that $\overline{\partial} f = 0$.

EXAMPLE 2.16. Similarly, the sequence
$$0 \to \Omega^1 \to \mathcal{E}^{1,0} \stackrel{\overline{\partial}}{\to} \mathcal{E}^2 \to 0$$
is exact.

EXAMPLE 2.17. The sequence
$$0 \to \mathcal{H} \to \mathcal{C}^\infty \stackrel{\partial\overline{\partial}}{\to} \mathcal{E}^2 \to 0$$
is exact, by the definition of the harmonic functions.

EXAMPLE 2.18. The sequence
$$0 \to \underline{\mathbb{C}} \to \mathcal{H} \stackrel{d}{\to} \Omega^1 \oplus \overline{\Omega}^1 \to 0$$
is exact.

EXAMPLE 2.19. The sequence
$$0 \to \underline{\mathbb{C}} \to \mathcal{C}^\infty \stackrel{d}{\to} \ker(d : \mathcal{E}^1 \to \mathcal{E}^2) \to 0$$
is exact, since the kernel of d in this setting is exactly the constant functions.

EXAMPLE 2.20. For any divisor D and any point p, the sequence
$$0 \to \mathcal{O}[D-p] \to \mathcal{O}[D] \stackrel{\mathrm{eval}_p}{\to} \mathbb{C}_p \to 0$$
is exact.

EXAMPLE 2.21. For any divisor D and any point p with $D(p) = 1$, the sequence
$$0 \to \Omega^1[D-p] \to \Omega^1[D] \xrightarrow{\text{Res}} \mathbb{C}_p \to 0$$
is exact.

EXAMPLE 2.22. Since a meromorphic function whose divisor is trivial has no zeroes or poles, the sequence
$$0 \to \mathcal{O}_X^* \to \mathcal{M}_X^* \xrightarrow{\text{div}} \mathcal{D}iv_X \to 0$$
is exact.

EXAMPLE 2.23. For any divisor D, the sequence
$$0 \to \mathcal{O}_X[D] \to \mathcal{M}_X \xrightarrow{\alpha_D} \mathcal{T}_X[D] \to 0$$
is exact. If $D_1 \leq D_2$, then the sequence
$$0 \to \mathcal{O}_X[D_1] \to \mathcal{O}_X[D_2] \xrightarrow{\alpha_{D_1/D_2}} \mathcal{T}_X[D_1/D_2] \to 0$$
is exact.

EXAMPLE 2.24. For any two divisors $D_1 \leq D_2$, the sequence
$$0 \to \mathcal{T}_X[D_1/D_2] \to \mathcal{T}_X[D_1] \xrightarrow{t} \mathcal{T}_X[D_2] \to 0$$
is exact.

Exact Sequences of Sheaves. The general definition of an exact sequence of sheaves will really not arise too often, but it is worth remarking on. We take our cue from the definition of a 1-1 or onto sheaf map.

DEFINITION 2.25. Let $\mathcal{A} \xrightarrow{\alpha} \mathcal{B} \xrightarrow{\beta} \mathcal{C}$ be a sequence of sheaf maps. This sequence is *exact at* $\mathcal{B}$ if, firstly, the composition of the maps is zero; and secondly, for every open set U and every point $p \in U$ and every section $b \in \mathcal{B}(U)$ which is in the kernel of ϕ_U, there is an open subset $V \subset U$ containing p such that $\rho_V^U(b)$ is in the image of α_V.

The reader should check that a short exact sequence of sheaf maps is exact at the three possible positions.

Sheaf Isomorphisms. Before leaving the subject of sheaf maps, we must make sure we know what the isomorphisms are in this category. Let $\phi : \mathcal{F} \to \mathcal{G}$ be a sheaf map. Taking our cue from algebra, we would declare ϕ to be an isomorphism if it is both 1-1 and onto; taking our cue from category theory, we would require ϕ to have an inverse sheaf map $\psi : \mathcal{G} \to \mathcal{F}$ (such that the composition either way is the identity sheaf map).

Luckily, these give equivalent definitions.

LEMMA 2.26. *A sheaf map $\phi : \mathcal{F} \to \mathcal{G}$ is both 1-1 and onto if and only if it has an inverse sheaf map.*

PROOF. Suppose that an inverse sheaf map ψ exists. Then ψ_U is an inverse to ϕ_U for every open set U; hence each ϕ_U is an isomorphism, and clearly then ϕ is both 1-1 and onto.

Suppose then that ϕ is both 1-1 and onto. It suffices to show that for every open set, the map ϕ_U is an isomorphism; then the collection of inverses will define the inverse sheaf map ψ to ϕ. Since ϕ is 1-1, every ϕ_U is 1-1 by Lemma 2.4.

Fix an open set U. We must now show ϕ_U is onto. To this end fix a section $g \in \mathcal{G}(U)$. Since the sheaf map ϕ is onto, for every point $p \in U$, there is an open neighborhood V_p of p contained in U such that $g_p = \rho^U_{V_p}(g)$ is in the image of ϕ_{V_p}. Let $f_p \in \mathcal{F}(V_p)$ be a preimage of g_p for each p.

The open subsets $\{V_p\}$ cover U. We will use the sheaf axiom to patch together the local sections $\{f_p\}$ to obtain a section f, which will map to the original section g. The first step is to check that the restrictions of different f_p's agree on the intersections. Therefore fix $p \neq q$, and let $W = V_p \cap V_q$; we must show that $\rho^{V_p}_W(f_p) = \rho^{V_q}_W(f_q)$. Since ϕ is 1-1, so is ϕ_W; hence to check this equality, it suffices to check it after applying ϕ_W. But

$$\phi_W(\rho^{V_p}_W(f_p)) = \rho^{V_p}_W(\phi_{V_p}(f_p)) = \rho^{V_p}_W(g_p) = \rho^{V_p}_W(\rho^U_{V_p}(g)) = \rho^U_W(g)$$

and similarly with q: $\phi_W(\rho^{V_q}_W(f_q)) = \rho^U_W(g)$ also. Hence they are equal to each other, and the collection of sections $\{f_p\}$ patch together (using the sheaf axiom for $\mathcal{F}$) to give a section $f \in \mathcal{F}(U)$.

Finally we must check that $\phi_U(f) = g$. Again by the sheaf axiom, this can be checked locally, on each V_p. Making the computation

$$\rho^U_{V_p}(\phi_U(f)) = \phi_{V_p}(f_p) = g_p = \rho^U_{V_p}(g)$$

shows that $\phi_U(f)$ and g do agree on each member of the open cover $\{V_p\}$ of U, and hence are equal by the sheaf axiom. □

Note that in the above proof of the surjectivity of ϕ_U, we needed to use the injectivity of ϕ; as we remarked earlier, it is not the case in general that a sheaf map is onto only when all of the homomorphisms ϕ_U are onto.

DEFINITION 2.27. A sheaf map $\phi : \mathcal{F} \to \mathcal{G}$ is an *isomorphism* if it is both 1-1 and onto, or, equivalently, if it has an inverse sheaf map.

There are several natural isomorphisms among the sheaves we have introduced so far. These are usually induced by multiplication. In particular, if f is a global meromorphic function on a Riemann surface X which is not identically zero, then

$$\mu_f : \mathcal{M}_X \to \mathcal{M}_X, \quad \mu_f : \mathcal{M}^{(1)}_X \to \mathcal{M}^{(1)}_X,$$

and for a fixed divisor D

$$\mu_f : \mathcal{O}_X[D] \to \mathcal{O}_X[D - \operatorname{div}(f)] \quad \text{and} \quad \mu_f : \Omega^1_X[D] \to \Omega^1_X[D - \operatorname{div}(f)]$$

are all sheaf isomorphisms; the inverse in each case is given by multiplication by $1/f$.

If ω is a nonzero global meromorphic 1-form on X, then the sheaf map
$$\mu_\omega : \mathcal{M}_X \to \mathcal{M}_X^{(1)}$$
and (for any divisor D) the sheaf map
$$\mu_\omega : \mathcal{O}_X[D] \to \Omega_X^1[D - \operatorname{div}(\omega)]$$
are both sheaf isomorphisms. Note that the first map $\mu_\omega : \mathcal{M}_X \to \mathcal{M}_X^{(1)}$ is only an isomorphism of sheaves of abelian groups; the sheaf $\mathcal{M}_X$ is also a sheaf of rings, which $\mathcal{M}_X^{(1)}$ is not.

Using Sheaves to Define the Category. Our first application of the language of sheaves is to use them to define a Riemann surface. Recall what a Riemann surface is: a second countable connected Hausdorff topological space with an atlas of compatible complex charts. We've seen above that if X is a Riemann surface, then X has a natural sheaf of rings on it, namely the sheaf $\mathcal{O}_X$ of holomorphic functions.

Now suppose that X is any second countable connected Hausdorff topological space, with a sheaf $\mathcal{R}$ of rings of complex-valued functions on X (that is, for each open subset $U \subset X$, the ring $\mathcal{R}(U)$ consists of a ring of functions from U to $\mathbb{C}$).

Fix any open set $U \subset X$, and a function $f : U \to V$, where $V \subset \mathbb{C}$ is an open subset of the complex plane. Composition with f then gives a ring homomorphism
$$\{\mathbb{C}\text{-valued functions on } A\} \xrightarrow{-\circ f} \{\mathbb{C}\text{-valued functions on } f^{-1}(A)\}$$
for any open subset $A \subset V$. Note that
$$\mathcal{O}_V(A) \subset \{\mathbb{C}\text{-valued functions on } A\}$$
and by assumption
$$\mathcal{R}(f^{-1}(A)) \subset \{\mathbb{C}\text{-valued functions on } f^{-1}(A)\}.$$

It is in this setting that we can formulate the definition of a Riemann surface in terms of sheaves.

PROPOSITION 2.28. *Let X be a second countable connected Hausdorff topological space, with a sheaf $\mathcal{R}$ of rings of complex-valued functions on X. Suppose an open cover $\{U_i\}$ of X is given, and homeomorphisms $f_i : U_i \to V_i \subset \mathbb{C}$ are given for each i, satisfying the following two conditions:*

(a) As complex-valued functions on U_i, each f_i is a section of the sheaf $\mathcal{R}$:
$$f_i \in \mathcal{R}(U_i) \quad \text{for each} \quad i;$$

(b) *Composition with f_i carries $\mathcal{O}_{V_i}$ isomorphically as a sheaf onto $\mathcal{R}|_{U_i}$ for each i, in the sense that*

$$- \circ f_i : \mathcal{O}_{V_i}(A) \xrightarrow{\cong} \mathcal{R}(f_i^{-1}(A))$$

is a ring isomorphism for every open $A \subset V_i$.

Then the homeomorphisms f_i are compatible chart maps on the space X and define a complex structure on X, making X into a Riemann surface.

PROOF. Since the f_i's are already homeomorphisms, they are certainly chart maps; the only thing to check is the compatibility. Fix two such, say $f_1 : U_1 \to V_1$ and $f_2 : U_2 \to V_2$. Let $W = U_1 \cap U_2$. We must show that $h = f_2 \circ f_1^{-1} : f_1(W) \to f_2(W)$ is holomorphic. The function h is defined on $f_1(W)$, which is a subset of V_1. Therefore to check that it is holomorphic, we must show that it is in $\mathcal{O}_{V_1}(f_1(W))$; by condition (b) above it then suffices to check that after composing with f_1 we obtain a function in $\mathcal{R}(W)$. But $h \circ f_1 = f_2|_W$, which by condition (a) is a section of $\mathcal{R}(W)$. □

Thus we see that a Riemann surface can be defined as a certain type of space with a special sheaf of functions defined on it. It would be hard to argue that this approach is simpler than the original. However this same approach may be generalized to many other categories, and is nowadays the "highbrow" approach to defining lots of geometric categories, including topological manifolds, differentiable manifolds, projective varieties, and schemes. We will not pursue this point of view much further, but it is worth being aware of.

Problems IX.2

A. Prove that the identity map is a sheaf map, and that the composition of two sheaf maps is a sheaf map.

B. Check that the sheaf maps described in Examples 2.5 through 2.11 are indeed onto as claimed.

C. Check that the sequences of sheaf maps described in Examples 2.13 through 2.24 are indeed short exact sequences of sheaves as claimed.

D. Check that a short exact sequence of sheaves is exact at the three possible positions. Check that a sheaf map $\phi : \mathcal{F} \to \mathcal{G}$ is 1-1 if and only if the sequence

$$0 \to \mathcal{F} \xrightarrow{\phi} \mathcal{G}$$

is exact at $\mathcal{F}$; check that ϕ is onto if and only if the sequence

$$\mathcal{F} \xrightarrow{\phi} \mathcal{G} \to 0$$

is exact at $\mathcal{G}$.

E. Suppose that a global $\mathcal{C}^\infty$ 2-form η is given on a Riemann surface X. Show that multiplication by η is a sheaf map from the sheaf of $\mathcal{C}^\infty$ functions $\mathcal{C}_X^\infty$ on X to the sheaf of $\mathcal{C}^\infty$ 2-forms $\mathcal{E}_X^2$ on X. Show that this multiplication map is an isomorphism if and only if η is nowhere zero.

F. Let ω be a global meromorphic 1-form on X, which is not identically zero. Show that for any divisor D on X, multiplication by ω
$$\mu_\omega : \mathcal{O}_X[D] \to \Omega^1_X[D - \text{div}(\omega)]$$
is a sheaf isomorphism.

G. Let $\phi : \mathcal{F} \to \mathcal{G}$ be a sheaf map, and fix a point $p \in X$. Define a map $\phi_p : \mathcal{F}_p \to \mathcal{G}_p$ on the stalks as follows. For an element $f \in \mathcal{F}_p$, represent f by a section $s \in \mathcal{F}(U)$ for some neighborhood U of p, and define $\phi_p(f) = g$, where g is represented by $\phi_U(s) \in \mathcal{G}(U)$. Show that this map ϕ_p is well defined, and is a homomorphism.

H. Show that a sheaf map ϕ is onto if and only if every stalk map ϕ_p is onto. Prove the same statement for 1-1 sheaf maps. Formulate and prove a criterion for a sequence of sheaf maps to be exact in terms of the stalk maps.

3. Čech Cohomology of Sheaves

Cohomology is a way of attaching ordinary groups to sheaves of groups (or rings to sheaves of rings, etc.) which measure the more global aspects of a sheaf. Sheaves are designed to make all local statements easy to formulate, and this is because in many instances local statements are easy to come by. However in geometry one is usually interested in global information.

The prototype for this phenomenon is the sheaf $\mathcal{O}_X[D]$ on a Riemann surface X. Locally, it is trivial to find meromorphic functions with poles bounded by a divisor D; if $D(p) = k$, and z is a local coordinate centered at p, then $z^{-k}h(z)$ for any holomorphic function h will be a local section of $\mathcal{O}_X[D]$. However, asking for a *global* meromorphic function with this property is a quite different matter; we have seen that this is exactly the content of the Riemann-Roch problem, and its solution may be rather tricky.

So in general we are left with the following situation: we can solve our problems locally easily, by finding sections of some sheaf; but we really want to solve problems globally, by finding global sections of that sheaf. The sheaf construction does offer some help in this direction, because the sheaf axiom itself insures that global sections will exist if local sections exist which agree on the overlap domains.

The cohomology construction turns the agreement condition into an algebraic one, by writing down a suitable homomorphism whose kernel is the set of sections which agree on their overlap domains. It not only provides a space of solutions, but in many cases a space of obstructions to finding solutions.

The technicalities of properly constructing the cohomology groups of a sheaf and the rather formidable notation involved should not be taken too seriously at one's first encounter with the theory. For many geometers, cohomology theory is applied mathematics: something to be appreciated and actively used more than studied and analyzed in its own right. Its use may be likened to driving a car: you can easily get a license to drive (and most people do) without being able to

3. ČECH COHOMOLOGY OF SHEAVES

build an engine.

For this reason we will take an abbreviated approach to cohomology, checking some but not all of the statements necessary to make a complete theory.

Čech Cochains. Let $\mathcal{F}$ be a sheaf of abelian groups on a topological space X. Let $\mathcal{U} = \{U_i\}$ be an open covering of X, and fix an integer $n \geq 0$. For every collection of indices $(i_0, i_1, \ldots, i_n)$, we denote the intersection of the corresponding open sets by

$$U_{i_0, i_1, \ldots, i_n} = U_{i_0} \cap U_{i_1} \cap \cdots \cap U_{i_n}.$$

The deletion of one of the indices is indicated with the use of a "$\widehat{i_k}$": the open set $U_{i_0, i_1, \ldots, \widehat{i_k}, \ldots, i_n}$ is exactly $U_{i_0, i_1, \ldots, i_{k-1}, i_{k+1}, \ldots, i_n}$. Note that we always have

$$U_{i_0, i_1, \ldots, i_n} \subseteq U_{i_0, i_1, \ldots, \widehat{i_k}, \ldots, i_n}.$$

DEFINITION 3.1. A *Čech n-cochain* (or simply *n-cochain*) for the sheaf $\mathcal{F}$ over the open cover $\mathcal{U}$ is a collection of sections of $\mathcal{F}$, one over each $U_{i_0, i_1, \ldots, i_n}$. The space of Čech n-cochains for $\mathcal{F}$ over $\mathcal{U}$ is denoted by $\check{C}^n(\mathcal{U}, \mathcal{F})$; thus

$$\check{C}^n(\mathcal{U}, \mathcal{F}) = \prod_{(i_0, i_1, \ldots, i_n)} \mathcal{F}(U_{i_0, i_1, \ldots, i_n}).$$

Thus a Čech 0-cochain is simply a collection $(f_i \in \mathcal{F}(U_i))$; that is, one gives a section of $\mathcal{F}$ over each open set in the cover. Similarly, a 1-cochain is a collection of sections of $\mathcal{F}$ over every double intersection of open sets in the cover; typical notation for a 1-cochain is (f_{ij}), where $f_{ij} \in \mathcal{F}(U_i \cap U_j)$ for every pair of indices i and j.

In general, an n-cochain would be denoted by $(f_{i_0, \ldots, i_n})$.

We note that if $\phi : \mathcal{F} \to \mathcal{G}$ is a sheaf map, then there is an induced map on cochains

$$\phi : \check{C}^n(\mathcal{U}, \mathcal{F}) \to \check{C}^n(\mathcal{U}, \mathcal{G})$$

for any open covering $\mathcal{U}$, sending a cochain $(f_{i_0, \ldots, i_n})$ to $(\phi(f_{i_0, \ldots, i_n}))$.

Čech Cochain Complexes. Define a "coboundary operator"

$$d : \check{C}^n(\mathcal{U}, \mathcal{F}) \to \check{C}^{n+1}(\mathcal{U}, \mathcal{F})$$

by setting

$$d((f_{i_0, \ldots, i_n})) = (g_{i_0, \ldots, i_{n+1}}),$$

where

$$g_{i_0, \ldots, i_{n+1}} = \sum_{k=0}^{n+1} (-1)^k \rho(f_{i_0, i_1, \ldots, \widehat{i_k}, \ldots, i_{n+1}}).$$

In the above formula ρ denotes the restriction map for the sheaf $\mathcal{F}$ corresponding to the subset $U_{i_0, \ldots, i_{n+1}} \subset U_{i_0, i_1, \ldots, \widehat{i_k}, \ldots, i_{n+1}}$.

At the 0 level, d sends a 0-cochain (f_i) to the 1-cochain (g_{ij}) where

$$g_{ij} = f_j - f_i,$$

suitably restricted of course. (We will often abuse the notation and drop the explicit mention of the suitable restriction map ρ.) We see here the algebraic patching condition: if $d(f_i) = 0$ as a 1-cochain, i.e., if every $g_{ij} = f_j - f_i$ s zero, then the f_i's agree on the overlap of their domains, and will patch together by the sheaf axiom to give a global section of the sheaf.

At the 1 level, d sends a 1-cochain (f_{ij}) to the 2-cochain (g_{ijk}), where

$$g_{ijk} = f_{jk} - f_{ik} + f_{ij}.$$

Any n-cochain c with $dc = 0$ is called an n-$cocycle$; the space of n-cocycles is denoted by $\check{Z}^n(\mathcal{U}, \mathcal{F})$. This is simply the kernel of d at the nth level.

Any n-cochain which is in the image of d (coming out of the space of $(n-1)$-cochains) is called an n-$coboundary$; the space of n-coboundaries is denoted by $\check{B}^n(\mathcal{U}, \mathcal{F})$. Note that since there are no (-1)-cochains, there are no 0-coboundaries; i.e., $\check{B}^0(\mathcal{U}, \mathcal{F}) = 0$ always.

It is elementary to check that $d \circ d = 0$; you should at least check this at the 0 level. Thus we have a *Čech cochain complex*

$$0 \to \check{C}^0(\mathcal{U}, \mathcal{F}) \xrightarrow{d} \check{C}^1(\mathcal{U}, \mathcal{F}) \xrightarrow{d} \check{C}^2(\mathcal{U}, \mathcal{F}) \xrightarrow{d} \ldots$$

(In general, a *complex* is a sequence of homomorphisms such that the composition of any two in a row gives the zero map.)

The coboundary operator commutes with the map on cochains coming from a map of sheaves.

Cohomology with respect to a Cover. The fact that $d \circ d = 0$ implies that every n-coboundary is an n-cocycle:

$$\check{B}^n(\mathcal{U}, \mathcal{F}) \subseteq \check{Z}^n(\mathcal{U}, \mathcal{F}).$$

DEFINITION 3.2. *The nth cohomology group $\check{H}^n(\mathcal{U}, \mathcal{F})$ of $\mathcal{F}$ with respect to the open cover $\mathcal{U}$ is the quotient group*

$$\check{H}^n(\mathcal{U}, \mathcal{F}) = \check{Z}^n(\mathcal{U}, \mathcal{F}) / \check{B}^n(\mathcal{U}, \mathcal{F})$$

of n-cocycles modulo n-coboundaries.

We have an immediate interpretation of $\check{H}^0$:

LEMMA 3.3. *For any open covering $\mathcal{U}$, the 0th cohomology group of a sheaf $\mathcal{F}$ is isomorphic to the group of global sections of $\mathcal{F}$:*

$$\check{H}^0(\mathcal{U}, \mathcal{F}) \cong \mathcal{F}(X).$$

PROOF. Since there are no 0-coboundaries, $\check{H}^0$ is simply $\check{Z}^0$, the space of 0-cocycles. Define the map $\alpha : \mathcal{F}(X) \to \check{C}^0$ by sending a global section f to the 0-cochain (f_i), where $f_i = \rho_{U_i}^X(f)$ is the restriction of f to U_i. Then $d(f_i) = (g_{ij})$, where $g_{ij} = f_j - f_i$; this is zero for every i and j, since both f_i and f_j are just the restrictions of f. Therefore α maps $\mathcal{F}(X)$ to the space of 0-cocycles $\check{Z}^0$, and hence to $\check{H}^0$.

3. ČECH COHOMOLOGY OF SHEAVES

That α is 1-1 and onto is exactly the content of the sheaf axiom. □

Since the coboundary map commutes with any map induced by a map of sheaves, such a map sends cocycles to cocycles and coboundaries to coboundaries; hence there is an induced map on cohomology groups:

$$\phi : \mathcal{F} \to \mathcal{G} \quad \text{induces} \quad \phi_* : \check{H}^n(\mathcal{U}, \mathcal{F}) \to \check{H}^n(\mathcal{U}, \mathcal{G}).$$

Refinements. We would like to associate, to every sheaf $\mathcal{F}$, a cohomology group which does not depend on the choice of any open covering. For this we need to compare the cohomology groups defined above for different coverings; this requires the following notion.

DEFINITION 3.4. Let $\mathcal{U} = \{U_i\}_{i \in I}$ and $\mathcal{V} = \{V_j\}_{j \in J}$ be two open coverings of X. We say that $\mathcal{V}$ is a *refinement* of $\mathcal{U}$, denoted by $\mathcal{V} \prec \mathcal{U}$, if for every open set V_j from the covering $\mathcal{V}$ there is an open set U_i from the covering $\mathcal{U}$ with $V_j \subseteq U_i$.

One also says that $\mathcal{V}$ is *finer than* $\mathcal{U}$ if it is a refinement. Any choice of such a U_i for every V_j can be viewed as a function $r : J \to I$ on the index sets for the two coverings, such that $V_j \subset U_{r(j)}$ for every j. Such a function is called a *refining map* for the coverings. The refining map is not unique.

Note that the concept of refining gives a partial ordering on the set of all coverings of X.

We leave to the reader to check the following examples.

EXAMPLE 3.5. Any subcovering of an open covering is a refinement. In particular, note that if one takes an open covering and simply adds more open sets, the original covering is finer than the one with the added open sets. This is a bit counterintuitive, so beware.

EXAMPLE 3.6. Let X be any topological space in which points are closed, and let $\mathcal{U}$ be an open cover of X. Then for any point p in X there is a refinement $\mathcal{V}$ of $\mathcal{U}$ such that p is in only one open set of $\mathcal{V}$.

EXAMPLE 3.7. Let X be any Riemann surface. Then any open covering has a refinement consisting entirely of chart domains.

EXAMPLE 3.8. Any two open coverings have a common refinement.

Suppose now that $\mathcal{V} = \{V_j\}_{j \in J}$ is a refinement of $\mathcal{U} = \{U_i\}_{i \in I}$. Let $r : J \to I$ be a refining map. Then r induces a map on n-cochains $\tilde{r} : \check{C}^n(\mathcal{U}, \mathcal{F}) \to \check{C}^n(\mathcal{V}, \mathcal{F})$ by the formula

$$\tilde{r}((f_{i_0,\ldots,i_n})) = (g_{j_0,\ldots,j_n}),$$

where

$$g_{j_0,\ldots,j_n} = f_{r(j_0),\ldots,r(j_n)},$$

restricted to $V_{j_0,\ldots,j_n}$.

The following is an immediate check which we leave to the reader.

LEMMA 3.9. *With the above notations, the map $\tilde{r}$ on n-cochains sends n-cocycles to n-cocycles and n-coboundaries to n-coboundaries. Therefore $\tilde{r}$ induces a map*
$$H(r) : \check{H}^n(\mathcal{U}, \mathcal{F}) \to \check{H}^n(\mathcal{V}, \mathcal{F})$$
for every n.

We remark that $H(r)$ on the 0 level is the identity, once $\check{H}^n(\mathcal{U}, \mathcal{F})$ and $\check{H}^n(\mathcal{V}, \mathcal{F})$ are identified with the space of global sections $\mathcal{F}(X)$.

We see that having a refining map r gives a way of comparing the cohomology spaces associated to two covers, one of which is finer than the other. In fact the particular refining map is irrelevant:

LEMMA 3.10. *The map $H(r)$ on the cohomology groups is independent of the refining map r, and depends only on the two coverings $\mathcal{U}$ and $\mathcal{V}$.*

PROOF. Suppose that r and r' are both refining maps for the refinement $\mathcal{V} \prec \mathcal{U}$. We must show that $H(r) = H(r')$. This is clear for $n = 0$, so assume that $n \geq 1$.

Fix a cohomology class $h \in \check{H}^n(\mathcal{U}, \mathcal{F})$, and represent h by an n-cocycle $(f_{i_0,\ldots,i_n})$. Then $H(r)(h)$ is represented by the n-cocycle $(g_{j_0,\ldots,j_n})$, and $H(r')(h)$ is represented by the n-cocycle $(g'_{j_0,\ldots,j_n})$, where
$$g_{j_0,\ldots,j_n} = f_{r(j_0),\ldots,r(j_n)} \quad \text{and} \quad g'_{j_0,\ldots,j_n} = f_{r'(j_0),\ldots,r'(j_n)}$$
for every set of $n+1$ indices $j_0, \ldots, j_n$. We must show that the difference $(g'_{j_0,\ldots,j_n} - g_{j_0,\ldots,j_n})$ is zero in cohomology, i.e., that it is a coboundary.

Form the $(n-1)$-cochain $(h_{\ell_0,\ldots,\ell_{n-1}})$ defined by
$$h_{\ell_0,\ldots,\ell_{n-1}} = \sum_{k=0}^{n-1} (-1)^k f_{r(\ell_0),\ldots,r(\ell_k),r'(\ell_k),\ldots,r'(\ell_{n-1})}.$$

A computation (using the fact that $(f_{i_0,\ldots,i_n})$ is a cocycle) yields that
$$d((h_{\ell_0,\ldots,\ell_{n-1}})) = (g'_{j_0,\ldots,j_n} - g_{j_0,\ldots,j_n});$$
Therefore these two cocycles differ by a coboundary, and we conclude that $H(r)(h) = H(r')(h)$. □

We will therefore denote this refining map on the cohomology level simply by $H_\mathcal{V}^\mathcal{U}$.

Note that if $\mathcal{W} \prec \mathcal{V} \prec \mathcal{U}$ are three covers, each finer than the next, then
$$H_\mathcal{W}^\mathcal{V} \circ H_\mathcal{V}^\mathcal{U} = H_\mathcal{W}^\mathcal{U}.$$

Also, these refining maps commute with any map ϕ_* induced by a map of sheaves.

LEMMA 3.11. *The map $H_\mathcal{V}^\mathcal{U}$ on the cohomology groups is 1-1 at the $\check{H}^1$ level.*

3. ČECH COHOMOLOGY OF SHEAVES

PROOF. Let $(f_{ab}) \in \check{Z}^1(\mathcal{U}, \mathcal{F})$ represent a class in $\check{H}^1(\mathcal{U}, \mathcal{F})$ which goes to zero in $\check{H}^1(\mathcal{V}, \mathcal{F})$ under the map $H_{\mathcal{V}}^{\mathcal{U}}$. This means that the 1-cocycle $(f_{r(i)r(j)})$ is a coboundary; so there is a 0-chain $(g_k) \in \check{C}^0(\mathcal{V}, \mathcal{F})$ such that $f_{r(i)r(j)} = g_j - g_i$ for every index i and j for the $\mathcal{V}$ covering. To prove that $H_{\mathcal{V}}^{\mathcal{U}}$ is 1-1, we must show that in fact (f_{ab}) is a coboundary.

Fix an index k for the $\mathcal{U}$ covering, and note that since (f_{ab}) is a cocycle,
$$g_j - g_i = f_{r(i)r(j)} = f_{kr(j)} - f_{kr(i)} \text{ on } V_i \cap V_j \cap U_k;$$
therefore for every i and j,
$$g_j - f_{kr(j)} = g_i - f_{kr(i)} \text{ on } V_i \cap V_j \cap U_k.$$
As we vary i, we obtain a covering of U_k by the sets $V_i \cap U_k$. Hence the sheaf axiom for $\mathcal{F}$ implies that there is a unique section $h_k \in \mathcal{F}(U_k)$ such that $h_k = g_i - f_{kr(i)}$ on $V_i \cap U_k$.

We claim that $d((h_k)) = (f_{ab})$, so that (f_{ab}) is in fact a coboundary and its cohomology class is zero. For this we must show that $f_{ab} = h_b - h_a$ on $U_a \cap U_b$ for every pair of indices a, b for the $\mathcal{U}$ covering.

If we fix an index k for the $\mathcal{V}$ covering, we see that on $U_a \cap U_b \cap V_k$
$$f_{ab} = f_{ar(k)} - f_{br(k)} = f_{ar(k)} - g_k - f_{br(k)} + g_k = h_b - h_a.$$
As we vary k, we obtain a covering of $U_a \cap U_b$ by the sets $U_a \cap U_b \cap V_k$; hence the sheaf axiom allows us to conclude that $f_{ab} = h_b - h_a$ as required. □

Čech Cohomology Groups. We are now in a position to define a series of cohomology groups which do not depend on a covering. This involves the direct limit concept. Suppose a partially ordered index set A is given, and a group G_a is given for each $a \in A$. (In our application the index set is the set of all coverings, and to each covering $\mathcal{U}$ we will associate the group $\check{H}^n(\mathcal{U}, \mathcal{F})$.) Suppose that a map $H_a^b : G_b \to G_a$ is given for every pair of comparable indices $a < b$, satisfying $H_a^b \circ H_b^c = H_a^c$ if $a < b < c$. Finally assume that for every pair of indices a and b there is an index c with $c < a$ and $c < b$. (Such a collection of groups and maps are called a *direct system* of groups.)

In this situation the *direct limit* of the system of groups $\{G_a \mid a \in A\}$ exists; this is a group, denoted by
$$L = \varinjlim_{a \in A} G_a,$$
together with maps $h_a : G_a \to L$ for every a, such that $h_a \circ H_a^b = h_b$ for every $a < b$. Moreover L is universal with respect to this property, in the sense that if any other group L' receives a map $h_a' : G_a \to L'$ from every group G_a, such that $h_a' \circ H_a^b = h_b'$ for every $a < b$, then there is a unique homomorphism $f : L \to L'$ such that $f \circ h_a = h_a'$ for every a.

In case all of the groups G_a are subgroups of some fixed larger group G, and all of the maps are inclusions, then the direct limit is simply the union of the subgroups G_a.

DEFINITION 3.12. Fix a sheaf $\mathcal{F}$ on X and an integer $n \geq 0$. The nth Čech cohomology group of $\mathcal{F}$ on X is the group

$$\check{H}^n(X, \mathcal{F}) = \varinjlim_{\mathcal{U}} \check{H}^n(\mathcal{U}, \mathcal{F}).$$

Since at the $\check{H}^0$ level, all of the groups are isomorphic to $\mathcal{F}(X)$ and all of the maps $H_\mathcal{V}^\mathcal{U}$ are compatible isomorphisms, the direct limit is also isomorphic to $\mathcal{F}(X)$:

$$\check{H}^0(X, \mathcal{F}) \cong \mathcal{F}(X).$$

At the $\check{H}^1$ level, we have seen that all of the maps are 1-1; in this case one can imagine that all of the cohomology groups are subgroups of a fixed space with the maps being inclusions. If this were true then

$$\check{H}^1(X, \mathcal{F}) = \bigcup_\mathcal{U} \check{H}^1(\mathcal{U}, \mathcal{F}).$$

We can achieve this with a bit of set-theoretic trickery as follows. Form the disjoint union A of all the Čech cohomology groups

$$A = \bigsqcup_\mathcal{U} \check{H}^1(\mathcal{U}, \mathcal{F}).$$

Define an equivalence relation $\approx$ on A by declaring $h_1 \in \check{H}^1(\mathcal{U}_1, \mathcal{F})$ equivalent to $h_2 \in \check{H}^1(\mathcal{U}_2, \mathcal{F})$ if there is a common refinement $\mathcal{V}$ of $\mathcal{U}_1$ and $\mathcal{U}_2$ such that $H_\mathcal{V}^{\mathcal{U}_1}(h_1) = H_\mathcal{V}^{\mathcal{U}_2}(h_2)$ in $\check{H}^1(\mathcal{V}, \mathcal{F})$. The set of equivalence classes $B = A/\approx$ has as natural subsets the isomorphic images $B_\mathcal{U}$ of the groups $\check{H}^1(\mathcal{U}, \mathcal{F})$, and inherits a group structure from them; moreover B is their union. Thus we have artificially constructed a single group B with the cohomology groups as subgroups, and it is not hard to see that B is the direct limit.

In this case the following is clear:

COROLLARY 3.13. For any sheaf $\mathcal{F}$ on X, $\check{H}^1(X, \mathcal{F}) = 0$ if and only if $\check{H}^1(\mathcal{U}, \mathcal{F}) = 0$ for every open covering $\mathcal{U}$.

Using the maps given in the direct limit, we have natural maps from every $\check{H}^n(\mathcal{U}, \mathcal{F})$ to $\check{H}^n(X, \mathcal{F})$. Thus every n-cocycle for any covering gives a class in $\check{H}^n(X, \mathcal{F})$. This class is zero if and only if there is a refinement of the covering such that the class is zero in the cohomology group for the refinement; that is, it is zero if and only if it is a coboundary after some refinement.

By the universal property of direct limits, if $\phi: \mathcal{F} \to \mathcal{G}$ is a sheaf map, then the collection of ϕ_*'s on the cohomology groups for the coverings induces a map on the limit group, which is also called ϕ_*:

$$\phi_*: \check{H}^n(X, \mathcal{F}) \to \check{H}^n(X, \mathcal{G}).$$

This map is functorial in the sense that $\mathrm{id}_* = \mathrm{id}$ and $(\phi \circ \psi)_* = \phi_* \circ \psi_*$.

To read more about the direct limit construction, see [**AM69**] or [**Lang84**].

3. ČECH COHOMOLOGY OF SHEAVES

The Connecting Homomorphism. Suppose that $\phi : \mathcal{F} \to \mathcal{G}$ is an onto map of sheaves. Let $\mathcal{K}$ be the kernel sheaf for ϕ. Let us define a map, called the *connecting homomorphism*

$$\Delta : \check{H}^0(X, \mathcal{G})(\cong \mathcal{G}(X)) \to \check{H}^1(X, \mathcal{K})$$

as follows. Take $g \in \mathcal{G}(X)$. Since ϕ is onto, for every point p there is a neighborhood U_p of p such that $g = \phi(f_p)$ on U_p for some $f_p \in \mathcal{F}(U_p)$. Note that the collection $\mathcal{U} = \{U_p\}$ is an open cover of X; let $h_{pq} = f_q - f_p \in \mathcal{F}(U_p \cap U_q)$. It is clear that (h_{pq}) is a 1-cocycle for the sheaf $\mathcal{F}$; moreover $\phi(h_{pq}) = 0$, since the difference is essentially $g - g$. Therefore (h_{pq}) is a 1-cocycle for the kernel sheaf $\mathcal{K}$, and represents a cohomology class in $\check{H}^1(\mathcal{U}, \mathcal{K})$. Its image in $\check{H}^1(X, \mathcal{K})$ will be denoted by $\Delta(g)$.

LEMMA 3.14. *This construction of $\Delta(g)$ is independent of the choice of covering $\mathcal{U}$ and the choice of preimages f_p.*

PROOF. Fix an open covering $\mathcal{U} = \{U_i\}$ and let us first check the independence of the choice of preimages. Suppose that on each U_i, there are two sections f_i and f'_i in $\mathcal{F}(U_i)$ such that $\phi(f_i) = \phi(f'_i) = g|_{U_i}$ for every i. Set $h_{pq} = f_q - f_p$ and $h'_{pq} = f'_q - f'_p$; we must show that (h_{pq}) and (h'_{pq}), which are both 1-cocycles for the kernel sheaf $\mathcal{K}$, differ by a 1-coboundary. Define $k_i = f_i - f'_i \in \mathcal{F}(U_i)$; note that in fact $k_i \in \mathcal{K}(U_i)$ for every i, so that (k_i) is a 0-cochain for the kernel sheaf $\mathcal{K}$. Moreover $d(k_i) = (\ell_{pq})$, where

$$\ell_{pq} = k_q - k_p = (f_q - f'_q) - (f_p - f'_p) = h_{pq} - h'_{pq}.$$

Therefore the difference $(h_{pq}) - (h'_{pq}) = (\ell_{pq})$ is a coboundary, and we have produced the same element in the cohomology group $\check{H}^1(\mathcal{U}, \mathcal{K})$; in particular, we have the same element in the direct limit group $\check{H}^1(X, \mathcal{K})$.

To check the independence of the choice of coverings, we may assume that one of the coverings is finer than the other (since any two coverings have a common refinement). Suppose then that $\mathcal{V} \prec \mathcal{U}$, and let $f_i \in \mathcal{F}(U_i)$ be preimages of $g|_{U_i}$ for each i. If r is a refining map for the comparison between $\mathcal{V}$ and $\mathcal{U}$, then note that on V_j, we may set $f'_j = f_{r(j)}|_{V_j}$ and obtain preimages of g on the sets V_j. Since we have already proven that for a fixed covering, the construction of $\Delta(g)$ does not depend on the choice of preimage, we may compare the use of the coverings $\mathcal{U}$ and $\mathcal{V}$ using the preimages f_i and f'_j.

But in this case, the 1-cocycle (f_i) exactly maps to the 1-cocycle (f'_j) under the map $H(r)$ which induces the map $H^{\mathcal{U}}_{\mathcal{V}}$ in cohomology. Therefore in the direct limit group $\check{H}^1(X, \mathcal{K})$, these two cocycles are the same. □

The purpose of the connecting homomorphism Δ is to give a criterion for when a given global section $g \in \mathcal{G}(X)$ is hit by a global section of $\mathcal{F}$:

LEMMA 3.15. *Suppose $g \in \mathcal{G}(X)$ is a global section of $\mathcal{G}$. Then there is a global section $s \in \mathcal{F}(X)$ of $\mathcal{F}$ such that $\phi(s) = g$ if and only if $\Delta(g) = 0$.*

PROOF. Suppose that $\phi(s) = g$ for some $s \in \mathcal{F}(X)$. Then, in the definition of the connecting homomorphism given above, We may choose $U_p = X$ for every point p, and $f_p = s$. Using the notation above, we then have $h_{pq} = 0$ for every p and q, so this is the identically zero 1-cocycle, which of course induces the 0 element in cohomology.

Conversely, suppose that $\Delta(g) = 0$ in $\check{H}^1(X, \mathcal{K})$. Using the above notation, this means that the 1-cocycle (h_{pq}) is a coboundary, and we may write $h_{pq} = k_q - k_p$ for some 0-cochain (k_p) for the kernel sheaf $\mathcal{K}$. Set $s_p = f_p - k_p$, where f_p is the preimage of g under ϕ locally on the set U_p.

On the overlap $U_p \cap U_q$, we have

$$s_p - s_q = (f_p - k_p) - (f_q - k_q) = (k_q - k_p) - (f_q - f_p) = k_q - k_p - h_{pq} = 0,$$

and so by the sheaf axiom the sections $\{s_p\}$ patch together to give a global section $s \in \mathcal{F}(X)$.

We claim that $\phi(s) = g$. By the sheaf axiom, it is enough to see this locally on each U_p. But on each U_p, we have

$$g|_{U_p} = \phi(f_p) = \phi(f_p - k_p) = \phi(s_p) = \phi(s|_{U_p}) = \phi(s)|_{U_p},$$

so we are done. $\square$

COROLLARY 3.16. *Let $\phi : \mathcal{F} \to \mathcal{G}$ be an onto map of sheaves with kernel sheaf $\mathcal{K}$. Then the map on global sections $\phi : \mathcal{F}(X) \to \mathcal{G}(X)$ is onto if $\check{H}^1(X, \mathcal{K}) = 0$.*

One should view this as follows: if this $\check{H}^1$ vanishes, then we can solve an equation globally on X. This is always important information.

The Long Exact Sequence of Cohomology. The property expressed in Lemma 3.15 can be viewed as saying that the sequence of maps

$$\mathcal{F}(X) \xrightarrow{\phi_X} \mathcal{G}(X) \xrightarrow{\Delta} \check{H}^1(X, \mathcal{K})$$

is an exact sequence of groups. This little exact sequence is part of a long exact sequence of cohomology groups:

PROPOSITION 3.17 (THE LONG EXACT SEQUENCE IN COHOMOLOGY). *Let $\phi : \mathcal{F} \to \mathcal{G}$ be an onto map of sheaves with kernel sheaf $\mathcal{K}$. Then the sequence*

$$0 \to \mathcal{K}(X) \xrightarrow{\text{inc}} \mathcal{F}(X) \xrightarrow{\phi_X} \mathcal{G}(X) \xrightarrow{\Delta} \check{H}^1(X, \mathcal{K}) \xrightarrow{\text{inc}_*} \check{H}^1(X, \mathcal{F}) \xrightarrow{\phi_*} \check{H}^1(X, \mathcal{G})$$

is exact at every step.

Here "inc" is the inclusion map of the kernel sheaf $\mathcal{K}$ into $\mathcal{F}$.

PROOF. The exactness at $\mathcal{K}(X)$ and at $\mathcal{F}(X)$ is just the definition of the kernel sheaf. The exactness at $\mathcal{G}(X)$ is, as mentioned above, exactly the content of Lemma 3.15.

To see that $\text{image}(\Delta) \subset \ker(\text{inc}_*)$, suppose that $g \in \mathcal{G}(X)$. The first step in defining $\Delta(g)$ is to choose an open covering $\{U_i\}$ and find elements $f_i \in \mathcal{F}(U_i)$

with $\phi_{U_i}(f_i) = g|_{U_i}$; then $\Delta(g)$ is defined by the 1-cocycle $f_i - f_j$ for the sheaf $\mathcal{K}$. But this cocycle is obviously a coboundary in the sheaf $\mathcal{F}$.

To finish the exactness at $\check{H}^1(X, \mathcal{K})$, we must check that $\ker(\text{inc}_*) \subset \text{image}(\Delta)$. Suppose that (k_{ij}) is a 1-cocycle for the sheaf $\mathcal{K}$ which represents a class in the kernel of inc_*. Then (k_{ij}) is a coboundary, considered as a 1-cocycle for the sheaf $\mathcal{F}$, and so there is a 0-cochain (f_i) such that $k_{ij} = f_j - f_i$ on $U_i \cap U_j$ for every i and j. Consider the 0-cochain (g_i) for $\mathcal{G}$, where $g_i = \phi(f_i)$. Note that

$$g_i - g_j = \phi(f_i - f_j) = \phi(k_{ji}) = 0$$

on $U_i \cap U_j$, so by the sheaf axiom for $\mathcal{G}$ there is a global section $g \in \mathcal{G}(X)$ such that $g|_{U_i} = g_i$ for every i. It is clear from the definition of Δ that $\Delta(g)$ is the class of (k_{ij}).

Finally we must check the exactness at $\check{H}^1(X, \mathcal{F})$. It is clear that $\text{inc}_* \circ \phi_* = 0$, so we only need to check that $\ker(\phi_*) \subset \text{image}(\text{inc}_*)$. Let c be a class in $\ker(\phi_*)$, and represent c by a 1-cocycle (f_{ij}) with respect to some open covering $\mathcal{U}$. Since $\phi_*(c) = 0$, we have that the 1-cocycle $(\phi(f_{ij}))$ represents 0 in $\check{H}^1(X, \mathcal{G})$. Therefore it is a coboundary; there is a 0-cochain (g_i) with respect to the cover $\mathcal{U}$ such that $\phi(f_{ij}) = g_j - g_i$ for every i and j in J. After refining $\mathcal{U}$ further we may assume, since ϕ is an onto map of sheaves, that each g_i is equal to $\phi(f_i)$ for some element $f_i \in \mathcal{F}(U_i)$.

Let $h_{ij} = f_{ij} - f_j + f_i \in \mathcal{F}(U_i \cap U_j)$; this is clearly a 1-cocycle since (f_{ij}) is. Applying ϕ, we see that

$$\phi(h_{ij}) = \phi(f_{ij}) - g_j + g_i = 0,$$

so that (h_{ij}) is actually a 1-cocycle for the kernel sheaf $\mathcal{K}$. Since it differs from the cocycle (f_{ij}) by the coboundary of the 0-cochain (f_i), it also gives the original class c in cohomology. Thus c is in the image of inc_*. □

The above proposition is usually expressed as saying that "a short exact sequence of sheaves gives a long exact sequence in cohomolomogy". The sequence continues under certain hypotheses. An open covering of a space is *locally finite* if every point has a neighborhood which intersects only finitely many of the open sets in the covering. A space is *paracompact* if it is Hausdorff and every open covering has a locally finite refinement. Every Riemann surface is paracompact; indeed, any manifold is paracompact (see [**Munkres75**]). Paracompactness is the property which insures that the long exact sequence of cohomology continues past the $\check{H}^1$ level. We will not prove this here; see [**Serre55**, section 25].

THEOREM 3.18. *Let X be a paracompact space (e.g., a Riemann surface) and let*

$$0 \to \mathcal{K} \to \mathcal{F} \to \mathcal{G} \to 0$$

be a short exact sequence of sheaves on X. Then there are connecting homomorphisms $\Delta : \check{H}^n(X, \mathcal{G}) \to \check{H}^{n+1}(X, \mathcal{K})$ for every $n \geq 0$ such that the sequence of *cohomology groups*

$$\begin{aligned}
0 &\to \check{H}^0(X,\mathcal{K}) \xrightarrow{\mathrm{inc}_*} \check{H}^0(X,\mathcal{F}) \xrightarrow{\phi_*} \check{H}^0(X,\mathcal{G}) \xrightarrow{\Delta} \\
&\to \check{H}^1(X,\mathcal{K}) \xrightarrow{\mathrm{inc}_*} \check{H}^1(X,\mathcal{F}) \xrightarrow{\phi_*} \check{H}^1(X,\mathcal{G}) \xrightarrow{\Delta} \\
&\to \check{H}^2(X,\mathcal{K}) \xrightarrow{\mathrm{inc}_*} \check{H}^2(X,\mathcal{F}) \xrightarrow{\phi_*} \check{H}^2(X,\mathcal{G}) \xrightarrow{\Delta} \cdots
\end{aligned}$$

is exact.

Problems IX.3

A. Show that the coboundary operator d commutes with the map on cochains coming from a map of sheaves.

B. Check that the map $\alpha : \mathcal{F}(X) \to \check{Z}^0$ defined by sending a global section f of $\mathcal{F}$ to the 0-cocycle (f_i) (where each f_i is simply the restriction of f to U_i) is a bijection, by using the sheaf axiom for $\mathcal{F}$.

C. Show that refinement gives a partial ordering on the set of all open coverings of a space X.

D. Verify Example 3.6: for any space X in which points are closed, and any open cover $\mathcal{U}$ of X, and for any point p in X, there is a refinement $\mathcal{V}$ of $\mathcal{U}$ such that p is in only one open set of $\mathcal{V}$.

E. Verify Example 3.7: any open covering of a Riemann surface X has a refinement consisting entirely of chart domains.

F. Show that any two open coverings of a space X have a common refinement, by taking intersections.

G. Show that a refining map r for comparing two coverings induces a map $\tilde{r}$ on cochains which sends cocycles to cocycles and coboundaries to coboundaries. Conclude that $H(r)$ is a well defined map on cohomology. Check that on the 0 level, $H(r)$ is always an isomorphism, which is the identity on global sections after making the identification of the two $\check{H}^0$'s with the group of global sections.

H. Show that the $(n-1)$-cochain $(h_{i_0,\ldots,i_{n-1}})$ defined in the proof of Lemma 3.10 does indeed satisfy

$$d((h_{i_0,\ldots,i_{n-1}})) = (g'_{j_0,\ldots,j_n} - g_{j_0,\ldots,j_n})$$

as claimed.

I. Show that if $\mathcal{W} \prec \mathcal{V} \prec \mathcal{U}$ are three covers, each finer than the next, then

$$H^{\mathcal{V}}_{\mathcal{W}} \circ H^{\mathcal{U}}_{\mathcal{V}} = H^{\mathcal{U}}_{\mathcal{W}}.$$

J. Show that the refining maps $H^{\mathcal{U}}_{\mathcal{V}}$ commute with any map ϕ_* induced by a map of sheaves, in the following sense. Suppose that $\phi : \mathcal{F} \to \mathcal{G}$ is a map of sheaves on X, and $\mathcal{U}$ and $\mathcal{V}$ are two open coverings of X with $\mathcal{V}$ finer than

$\mathcal{U}$. Then for any n, the diagram

$$\begin{array}{ccc} \check{H}^n(\mathcal{U},\mathcal{F}) & \xrightarrow{\phi_*} & \check{H}^n(\mathcal{U},\mathcal{G}) \\ H^{\mathcal{U}}_{\mathcal{V}} \downarrow & & \downarrow H^{\mathcal{U}}_{\mathcal{V}} \\ \check{H}^n(\mathcal{V},\mathcal{F}) & \xrightarrow{\phi_*} & \check{H}^n(\mathcal{V},\mathcal{G}) \end{array}$$

commutes.

K. Show that if a family of subgroups $\{G_a\}$ of a fixed group G is given, with the property that any two are both contained in a third, then the union $L = \cup_a G_a$ is a subgroup of G, which satisfies the universal property for the direct limit of the subgroups. (Here the maps between the subgroups are the inclusion maps when one is contained in another.)

L. Show that if a direct system of groups $\{G_a\}$ and maps H_a^b are given, such that every map H_a^b is an isomorphism, then the direct limit L of the system of groups is also isomorphic to each, and in fact the natural map $h_a : G_a \to L$ is an isomorphism. (Use the universal property of the direct limit.)

M. Let X be the Riemann Sphere $\mathbb{C}_\infty$, and let $U_0 = X - \{0\}$ and $U_1 = X - \{\infty\}$ be the standard open covering $\mathcal{U}$ of X. Compute $\check{H}^1(\mathcal{U}, \mathcal{O}_X[n \cdot \infty])$ for all n explicitly by writing down the spaces of relevant cochains, computing the 1-cocycles and 1-coboundaries, and taking the quotient group. Show that this cohomology group is a complex vector space.

N. Let $(f_{i_0,\ldots,i_n})$ be an n-cocycle for a sheaf $\mathcal{F}$. Show that if any two of the indices are equal, then $f_{i_0,\ldots,i_n} = 0$. Show that if all of the indices are distinct, and σ is a permutation of the indices, then $f_{\sigma(i_0),\ldots,\sigma(i_n)} = \text{sign}(\sigma) f_{i_0,\ldots,i_n}$.

4. Cohomology Computations

As mentioned in the previous sections of this chapter, most of the time one is primarily interested in computing the group of global functions or forms satisfying some local conditions. Cohomologically speaking, this is always some $\check{H}^0$ of a sheaf on the space in question.

Short exact sequences of sheaves give precise relationships between different sheaves, and the computation of global sections can, by appealing to the long exact sequence in cohomology, often be reduced to some computation of an $\check{H}^1$. These in turn can be related to $\check{H}^2$'s, etc. So eventually all the cohomology groups can get involved.

It is most useful to have general statements that with certain sheaves or types of sheaves, higher cohomology groups automatically vanish. If so, then whenever such sheaves appear in a short exact sequence of sheaves, we will have that every third term of the long exact sequence will vanish, which is great information relating the cohomology of the other two sheaves.

We will begin this section by proving that the higher cohomology of the $\mathcal{C}^\infty$ sheaves and the skyscraper sheaves do vanish, and then proceed to draw conclusions concerning cohomology of other sheaves by using the long exact sequence.

There are at least three basic ways to use vanishing of cohomology groups to make conclusions about other cohomology groups, using the long exact sequence. The most trivial is if, in the long exact sequence, one has two vanishing groups A and C separated by a single group B:
$$0 = A \to B \to C = 0.$$
One concludes that $B = 0$ in this situation.

A second is if one has two vanishing groups A and D separated by two groups B and C:
$$0 = A \to B \to C \to D = 0.$$
In this case one concludes that the map from B to C is an isomorphism, and in particular that $B \cong C$.

A third is if one shows that in a short exact sequence of sheaves
$$0 \to \mathcal{K} \to \mathcal{F} \xrightarrow{\phi} \mathcal{G} \to 0,$$
the $\check{H}^1(X, \mathcal{F})$ of the middle sheaf is zero. One then concludes that
$$\check{H}^1(X, \mathcal{K}) \cong \frac{\mathcal{G}(X)}{\phi(\mathcal{F}(X))}.$$

A remark on notation: often one simply writes $\check{H}^n(\mathcal{F})$ for the cohomology group of a sheaf $\mathcal{F}$ on a space X, when the space X is by the context obvious.

The Vanishing of $\check{H}^1$ for $\mathcal{C}^\infty$ Sheaves. Recall that the *support* of a continuous function ϕ on a topological space X is the closure of the subset $\{x \in X \mid \phi(x) \neq 0\}$. On any paracompact space, such as a Riemann surface, one has *partitions of unity* for any open covering $\mathcal{U} = \{U_i\}$. This is a set of $\mathcal{C}^\infty$ functions $\{\varphi_i\}$ such that
- every point in X has a neighborhood meeting only finitely many of the support sets of the φ_i,
- for every point $p \in X$, $\sum_i \varphi_i(p) = 1$, and
- $\text{Supp}(\varphi_i) \subseteq U_i$ for every i.

The existence of partitions of unity is the key ingredient in proving the following vanishing result.

PROPOSITION 4.1. *Let X be a Riemann surface. Then for any $n \geq 1$,*
 a. $\check{H}^n(X, \mathcal{C}^\infty) = 0$,
 b. $\check{H}^n(X, \mathcal{E}^1) = 0$,
 c. $\check{H}^n(X, \mathcal{E}^{1,0}) = 0$,
 d. $\check{H}^n(X, \mathcal{E}^{0,1}) = 0$, and
 e. $\check{H}^n(X, \mathcal{E}^2) = 0$.

PROOF. We will only show that $\check{H}^1(\mathcal{U}, \mathcal{C}^\infty) = 0$ for every open covering $\mathcal{U} = \{U_i\}$ on X, by showing that every 1-cocycle is a coboundary. This will imply that $\check{H}^1(X, \mathcal{C}^\infty) = 0$, by Corollary 3.13. The vanishing of the $\check{H}^1$ for the other

4. COHOMOLOGY COMPUTATIONS

$\mathcal{C}^\infty$ sheaves is proved in an identical manner. Also, the proof for the higher cohomology groups varies only in that there are more indices to keep track of.

Fix a covering $\mathcal{U}$, and let (f_{ij}) be a 1-cocycle for the sheaf $\mathcal{C}^\infty$ on this covering. Consider the $\mathcal{C}^\infty$ function $\varphi_j f_{ij}$; extend it by zero outside of $\text{Supp}(\varphi_i)$ and consider it as a $\mathcal{C}^\infty$ function defined on all of U_i. Set $g_i = -\sum_j \varphi_j f_{ij}$; this is also a $\mathcal{C}^\infty$ function defined on U_i. (The sum is finite for any point, by the local finiteness of the partition of unity.) Then, using that (f_{ij}) is a 1-cocycle, we have

$$g_j - g_i = -\sum_k \varphi_k f_{jk} + \sum_k \varphi_k f_{ik} = \sum_k \varphi_k (f_{ik} - f_{jk}) = \sum_k \varphi_k f_{ij} = f_{ij}$$

so that $(f_{ij}) = d(g_i)$ is a coboundary. $\square$

The Vanishing of $\check{H}^1$ for Skyscraper Sheaves. A variant of the partition of unity argument given above for the $\mathcal{C}^\infty$ sheaves can be used to show that any skyscraper sheaf has a vanishing $\check{H}^1$.

This is based on an integer-valued version of a partition of unity:

LEMMA 4.2. *Let X be a space, and let $\mathcal{U} = \{U_i\}$ be an open cover of X. Then there is a collection of integer-valued functions $\{\varphi_i\}$ on X satisfying*
- *every point p in X lies in only finitely many of the support sets of the φ_i,*
- *for every point $p \in X$, $\sum_i \varphi_i(p) = 1$, and*
- *$\text{Supp}(\varphi_i) \subseteq U_i$ for every i.*

PROOF. Take the open covering $\{U_i\}$ and totally order the index set. Then define

$$\varphi_i(p) = \begin{cases} 1 & \text{if } p \in U_i - \cup_{j<i} U_j, \\ 0 & \text{otherwise.} \end{cases}$$

This collection of functions works. $\square$

These functions are generally discontinuous of course. But if f is a section of a skyscraper sheaf $\mathcal{F}$ on an open set U, and φ is any $\mathbb{Z}$-valued function defined on U, then φf is also a section of $\mathcal{F}$ on U. Hence these are possible functions to use in partition of unity arguments involving skyscraper sheaves.

PROPOSITION 4.3. *Let X be a space, and let $\mathcal{F}$ be a skyscraper sheaf on X. Then for any $n \geq 1$, $\check{H}^n(X, \mathcal{F}) = 0$.*

PROOF. Again we will only give the proof for the $\check{H}^1$. It suffices to show that $\check{H}^1(\mathcal{U}, \mathcal{F}) = 0$ for every open covering $\mathcal{U} = \{U_i\}$ on X. We mimic in every detail the proof of Proposition 4.1.

Fix a covering $\mathcal{U}$, and let $\{\varphi_i\}$ be an integer-valued partition of unity as described above for $\mathcal{U}$. Let (f_{ij}) be a 1-cocycle for the sheaf $\mathcal{F}$ on this covering. Consider the section $\varphi_j f_{ij}$; extend it by zero outside of $\text{Supp}(\varphi_i)$ and consider it as a section of $\mathcal{F}$ defined on U_i. Set $g_i = -\sum_j \varphi_j f_{ij}$; this is also a section of $\mathcal{F}$ defined on U_i. Then, using that (f_{ij}) is a 1-cocycle, we have $g_j - g_i = f_{ij}$ exactly as before, so that $(f_{ij}) = d(g_i)$ is a coboundary. $\square$

COROLLARY 4.4. *Let X be a Riemann surface. Then:*
   a. *for any point $p \in X$, $\check{H}^n(X, \mathbb{C}_p) = 0$ for $n \geq 1$;*
   b. *$\check{H}^n(X, \mathcal{D}iv_X) = 0$ for $n \geq 1$;*
   c. *for any divisor D on X, $\check{H}^n(X, \mathcal{T}_X[D]) = 0$ for $n \geq 1$;*
   d. *for any pair of divisors $D_1 \leq D_2$, $\check{H}^n(X, \mathcal{T}_X[D_1/D_2]) = 0$ for $n \geq 1$.*

Cohomology of Locally Constant Sheaves. Suppose we have a group G and we consider the locally constant sheaf $\underline{G}$ of locally constant functions from X to G. Clearly all of the cohomological constructions made for this sheaf depend only on the topology of X; if X is a Riemann surface, then the $\mathcal{C}^\infty$ structure on X and certainly the complex structure is irrelevant.

It is a basic result in algebraic topology that the Čech cohomology groups for the locally constant sheaves agree with the simplicial cohomology for any triangulable space. (See [**Munkres84**, Section 73].) For contractible spaces, these groups are mostly zero; for compact Riemann surfaces, these are in any case well-known groups. Thus we obtain the following computations from algebraic topology:

PROPOSITION 4.5. *Let X be a contractible Riemann surface (e.g., the disc or the plane $\mathbb{C}$), and let G be an abelian group. Then*
   a. *$\check{H}^0(X, \underline{G}) \cong G$, and*
   b. *$\check{H}^n(X, \underline{G}) = 0$ for $n \geq 1$.*

PROPOSITION 4.6. *Let X be a compact Riemann surface of genus g. Let G be an abelian group. Then*
   a. *$\check{H}^0(X, \underline{G}) \cong G$,*
   b. *$\check{H}^1(X, \underline{G}) \cong G^{2g}$,*
   c. *$\check{H}^2(X, \underline{G}) \cong G$, and*
   d. *$\check{H}^n(X, \underline{G}) = 0$ for $n \geq 3$.*

The Vanishing of $\check{H}^2(X, \mathcal{O}_X[D])$. We may use the long exact sequence of cohomology and the vanishing results above to prove the following.

PROPOSITION 4.7. *Let X be a Riemann surface and let D be a divisor on X. Then $\check{H}^n(X, \mathcal{O}_X[D]) = 0$ for $n \geq 2$.*

PROOF. First let us check this for $D = 0$. The short exact sequence
$$0 \to \mathcal{O} \to \mathcal{C}^\infty \xrightarrow{\bar{\partial}} \mathcal{E}^{0,1} \to 0$$
gives exact sequences
$$\check{H}^n(\mathcal{E}^{0,1}) \xrightarrow{\Delta} \check{H}^{n+1}(\mathcal{O}) \to \check{H}^{n+1}(\mathcal{C}^\infty)$$

for every $n \geq 0$. Since the two spaces at the ends vanish for $n \geq 1$, the result follows.

The general case follows by using the sequence
$$0 \to \mathcal{O}_X[D_1] \to \mathcal{O}_X[D_2] \stackrel{\alpha_{D_1/D_2}}{\to} \mathcal{T}_X[D_1/D_2] \to 0$$
which exists and is exact whenever $D_1 \leq D_2$. This induces in the long exact sequence
$$\check{H}^{n-1}(\mathcal{T}_X[D_1/D_2]) \to \check{H}^n(\mathcal{O}_X[D_1]) \to \check{H}^n(\mathcal{O}_X[D_2]) \to \check{H}^n(\mathcal{T}_X[D_1/D_2])$$
and the two spaces at the ends vanish for $n \geq 2$ since the sheaf $\mathcal{T}_X[D_1/D_2]$ is a discrete skyscraper sheaf. Therefore
$$\check{H}^n(\mathcal{O}_X[D_1]) \cong \check{H}^n(\mathcal{O}_X[D_2]) \text{ for } n \geq 2$$
if $D_1 \leq D_2$. Writing a divisor D as $D = P - N$ with $P, N \geq 0$, we see that
$$\begin{aligned} \check{H}^n(\mathcal{O}_X[D]) &\cong \check{H}^n(\mathcal{O}_X[P]) &&\text{since } D \leq P \\ &\cong \check{H}^n(\mathcal{O}_X) &&\text{since } 0 \leq P \\ &= 0 \end{aligned}$$
by the result for $D = 0$, i.e., for $\mathcal{O}_X$. $\square$

Let ω be a meromorphic 1-form on a Riemann surface, with canonical divisor K. Then for any divisor D on X, we have an isomorphism of sheaves
$$\mathcal{O}_X[K+D] \to \Omega^1_X[D]$$
given by multiplication by ω. This remark, with the above Proposition, gives the following.

COROLLARY 4.8. *Let X be a Riemann surface and let D be a divisor on X. Then $\check{H}^n(X, \Omega^1_X[D]) = 0$ for $n \geq 2$.*

De Rham Cohomology. The De Rham cohomology groups are defined using $\mathcal{C}^\infty$ forms and noting that the operator d satisfies $d \circ d = 0$. Therefore any $\mathcal{C}^\infty$ k-form ω which is d-exact, i.e., which is $d\eta$ for some $\mathcal{C}^\infty$ $(k-1)$-form η certainly is d-closed: $d\omega = 0$.

DEFINITION 4.9. *Let X be a differentiable manifold. The kth De Rham cohomology group, denoted by $H^k_d(X)$, is the quotient space of d-closed $\mathcal{C}^\infty$ k-forms modulo the image of d:*
$$H^k_d(X) = \frac{\{\mathcal{C}^\infty \text{ } k\text{-forms } \omega \mid d\omega = 0\}}{\{d\eta \mid \eta \text{ is a } \mathcal{C}^\infty \text{ } (k-1)\text{-form}\}}.$$

Note that $H^0_d(X) \cong \mathbb{C}$ is the space of constant functions on X.

PROPOSITION 4.10. *Let X be a Riemann surface. Then for any $n \geq 0$,*
$$H^n_d(X) \cong \check{H}^n(X, \underline{\mathbb{C}}).$$

PROOF. The result is clear for $n = 0$ (both are the space of constant functions on X) and for $n \geq 3$ (both spaces are 0).

To understand $H_d^1(X)$, recall the short exact sequence

(4.11) $$0 \to \underline{\mathbb{C}} \to \mathcal{C}^\infty \xrightarrow{d} \mathcal{K} \to 0$$

where $\mathcal{K}$ is the kernel sheaf for the sheaf map $d : \mathcal{E}^1 \to \mathcal{E}^2$. We see that $H_d^1(X)$ is exactly the cokernel of the map d on global sections:

$$H_d^1(X) \cong \mathcal{K}(X)/d(\mathcal{C}^\infty(X)).$$

Since $\check{H}^1(X, \mathcal{C}^\infty) = 0$, this cokernel is isomorphic to $\check{H}^1(X, \underline{\mathbb{C}})$, using the long exact sequence. Note also that

$$\check{H}^n(X, \mathcal{K}) \cong \check{H}^{n+1}(X, \underline{\mathbb{C}})$$

for every $n \geq 1$, since the higher cohomology groups for the sheaf $\mathcal{C}^\infty$ vanish.

The analysis of the H_d^2 is similar. By Poincaré's Lemma, the sheaf map $d : \mathcal{E}^1 \to \mathcal{E}^2$ is onto with kernel $\mathcal{K}$. We then have a long exact sequence in cohomology; this gives that

$$\check{H}^n(X, \mathcal{K}) = 0 \text{ for } n \geq 2$$

and

$$0 \to \mathcal{K}(X) \to \mathcal{E}^1(X) \xrightarrow{d} \mathcal{E}^2(X) \to \check{H}^1(X, \mathcal{K}) \to 0$$

since the higher cohomology groups of the two $\mathcal{C}^\infty$ sheaves are 0. Thus we have that

$$H_d^2(X) \cong \check{H}^1(X, \mathcal{K}) \cong \check{H}^2(X, \underline{\mathbb{C}}).$$

□

Note that by Propositions 4.5 and 4.6, we have that $H_d^1(X)$ and $H_d^2(X)$ are both zero if X is contractible; if X is a compact Riemann surface of genus g, then $\dim H_d^1(X) = 2g$ and $\dim H_d^2(X) = 1$.

Dolbeault Cohomology. Let X be a Riemann surface. The Dolbeault cohomology groups are defined similarly to the De Rham groups, using the operator $\bar{\partial}$ instead of d.

DEFINITION 4.12. The *Dolbeault cohomology group of X* (of type (p,q)) is the group $H_{\bar{\partial}}^{p,q}(X)$ defined by

$$H_{\bar{\partial}}^{p,q}(X) = \frac{\ker \bar{\partial} : \mathcal{E}^{p,q}(X) \to \mathcal{E}^{p,q+1}(X)}{\operatorname{image} \bar{\partial} : \mathcal{E}^{p,q-1}(X) \to \mathcal{E}^{p,q}(X)}.$$

4. COHOMOLOGY COMPUTATIONS

In the above definition we set $\mathcal{E}^{0,0}$ to be the sheaf $\mathcal{C}^\infty$ of C^∞ functions, $\mathcal{E}^{1,1} = \mathcal{E}^2$ to be the sheaf of C^∞ 2-forms, and $\mathcal{E}^{p,q} = 0$ if $p+q > 2$ or $p = 2$ or $q = 2$. We have in particular only 4 possible groups here:

$$\begin{aligned} H_{\bar{\partial}}^{0,0}(X) &= \mathcal{O}(X), \\ H_{\bar{\partial}}^{1,0}(X) &= \Omega^1(X), \\ H_{\bar{\partial}}^{0,1}(X) &= \frac{\mathcal{E}^{0,1}(X)}{\text{image}\,\bar{\partial}: C^\infty(X) \to \mathcal{E}^{0,1}(X)}, \text{ and} \\ H_{\bar{\partial}}^{1,1}(X) &= \frac{\mathcal{E}^2(X)}{\text{image}\,\bar{\partial}: \mathcal{E}^{1,0}(X) \to \mathcal{E}^2(X)}. \end{aligned}$$

The first two groups are adequately described above. The second two have interpretations using Čech cohomology. Consider the short exact sequence

$$0 \to \mathcal{O} \to \mathcal{C}^\infty \xrightarrow{\bar{\partial}} \mathcal{E}^{0,1} \to 0$$

which gives the long exact sequence

$$0 \to \mathcal{O}(X) \to \mathcal{C}^\infty(X) \xrightarrow{\bar{\partial}} \mathcal{E}^{0,1}(X) \to \check{H}^1(X, \mathcal{O}) \to 0.$$

We see immediately that

$$H_{\bar{\partial}}^{0,1}(X) \cong \check{H}^1(X, \mathcal{O}).$$

Similarly, consider the short exact sequence

$$0 \to \Omega^1 \to \mathcal{E}^{1,0} \xrightarrow{\bar{\partial}} \mathcal{E}^2 \to 0$$

which gives the long exact sequence

$$0 \to \Omega^1(X) \to \mathcal{E}^{0,1}(X) \xrightarrow{\bar{\partial}} \mathcal{E}^2(X) \to \check{H}^1(X, \Omega^1) \to 0.$$

Therefore we have

$$H_{\bar{\partial}}^{1,1}(X) \cong \check{H}^1(X, \Omega^1).$$

If we adopt the notation that a function is a 0-form, and use Ω_X^0 for the sheaf of holomorphic functions $\mathcal{O}_X$, then all four Dolbeault groups can be expressed via Čech cohomology as

(4.13) $$H_{\bar{\partial}}^{p,q}(X) \cong \check{H}^q(X, \Omega_X^p).$$

Problems IX.4

A. Show that if $\mathcal{F}$ is a skyscraper sheaf on X, U is an open set in X, and φ is any integer-valued function on U, then for every $f \in \mathcal{F}(U)$, we have $\varphi f \in \mathcal{F}(U)$ also.

B. Let X be a Riemann surface. Analyze the long exact sequence in cohomology for the short exact sequence

$$0 \to \mathbb{C} \to \mathcal{O} \stackrel{d=\partial}{\to} \Omega^1 \to 0$$

and say as much as you can about the terms involved.

C. Show that on a Riemann surface X, we have

$$\check{H}^1(X, \mathcal{H}) \cong \frac{\mathcal{E}^2(X)}{\partial\bar{\partial}(\mathcal{C}^\infty(X))}.$$

Show that $\check{H}^n(X, \mathcal{H}) = 0$ for $n \geq 2$.

D. Let D be a divisor on a Riemann surface X, and let p be a point on X. Show that if $\check{H}^1(X, \mathcal{O}_X[D-p]) = 0$, then $\check{H}^1(X, \mathcal{O}_X[D]) = 0$. Use induction to show that if $D_1 \leq D_2$ and $\check{H}^1(X, \mathcal{O}_X[D_1]) = 0$, then $\check{H}^1(X, \mathcal{O}_X[D_2]) = 0$.

Further Reading

There are more complete treatments of sheaf and Čech cohomology theory in [**Hirzebruch66**], [**Warner71**], and [**G-H78**]. Serre's monograph [**Serre55**] is still hard to beat, and there the point of view of the étale space is brought forward. In [**Hartshorne77**] the construction of the derived functor cohomology theory is explained. The reader may also profit from [**Gomez-Mont89**] and [**Godement58**], which is probably the most complete.

For De Rham and Dolbeault cohomology, the reader may consult [**K-M71**], [**Warner71**], [**B-T82**], [**Wells73**], [**Hartshorne76**], and [**G-H78**].

Chapter X. Algebraic Sheaves

1. Algebraic Sheaves of Functions and Forms

Algebraic Curves. Recall that an *algebraic curve* is a compact Riemann surface whose field of meromorphic functions separates points and tangents. Any projective curve is algebraic, as we have noted. Indeed, it is a fundamental theorem in the analytical part of the theory that any compact Riemann surface is an algebraic curve.

Since any projective curve X is algebraic, and the global meromorphic functions on X are all rational functions, the field $\mathcal{M}(X)$ of global meromorphic functions on X is often called the *rational function field* of X. Similarly, the vector space $\mathcal{M}^{(1)}(X)$ of global meromorphic 1-forms on X is called the *space of rational 1-forms*.

We have introduced sheaf theory as a tool to organize functions and forms satisfying local properties, and Čech cohomology to extract global information from local (i.e. sheaf) data. The usual definition of the important sheaves of functions and forms associates to every open set U in X the set of all holomorphic or meromorphic functions or forms of the desired type, defined on that open set U. Most of these functions will not extend to all of X. Hence it seems inefficient to use them at every stage of the constructions, only to have them disappear at the end when one tries to draw global conclusions.

This is the main idea of introducing the sheaves of algebraic functions and forms: stay with the globally defined objects all the way.

Algebraic Sheaves of Functions. Let X be an algebraic curve, and let $\mathcal{M}(X)$ be the field of rational functions on X. Define a sheaf $\mathcal{O}_{X,alg}$ on X by setting

$$\mathcal{O}_{X,alg}(U) = \{f \in \mathcal{M}(X) \mid f \text{ is holomorphic at all points of } U\}.$$

Thus for every U, we have $\mathcal{O}_{X,alg}(U)$ is a subring of the fixed rational function field $\mathcal{M}(X)$. This sheaf is called the *sheaf of regular functions* on X.

Note that there is a natural inclusion map

$$\mathcal{O}_{X,alg} \hookrightarrow \mathcal{O}_X$$

since every one of these functions is by definition holomorphic.

The use of the word "regular" is simply meant to replace "holomorphic", when this whole idea is generalized to fields other than $\mathbb{C}$. Similarly, the word "rational" is more commonly used than "meromorphic", when working with the algebraic sheaves. This, as noted above, is because for a projective curve, every global meromorphic function is the restriction of a rational function.

More generally, given a divisor D on X, we construct the algebraic analogue of the sheaf of meromorphic functions with poles bounded by D:

$$\mathcal{O}_{X,alg}[D](U) = \{f \in \mathcal{M}(X) \mid \operatorname{div}(f) \geq -D \text{ at all points of } U\}.$$

This is called the *sheaf of rational functions with poles bounded by D on X*. Note that of course $\mathcal{O}_{X,alg}[0] = \mathcal{O}_{X,alg}$. Again there is a natural inclusion

$$\mathcal{O}_{X,alg}[D] \hookrightarrow \mathcal{O}_X[D].$$

Finally we have the analogue of the sheaf of meromorphic functions itself. This is the *sheaf of rational functions* on X, defined by

$$\mathcal{M}_{X,alg}(U) = \underline{\mathcal{M}(X)}(U)$$

for every U, i.e., the locally constant sheaf of locally constant functions from U to the discrete group $\mathcal{M}(X)$. Of course we again have an inclusion of sheaves

$$\mathcal{M}_{X,alg} \hookrightarrow \mathcal{M}_X.$$

We feel free to drop the subscript X in the notation when there is no chance of confusion.

Algebraic Sheaves of Forms. The same constructions made above for functions can be made for 1-forms. We let $\mathcal{M}^{(1)}(X)$ denote the group of rational 1-forms on X. It is a 1-dimensional vector space over the rational function field $\mathcal{M}(X)$, generated by any nonzero rational 1-form.

We have the *sheaf of regular 1-forms* on X, defined by

$$\Omega^1_{X,alg}(U) = \{\omega \in \mathcal{M}^{(1)}(X) \mid \omega \text{ is holomorphic at all points of } U\}.$$

Given a divisor D, we have the *sheaf of rational 1-forms with poles bounded by D*, defined by

$$\Omega^1_{X,alg}[D](U) = \{\omega \in \mathcal{M}^{(1)}(X) \mid \operatorname{div}(\omega) \geq -D \text{ at all points of } U\};$$

again we have $\Omega^1_{X,alg}[0] = \Omega^1_{X,alg}$.

Finally we have the locally constant *sheaf of rational 1-forms* on X:

$$\mathcal{M}^{(1)}_{X,alg}(U) = \underline{\mathcal{M}^{(1)}(X)}(U)$$

for every U, i.e., the sheaf of locally constant functions from U to the discrete group $\mathcal{M}^{(1)}(X)$.

1. ALGEBRAIC SHEAVES OF FUNCTIONS AND FORMS

We also have the corresponding inclusions of sheaves:

$$\Omega^1_{X,alg} \hookrightarrow \Omega^1_X, \ \ \Omega^1_{X,alg}[D] \hookrightarrow \Omega^1_X[D], \ \text{ and } \ \mathcal{M}^{(1)}_{X,alg} \hookrightarrow \mathcal{M}^{(1)}_X.$$

Again the subscript X is often omitted from the notation.

The Zariski Topology. The construction of the algebraic sheaves above is all well and good, but the real utility of these sheaves comes when we not only restrict the sheaves but also restrict the open sets on which we build the sheaves. Recall that a *cofinite* subset of X is a subset whose complement is finite. All such sets are of course open sets of X, and these are the open sets to which we will focus our attention. The motivation for looking at the cofinite sets comes from the following observation.

LEMMA 1.1. *For any divisor D, consider the algebraic sheaf $\mathcal{O}_{X,alg}[D]$ defined above. Then for any open set U, and any $f \in \mathcal{O}_{alg}[D](U)$, there is cofinite set V with $U \subseteq V \subseteq X$ such that the restriction map*

$$\rho^V_U : \mathcal{O}_{alg}[D](V) \to \mathcal{O}_{alg}[D](U)$$

hits the function f. The same statement is true for any of the sheaves of rational 1-forms $\Omega^1_{X,alg}[D]$.

PROOF. Since f is a global meromorphic function, it has a finite number of poles, and in particular a finite number of poles outside of U. Let $p_1, \ldots, p_n$ be the poles of f outside of U. Since D has finite support, there are finitely many points $q_1, \ldots, q_m$ with $D(q_i) < 0$ outside of U.

Let V be the complement of the sets of p's and q's. Then $\text{div}(f) \geq -D$ on all of V, since it is on all of U, and at any point p of $V - U$, $\text{div}(f)(p) \geq 0$ and $D(p) \geq 0$. Hence $f \in \mathcal{O}_{alg}[D](V)$.

The same proof works for the sheaf $\Omega^1_{X,alg}[D]$ of rational 1-forms with poles bounded by D. $\square$

The moral of the above lemma is that, as far as the sections of the algebraic sheaves go, one does not need all of the open sets, just the cofinite sets. In effect, every section of every one of these sheaves over any open set actually lives as a section over a cofinite open set. Luckily, these sets are enough to define a topology on X.

DEFINITION 1.2. Let X be a compact Riemann surface. The *Zariski topology* on X is the topology whose open sets are the cofinite sets (and the empty set, of course).

When we explicitly want to refer to X with its Zariski topology, we will write X_{Zar}. We note several immediate points:
- X_{Zar} is not a Hausdorff space.

- X_{Zar} is compact, in the sense that every open cover has a finite subcover. (This property is sometimes referred to as *quasi-compactness*, when the space is not Hausdorff; some authors reserve compactness to imply also that the space is Hausdorff.)
- Any two nonempty open sets of X_{Zar} intersect. (A space with this property is said to be *irreducible*.)

Since any Zariski open set is a classical open set, the Zariski topology is a subtopology of the classical topology. In particular, every sheaf on X (using the classical topology) induces a sheaf on X_{Zar}, simply by only considering the sheaf on the Zariski open sets.

This is usually only done for the algebraic sheaves, in which case one obtains sheaves

$$\mathcal{O}_{X_{Zar},alg}[D], \quad \Omega^1_{X_{Zar},alg}[D], \quad \mathcal{M}_{X_{Zar},alg}, \quad \text{and} \quad \mathcal{M}^{(1)}_{X_{Zar},alg}$$

of algebraic functions and forms on X_{Zar}. Note that the sheaves

$$\mathcal{M}_{X_{Zar},alg} \quad \text{and} \quad \mathcal{M}^{(1)}_{X_{Zar},alg}$$

are actually *constant* sheaves, since every two open sets intersect in X_{Zar}. The first has sections equal to the field of global meromorphic functions $\mathcal{M}(X)$ for every open set; the second has the group $\mathcal{M}^{(1)}(X)$ of global meromorphic 1-forms as sections over each open set.

Problems X.1

A. Note that in the text we defined the various algebraic sheaves of functions and forms by giving only a presheaf. Check that the presheaf $\mathcal{O}_{X,alg}$ of regular functions on an algebraic curve X (and more generally the presheaves $\mathcal{O}_{X,alg}[D]$) satisfy the sheaf axiom (with the classical topology).

B. Repeat Problem A. for the algebraic sheaves of forms.

C. Verify that the Zariski topology is a topology.

D. Show that the stalk $\mathcal{O}_{X,alg,p}$ of the sheaf $\mathcal{O}_{X,alg}$ at a point $p \in X$ is the subring of the rational function field $\mathcal{M}(X)$ consisting of those rational functions which are holomorphic at the point p.

E. Show that if K is a canonical divisor on an algebraic curve X, then

$$\Omega^1_{X,alg}[D] \cong \mathcal{O}_{X,alg}[K+D].$$

2. Zariski Cohomology

Since the Zariski topology is a topology on X, one can use this topology to define Čech cohomology groups on X_{Zar} for any sheaf on X_{Zar}. The same identical formalism is used as in the construction of the Čech cohomology groups for sheaves on X with the classical topology.

2. ZARISKI COHOMOLOGY

For every Zariski open covering $\mathcal{U} = \{U_i\}$, we may form the Čech cochain group

$$\check{C}^n(\mathcal{U}, \mathcal{F}) = \prod_{(i_0, i_1, \ldots, i_n)} \mathcal{F}(U_{i_0, i_1, \ldots, i_n}),$$

and there are coboundary maps

$$d : \check{C}^n(\mathcal{U}, \mathcal{F}) \to \check{C}^{n+1}(\mathcal{U}, \mathcal{F})$$

defined by setting

$$d((f_{i_0,\ldots,i_n})) = (g_{i_0,\ldots,i_{n+1}})$$

where

$$g_{i_0,\ldots,i_{n+1}} = \sum_{k=0}^{n+1} (-1)^k \rho(f_{i_0, i_1, \ldots, \widehat{i_k}, \ldots, i_{n+1}})$$

just as before. We still have $d \circ d = 0$, and so we obtain the Čech groups with respect to the cover $\mathcal{U}$, defined by taking the kernel of d modulo the image of d. Again taking the limit over all covers gives the Čech cohomology group

$$\check{H}^n(X_{Zar}, \mathcal{F})$$

for any sheaf on X_{Zar}.

The Vanishing of $\check{H}^1(X_{Zar}, \mathcal{F})$ for a Constant Sheaf $\mathcal{F}$. As we have seen, both of the sheaves $\mathcal{M}_{alg}$ and $\mathcal{M}_{alg}^{(1)}$ are constant sheaves on X_{Zar}. A general result allows us to conclude that the higher cohomology of these sheaves vanish.

PROPOSITION 2.1. *Let $\underline{G}$ be a constant sheaf (constantly equal to the group G) on X_{Zar}. Then for every $n \geq 1$, $\check{H}^n(X_{Zar}, \underline{G}) = 0$.*

PROOF. We will give the proof for $\check{H}^1$; the proof in general only involves more indices, not more ideas. Let c denote a cohomology class in $\check{H}^1(X_{Zar}, \underline{G})$, and represent c as a 1-cocycle (f_{ij}) with respect to some covering $\mathcal{U} = \{U_i\}$, which we may take to be finite since X_{Zar} is quasi-compact. Choose an ordering of the open sets U_i, so that we have $U_0, U_1, \ldots, U_n$ for some n. The cocycle condition on the f_{ij}'s implies that $f_{ii} = 0$ for every i and $f_{ij} = -f_{ji}$ for every $i \neq j$. Therefore the cocycle is determined by the elements f_{ij} with $i < j$.

In fact, the cocycle is determined by the elements $f_{i,i+1}$ for each i. This is because if $i < j < k$, then $f_{ik} = f_{ij} + f_{jk}$ by the cocycle condition, and all of these elements make sense since every two open sets intersect. Moreover it is easy to see that if $f_{i,i+1}$ are chosen arbitrarily in G, then one recovers the cocycle by setting

$$f_{ij} = \sum_{k=i}^{j-1} f_{k,k+1}$$

for every $i < j$.

Now set $g_0 = 0$ and for $i \geq 1$ set $g_i = \sum_{k=0}^{i-1} f_{k,k+1}$. Then (g_i) is a 0-cochain for the sheaf $\underline{G}$, and clearly $f_{ij} = g_j - g_i$ for $i < j$. Therefore the 1-cocycle (f_{ij}) is a coboundary, so its cohomology class c is zero in $\check{H}^1$. $\square$

COROLLARY 2.2. *If X is a compact Riemann surface, then for $n \geq 1$,*
$$\check{H}^n(X_{Zar}, \mathcal{M}_{alg}) = \check{H}^n(X_{Zar}, \mathcal{M}_{alg}^{(1)}) = 0.$$

The Interpretation of $H^1(D)$. If one has a short exact sequence of sheaves on X_{Zar}, one may not have a complete long exact sequence in cohomology, since X_{Zar} is not paracompact. However one always has the long exact sequence up through the $\check{H}^1$ level, and this is enough for our purposes at the moment.

Recall the sheaf of Laurent tail divisors $\mathcal{T}_X[D]$ whose sections over a classical open set U is the set of Laurent tail divisors with terms bounded above by $-D$. Using the classical topology, the definition of a Laurent tail divisor implied that the set of points where the divisor was nontrivial was a discrete subset of the open set. Therefore on the entire compact Riemann surface, a Laurent tail divisor had to have finite support.

In the algebraic setting, we simply require every Laurent tail divisor to have finite support, i.e., it must be a global Laurent tail. This exactly mimics the definition of the algebraic sheaves: take only the global objects satisfying the required condition. This give us a sheaf $\mathcal{T}_{X,alg}[D]$ for every divisor D on X.

We still have a map α_D in the algebraic setting, sending a global meromorphic function to its suitably truncated Laurent tail. This gives now a map of sheaves on X_{Zar}:
$$\alpha_{D,alg} : \mathcal{M}_{alg} \to \mathcal{T}_{X,alg}[D].$$

LEMMA 2.3. *For every divisor D on X, the map $\alpha_{D,alg}$ is an onto map of sheaves on X_{Zar} with kernel $\mathcal{O}_{X,alg}[D]$. Hence we have a short exact sequence*
$$0 \to \mathcal{O}_{X,alg}[D] \to \mathcal{M}_{alg} \stackrel{\alpha_{D,alg}}{\to} \mathcal{T}_{X,alg}[D] \to 0.$$

PROOF. Clearly the algebraic sheaf $\mathcal{O}_{X,alg}[D]$ is the kernel of $\alpha_{D,alg}$; the surjectivity of $\alpha_{D,alg}$ is the real question. We will show that $\alpha_{D,alg}$ is surjective on any open set U which is not the entire surface X. This will suffice to show that $\alpha_{D,alg}$ is an onto map of sheaves.

Fix a point p in the complement of U. Let Z be a finite Laurent tail divisor supported on U, with terms bounded above by $-D$, so that $Z \in \mathcal{T}_{X,alg}[D](U)$. Consider the divisor $D_n = D + n \cdot p$. For large n we have $H^1(D_n) = 0$ (recall that this is the cokernel of the global map $\alpha_{D_n} : \mathcal{M}(X) \to \mathcal{T}[D_n](X)$, see Chapter VI, Section 2). Indeed, using Corollary 3.12 of Chapter VI, it is enough to have $\deg(D_n) \geq 2g - 1$ where g is the genus of X.

Therefore for large n the global map α_{D_n} is surjective. Note that the finite Laurent tail divisor Z is in the space of global Laurent tail divisors $\mathcal{T}[D_n](X)$, since Z does not have p in its support. Hence there is a global meromorphic function f with $\alpha_{D_n}(f) = Z$. This function f, when restricted to the original open set U, is a preimage of Z also. $\square$

The above proof amounts to saying that if one allows an arbitrarily bad pole at some point outside the set U, then one can arrange any finite set of Laurent tails inside U.

By Corollary 2.2, the long exact sequence for this short exact sequence starts out as

$$0 \to L(D) \to \mathcal{M}(X) \xrightarrow{\alpha_D} \mathcal{T}[D](X) \to \check{H}^1(X_{Zar}, \mathcal{O}_{X,alg}[D]) \to 0.$$

We therefore conclude that $\check{H}^1(X_{Zar}, \mathcal{O}_{X,alg}[D])$ is isomorphic to the cokernel of the global map α_D. In other words, we have proved the following.

PROPOSITION 2.4. *If D is any divisor on an algebraic curve X, then*

$$H^1(D) \cong \check{H}^1(X_{Zar}, \mathcal{O}_{X,alg}[D]).$$

Finally the mystery behind the H^1 notation for this cokernel is solved!

GAGA Theorems. With now two different ways of taking cohomology, a natural problem arises, namely to compare them. For sanity we will write X_{an} for the Riemann surface X with the classical, or analytic, topology, and also we will write $\mathcal{O}_{X,an}[D]$ for the sheaves of meromorphic functions on X_{an} with poles bounded by D. Specifically one wants to compare the groups $\check{H}^n(X_{an}, \mathcal{O}_{X,an}[D])$ and $\check{H}^n(X_{Zar}, \mathcal{O}_{X,alg}[D])$.

Recall that the algebraic sheaf $\mathcal{O}_{X,alg}[D]$ is a subsheaf of the analytic sheaf $\mathcal{O}_X[D]$, both considered as sheaves on X with the classical topology. Hence for every n there is an induced map

$$j_1 : \check{H}^n(X_{an}, \mathcal{O}_{X,alg}[D]) \to \check{H}^n(X_{an}, \mathcal{O}_{X,an}[D]).$$

Furthermore, since any Zariski open set is classically open, any Zariski open covering is a classical open cover; hence any cochain for the Zariski topology is a cochain for the classical topology. This induces a map on cohomology

$$j_2 : \check{H}^n(X_{Zar}, \mathcal{O}_{X,alg}[D]) \to \check{H}^n(X_{an}, \mathcal{O}_{X,alg}[D]).$$

The composition j_2 with j_1 gives a natural comparison map

$$j : \check{H}^n(X_{Zar}, \mathcal{O}_{X,alg}[D]) \to \check{H}^n(X_{an}, \mathcal{O}_{X,an}[D]).$$

The same constructions may be used to obtain a comparison map for the sheaves of 1-forms also. This is then a map

$$j^1 : \check{H}^n(X_{Zar}, \Omega^1_{X,alg}[D]) \to \check{H}^n(X_{an}, \Omega^1_{X,an}[D]).$$

Now it is a fundamental result that these comparison maps j and j^1 are isomorphisms of groups. This type of theorem is called a GAGA theorem, after the article of Serre [**Serre56**] in which such theorems are first proved. "GAGA" is an acronym for "Geometrie Analytique et Geometrie Algebrique", which is the title of Serre's paper. The proof of the GAGA theorems are beyond us at this moment; but for the sheaves we have seen the statement is the following.

THEOREM 2.5 (GAGA). *Let X be an algebraic curve. Then for any divisor D, the comparison maps*
$$j : \check{H}^n(X_{Zar}, \mathcal{O}_{X,alg}[D]) \to \check{H}^n(X_{an}, \mathcal{O}_{X,an}[D])$$
and
$$j^1 : \check{H}^n(X_{Zar}, \Omega^1_{X,alg}[D]) \to \check{H}^n(X_{an}, \Omega^1_{X,an}[D])$$
are group isomorphisms for all n.

In addition to Serre's original paper, one may consult [**GA74**] for the full statements and proofs.

Note that it is not being claimed that *any* sheaf has the same cohomology using either of the topologies. A good example where they diverge is the locally constant sheaves. By Proposition 2.1, $\check{H}^1(X_{Zar}, \underline{G}) = 0$; but we have seen that $\check{H}^1(X_{an}, \underline{G})$ is isomorphic to G crossed with itself $2g$ times. The GAGA theorem therefore is more subtle than simply a statement about comparing cohomology with different topologies; the sheaf counts also.

Further Computations. The GAGA Theorem, which relates the Zariski cohomology groups of the algebraic sheaves to the cohomology of the analytic sheaves, and the interpretation of the cokernel space $H^1(D)$ as a Zariski cohomology group, allows us to give some precise computations of the cohomology of the analytic sheaves. This in turn gives us some insight into several analytic theorems.

The following is a direct consequence of the GAGA Theorem, Proposition 2.4, and the results on $H^1(D)$ in Chapter VI, Section 3.

PROPOSITION 2.6. *Let X be an algebraic curve of genus g. Let D be a divisor on X. Then the spaces $\check{H}^1(X_{an}, \mathcal{O}_{an}[D])$ and $\check{H}^1(X_{an}, \Omega^1_{an}[D])$ are finite-dimensional. Moreover,*
$$\dim \check{H}^1(X_{an}, \mathcal{O}_{an}) = g$$
and
$$\dim \check{H}^1(X_{an}, \Omega^1_{an}) = 1.$$
If $\deg(D) \geq 2g - 1$, then $\check{H}^1(X_{an}, \mathcal{O}_{an}[D]) = 0$.

Let us now turn to the short exact sequence
$$0 \to \underline{\mathbb{C}} \to \mathcal{O} \xrightarrow{d} \Omega^1 \to 0$$
which gives the long exact sequence
$$\begin{aligned} 0 \to \quad & \underline{\mathbb{C}}(X) & \to & \quad \mathcal{O}(X) & \to & \quad \Omega^1(X) & \to \\ \to \quad & \check{H}^1(X, \underline{\mathbb{C}}) & \to & \quad \check{H}^1(X, \mathcal{O}) & \to & \quad \check{H}^1(X, \Omega^1) & \to \\ \to \quad & \check{H}^2(X, \underline{\mathbb{C}}) & \to & \quad 0. \end{aligned}$$

2. ZARISKI COHOMOLOGY

If we assume that X is an algebraic curve of genus g, then the first two spaces are 1-dimensional, as are the last two spaces. Therefore the sequence breaks, leaving a short exact sequence

$$0 \to \Omega^1(X) \to \check{H}^1(X, \mathbb{C}) \to \check{H}^1(X, \mathcal{O}) \to 0.$$

The first and last space here have dimension g, and the middle space has dimension $2g$. This sequence is called the *Hodge filtration* on the middle space $\check{H}^1(X, \mathbb{C})$.

Now consider the sheaf sequence

$$0 \to \mathcal{O}_X[D] \to \mathcal{M}_X \overset{\alpha_D}{\to} \mathcal{T}_X[D] \to 0$$

which gives the long exact sequence

$$\begin{aligned} 0 \to\ & \mathcal{O}[D](X) \to \mathcal{M}(X) \overset{\alpha_D(X)}{\to} \mathcal{T}_X[D](X) \to \\ \to\ & \check{H}^1(X, \mathcal{O}_X[D]) \to \check{H}^1(X, \mathcal{M}_X) \to 0 \end{aligned},$$

since $\mathcal{T}_X[D]$ is a skyscraper sheaf and hence $\check{H}^1(X, \mathcal{T}_X[D]) = 0$. Now we have seen above that when the degree of D is large enough, $\check{H}^1(X, \mathcal{O}_X[D]) = 0$. Hence we conclude, independently of D, that

$$\check{H}^1(X, \mathcal{M}_X) = 0,$$

which is the same result as we had for the Zariski cohomology of the corresponding algebraic sheaf in Corollary 2.2.

The Zero Mean Theorem. Let X be an algebraic curve. Consider the short exact sequence of sheaves

(2.7) $$0 \to \mathcal{H} \to \mathcal{C}^\infty \overset{\partial\overline{\partial}}{\to} \mathcal{E}^2 \to 0;$$

from this we see that

$$\check{H}^n(X, \mathcal{H}) = 0 \text{ for } n \geq 2.$$

We also have the sequence

$$0 \to \mathbb{C} \to \mathcal{H} \overset{d}{\to} \Omega^1 \oplus \overline{\Omega}^1 \to 0.$$

The long exact sequence of cohomology groups is then

$$\begin{aligned} 0 \to\ & \mathbb{C} \to \mathcal{H}(X) \to \Omega^1(X) \oplus \overline{\Omega}^1(X) \to \\ \to\ & \check{H}^1(X, \mathbb{C}) \to \check{H}^1(X, \mathcal{H}) \to \check{H}^1(X, \Omega^1) \oplus \check{H}^1(X, \overline{\Omega}^1) \to \\ \to\ & \check{H}^2(X, \mathbb{C}) \to 0. \end{aligned}$$

The first two spaces are simply the constant functions, by the maximum modulus theorem for harmonic functions. Therefore these two break off, giving the exact sequence

$$
\begin{array}{rcccc}
& & 0 & \to & \Omega^1(X) \oplus \overline{\Omega}^1(X) & \to \\
\to & \check{H}^1(X,\underline{\mathbb{C}}) & \to & \check{H}^1(X,\mathcal{H}) & \to & \check{H}^1(X,\Omega^1) \oplus \check{H}^1(X,\overline{\Omega}^1) & \to \\
\to & \check{H}^2(X,\underline{\mathbb{C}}) & \to & 0.
\end{array}
$$

Now $\Omega^1(X)$ has dimension g over $\mathbb{C}$, hence dimension $2g$ over $\mathbb{R}$; its conjugate space also has dimension $2g$ over $\mathbb{R}$, since they are isomorphic (via conjugation of course) over $\mathbb{R}$. The second space in the sequence above has dimension $2g$ over $\mathbb{C}$, hence dimension $4g$ over $\mathbb{R}$. Therefore the first map above is an isomorphism, and these two spaces also break off the sequence, giving the short exact sequence

$$0 \to \check{H}^1(X,\mathcal{H}) \to \check{H}^1(X,\Omega^1) \oplus \check{H}^1(X,\overline{\Omega}^1) \to \check{H}^2(X,\underline{\mathbb{C}}) \to 0.$$

We have that $\check{H}^1(X,\Omega^1)$ is 1-dimensional over $\mathbb{C}$, hence 2-dimensional over $\mathbb{R}$; the same is true of $\check{H}^1(X,\overline{\Omega}^1)$. Therefore the middle space here is 4-dimensional over $\mathbb{R}$. The last space $\check{H}^2(X,\underline{\mathbb{C}})$ is 1-dimensional over $\mathbb{C}$, hence 2-dimensional over $\mathbb{R}$; we conclude that $\check{H}^1(X,\mathcal{H})$ is 2-dimensional over $\mathbb{R}$. Since it is a complex space, we see that

$$\dim_{\mathbb{C}} \check{H}^1(X,\mathcal{H}) = 1.$$

Returning to the short exact sequence (2.7), we have a long exact sequence

$$0 \to \mathcal{H}(X) \to \mathcal{C}^\infty(X) \xrightarrow{\partial\overline{\partial}} \mathcal{E}^2(X) \xrightarrow{\Delta} \check{H}^1(X,\mathcal{H}) \to 0,$$

which shows that

$$\mathcal{E}^2(X)/\partial\overline{\partial}(\mathcal{C}^\infty(X)) \cong \check{H}^1(X,\mathcal{H}) \cong \mathbb{C}.$$

One interprets this as saying that there is one linear condition on a $\mathcal{C}^\infty$ 2-form η for it to be $\partial\overline{\partial}f$ for some $\mathcal{C}^\infty$ function f. One can write down this linear condition immediately in terms of integration: by Stoke's theorem, we have

$$\iint_X \partial\overline{\partial} f = 0$$

for any $\mathcal{C}^\infty$ function f. Therefore:

PROPOSITION 2.8 (THE ZERO MEAN THEOREM). *Let η be a $\mathcal{C}^\infty$ 2-form on an algebraic curve X. Then there exists a $\mathcal{C}^\infty$ function f on X such that $\partial\overline{\partial} f = \eta$ if and only if*

$$\iint_X \eta = 0.$$

M. Cornalba refers to this theorem as the "cornerstone of the theory of Riemann surfaces" in his highly recommended notes on the Riemann-Roch and Abel Theorems [**Cornalba89**]. It can be proved without resorting to the GAGA theorems and the algebraic computation of $H^1(D)$ using a bit of functional analysis. The reader may consult Cornalba's notes or Warner's text [**Warner71**] for a proof in this spirit. If one does this, the entire theory may be reversed in a

2. ZARISKI COHOMOLOGY

sense, and one can recover all of the major theorems simply from the Zero Mean Theorem.

The High Road to Abel's Theorem. Consider the exponential sequence

$$0 \to \underline{\mathbb{Z}} \to \mathcal{O}_X \to \mathcal{O}_X^* \to 0$$

for an algebraic curve X, which induces the long exact sequence

$$0 \to \check{H}^1(X, \underline{\mathbb{Z}}) \to \check{H}^1(X, \mathcal{O}_X) \to \check{H}^1(X, \mathcal{O}_X^*) \to \check{H}^2(X, \underline{\mathbb{Z}}).$$

Also consider the sequence

$$0 \to \mathcal{O}_X^* \to \mathcal{M}_X^* \to \mathcal{D}iv_X \to 0$$

which, on the right side, sends a meromorphic function to its divisor. The long exact sequence here starts out as

$$0 \to \mathbb{C}^* \to \mathcal{M}(X)^* \to \text{Div}(X) \to \check{H}^1(X, \mathcal{O}_X^*).$$

Now recall that $\check{H}^2(X, \underline{\mathbb{Z}}) \cong \mathbb{Z}$; in fact there is an isomorphism such that the composition map

$$\text{Div}(X) \to \check{H}^1(X, \mathcal{O}_X^*) \to \check{H}^2(X, \underline{\mathbb{Z}}) \cong \mathbb{Z}$$

is exactly the degree mapping, sending a divisor D to its degree. Moreover the image of $\mathcal{M}(X)^* \to \text{Div}(X)$ is exactly the subgroup $\text{PDiv}(X)$ of principal divisors on X. Therefore we may build the diagram

$$\begin{array}{ccccccccc}
& & 0 & = & 0 & & & & \\
& & \downarrow & & \downarrow & & & & \\
& & \text{PDiv}(X) & = & \text{PDiv}(X) & & & & \\
& & \downarrow & & \downarrow & & & & \\
0 & \to & \text{Div}_0(X) & \to & \text{Div}(X) & \stackrel{\deg}{\to} & \mathbb{Z} & \to & 0 \\
& & \downarrow & & \downarrow & & \| & & \\
0 & \to & \check{H}^1(X, \mathcal{O}_X)/\check{H}^1(X, \underline{\mathbb{Z}}) & \to & \check{H}^1(X, \mathcal{O}_X^*) & \stackrel{\deg}{\to} & \check{H}^2(X, \underline{\mathbb{Z}}) & &
\end{array}$$

The vertical sequence on the left side of this diagram shows that we have an alternate criterion for when a divisor of degree 0 is a principal divisor: D is principal if and only if D goes to 0 in the quotient group $\check{H}^1(X, \mathcal{O}_X)/\check{H}^1(X, \underline{\mathbb{Z}})$.

Actually, this quotient group is isomorphic to the Jacobian $\text{Jac}(X)$ of X, and the map is of course the Abel-Jacobi mapping. At this point let us be content with remarking that, due to Serre Duality and the GAGA theorems, we have a natural isomorphism

$$\check{H}^1(X, \mathcal{O}_X) \cong \Omega^1(X)^*$$

between the numerator of this quotient group and the dual space to the space of holomorphic 1-forms. Moreover, the cohomology group $\check{H}^1(X, \underline{\mathbb{Z}})$ is a free abelian group of rank $2g$, where g is the genus of X. Its image in $\Omega^1(X)^*$ is exactly the period lattice, and we have the Jacobian of X.

This cohomological point of view can be taken from the beginning to prove Abel's theorem. One gets the criterion immediately, and then one has to identify all the maps and the spaces. The more pedestrian approach taken in Chapter VIII defines the maps and spaces at the outset, and then the work is done in verifying the theorem.

Problems X.2

A. Give a proof that $\check{H}^2(X_{Zar}, \underline{G}) = 0$ for a constant sheaf $\underline{G}$ on an algebraic curve X.

B. Verify that $\alpha_{D,alg} : \mathcal{M}_{alg} \to \mathcal{T}_{X,alg}[D]$ is a map of sheaves on X_{Zar}.

C. Why is it enough to show that if a map α of sheaves is onto for every open set which is not the whole space X, then α is an onto map of sheaves? Generalize this statement.

D. By analyzing the long exact sequence associated to the short exact sequence
$$0 \to \mathcal{O}_X[D_1] \to \mathcal{O}_X[D_2] \to \mathcal{T}_X[D_1/D_2] \to 0$$
whenever $D_1 \leq D_2$, prove the first form of the Riemann-Roch Theorem.

E. Assuming that the Zero Mean Theorem holds for any compact Riemann surface X (which it does of course), show that if ω is any $\mathcal{C}^\infty$ 1-form on X, then there exists a $\mathcal{C}^\infty$ function f on X such that $\partial\bar{\partial} f = d\omega$. Conclude that if α is a $(0,1)$-form on X, then there exists a $\mathcal{C}^\infty$ function f on X such that $\partial\bar{\partial} f = \partial\alpha$.

F. Let X be a compact Riemann surface. Define a mapping $\text{Bar} : \Omega^1(X) \to H^{0,1}_{\bar{\partial}}(X)$ by sending a holomorphic 1-form ω to the Dolbeault cohomology class of the conjugate form $\bar{\omega}$. Show that Bar is $\mathbb{C}$-anti-linear, and is 1-1. Show that if one assumes that the Zero Mean Theorem holds for X, then Bar is onto.

G. Let X be a compact Riemann surface. Show that the natural map from $H^2_{\bar{\partial}}(X)$ to $H^{1,1}_d(X)$, sending the Dolbeault cohomology class of a $\mathcal{C}^\infty$ 2-form η to its De Rham cohomology class, is an isomorphism of complex vector spaces, using the Zero Mean Theorem.

H. Show that for a compact Riemann surface X, the following sequence is exact:
$$0 \to \Omega^1(X) \xrightarrow{A} H^1_d(X) \xrightarrow{B} H^{0,1}_{\bar{\partial}}(X) \xrightarrow{D} H^{1,1}_{\bar{\partial}}(X) \xrightarrow{C} H^2_d(X) \to 0,$$
where A and C are the obvious maps, B is induced by projection of a 1-form onto its $(0,1)$ part, and D is induced by the differentiation operator d.

I. Continuing with the sequence of the previous problem, now assume that the Zero Mean Theorem holds for X. Conclude that $\dim \Omega^1(X) = \frac{1}{2} \dim H^1_d(X)$. Finally note that since the operator d is defined purely in terms of the $\mathcal{C}^\infty$ structure, the De Rham cohomology group $H^1_d(X)$ has the same dimension for all X of the same topological genus; all such X are diffeomorphic. Therefore we conclude that $\dim \Omega^1(X)$ depends only on the genus g of X, and not on X itself. One concludes that $\dim \Omega^1(X) = g$ by explicitly computing the space of holomorphic 1-forms on a single surface of genus g, say a hyperelliptic surface.

This series of problems illustrate how the Zero Mean Theorem is used to begin the computations of cohomology spaces in the theory of compact Riemann surfaces.

J. Show that the exact sequence of the previous problem corresponds to part of the long exact sequence in cohomology for the short exact sequence of sheaves
$$0 \to \underline{\mathbb{C}} \to \mathcal{O}_X \xrightarrow{d} \Omega^1 \to 0$$
using the De Rham and Dolbeault isomorphisms.

Further Reading

The final algebraic nail has now been driven in, by introducing the algebraic sheaves and the Zariski topology. The move to working over an arbitrary field is almost complete; a text taking this point of view from the beginning is [**Hartshorne77**].

The Zariski topology may be defined for any algebraic variety of arbitrary dimension: the closed sets are the algebraic subsets. This has been generalized further to the spectrum of commutative rings (see [**AM69**]), and then to schemes (see [**E-H92**] for an introduction). A professional will eventually have need of the encyclopedic [**Grothendieck**].

The GAGA theorems appear in [**Serre56**]; see also [**GA74**].

Chapter XI. Invertible Sheaves, Line Bundles, and $\check{H}^1$

The Picard group $\text{Pic}(X)$ of an algebraic curve X has been defined as the group of divisors on X modulo the subgroup of principal divisors; in other words, it classifies divisors up to linear equivalence. It turns out that the Picard group classifies many things, which a priori are unrelated, but in the end share a very close relationship. These are "invertible" sheaves, defined to be sheaves on X which are locally isomorphic to $\mathcal{O}_X$, and "line bundles", about which more will be said later. In addition, the Picard group can also be represented as a cohomology group.

These four aspects of the Picard group bind the theory together in an especially intricate way, and the language and notation used by working algebraic geometers reflects this; it is common parlance that a linear equivalence class of divisors is called a "bundle", and a cohomology class will be referred to as an "invertible sheaf". It is part of the study to get used to these abuses.

In this chapter we will lay out these other ways of thinking about $\text{Pic}(X)$, and work through all the relationships.

We will work in this chapter in the algebraic category, taking as the underlying topology the Zariski topology and taking the algebraic sheaves as the fundamental tools. The interested reader will have no trouble translating the definitions into the analytic category if it is desired.

1. Invertible Sheaves

Sheaves of $\mathcal{O}$-Modules. The importance of the Picard group comes from its ability to classify *locally trivial* objects of a special type; this is not so clear from the divisor-mod-linear equivalence viewpoint, but it is obvious from the invertible-sheaf and line-bundle viewpoints. Before defining an invertible sheaf, we must understand the concept of a sheaf of modules.

DEFINITION 1.1. Let X be an algebraic curve (with its Zariski topology), and denote simply by $\mathcal{O}$ the sheaf $\mathcal{O}_{X,alg}$ of regular functions on X. A sheaf $\mathcal{F}$ on X is a *sheaf of $\mathcal{O}$-modules* (or simply an $\mathcal{O}$-*module*) if
  (i) for every open set $U \subseteq X$, the group $\mathcal{F}(U)$ is an $\mathcal{O}(U)$-module;

(ii) whenever $V \subseteq U$, then the restriction map $\rho_V^U : \mathcal{F}(U) \to \mathcal{F}(V)$ is $\mathcal{O}$-linear in the sense that if $r \in \mathcal{O}(U)$ and $f \in \mathcal{F}(U)$ then $\rho_V^U(r \cdot f) = \rho_V^U(r) \cdot \rho_V^U(f)$.

Of course in the algebraic category, where every ring $\mathcal{O}(U)$ is a subring of the rational function field $\mathcal{M}(X)$, the restriction maps are all inclusions, and making the appropriate identifications the last condition is more naturally written simply as $\rho_V^U(r \cdot f) = r \cdot \rho_V^U(f)$, i.e., that ρ_V^U is $\mathcal{O}$-linear.

A sheaf map $\phi : \mathcal{F} \to \mathcal{G}$ between two $\mathcal{O}$-modules is simply a sheaf map such that for every U, $\phi_U : \mathcal{F}(U) \to \mathcal{G}(U)$ is a homomorphism of $\mathcal{O}(U)$-modules. The kernel of a sheaf map of $\mathcal{O}$-modules is a sheaf of $\mathcal{O}$-modules.

The main examples of $\mathcal{O}$-modules are the sheaves $\mathcal{O}_{X,alg}[D]$ of rational functions with poles bounded by a divisor D, the sheaves $\Omega^1_{X,alg}[D]$ of rational 1-forms with poles bounded by D, and the (constant) sheaf $\mathcal{M}_{X,alg}$ of rational functions on X.

Definition of an Invertible Sheaf. We can now state the definition of an invertible sheaf, which is expressed in terms of restricted sheaves. Recall that if $\mathcal{F}$ is a sheaf on X, and $Y \subset X$ is an open set, then we may define a sheaf $\mathcal{F}|_Y$ on Y by setting $\mathcal{F}|_Y(U) = \mathcal{F}(U)$ for any open subset $U \subseteq Y$ (which is then also an open subset of X).

DEFINITION 1.2. Let X be an algebraic curve, and let $\mathcal{F}$ be a sheaf of $\mathcal{O}$-modules. We say that $\mathcal{F}$ is *invertible* if for every $p \in X$ there is an open neighborhood U of p, such that $\mathcal{F}|_U \cong \mathcal{O}|_U$ as sheaves of $\mathcal{O}|_U$-modules on the space U.

Thus invertible sheaves are "locally free rank one" $\mathcal{O}$-modules. An isomorphism $\phi_U : \mathcal{O}|_U \to \mathcal{F}|_U$ is called a *trivialization* of $\mathcal{F}$ over U.

An equivalent way of giving the invertible definition is to require that there is an open cover $\{U_i\}$ of X such that for each i, $\mathcal{F}|_{U_i} \cong \mathcal{O}|_{U_i}$ as sheaves of $\mathcal{O}|_{U_i}$-modules on U_i.

It is sometimes convenient to express the invertibility property in terms of generators for the modules $\mathcal{F}(V)$. Suppose that U is an open subset of X on which $\mathcal{F}|_U \cong \mathcal{O}|_U$ as sheaves of $\mathcal{O}$-modules. In this case we have an isomorphism of sheaves $\phi : \mathcal{O}|_U \to \mathcal{F}|_U$, which exactly means that there are isomorphisms $\phi_V : \mathcal{O}(V) \to \mathcal{F}(V)$ for all open $V \subseteq U$; moreover each map ϕ_V is a map of $\mathcal{O}(V)$-modules, and these isomorphisms commute with the restriction maps. In particular, there is an isomorphism $\phi_U : \mathcal{O}(U) \to \mathcal{F}(U)$ on the entire open subset U.

Let $f_U \in \mathcal{F}(U)$ be the image of $1 \in \mathcal{O}(U)$, which is then a generator for the free module $\mathcal{F}(U)$ over $\mathcal{O}(U)$. Note that if we define $f_V = \rho_V^U(f_U)$ for every open $V \subseteq U$, then f_V is also a generator of the free module $\mathcal{F}(V)$, since $f_V = \rho_V^U(f_U) = \rho_V^U(\phi_U(1)) = \phi_V(\rho_V^U(1)) = \phi_V(1)$.

1. INVERTIBLE SHEAVES

Thus the element f_U is not only a generator for $\mathcal{F}(U)$, but it restricts to generators f_V for $\mathcal{F}(V)$ for every open $V \subset U$.

It is easy to express the fact that $\mathcal{F}(V)$ is free of rank one over $\mathcal{O}(V)$ using a generator f_V; we simply require that f_V generate $\mathcal{F}(V)$ over $\mathcal{O}(V)$, and that it has a trivial annihilator in $\mathcal{O}(V)$; that is, if $r \in \mathcal{O}(V)$ and $r \cdot f_V = 0$ then $r = 0$.

So we are led to the following criterion for $\mathcal{F}|_U$ to be isomorphic to $\mathcal{O}|_U$: there should be a section $f_U \in \mathcal{F}(U)$ such that for every open $V \subseteq U$, the restriction $f_V = \rho_V^U(f_U)$ generates $\mathcal{F}(V)$ and has a trivial annihilator. This proves the following:

LEMMA 1.3. *Let X be an algebraic curve, and let $\mathcal{F}$ be a sheaf of $\mathcal{O}$-modules. Then $\mathcal{F}$ is invertible if and only if for every $p \in X$ there is an open neighborhood U of p and a section $f_U \in \mathcal{F}(U)$ such that for all $V \subseteq U$, the restricted section $f_V = \rho_V^U(f_U)$ generates the module $\mathcal{F}(V)$ over $\mathcal{O}(V)$, and has a trivial annihilator.*

Such an element f_U will be called a *local generator* for the invertible sheaf $\mathcal{F}$ at the point p. Hence we may loosely say that a sheaf $\mathcal{F}$ is invertible if it has a local generator at every point of X.

EXAMPLE 1.4. Let X be an algebraic curve. Then the sheaf $\Omega^1 = \Omega^1_{X,alg}$ of regular 1-forms on X is invertible. A local generator for Ω^1 in a neighborhood of a point p is the 1-form dz, where z is any rational local coordinate for X centered at p (that is, z is a rational function with a simple zero at p).

Invertible Sheaves associated to Divisors. The first (and as we will see, the "only") example of an invertible sheaf is afforded by the sheaf of rational functions whose poles are bounded by a divisor D on X.

LEMMA 1.5. *Let X be an algebraic curve, and let D be a divisor on X. Then the sheaf $\mathcal{O}_{X,alg}[D]$ is an invertible sheaf. Moreover, a local generator for $\mathcal{O}_{X,alg}[D]$ at a point $p \in X$ is $z^{-D(p)}$, where z is any rational function on X with a simple zero at p.*

PROOF. Fix a point $p \in X$, and let $z \in \mathcal{M}(X)$ be a rational function with a simple zero at p. (Such a global function exists since X is an algebraic curve.) Then z can be taken as a local complex coordinate on X at p. Let U be the Zariski open subset of X defined by removing all zeroes and poles of z (except for p) and all points q with $D(q) \neq 0$ (except for p). This is a cofinite subset of X, and for every $q \in U$ with $q \neq p$ we have $\mathrm{ord}_q(z) = D(q) = 0$.

Let $n = D(p)$; we claim that $f_U = z^{-n}$ is a local generator for $\mathcal{O}_{X,alg}[D]$ at p. Firstly, by the choice of U, we have that $f_U \in \mathcal{O}_{X,alg}[D](U)$, and so we may define $f_V \in \mathcal{O}_{X,alg}[D](V)$ for every open subset $V \subseteq U$. Of course each of these sections f_V has trivial annihilator; it is nonzero, and the multiplication is all happening in the rational function field $\mathcal{M}(X)$.

To finish the proof, using Lemma 1.3, we must check that f_V generates $\mathcal{O}_{X,alg}[D](V)$ over $\mathcal{O}(V)$ for every open $V \subseteq U$. Let $g \in \mathcal{O}_{X,alg}[D](V)$; since $V \subseteq U$, we have that at all points $q \in V$ with $q \neq p$, $\text{ord}_q(g) \geq 0$, and at p we have $\text{ord}_p(g) \geq -n$. Now the purported generator $f_V = z^{-n}$ satisfies $\text{ord}_q(f_V) = 0$ for all $q \neq p$, and $\text{ord}_p(f_V) = -n$. Therefore the ratio $r = g/f_V$ has $\text{ord}_q(r) \geq 0$ for every $q \in V$ (including $q = p$), and hence $r \in \mathcal{O}(V)$. Hence $g = r \cdot f_V$, and so f_V does generate $\mathcal{O}_{X,alg}[D](V)$ over $\mathcal{O}(V)$. □

The same proof, using rational 1-forms instead of rational functions, gives the following.

LEMMA 1.6. *Let X be an algebraic curve, and let D be a divisor on X. Then the sheaf $\Omega^1_{X,alg}[D]$ is an invertible sheaf. Moreover, a local generator for $\Omega^1_{X,alg}[D]$ at a point $p \in X$ is $z^{-D(p)}dz$, where z is any rational function on X with a simple zero at p.*

Instead of copying the proof, one could instead appeal to the isomorphism of sheaves $\Omega^1_{X,alg}[D] \cong \mathcal{O}_{X,alg}[K + D]$ where K is a canonical divisor on X; then the above lemma becomes a direct corollary of Lemma 1.5.

The Tensor Product of Invertible Sheaves. If F and G are two free modules of rank one over a ring R, generated by f and g respectively, then the tensor product $F \otimes_R G$ is also free of rank one over R, generated by $f \otimes g$. It is possible to "sheafify" this remark and define a tensor product operation on invertible sheaves.

Let $\mathcal{F}$ and $\mathcal{G}$ be two invertible sheaves on an algebraic curve X. One's first instinct in trying to define $\mathcal{F} \otimes_\mathcal{O} \mathcal{G}$ is to set

$$(1.7) \qquad \mathcal{F} \otimes \mathcal{G}(U) = \mathcal{F}(U) \otimes_{\mathcal{O}(U)} \mathcal{G}(U),$$

which does produce a presheaf of $\mathcal{O}$-modules on X. However it does *not* in general satisfy the sheaf axiom. This is a big disappointment, since (1.7) is an attractively simple definition.

The sheaf axiom states that if $\{U_i\}$ is an open cover of X, and $\mathcal{F}$ is any sheaf on X, then the natural map

$$\mathcal{F}(U) \to \{(s_i) \in \prod_i \mathcal{F}(U \cap U_i) \mid s_i|_{U \cap U_i \cap U_j} = s_j|_{U \cap U_i \cap U_j} \text{ for all } i, j\}$$

is an isomorphism of groups. (The map sends a section s to the collection of restrictions of s to $U \cap U_i$.) We can reverse our point of view and take this property as the *definition* of $\mathcal{F}(U)$ if we know enough about the groups $\mathcal{F}(U \cap U_i)$.

Suppose then that U is an open set on which both invertible sheaves $\mathcal{F}$ and $\mathcal{G}$ have trivializations. In other words, we have compatible isomorphisms from $\mathcal{O}(V)$ to both $\mathcal{F}(V)$ and $\mathcal{G}(V)$ for every open $V \subseteq U$. In this case we should be satisfied with the definition of $\mathcal{F} \otimes_\mathcal{O} \mathcal{G}$ given in (1.7), not only for U but for every

1. INVERTIBLE SHEAVES

open subset of U; this produces a free rank one module for every open subset, and the compatibility conditions ensure that in fact with this definition we have

$$(\mathcal{F} \otimes_{\mathcal{O}} \mathcal{G})|_U \cong \mathcal{O}|_U.$$

With this in mind we follow the dictates of the sheaf axiom to define $\mathcal{F} \otimes_{\mathcal{O}} \mathcal{G}(U)$ for any open U, as follows. Let $\{U_i\}$ be the collection of all open subsets of X on which both $\mathcal{F}$ and $\mathcal{G}$ may be trivialized. This forms an open covering of X. For any open subset U of X, define

$$(1.8) \quad \mathcal{F} \otimes_{\mathcal{O}} \mathcal{G}(U) = \{(s_i) \in \prod_i \mathcal{F}(U \cap U_i) \otimes_{\mathcal{O}(U \cap U_i)} \mathcal{G}(U \cap U_i) \mid$$
$$s_i|_{U \cap U_i \cap U_j} = s_j|_{U \cap U_i \cap U_j} \text{ for all } i, j\}.$$

LEMMA 1.9. *The definition given in (1.8) defines an invertible sheaf on X, denoted by $\mathcal{F} \otimes_{\mathcal{O}} \mathcal{G}$, and is called the* tensor product *of the invertible sheaves $\mathcal{F}$ and $\mathcal{G}$.*

PROOF. It is clear that (1.8) defines a presheaf. Let us check the sheaf axiom. For this choose an open set U and an open covering $\{V_k\}$ of U. Suppose that sections $(s_i^{(k)})$ in $\mathcal{F} \otimes_{\mathcal{O}} \mathcal{G}(V_k)$ are given for each k, which agree on $V_k \cap V_\ell$ for every k, ℓ. We must show that this collection of sections comes from a unique section of $\mathcal{F} \otimes_{\mathcal{O}} \mathcal{G}(U)$.

Note that for every i and k, we have $s_i^{(k)} \in \mathcal{F}(V_k \cap U_i) \otimes_{\mathcal{O}(V_k \cap U_i)} \mathcal{G}(V_k \cap U_i)$.

Let $f_i \in \mathcal{F}(U_i)$ be a generator over $\mathcal{O}(U_i)$, and let $g_i \in \mathcal{G}(U_i)$ be a generator over $\mathcal{O}(U_i)$. We note that for every open subset of U_i, f_i and g_i are generators of the corresponding modules, by restriction. In particular, we have that $f_i \otimes g_i$ is a generator of $\mathcal{F}(U_i) \otimes_{\mathcal{O}(U_i)} \mathcal{G}(U_i)$ and of the corresponding module over any open subset of U_i. Hence we may write

$$s_i^{(k)} = r_i^{(k)} f_i \otimes g_i$$

for some unique $r_i^{(k)} \in \mathcal{O}(V_k \cap U_i)$, for every k and i.

Since the collections $(s_i^{(k)})$ agree on the overlaps, we have for every k, ℓ, and i that $r_i^{(k)}|_{V_k \cap V_\ell \cap U_i} = r_i^{(\ell)}|_{V_k \cap V_\ell \cap U_i}$. Therefore since $\mathcal{O}$ satisfies the sheaf axiom these ring elements patch together to give a unique $r_i \in \mathcal{O}(U \cap U_i)$ such that $r_i|_{V_k \cap U_i} = r_i^{(k)}$ for every k.

In this case we set $s_i = r_i \cdot f_i \otimes g_i \in \mathcal{F}(U \cap U_i) \otimes_{\mathcal{O}(U \cap U_i)} \mathcal{G}(U \cap U_i)$. We then have that the collection (s_i) is a section of $\mathcal{F} \otimes_{\mathcal{O}} \mathcal{G}$ over U, and restricts to the collections $(s_i^{(k)})$ for every k.

The uniqueness of the section (s_i) comes from the uniqueness of the r_i's. This finishes the proof that $\mathcal{F} \otimes_{\mathcal{O}} \mathcal{G}$ is a sheaf.

To see that it is invertible, fix an index i, and consider the local generators f_i and g_i for $\mathcal{F}|_{U_i}$ and $\mathcal{G}|_{U_i}$ respectively. If for every j we set $s_j = f_i|_{U_j \cap U_i} \otimes g_i|_{U_j \cap U_i}$, we see that the collection (s_j) is a section of $\mathcal{F} \otimes_{\mathcal{O}} \mathcal{G}$ over U_i. Moreover it is clear that it is a local generator, since f_i and g_i are local generators. This finishes the proof of the lemma. □

The last paragraph of the proof shows that if $\mathcal{F}$ and $\mathcal{G}$ are both trivializable over an open set U, then so is $\mathcal{F} \otimes_{\mathcal{O}} \mathcal{G}$. Indeed, a local generator is induced by the tensor product of the local generators for $\mathcal{F}$ and $\mathcal{G}$.

In the definition of $\mathcal{F} \otimes_{\mathcal{O}} \mathcal{G}$, we used every open set U_i on which both $\mathcal{F}$ and $\mathcal{G}$ could be trivialized. We remark here that this is not really necessary; if we use any collection of such U_i's which cover X, we will obtain a sheaf isomorphic to the one above. We leave it to the reader to check this statement.

Using all of these open sets was only done to avoid having to make a choice of such a covering, which would have necessitated showing that the result was independent of the choice of covering, up to isomorphism.

The proof of the following easy lemma is left to the reader.

LEMMA 1.10. *Let X be an algebraic curve.*
  (i) *If $\mathcal{F}$ is an invertible sheaf on X, then $\mathcal{O} \otimes_{\mathcal{O}} \mathcal{F} \cong \mathcal{F}$.*
  (ii) *If $\mathcal{F}$ and $\mathcal{G}$ are invertible sheaves on X, then $\mathcal{F} \otimes_{\mathcal{O}} \mathcal{G} \cong \mathcal{G} \otimes_{\mathcal{O}} \mathcal{F}$.*
  (iii) *If $\mathcal{F}$, $\mathcal{G}$, and $\mathcal{H}$ are invertible sheaves on X, then $(\mathcal{F} \otimes_{\mathcal{O}} \mathcal{G}) \otimes_{\mathcal{O}} \mathcal{H} \cong \mathcal{F} \otimes_{\mathcal{O}} (\mathcal{G} \otimes_{\mathcal{O}} \mathcal{H})$.*

The Inverse of an Invertible Sheaf. If F is a free module of rank one over a ring R, generated by f, then the dual module $F^{-1} = \operatorname{Hom}_R(F, R)$ is also free of rank one over R, generated by the functional which sends f to 1. It is possible to "sheafify" this remark and define a dualizing operation on invertible sheaves.

Let $\mathcal{F}$ be an invertible sheaf on an algebraic curve X. One's first instinct in trying to define $\mathcal{F}^{-1}$ is to set

$$(1.11) \qquad \mathcal{F}^{-1}(U) = \operatorname{Hom}_{\mathcal{O}(U)}(\mathcal{F}(U), \mathcal{O}(U)),$$

which does produce a presheaf of $\mathcal{O}$-modules on X. However again it does *not* in general satisfy the sheaf axiom.

One can recover in the same manner as above; this time we will leave all the details to the reader. Let $\{U_i\}$ be the collection of all open subsets of X on which $\mathcal{F}$ may be trivialized. This is an open covering of X. For any open subset U of X, define

$$(1.12) \quad \mathcal{F}^{-1}(U) = \{(s_i) \in \prod_i \operatorname{Hom}_{\mathcal{O}(U \cap U_i)}(\mathcal{F}(U \cap U_i), \mathcal{O}(U \cap U_i)) \mid$$
$$s_i|_{U \cap U_i \cap U_j} = s_j|_{U \cap U_i \cap U_j} \text{ for all } i, j\}.$$

LEMMA 1.13. *The definition given in (1.12) defines an invertible sheaf on X, denoted by $\mathcal{F}^{-1}$, and is called the* inverse *of the invertible sheaf $\mathcal{F}$.*

1. INVERTIBLE SHEAVES

In addition, if $\mathcal{F}$ is trivializable over an open set U, then so is the inverse sheaf $\mathcal{F}^{-1}$. If f is a local generator for $\mathcal{F}$ on U, then a local generator for $\mathcal{F}^{-1}$ is induced by the functional sending f to 1.

One of the most important examples of an inverse sheaf construction is to take the inverse sheaf to the invertible sheaf Ω^1 of regular 1-forms on X. This sheaf is called the *tangent sheaf*, and is often denoted by Θ_X (or simply Θ). More will be said later concerning the tangent sheaf.

The Group of Isomorphism Classes of Invertible Sheaves. For an algebraic curve X, we have seen that the tensor product gives a binary operation on the class of invertible sheaves. We form a group from the isomorphism classes as follows.

Denote by $\mathrm{Inv}(X)$ the set of isomorphism classes of invertible sheaves on X. If $\mathcal{F}$ is an invertible sheaf, its isomorphism class will be denoted by $[\mathcal{F}]$. Define

$$[\mathcal{F}] \otimes [\mathcal{G}] := [\mathcal{F} \otimes_{\mathcal{O}} \mathcal{G}].$$

It is clear that this is well defined, depending only on the isomorphism classes of the invertible sheaves and not on the sheaves themselves. Moreover by Lemma 1.10, this binary operation on $\mathrm{Inv}(X)$ has $\mathcal{O}$ as an identity, and is commutative and associative.

We claim that the class of the inverse invertible sheaf gives an inverse for this operation. To see this, we must check that for an invertible sheaf $\mathcal{F}$ on X, we have

$$\mathcal{F} \otimes_{\mathcal{O}} \mathcal{F}^{-1} \cong \mathcal{O}.$$

Such an isomorphism is to be expected, by analogy with the free rank one module situation: if F is a free rank one module over a commutative ring R with identity, then $F \otimes_R \mathrm{Hom}_R(F, R) \cong R$. Moreover the isomorphism sends $f \otimes \phi$ to $\phi(f)$. We need to "sheafify" this isomorphism.

Let $\{U_i\}$ be an open cover of X such that over each U_i there is a local generator f_i for the invertible sheaf $\mathcal{F}$. Then by definition we have for any open subset U of X,

$$\mathcal{F} \otimes_{\mathcal{O}} \mathcal{F}^{-1}(U) = \{(s_i) \in \prod_i \mathcal{F}(U \cap U_i) \otimes_{\mathcal{O}(U \cap U_i)} \mathcal{F}^{-1}(U \cap U_i) \mid$$
$$s_i|_{U \cap U_i \cap U_j} = s_j|_{U \cap U_i \cap U_j} \text{ for all } i,j\}.$$

On $U \cap U_i$, $\mathcal{F}(U \cap U_i)$ is a free rank one module over $\mathcal{O}(U \cap U_i)$, generated by $f_i|_{U \cap U_i}$. Similarly, $\mathcal{F}^{-1}(U \cap U_i)$ is also free of rank one, generated by the functional ϕ_i which sends f_i to 1. Therefore for every i we may write $s_i = r_i \cdot f_i \otimes \phi_i$, for some unique $r_i \in \mathcal{O}(U \cap U_i)$. The compatibility condition for the s_i's ensures that $r_i|_{U \cap U_i \cap U_j} = r_j|_{U \cap U_i \cap U_j}$; therefore this collection of r_i's patch together to give a section $r \in \mathcal{O}(U)$. This prescription defines a sheaf map

$$\Phi : \mathcal{F} \otimes_{\mathcal{O}} \mathcal{F}^{-1} \to \mathcal{O},$$

and we must now check that Φ is an isomorphism.

The inverse Ψ to Φ is readily defined. Fix an open set U, and a section $r \in \mathcal{O}(U)$. Define $\Psi(r)$ to be the collection (s_i), where $s_i = r|_{U \cap U_i} \cdot f_i \otimes \phi_i$. This defines a sheaf map, and we leave it to the reader to check that Φ and Ψ are inverses of one another.

This proves the following.

PROPOSITION 1.14. *Let X be an algebraic curve. Then the set $\text{Inv}(X)$ of isomorphism classes of invertible sheaves on X forms an abelian group whose operation is induced by the tensor product. The identity is the class of the sheaf $\mathcal{O}$ of regular functions on X. The inverse of the class of an invertible sheaf is the class of the inverse invertible sheaf.*

We will see in Section 3 that the map sending a divisor D to the invertible sheaf $\mathcal{O}[D]$ induces an isomorphism from the Picard group $\text{Pic}(X)$ of divisors modulo linear equivalence, to the group $\text{Inv}(X)$ of invertible sheaves. For this reason all algebraic geometers refer to $\text{Inv}(X)$ as the Picard group of X.

Problems XI.1

A. Show that the kernel of a sheaf map of $\mathcal{O}$-modules is an $\mathcal{O}$-module.
B. Show that $\mathcal{O}_{X,alg}[D]$, $\Omega^1_{X,alg}[D]$, and $\mathcal{M}_{X,alg}$ are sheaves of $\mathcal{O}$-modules.
C. Show that if $\mathcal{F}$ is a sheaf of $\mathcal{O}$-modules, then for every point $p \in X$, the stalk $\mathcal{F}_p$ is a module over the stalk $\mathcal{O}_p$.
D. Prove or disprove: a sheaf $\mathcal{F}$ of $\mathcal{O}$-modules is invertible if and only if for every point $p \in X$, the stalk $\mathcal{F}_p$ is a free rank one $\mathcal{O}_p$-module.
E. Show that the presheaf defined by (1.7) may not satisfy the sheaf axiom, by considering the Riemann Sphere X, a point $p \in X$ (say $p = \infty$), and the two invertible sheaves $\mathcal{O}_{X,alg}[-p]$ and $\mathcal{O}_{X,alg}[p]$. Specifically, show that there is an open cover of X and sections of the presheaf over each open set in the cover, which agree on the intersections, but which do not come from a global section of the presheaf.
F. Show that if $\{U_i\}$ is any open covering of X such that both $\mathcal{F}$ and $\mathcal{G}$ are trivialized over each U_i, and we define a sheaf $\mathcal{T}$ on X by setting

$$\mathcal{T}(U) = \{(s_i) \in \prod_i \mathcal{F}(U \cap U_i) \otimes_{\mathcal{O}(U \cap U_i)} \mathcal{G}(U \cap U_i) \mid$$
$$s_i|_{U \cap U_i \cap U_j} = s_j|_{U \cap U_i \cap U_j} \text{ for all } i,j\},$$

then the sheaf $\mathcal{T}$ is isomorphic to the tensor product sheaf $\mathcal{F} \otimes_{\mathcal{O}} \mathcal{G}$.
G. Prove Lemma 1.10.
H. Show that the presheaf defined by (1.11) may not satisfy the sheaf axiom, by considering the Riemann Sphere X, a point $p \in X$ (say $p = \infty$), and the invertible sheaf $\mathcal{O}_{X,alg}[-p]$. Specifically, show that there is an open cover of X and sections of the inverse presheaf over each open set in the cover, which agree on the intersections, but which do not come from a global section of the inverse presheaf.

I. Prove Lemma 1.13.

2. Line Bundles

A line bundle on an algebraic curve X is essentially the assignment of a complex line to each point $p \in X$. (By "complex line" we mean a one-dimensional complex vector space.) The prototype for this is projective space itself; a point $p \in \mathbb{P}^n$ is defined to be a 1-dimensional subspace L_p of $\mathbb{C}^{n+1}$. Hence assigning the subspace L_p (considered as a vector space) to itself (considered as the point p in $\mathbb{P}^n$) gives a line bundle on $\mathbb{P}^n$.

Since $\mathbb{P}^n$ has this most natural line bundle on it, anything that maps to $\mathbb{P}^n$ will have an induced line bundle: if $\phi : X \to \mathbb{P}^n$ is a function, we may assign to a point $x \in X$ the complex line given by $\phi(x)$.

In this way line bundles are ubiquitous, once you know what they are. In this section we will speak only of line bundles on algebraic curves, although the reader should have no trouble transferring the ideas to other categories.

The Definition of a Line Bundle. Let X be an algebraic curve. In assigning a one-dimensional complex vector space L_p (a complex line) to every point p of X, we want to include an extra condition that roughly speaking makes the complex lines vary continuously with the points of X. In fact, the best one can hope for is that the dependence on the point is locally trivial. The model for this should be $\mathbb{C} \times U$, for an open set $U \subset X$; the second projection map $pr_2 : \mathbb{C} \times U \to U$ sending (λ, p) to p has the property that $pr_2^{-1}(p) = \mathbb{C} \times \{p\}$ is naturally a complex line, and as one varies p the line varies "trivially".

The concept is somewhat similar to the definition of a Riemann surface itself. There we had a topological space which was locally biholomorphic to an open subset of $\mathbb{C}$; these isomorphisms were given by the complex charts.

Transferring this idea in our context leads to the following definition.

DEFINITION 2.1. Let X be an algebraic curve, let L be a set, and let $\pi : L \to X$ be a function. A *line bundle chart* for L (or for π) is a bijection $\phi : \pi^{-1}(U) \to \mathbb{C} \times U$ for some open set $U \subseteq X$, such that $pr_2 \circ \phi = \pi$ on $\pi^{-1}(U)$.

The open set U is called the *support* of the line bundle chart. Note that if $\phi : \pi^{-1}(U) \to \mathbb{C} \times U$ is a line bundle chart for L, then we automatically have that the fibers $L_p = \pi^{-1}(p)$ for each point p in U are complex lines, with a vector space structure: via ϕ, L_p is carried exactly onto $\mathbb{C} \times \{p\}$, and we can transport the addition and scalar multiplication via ϕ. Explicitly, we have

$$\lambda \cdot v = \phi^{-1}(\lambda pr_1(\phi(v)), p)$$

for $\lambda \in \mathbb{C}$ and $v \in L_p$, and

$$v + w = \phi^{-1}(pr_1(\phi(v)) + pr_1(\phi(w)), p)$$

for v and w in L_p.

If $\phi : \pi^{-1}(U) \to \mathbb{C} \times U$ is a line bundle chart for L, a complex coordinate z in the first component $\mathbb{C}$ of the target is called the *fiber coordinate* of the line bundle L with respect to this chart ϕ.

Now we will have to define what it means for two line bundle charts to be compatible. The reader will see that we are recapitulating the idea of a Riemann surface in its entirety.

The additional complication for line bundle charts is that we want to preserve the vector space structure in the fibers L_p of the map π.

Suppose that $\phi_1 : \pi^{-1}(U_1) \to U_1$ and $\phi_2 : \pi^{-1}(U_2) \to U_2$ are two line bundle charts on L, and $p \in U_1 \cap U_2$. Note that we have two separate identifications of the fiber $L_p = \pi^{-1}(p)$ with $\mathbb{C} \times \{p\}$, given by ϕ_1 and ϕ_2; we demand first that the composition $\phi_2 \circ \phi_1^{-1}$ be a linear isomorphism of vector spaces. Since these are just one-dimensional spaces, the map must be obtained by scaling by some nonzero number $r(p)$. Secondly, we should demand that this scaling factor vary holomorphically with the point p. Since we are in the algebraic category here, we in fact demand that it vary as a global meromorphic function. This gives the following.

DEFINITION 2.2. Let X be an algebraic curve, let L be a set, and let $\pi : L \to X$ be a function. Suppose that $\phi_1 : \pi^{-1}(U_1) \to \mathbb{C} \times U_1$ and $\phi_2 : \pi^{-1}(U_2) \to \mathbb{C} \times U_2$ are two line bundle charts on L. We say that ϕ_1 and ϕ_2 are *compatible* if either $U_1 \cap U_2 = \emptyset$ or the map

$$\phi_2 \circ \phi_1^{-1} : \mathbb{C} \times (U_1 \cap U_2) \to \mathbb{C} \times (U_1 \cap U_2)$$

has the form

$$(v, p) \mapsto (r(p) \cdot v, p)$$

for some regular nowhere zero function $r \in \mathcal{O}(U_1 \cap U_2)$.

The regular nowhere zero function r is called the *transition function* between the two line bundle charts.

The form of the map above is expressed conveniently in terms of the fiber coordinates of the two line bundle charts: if z_i is the fiber coordinate for L with respect to ϕ_i, then these fiber coordinates must be related by an equation of the form

$$z_2 = r \cdot z_1$$

for some regular nowhere zero function $r \in \mathcal{O}(U_1 \cap U_2)$.

This notion of compatibility ensures that the vector space structures induced on the fiber L_p via the two different line bundle charts are in fact exactly the same.

We can now finish the definition of a line bundle in a rather formal way, again taking our cue from the Riemann surface definition.

2. LINE BUNDLES

DEFINITION 2.3. Let X be an algebraic curve, let L be a set, and let $\pi : L \to X$ be a function. A *line bundle atlas* for L (or for π) is a collection of pairwise compatible line bundle charts for L whose supports cover X. Two line bundle atlases for L are *equivalent* if every line bundle chart of one is compatible with every line bundle chart of the other. A *line bundle structure* for L (or for π) is a maximal line bundle atlas for L, or, equivalently, an equivalence class of line bundle atlases for L. A *line bundle on* X is a map $\pi : L \to X$ which has a line bundle structure.

The first example of a line bundle on an algebraic curve X is the *trivial line bundle*, defined to be simply the product $L = \mathbb{C} \times X$, with the map π being the second projection. Here a line bundle atlas consists of a single line bundle chart.

The Tautological Line Bundle for a Map to $\mathbb{P}^n$. As mentioned above, one of the most important examples of a line bundle on an algebraic curve X comes from a holomorphic map to a projective space. Choose $n+1$ rational functions $f_0, f_1, \ldots, f_n$ on X which are not all identically zero. These functions define a mapping

$$\Phi : X \to \mathbb{P}^n$$

by sending $p \in X$ to $[f_0(p) : f_1(p) : \cdots : f_n(p)]$; see Chapter V, Section 4. Consider the subset $L \subset \mathbb{C}^{n+1} \times X$ defined by

$$L = \{(v, p) \in \mathbb{C}^{n+1} \times X \mid v \in \Phi(p)\} = \bigcup_{p \in X} (\Phi(p) \times \{p\}),$$

where we of course are identifying the point $\Phi(p)$ in projective space with the corresponding one-dimensional subspace of $\mathbb{C}^{n+1}$. Note that the second projection gives a function $\pi : L \to X$, and the fiber $L_p = \pi^{-1}(p)$ is naturally isomorphic to the complex line associated to $\Phi(p)$ (it is in fact $\Phi(p) \times \{p\}$).

Let $U_i \subset X$ be the open subset of X where the ith coordinate of Φ is not zero; hence

$$U_i = \Phi^{-1}(\{[z_0 : \cdots : z_{i-1} : 1 : z_{i+1} : \cdots : z_n] \mid z_i \in \mathbb{C}\}).$$

For any point $p \in U_i$, we may write $\Phi(p)$ uniquely as $[z_0 : \cdots : z_{i-1} : 1 : z_{i+1} : \cdots : z_n]$; these coordinate functions $z_k = f_k/f_i$ (which are of course functions of p) are regular functions on U_i. (In fact, U_i may be defined to be the Zariski open subset of X obtained by removing all the poles of the rational functions f_k/f_i, as k varies.)

Hence for any $(v, p) \in \pi^{-1}(U_i)$, define $\phi_i : \pi^{-1}(U_i) \to \mathbb{C} \times U_i$ by setting

$$\phi_i(v, p) = (v_i, p),$$

where v_i is just the ith coordinate of the vector v. The inverse mapping sends a pair (s, p) to the pair (v, p), where $v = (sz_0, \ldots, sz_{i-1}, s, sz_{i+1}, \ldots, sz_n)$. It is an exercise to check that ϕ_i is a line bundle chart for L.

Let us check that ϕ_i and ϕ_j are compatible. For a point $p \in U_i \cap U_j$, write $\Phi(p) = [z_0 : \cdots z_{i-1} : 1 : z_{i+1} : \cdots : z_n]$ as above. Then if $(s,p) \in \mathbb{C} \times (U_i \cap U_j)$, we have

$$\begin{aligned}\phi_j \circ \phi_i^{-1}(s,p) &= \phi_j((sz_0, \ldots, sz_{i-1}, s, sz_{i+1}, \ldots, sz_n), p) \\ &= (sz_j, p) = (sf_j/f_i, p);\end{aligned}$$

since f_j/f_i is a regular function on U_i, this has the correct form for the compatibility condition.

Since the U_i's cover X, the ϕ_i's give a line bundle atlas for L, and hence induce a line bundle structure on L. This line bundle is called the *tautological line bundle* for the map Φ.

Line Bundle Homomorphisms. In any algebraic construction, it is never enough to give the objects without giving the maps between the objects. The case of line bundles is no exception, if for no other reason than to be able to speak of isomorphisms between different line bundles.

Essentially, a line bundle homomorphism should be a map between the line bundles which sends fibers to fibers and is linear on each fiber. Moreover the linear map should vary in a regular way. This leads to the following.

DEFINITION 2.4. Let X be an algebraic curve and suppose that $\pi_1 : L_1 \to X$ and $\pi_2 : L_2 \to X$ are two line bundles on X. A function $\alpha : L_1 \to L_2$ is a *line bundle homomorphism* if
  (i) $\pi_2 \circ \alpha = \pi_1$, and
  (ii) for every pair of line bundle charts $\phi_1 : \pi_1^{-1}(U_1) \to \mathbb{C} \times U_1$ and $\phi_2 : \pi_2^{-1}(U_2) \to \mathbb{C} \times U_2$ for L_1 and L_2 respectively, the composition

$$\phi_2 \circ \alpha \circ \phi_1^{-1} : \mathbb{C} \times (U_1 \cap U_2) \to \mathbb{C} \times (U_1 \cap U_2)$$

has the form

$$(s,p) \mapsto (f(p)s, p)$$

for some regular function f on $U_1 \cap U_2$.

It is an exercise to check that it suffices to check condition (ii) only for pairs of line bundles charts coming from two line bundle atlases for L_1 and L_2.

The composition of two line bundle homomorphisms is a line bundle homomorphism; the identity map for a line bundle is of course a line bundle homomorphism. A *line bundle isomorphism* is a line bundle homomorphism which has an inverse. Two line bundles are said to be *isomorphic* if there is a line bundle isomorphism between them.

We denote by $\mathrm{LB}(X)$ the set of isomorphism classes of line bundles on X.

2. LINE BUNDLES

Defining a Line Bundle via Transition Functions. Recall in Chapter III, Section 1 we introduced the notion of *glueing* Riemann surfaces together to obtain other Riemann surfaces. A similar idea can be developed to define line bundles on an algebraic curve X, by "glueing" trivial line bundles on open sets $\{U_i\}$ together. The required glueing data is exactly the set of transition functions.

Suppose then that an open covering $\{U_i\}$ of an algebraic curve X is given. Let us give, for each pair (i,j) of indices, a regular nowhere zero function t_{ij} on the intersection $U_i \cap U_j$. We consider the trivial line bundle $\mathbb{C} \times U_i$ over each U_i. In order to "glue" these line bundles together, whenever a point p is in both U_i and U_j, we need to identify the complex line $\mathbb{C} \times \{p\} \subset \mathbb{C} \times U_i$ with the complex line $\mathbb{C} \times \{p\} \subset \mathbb{C} \times U_j$. This is of course done via the transition function t_{ij}: we identify $(s, p) \in \mathbb{C} \times U_j$ with $(t_{ij}s, p) \in \mathbb{C} \times U_i$.

To be more precise and explicit, let $\tilde{L}$ be the disjoint union

$$\tilde{L} = \coprod_i (\mathbb{C} \times U_i).$$

Define a partition of $\tilde{L}$ by declaring $(s, p) \in \mathbb{C} \times U_j$ to be in the same partition subset as $(t_{ij}s, p) \in \mathbb{C} \times U_i$ whenever $p \in U_i \cap U_j$. Let L be the set of partition subsets; there is a natural map from $\tilde{L}$ to L, sending an element to its partition subset.

Under some mild hypotheses, L will be a line bundle over X, and the natural maps from $\mathbb{C} \times U_i$ to L will be inverses of line bundle charts for L. What is required is essentially that we not identify too much. For example, if $i = j$, we do not want to make any nontrivial identifications at all; hence we will insist that

(2.5) $$t_{ii} \equiv 1 \text{ on } U_i$$

for each i.

Similarly, if $i \neq j$, we will have that $(s, p) \in \mathbb{C} \times U_j$ is identified with $(t_{ij}s, p) \in \mathbb{C} \times U_i$, which is in turn identified back with $(t_{ji}t_{ij}s, p) \in \mathbb{C} \times U_j$; hence we will demand that

(2.6) $$t_{ji}t_{ij} \equiv 1 \text{ on } U_i \cap U_j$$

for every i and j.

Finally, if i, j, and k are three indices, by starting at $(s, p) \in \mathbb{C} \times U_k$ and moving to the U_j, then to the U_i, and back to the U_k products, we see that

(2.7) $$t_{ki}t_{ij}t_{jk} \equiv 1 \text{ on } U_i \cap U_j \cap U_k$$

is necessary for every i, j, and k.

These three properties (2.5)-(2.7) are called the *cocycle conditions* on the transition functions t_{ij}; compare them with the conditions for a 1-cochain to be a 1-cocycle. More will be said about this later.

Note that the cocycle conditions imply by induction that if $i_0, i_1, \ldots, i_n$ is any sequence of indices, then

$$(2.8) \qquad t_{i_0 i_n} \equiv t_{i_0 i_1} t_{i_1 i_2} \cdots t_{i_{n-1} i_n} \text{ on } U_{i_0} \cap \cdots \cap U_{i_n}.$$

Therefore any point $(s, p) \in \mathbb{C} \times U_j$ is identified exactly with the points $(t_{ij}s, p) \in \mathbb{C} \times U_i$ for those i such that $p \in U_i$, and with no other points. In particular, the composition

$$\mathbb{C} \times U_i \hookrightarrow \tilde{L} \to L$$

is injective.

Let L_i be the image of $\mathbb{C} \times U_i$ in L, so that the above composition gives a bijection between $\mathbb{C} \times U_i$ and L_i; denote by $\phi_i : L_i \to \mathbb{C} \times U_i$ the inverse of this bijection; it is a line bundle chart on L.

Finally we check the compatibility of these line bundle charts. Fix two indices i and j, and an ordered pair $(s, p) \in \mathbb{C} \times U_j$. Then $\phi_i(\phi_j^{-1}(s, p))$ is the point in $\mathbb{C} \times U_i$ which is identified with (s, p) in L; this is by our construction exactly the point $(t_{ij}s, p)$. Therefore since t_{ij} is a regular nowhere zero function on $U_i \cap U_j$ by assumption, the mapping $\phi_i \circ \phi_j^{-1}$ has the required form and the two line bundle charts are compatible.

Since the U_i's form an open cover of X, we have constructed a line bundle atlas for L, inducing a line bundle structure.

This proves the following.

PROPOSITION 2.9. *Let X be an algebraic curve, $\{U_i\}$ an open cover of X, and for each pair i, j of indices, suppose that t_{ij} is a nowhere zero regular function on $U_i \cap U_j$, such that the collection $\{t_{ij}\}$ satisfies the cocycle conditions (2.5)-(2.7). Then there is a line bundle L, unique up to isomorphism, with line bundle charts having the U_i's as supports, and having the functions t_{ij} as the transition functions. In terms of fiber coordinates z_i for these line bundle charts, we have the equation*

$$z_i = t_{ij} z_j$$

holding over $U_i \cap U_j$ for every i and j.

The only point which we leave to the reader is to check the uniqueness of L.

EXAMPLE 2.10. As an example of the previous construction, consider the algebraic curve $\mathbb{P}^1$, with its two charts $U_0 = \{[1 : z] \mid z \in \mathbb{C}\}$ and $U_1 = \{[w : 1] \mid w \in \mathbb{C}\}$. On the overlap of course we have $w = 1/z$. Define a set of transition functions by

$$t_{00} = t_{11} = 1, \quad t_{01} = z^n, \quad t_{10} = w^n.$$

This set satisfies the cocycle conditions, and defines a line bundle denoted by L_n on $\mathbb{P}^1$.

2. LINE BUNDLES

EXAMPLE 2.11. The above construction works both in the analytic setting (for a Riemann surface with the classical topology) and in the algebraic setting (for an algebraic curve with the Zariski topology). Let us give an example in the analytic setting.

Let X be any compact Riemann surface, and suppose that $\mathcal{A} = \{\phi_i : U_i \to V_i\}$ is an atlas on X. For every i and j, the charts ϕ_i and ϕ_j are compatible, and so the composition $T_{ij} = \phi_i \circ \phi_j^{-1}$ is a biholomorphic function from $\phi_j(U_i \cap U_j)$ to $\phi_i(U_I \cap U_J)$. Moreover by Lemma 1.7 of Chapter I, the derivative T'_{ij} is nowhere zero. Let $t_{ij} = T'_{ij} \circ \phi_i|_{U_i \cap U_j}$; this is a nowhere zero holomorphic function on $U_i \cap U_j$, and the collection (t_{ij}) satisfies the cocycle conditions.

The line bundle which this set of transition functions defines, constructed as above, is called the *tangent bundle* $\mathbb{T}_X$ (or simply $\mathbb{T}$) on X. We will see later that it depends only on the complex structure of X, not on the particular atlas chosen to define it.

EXAMPLE 2.12. Let X be any compact Riemann surface; with the same notation as above, we may use as transition functions not the derivatives T'_{ij}, but the reciprocal functions $1/T'_{ij}$. Since we have that $T_{ij} \circ T_{ji}$ is the identity, in fact $T'_{ji} = 1/T'_{ij}$. In any case if we let $t_{ij} = T'_{ji} \circ \phi_i|_{U_i \cap U_j}$, we obtain a nowhere zero holomorphic function on $U_i \cap U_j$, and the collection (t_{ij}) satisfies the cocycle conditions.

The line bundle which this set of transition functions defines, constructed as above, is called the *canonical bundle* $\mathbb{K}_X$ (or simply $\mathbb{K}$) on X. It also depends only on the complex structure of X, not on the particular atlas chosen to define it.

The Invertible Sheaf of Regular Sections of a Line Bundle. The way that line bundles and invertible sheaves are directly related is via the notion of a *section* of a line bundle.

DEFINITION 2.13. Let $\pi : L \to X$ be a line bundle over an algebraic curve X, and let $U \subset X$ be an open subset of X. A *regular section of L over U* is a function $s : U \to L$ such that
  (i) for every $p \in U$, $s(p)$ lies in the fiber of L over p, i.e.,
$$\pi \circ s = \mathrm{id}_U;$$
  and
  (ii) for every line bundle chart $\phi : \pi^{-1}(V) \to \mathbb{C} \times V$ for L, the composition
$$pr_1 \circ \phi \circ s|_{U \cap V} : U \cap V \to \mathbb{C}$$
  is a regular function on $U \cap V$.

It is an exercise to check that if a function s exists satisfying (i), then one does not need to check (ii) for every line bundle chart for L, but just for the line bundle charts in some line bundle atlas for L.

We denote by $\mathcal{O}_{X,alg}\{L\}(U)$ (or simply $\mathcal{O}\{L\}(U)$ if no confusion is possible) the set of regular sections of L over U. Our aim is to show that this construction gives an invertible sheaf on X.

First we note that whenever $V \subseteq U$ there is a natural restriction map from $\mathcal{O}\{L\}(U)$ to $\mathcal{O}\{L\}(V)$. Secondly, by using the addition in each fiber of L, we see immediately that the set of regular sections forms a group under pointwise addition: if s_1 and s_2 are regular sections over U, then $s_1 + s_2$ (defined by sending a point p to $s_1(p) + s_2(p)$, the sum being computed in the fiber of L over p) is a regular section of L over U. Thirdly, if s is a regular section of L over U and f is a regular function on U, then the product $f \cdot s$ (defined by sending a point $p \in U$ to the vector $f(p) \cdot s(p)$ in the fiber of L over p) is also a regular section of L over U; moreover it is clear that the required distributive laws hold. Therefore the set $\mathcal{O}\{L\}(U)$ is a module over the ring $\mathcal{O}(U)$; since the restriction map is compatible with this module structure, we have a presheaf of $\mathcal{O}$-modules.

To check the sheaf axiom, fix an open set U and an open covering $\{V_i\}$ of U. Suppose that regular sections s_i of L over V_i are given for each i, such that s_i and s_j agree on the intersections $V_i \cap V_j$ for every pair i, j. Define $s : U \to L$ by setting $s(p) = s_i(p)$ if $p \in V_i$; this is well defined, and satisfies $\pi \circ s = \mathrm{id}_U$. We must check the regularity condition.

Fix a line bundle chart $\phi : \pi^{-1}(W) \to \mathbb{C} \times W$ for L, and for each i let $f_i = pr_1 \circ \phi \circ s_i|_{V_i \cap W} : V_i \cap W \to \mathbb{C}$; by assumption, each f_i is a regular function on $V_i \cap W$. Since the s_i's agree on the double intersections, we have that $f_i = f_j$ on $V_i \cap V_j \cap W$ for every i, j. Hence since $\mathcal{O}$ is a sheaf, the f_i's patch together to give a regular function f on $U \cap W$. This regular function f is, by construction, equal to $pr_1 \circ \phi \circ s|_{U \cap W}$, and so s is a regular section of L over U.

This finishes the proof that $\mathcal{O}\{L\}$ is a sheaf of $\mathcal{O}$-modules. Finally to check that it is invertible, fix a line bundle chart $\phi : \pi^{-1}(U) \to \mathbb{C} \times U$ for L. Then for any regular function f on any open subset $V \subseteq U$ of U, we may define a regular section s_f for L over V by setting

$$s_f(p) = \phi^{-1}(f(p), p).$$

Conversely, given a regular section s for L over V, we obtain a regular function f_s on V by

$$f_s = pr_1 \circ \phi \circ s.$$

The reader can check that these correspondences give inverse $\mathcal{O}$-module isomorphisms between $\mathcal{O}(V)$ and $\mathcal{O}\{L\}(V)$ for every such V; moreover everything is compatible with the restriction maps and so we in fact obtain a sheaf isomorphism between $\mathcal{O}|_U$ and $\mathcal{O}\{L\}|_U$. Hence $\mathcal{O}\{L\}$ is invertible, and we have shown the following.

PROPOSITION 2.14. *Let X be an algebraic curve and $\pi : L \to X$ a line bundle on X. Then the presheaf $\mathcal{O}\{L\}$ of regular sections of L is an invertible sheaf on X.*

The construction above induces a function $\mathcal{O}\{-\}$ from the set $\mathrm{LB}(X)$ of isomorphism classes of line bundles on X to the group $\mathrm{Inv}(X)$ of isomorphism classes of invertible sheaves on X. We will see later that this is a bijection.

Regular sections can be conveniently described in terms of fiber coordinates and transition functions if one has a line bundle atlas for $\pi : L \to X$. Specifically, suppose that $\{\phi_i : \pi^{-1}(U_i) \to \mathbb{C} \times U_i\}$ is a line bundle atlas for L, with fiber coordinates z_i and transition functions t_{ij} satisfying the equations

$$z_i = t_{ij} z_j$$

for every i and j. Suppose that on each U_i a regular function s_i is given. We may view s_i as the first coordinate of a section S_i on the trivial bundle $\mathbb{C} \times U_i \to U_i$, by defining $S_i(p) = (s_i(p), p)$. Then we may transfer these sections S_i to $\pi^{-1}(U_i)$ via the line bundle chart maps, and define regular sections $\tilde{S}_i$ for L over U_i, by $\tilde{S}_i(p) = \phi^{-1}(S_i(p))$. If these agree on the intersections $U_i \cap U_j$, we will be able to define a global regular section S of L over X by setting $S(p) = \tilde{S}_i(p)$ if $p \in U_i$.

Note that the definition of S_i is essentially that $z_i = s_i(p)$; hence the compatibility condition for agreement of the sections is exactly that

$$s_i = t_{ij} s_j \text{ on } U_i \cap U_j$$

for every i and j. We may go through this construction for describing regular sections over any open set $U \subseteq X$, and we have therefore shown the following:

LEMMA 2.15. *Suppose that L is a line bundle on X having transition functions $\{t_{ij}\}$ with respect to a line bundle atlas having as support the open sets $\{U_i\}$. Then a regular section of L over an open set $U \subset X$ is given by a collection of regular functions $s_i \in \mathcal{O}(U \cap U_i)$ satisfying the compatibility condition*

$$s_i = t_{ij} s_j \text{ on } U \cap U_i \cap U_j$$

for every i and j.

EXAMPLE 2.16. Let us take the canonical bundle $\mathbb{K}$ defined in Example 2.12, via a complex atlas $\mathcal{A} = \{\phi_i : U_i \to V_i\}$, and transition functions t_{ij} defined there. Let z_i be the local coordinate for the chart ϕ_i. Suppose that ω is a holomorphic 1-form on an open set $U \subset X$. With respect to ϕ_i we may write ω locally as $s_i \mathrm{d} z_i$; we may consider s_i as a holomorphic function on $U \cap U_i$. If we have $z_i = T_{ij}(z_j)$ on $U_i \cap U_j$, then $\mathrm{d} z_i = T'_{ij} \mathrm{d} z_j$; hence the condition that ω be well defined on all of U is that $s_i T'_{ij} = s_j$, or, equivalently, that

$$s_i = t_{ij} s_j,$$

since with a mild abuse of notation we have $t_{ij} = 1/T'_{ij}$. This shows that ω induces a section of the canonical bundle over U. Conversely, a section of the canonical bundle over U induces a holomorphic 1-form ω, by reversing the construction.

In fact this correspondence holds over any open set; we leave the following to the reader:

LEMMA 2.17. *If $\mathbb{K}$ is the canonical bundle on X, then its sheaf of regular sections $\mathcal{O}\{\mathbb{K}\}$ is isomorphic to the sheaf Ω^1 of regular 1-forms on X.*

Sections of the Tangent Bundle and Tangent Vector Fields. We can also give some meaning to the sections of the tangent bundle, after introducing what a tangent vector field is. We will be brief; the analogue with the construction of forms should make it evident how to make rigorous definitions.

If a manifold is embedded in an ambient linear space, one may define the tangent vectors to the manifold at a point p to be the linear subspace of the ambient space most closely approximating the manifold. This is admittedly vague, but since we want to define tangent vectors at points of an arbitrary (unembedded) curve, we do not want to use this as a definition anyway.

Another approach to tangent vectors at a point p on a manifold is to take the space of all tangent vectors to curves passing through the given point p. Since we are trying to define tangent vectors to curves in the first place, this also seems circular (although it could be successfully pursued).

We prefer to think of a tangent vector as giving a *directional derivative* at a point. If one has local coordinates $(z_1, \ldots z_n)$ at a point, the tangent vector $\underline{v} = (v_1, \ldots, v_n)$ induces the directional derivative operator

$$D_{\underline{v}} = \sum_{i=1}^n v_i \frac{\partial}{\partial z_i}$$

on functions of the variables z_i. Note that the operator $D_{\underline{v}}$ is a *derivation*, in the sense that it satisfies the three conditions

  (i) $D_{\underline{v}}(c) = 0$ for all constant functions c;
  (ii) $D_{\underline{v}}(f + g) = D_{\underline{v}}(f) + D_{\underline{v}}(g)$ for all functions f and g; and
  (iii) $D_{\underline{v}}(fg) = fD_{\underline{v}}(g) + gD_{\underline{v}}(f)$ for all functions f and g.

Moreover it is not hard to check that all derivations have the directional derivative form given above.

In one dimension, which is our situation, a tangent vector will then correspond to a derivation of the the form $\lambda \partial/\partial z$ for a local coordinate z centered at p, and some constant λ. If we desire a *tangent vector field*, i.e., giving a tangent vector at each point in an open set with local coordinate z, then the constant λ would then vary with the point, giving an operator of the form $D = f(z)\partial/\partial z$. We take this to be our basic definition of a holomorphic tangent vector field (with respect to a local coordinate, i.e., with respect to a chart on X); we require the coefficient function f to be holomorphic.

Suppose that z and w are two local coordinates near p, and that $w = T(z)$ is the transition function. Let $f(z)\partial/\partial z$ and $g(w)\partial/\partial w$ be two holomorphic tangent vector fields with respect to these local coordinates. Since for any function h,

we have
$$f(z)\frac{\partial h}{\partial z} = f(z)\frac{\partial h}{\partial w}\frac{dw}{dz} = f(z)T'(z)\frac{\partial h}{\partial w},$$
these two derivations are equal exactly when $g(w) = f(z)T'(z)$; expressing this all in terms of the same variable z gives the compatibility condition

(2.18) $$g(T(z)) = f(z)T'(z).$$

With analogy to the definition of 1-forms, we therefore define a holomorphic tangent vector field on X to be a collection of holomorphic tangent vector fields $f_i \partial/\partial z_i$ with respect to local coordinates z_i on a complex atlas for X, which are pairwise compatible in the above sense. We can also of course define holomorphic tangent vector fields on any open set $U \subset X$ in the same way.

We leave it to the reader to check that if one has a holomorphic tangent vector field with respect to all the charts of one complex atlas for X, then one obtains unique holomorphic tangent vector fields with respect to every complex chart of X (this is the case for 1-forms, see Lemma 1.4 of Chapter IV).

The whole construction can be made in the algebraic category also; the local coordinates are taken to be rational, as are the coefficient functions. One then gets a definition of a *regular tangent vector field* on X.

A regular (or holomorphic) tangent vector field on an open set U can be naturally restricted to open subsets of U. In this way we obtain a sheaf of regular tangent vector fields, which in a natural way is a sheaf of $\mathcal{O}$-modules. In fact it is clearly invertible: a local generator is $\partial/\partial z$ if z is a local coordinate.

We now can state the analogue of Lemma 2.17.

LEMMA 2.19. *If $\mathbb{T}$ is the tangent bundle on X, then its sheaf of regular sections $\mathcal{O}\{\mathbb{T}\}$ is isomorphic to the invertible sheaf of regular tangent vector fields on X.*

PROOF. We will show that the local generators for the two sheaves have the same transition functions. Suppose that a complex atlas on X is given, with local coordinates z_i and transition functions T_{ij}, so that $z_i = T_{ij}(z_j)$. By the definition given above, a regular tangent vector field on X is given by operators $f_i(z_i)\partial/\partial z_i$ satisfying the compatibility condition (2.18), which is

$$f_i = f_j T'_{ij}.$$

This is precisely the condition that the local functions f_i patch together to give a section of the tangent bundle $\mathbb{T}$.

The same computation can be made over any open set U of X, and proves the lemma. □

We will see below that the sheaf of regular tangent vector fields on X is isomorphic to the tangent sheaf Θ, which by definition is the inverse sheaf to the sheaf Ω^1 of regular 1-forms on X.

Rational Sections of a Line Bundle. Just as there are regular and rational functions on an algebraic curve, so can there be regular and rational sections of a line bundle. These would be regular sections at all but finitely many points; at those point a "pole" is allowed.

DEFINITION 2.20. Let $\pi : L \to X$ be a line bundle over an algebraic curve X. A *rational section of L* is a regular section $s : U \to L$ on a Zariski open subset U of X. In other words, there is a finite set $P \subset X$, and a function $s : X - P \to L$, such that
(i) for every $p \in X - P$, $s(p)$ lies in the fiber of L over p, i.e.,

$$\pi \circ s = \mathrm{id}_{X-P};$$

and
(ii) for every line bundle chart $\phi : \pi^{-1}(V) \to \mathbb{C} \times V$ for L, the composition

$$pr_1 \circ \phi \circ s|_V : V \to \mathbb{C}$$

is a rational function.

As one might expect, rational sections are much easier to come by than regular sections. In fact, the rational sections of L are in 1-1 correspondence with the rational function field $\mathcal{M}(X)$. To see this, fix a line bundle chart $\phi : \pi^{-1}(V) \to \mathbb{C} \times V$ for L, and let f be any rational function on X. Then the formula

$$s(p) = \phi^{-1}(f(p), p)$$

defines a rational section of L. Conversely, if s is a rational section, then $pr_1 \circ \phi \circ s$ is a rational function.

This correspondence between rational sections and rational functions depends completely on the choice of the line bundle chart, and is not natural in any way. If we change line bundle charts, the rational function corresponding to a given rational section will of course get multiplied by the transition function (which is again a rational function); the resulting product is the rational function obtained using the new line bundle chart.

As with regular sections, rational sections may also be given locally, satisfying a compatibility condition identical to that given in Lemma 2.15. We leave the proof of the following to the reader.

LEMMA 2.21. *Suppose that L is a line bundle on X having transition functions $\{t_{ij}\}$ with respect to a line bundle atlas having support the open sets $\{U_i\}$. Then a rational section of L over X is given by a collection of rational functions $s_i \in \underline{\mathcal{M}}(U_i) = \mathcal{M}(X)$ satisfying the compatibility condition*

$$s_i = t_{ij} s_j \quad on \ U_i \cap U_j$$

for every i and j.

2. LINE BUNDLES

The Divisor of a Rational Section. The relationship between line bundles and divisors is afforded by defining the divisor of a rational section of a line bundle. We first define the order of a rational section.

DEFINITION 2.22. Let X be an algebraic curve, let $\pi : L \to X$ be a line bundle on X, and fix a rational section s of L. The *order* of s at a point $p \in X$, denoted by $\operatorname{ord}_p(s)$, is the order of the rational function $f = pr_1 \circ \phi \circ s$ where $\phi : \pi^{-1}(U) \to \mathbb{C} \times U$ is any line bundle chart for L whose support U contains p.

We must check that this is well defined, independent of the choice of line bundle chart. However if another line bundle chart $\phi' : \pi^{-1}(V) \to \mathbb{C} \times V$ is used, with a nowhere zero regular transition function t between the two line bundle charts, then the rational function $f' = pr_1 \circ \phi' \circ s$ is exactly tf; hence

$$\operatorname{ord}_p(f') = \operatorname{ord}_p(tf) = \operatorname{ord}_p(t) + \operatorname{ord}_p(f) = \operatorname{ord}_p(f)$$

since $\operatorname{ord}_p(t) = 0$. Therefore $\operatorname{ord}_p(s)$ is well defined.

Moreover since we are in the algebraic category, all but finitely many points p of X will have $\operatorname{ord}_p(s) = 0$. We define the *divisor* $\operatorname{div}(s)$ of the rational section s to be the divisor whose value at p is $\operatorname{ord}_p(s)$:

$$\operatorname{div}(s) = \sum_{p \in X} \operatorname{ord}_p(s) \cdot p.$$

If L has transition functions t_{ij}, and the rational section s is given by a collection of rational functions s_i satisfying $s_i = t_{ij} s_j$ (see Lemma 2.21), then the divisor of s restricted to U_i is the divisor of s_i restricted to U_i.

The fundamental remark which we are striving for here is that these divisors are all linearly equivalent:

PROPOSITION 2.23. *Let L be a line bundle on an algebraic curve X. Suppose that s_1 and s_2 are two rational sections of L. Then $\operatorname{div}(s_1) \sim \operatorname{div}(s_2)$.*

PROOF. Fix a line bundle chart $\phi : \pi^{-1}(U) \to \mathbb{C} \times U$ for L, and let $f_i = pr_1 \circ \phi \circ s_i$ be the corresponding rational functions to the two sections. Let $g = f_1/f_2$.

Now for any other line bundle chart $\phi' : \pi^{-1}(U') \to \mathbb{C} \times U'$ for L, denote the transition function from ϕ to ϕ' by t. Then the rational function $f'_i = pr_1 \circ \phi' \circ s_i$ is exactly tf_i for each i. In particular, we also have that $g = f'_1/f'_2$; so the rational function g obtained by taking the ratio of the local rational functions corresponding to the s_i's is independent of the choice of line bundle chart used to define the rational functions.

Hence if $p \in U \cap U'$, we have

$$\operatorname{ord}_p(s_2) = \operatorname{ord}_p(f'_2) = \operatorname{ord}_p(g) + \operatorname{ord}_p(f'_1) = \operatorname{ord}_p(g) + \operatorname{ord}_p(s_1).$$

Since as noted above the function g does not change when we change line bundle charts, this formula in fact holds for all p in X. Hence $\mathrm{div}(s_2) = \mathrm{div}(g) + \mathrm{div}(s_1)$, proving that $\mathrm{div}(s_2)$ is linearly equivalent to $\mathrm{div}(s_1)$. □

We therefore obtain a function from the set $\mathrm{LB}(X)$ of isomorphism classes of line bundles on X to the Picard group $\mathrm{Pic}(X)$ of divisors modulo linear equivalence, by sending the class of a line bundle L to the divisor of any of its rational sections. We will see in the next section that this is a bijection.

Problems XI.2

A. Check that the additions and scalar multiplications induced on the fiber L_p of a line bundle L over a point p by two different compatible line bundle charts are in fact the same.

B. Check that the functions ϕ_i used to define the tautological line bundle for a map to projective space are indeed line bundle charts.

C. As a projective space, the algebraic curve $\mathbb{P}^1$ has a tautological line bundle L. Show that L may be defined by an atlas with two line bundle charts, supported over the two standard charts of $\mathbb{P}^1$. Find the transition functions for L.

D. Show that the composition of two line bundle homomorphisms is a line bundle homomorphism.

E. Check that the line bundle L constructed in the proof of Proposition 2.9 is unique.

F. Show that the tautological line bundle on $\mathbb{P}^1$ is one of the line bundles L_n defined in Example 2.10. Which n is it?

G. Suppose that L is a line bundle on an algebraic curve X, U is an open subset of X, s is a function from U to L satisfying $\pi \circ s = \mathrm{id}_U$, and suppose that s satisfies condition (ii) of Definition 2.13 for all line bundle charts in some line bundle atlas for L. Show that s is a regular section of L over U, i.e., that s satisfies the condition (ii) for all line bundle charts for L.

H. Let L be a line bundle on an algebraic curve X. Show that a line bundle homomorphism $\alpha : \mathbb{C} \times X \to L$ from the trivial line bundle to L induces a global regular section s_α of L, by setting $s_\alpha(p) = \alpha(1, p)$ for each $p \in X$. Show that every global regular section of L is obtained from a unique such line bundle homomorphism α.

I. Show that if L is the trivial line bundle on X, then the invertible sheaf $\mathcal{O}\{L\}$ is isomorphic to the sheaf $\mathcal{O}$ of regular functions on X.

J. Let $\alpha : L_1 \to L_2$ be a line bundle homomorphism. Show that α induces a sheaf map from $\mathcal{O}\{L_1\}$ to $\mathcal{O}\{L_2\}$ by sending a regular section s of L over U to the composition $\alpha \circ s$.

K. Show that if $\mathbb{K}$ is the canonical bundle on X, then the sheaf $\mathcal{O}\{\mathbb{K}\}$ of regular sections of $\mathbb{K}$ is isomorphic to the sheaf of regular 1-forms Ω^1.

L. Prove Lemma 2.21.

M. Define the degree of a line bundle L on an algebraic curve to be the degree of a divisor of any rational section of L. Show that this is well defined. Compute the degree of the line bundles L_n on $\mathbb{P}^1$ defined in Example 2.10.

N. Show that a line bundle L on an algebraic curve X is isomorphic to the trivial line bundle if and only if there is a nowhere zero regular global section of L.

O. Suppose that D is the divisor of a rational section of a line bundle L on X. Show that any divisor D' linearly equivalent to D is the divisor of some rational section of L. In particular, if L has a nonzero regular section s_0, with $\mathrm{div}(s_0) = D_0$, show that

$$|D_0| = \{\mathrm{div}(s) \mid s \text{ is a regular section of } L\}.$$

P. Start with an algebraic curve X and a very ample divisor D on X. Embed X into projective space with the rational functions in $L(D)$. Consider the tautological bundle L for X with this map to $\mathbb{P}^n$. Compare the divisor of a rational section of L to the original divisor D; are they in the same linear equivalence class?

3. Avatars of the Picard Group

In the previous sections we have introduced invertible sheaves and line bundles on an algebraic curve X. We now want to explain how these ideas are related; we will see that there are in fact five different groups of objects which a priori look different, but which in the end turn out to be essentially the same. We begin with recalling the Picard group of an algebraic curve, and relate this to a cohomology group.

Divisors Modulo Linear Equivalence and Cocycles. Recall that the definition of the Picard Group $\mathrm{Pic}(X)$ for an algebraic curve X is

$$\mathrm{Pic}(X) = \mathrm{Div}(X)/\mathrm{PDiv}(X),$$

the group of divisors on X modulo the subgroup of principal divisors. We usually say that the Picard group is the group of divisors modulo linear equivalence, since two divisors are linearly equivalent exactly when their difference is a principal divisor.

Our first goal is to develop a cohomological interpretation for the Picard group. For this we need to introduce the sheaf of divisors on X, and since we are working in the algebraic category (where the philosophy is that all local objects should in fact be global) let us define the sheaf $\mathcal{D}iv_{X,alg}$ by

$$\mathcal{D}iv_{X,alg}(U) = \{ \text{ divisors with finite support contained in } U\}.$$

This contrasts with the analytic sheaf of divisors, which consists of all divisors supported on U; there the support simply had to be a discrete subset, not a finite subset.

There is a divisor map, taking a rational function on U (which is not identically zero) and sending it to the part of its divisor supported on U; this gives a sheaf

map
$$\text{div} : \mathcal{M}^*_{X,alg} \to \mathcal{D}iv_{X,alg}.$$

(Recall that $\mathcal{M}^*_{X,alg}$ is a constant sheaf on X, namely the constant sheaf of not-identically-zero rational functions.)

LEMMA 3.1. *For an algebraic curve X, with the Zariski topology, the divisor map* $\text{div} : \mathcal{M}^*_{X,alg} \to \mathcal{D}iv_{X,alg}$ *is an onto map of sheaves.*

PROOF. Fix a Zariski open set U containing a point $p \in X$, and a finitely supported divisor D on U. If $D(p) = 0$, consider the Zariski open subset V of U obtained by deleting the support of D; on V, the divisor D restricts to 0, and so is the divisor of the rational function 1. Hence we have lifted D after a restriction to a smaller neighborhood.

Suppose now that $n = D(p) \neq 0$. Since X is algebraic, we have a rational function z on X with $\text{ord}_p(z) = 1$. Now form the Zariski open subset V of U obtained by deleting the support of D, except for p, and the zeroes and poles of the rational function z, except for p; on V, the divisor D restricts to $n \cdot p$, and so is the divisor of the rational function z^n. Hence again we have lifted D after a restriction to a smaller neighborhood. □

What is the kernel of this sheaf map div? Clearly it is the sheaf of those rational functions which have a trivial divisor, and this is exactly the sheaf of rational functions which are both regular (no poles) and nowhere zero (no zeroes). We use the notation $\mathcal{O}^*_{X,alg}$ for this sheaf:

$$\mathcal{O}^*_{X,alg}(U) = \{f \in \mathcal{O}_{X,alg}(U) \mid f \text{ has no zeroes on } U\}.$$

This is a sheaf on X, with the group operation being multiplication of functions. In the rest of this chapter we will often write simply $\mathcal{O}^*$ for the sheaf $\mathcal{O}^*_{X,alg}$.

We therefore have a short exact sequence of sheaves

$$0 \to \mathcal{O}^*_{X,alg} \to \mathcal{M}^*_{X,alg} \xrightarrow{\text{div}} \mathcal{D}iv_{X,alg} \to 0$$

which induces the long exact sequence in cohomology

(3.2)
$$0 \to \mathcal{O}^*_{X,alg}(X) \to \mathcal{M}^*_{X,alg}(X) \xrightarrow{\text{div}} \mathcal{D}iv_{X,alg}(X) \to \check{H}^1(X_{Zar}, \mathcal{O}^*_{X,alg}) \to 0,$$

where the last term here is zero because $\mathcal{M}^*_{X,alg}$ is a constant sheaf on X and we are using the Zariski topology, so that $\check{H}^1(X_{Zar}, \mathcal{M}^*_{X,alg}) = 0$ by Proposition 2.1 of Chapter X.

Now $\mathcal{O}^*_{X,alg}(X)$ are the global regular nowhere zero functions on X; all such are constant, so this group is isomorphic to $\mathbb{C}^*$. Similarly $\mathcal{M}^*_{X,alg}(X)$ is just the multiplicative group of the rational function field $\mathcal{M}(X)$, which is simply

$\mathcal{M}(X) - \{0\}$. Finally the group $\mathcal{D}iv_{X,alg}(X)$ is the group $\mathrm{Div}(X)$ of global divisors on X. Therefore the sequence (3.2) is

$$0 \to \mathbb{C}^* \to \mathcal{M}(X) - \{0\} \xrightarrow{\mathrm{div}} \mathrm{Div}(X) \to \check{H}^1(X_{Zar}, \mathcal{O}^*_{X,alg}) \to 0,$$

which realizes the cohomology group $\check{H}^1(X_{Zar}, \mathcal{O}^*_{X,alg})$ as the cokernel of the global divisor map. Of course the image of the global divisor map is exactly the subgroup $\mathrm{PDiv}(X)$ of principal divisors on X. We conclude that the quotient is isomorphic to this $\check{H}^1$, and have proved the following.

PROPOSITION 3.3. *The Picard group* $\mathrm{Pic}(X)$ *of an algebraic curve* X *is isomorphic to the first cohomology group of the sheaf* $\mathcal{O}^*_{X,alg}$, *via the map* Δ *induced by the connecting homomorphism:*

$$\Delta : \mathrm{Pic}(X) \xrightarrow{\cong} \check{H}^1(X_{Zar}, \mathcal{O}^*_{X,alg}).$$

Let us be explicit and go through the exercise of computing a cocycle for the linear equivalence class of a divisor D. Write D as a finite sum

$$D = \sum_{i=1}^{N} n_i \cdot p_i$$

for integers n_i and points p_i on X. By Corollary 1.16 of Chapter VI, there is a rational function f on X with $\mathrm{ord}_{p_i} = n_i$ for each $i = 1, \ldots, N$. Let $q_1, \ldots, q_M$ be the set of zeroes and poles of f disjoint from the set of p_i's, and let $U_1 = X - \{q_1, \ldots, q_M\}$; let $U_2 = X - \{p_1, \ldots, p_N\}$. Then $\{U_1, U_2\}$ is a Zariski open covering of X, on which the divisor D lifts: on U_1, $D = \mathrm{div}(f_1)$ with $f_1 = f$, and on U_2, $D = 0$, so that if we set $f_2 = 1$, then $D = \mathrm{div}(f_2)$ on U_2.

The prescription for computing the connecting homomorphism says to lift locally, then take differences; in our multiplicative situation we would then take ratios. Hence we define a 1-cochain (t_{ij}) with respect to the cover $\{U_1, U_2\}$ by setting

$$t_{11} = t_{22} = 1, \quad t_{12} = f_2/f_1 = 1/f, \quad \text{and} \quad t_{21} = f_1/f_2 = f.$$

The element t_{11} is defined on U_1, and t_{22} is defined on U_2; the elements t_{12} and t_{21} are defined on $U_1 \cap U_2$. On these sets these elements are sections of the sheaf $\mathcal{O}^*$, and form a 1-cocycle. The class of this cocycle in $\check{H}^1(X_{Zar}, \mathcal{O}^*_{X,alg})$ is the image of the divisor class of D.

Beware: some authors define the cocycle using the convention that $t_{ij} = f_i/f_j$ (see for example [**Kodaira86**]); this differs from the above exactly in getting the inverse of the cocycle we defined.

The construction given above shows that the cohomology class for a divisor D can be represented using an open covering with only two open sets. But Proposition 3.3 implies that every cohomology class is the class corresponding to a divisor. Hence we obtain the following useful corollary:

COROLLARY 3.4. *Let X be an algebraic curve. Then every cohomology class in $\check{H}^1(X_{Zar}, \mathcal{O}_{X,alg}^*)$ can be represented by a cocycle (t_{ij}) with respect to an open covering $\mathcal{U} = \{U_1, U_2\}$ having only two open sets.*

Invertible Sheaves Modulo Isomorphism. Recall that given a divisor D on an algebraic curve X, the sheaf $\mathcal{O}[D]$ is invertible. Moreover, if D_1 and D_2 are linearly equivalent, then the sheaves $\mathcal{O}[D_1]$ and $\mathcal{O}[D_2]$ are isomorphic; the isomorphism is given by multiplication by the function whose divisor is $D_1 - D_2$.

Therefore we have a well defined map

$$\mathcal{O}[-] : \mathrm{Pic}(X) \to \mathrm{Inv}(X);$$

we will show that this is a group isomorphism.

First let us check that the map is a homomorphism; this is equivalent to checking the isomorphism

$$\mathcal{O}[D_1 + D_2] \cong \mathcal{O}[D_1] \otimes_{\mathcal{O}} \mathcal{O}[D_2]$$

for any two divisors D_1 and D_2 on X.

To check this, recall that $\mathcal{O}[D]$ is locally generated by a rational function f with $\mathrm{div}(f) = -D$ (see Lemma 1.5). Choose an open covering $\{U_i\}$ for X, such that on each U_i both of the sheaves $\mathcal{O}[D_1]$ and $\mathcal{O}[D_2]$ are trivialized; let $f_i^{(1)}$ and $f_i^{(2)}$ be local generators for these two sheaves on U_i respectively; hence $\mathrm{div}(f_i^{(j)}) = -D_j$ on U_i.

Therefore $\mathrm{div}(f_i^{(1)} f_i^{(2)}) = -D_1 - D_2$ on U_i for every i, and so $f_i^{(1)} f_i^{(2)}$ is a local generator for $\mathcal{O}[D_1 + D_2]$ on U_i.

Now there is a bilinear map induced by multiplication from $\mathcal{O}[D_1] \times \mathcal{O}[D_2]$ to $\mathcal{O}[D_1 + D_2]$; this bilinear map descends to the tensor product, giving a natural sheaf map

$$\mu : \mathcal{O}[D_1] \otimes_{\mathcal{O}} \mathcal{O}[D_2] \to \mathcal{O}[D_1 + D_2].$$

A local generator for the tensor product over U_i is $f_i^{(1)} \otimes f_i^{(2)}$; this local generator maps (via the multiplication map μ) to the product $f_i^{(1)} f_i^{(2)}$, which as we noted above is a local generator for $\mathcal{O}[D_1 + D_2]$. Hence the sheaf map μ is an isomorphism, since it sends the local generator to the local generator over each U_i.

This proves that $\mathcal{O}[-]$ is a homomorphism of groups.

PROPOSITION 3.5. *Let X be an algebraic curve. Then the map*

$$\mathcal{O}[-] : \mathrm{Pic}(X) \to \mathrm{Inv}(X)$$

is an isomorphism of groups.

PROOF. The identity for the group law in $\mathrm{Inv}(X)$ is the sheaf $\mathcal{O}$. Suppose then that $\mathcal{O}[D] \cong \mathcal{O}$ as sheaves of $\mathcal{O}$-modules. The invertible sheaf $\mathcal{O}$ has a global generator, namely the function 1; hence we conclude that $\mathcal{O}[D]$ will have a global generator. Call that global generator f; therefore f generates the free rank one module $\mathcal{O}[D](U)$ over $\mathcal{O}$ for every open set U. Hence $\mathcal{O}[D](U)$ consists exactly of the multiples of f by elements of $\mathcal{O}(U)$, for every U. We conclude that as a sheaf

$$\mathcal{O}[D] = \mathcal{O}[-\operatorname{div}(f)],$$

since both sides are the multiples of f by regular functions, locally. Note that this is an equality of sheaves, not an isomorphism.

This is enough to conclude that $D = -\operatorname{div}(f)$. The more general statement is that if $\mathcal{O}[D_1] = \mathcal{O}[D_2]$ as subsheaves of $\mathcal{M}$, then $D_1 = D_2$. To see this, suppose that $D_1(p) < D_2(p)$ for some point p. Let z be a rational function having order one at p. Then in a Zariski neighborhood U of p, $z^{-D_2(p)}$ will be in $\mathcal{O}[D_2](U)$ but not in $\mathcal{O}[D_1](U)$. This contradiction proves that $D_1(p) \geq D_2(p)$ for all p, so that $D_1 \geq D_2$; reversing the argument shows the other inequality and we conclude that $D_1 = D_2$.

Applying this in our case yields that $D = -\operatorname{div}(f)$, and is therefore principal, so that the class of D in $\operatorname{Pic}(X)$ is zero. Therefore we have proved that the kernel of the group homomorphism $\mathcal{O}[-]$ is trivial, so that $\mathcal{O}[-]$ is 1-1.

To show that the map is onto, we must recover a divisor class from an invertible sheaf. If the invertible sheaf was $\mathcal{O}[D]$, the way to recover D is to choose a covering $\{U_i\}$ on which the sheaf trivializes, and choose local generators f_i; then we have $D(p) = -\operatorname{ord}_p(f_i)$ if $p \in U_i$.

Let $\mathcal{F}$ be an arbitrary invertible sheaf on X, and let $\{U_i\}$ be a covering of X such that for every i, $\mathcal{F}|_{U_i}$ is trivial. Let f_i be a local generator for $\mathcal{F}|_{U_i}$, namely a generator of the module $\mathcal{F}(U_i)$ over $\mathcal{O}(U_i)$.

These elements f_i are not functions; we have no control over what they are. However for every pair i,j we have a regular nowhere zero function t_{ij} on $U_i \cap U_j$ such that $f_i = t_{ij} f_j$ on $U_i \cap U_j$, since over this set both f_i and f_j are local generators, and hence they "differ" by a unit in the ring $\mathcal{O}(U_i \cap U_j)$. (We see here the cocycle of the divisor before we are seeing the divisor!) Note that the collection (t_{ij}) satisfies the cocycle conditions.

Fix an index, say $i = 0$, and consider the functions t_{i0} for every i; this is regular and nowhere zero on $U_0 \cap U_i$. However it may have zeroes or poles on $U_i - (U_0 \cap U_i)$, and we will form a divisor D on X by setting

$$D(p) = -\operatorname{ord}_p(t_{i0}) \text{ if } p \in U_i.$$

Note that this is well defined: if $p \in U_i \cap U_j$, then $t_{i0} = t_{ij} t_{j0}$, so that $\operatorname{ord}_p(t_{i0}) = \operatorname{ord}_p(t_{j0})$ since $\operatorname{ord}_p(t_{ij}) = 0$ for any $p \in U_i \cap U_j$. Note that $D(p) = 0$ for all $p \in U_0$, since $t_{00} = 1$.

We remark at this point that $\mathcal{O}[D]$ is also trivializable over each U_i; in fact t_{i0} is a local generator.

Finally we must check that with this divisor D, we have $\mathcal{O}[D] \cong \mathcal{F}$. For this we must define isomorphisms $\mathcal{O}[D](U) \to \mathcal{F}(U)$ for every open set U, which are compatible with the restriction maps. However it suffices to define such isomorphisms for all U which are subsets of the U_i's; by the sheaf axiom we will obtain isomorphisms for all U.

Indeed, it suffices to define the isomorphisms only for the U_i, since both $\mathcal{O}[D]$ and $\mathcal{F}$ are trivializable over each U_i. The compatibility with the restriction maps will be ensured if we check that for an open subset $U \subset U_i \cap U_j$, the map for U defined by restricting the map on U_i is the same as that defined by restricting the map on U_j.

Now on U_i, simply send the local generator t_{i0} of $\mathcal{O}[D]$ to the local generator f_i of $\mathcal{F}$, and extend $\mathcal{O}(U_i)$-linearly. This is compatible with the restriction maps precisely because the t_{ij}'s satisfy the cocycle condition. If $U \subset U_i \cap U_j$, then using the U_i map we see that the local generator t_{i0} is sent to f_i; using the U_j map we see that the local generator t_{j0} is sent to f_j, These differ by t_{ij} both before and after applying the map, and so induce the same map on U.

This proves that $\mathcal{O}[D] \cong \mathcal{F}$, and hence the group homomorphism $\mathcal{O}[-]$ is onto, finishing the proof of the Proposition. □

We saw in the proof above the cocycle of the divisor rearing its head for a moment. Following that construction to its conclusion gives an isomorphism between $\mathrm{Inv}(X)$ and $\check{H}^1(X_{Zar}, \mathcal{O}^*)$; let us go through this for completeness.

Let $\mathcal{F}$ be an invertible sheaf on X, let $\mathcal{U} = \{U_i\}$ be an open cover of X such that $\mathcal{F}$ trivializes over each U_i, and let f_i be a local generator of $\mathcal{F}$ over U_i. For each pair i,j, define $t_{ij} \in \mathcal{O}^*(U_i \cap U_j)$ by writing $f_i = t_{ij} f_j$ in $\mathcal{F}(U_i \cap U_j)$. Then the collection (t_{ij}) is a 1-cocycle for the sheaf $\mathcal{O}^*$, with respect to the open covering $\mathcal{U}$, and hence induces a class in $\check{H}^1(\mathcal{U}, \mathcal{O}^*)$, and further in $\check{H}^1(X_{Zar}, \mathcal{O}^*)$. Call this class $H_I(\mathcal{F})$.

LEMMA 3.6. *The class $H_I(\mathcal{F}) \in \check{H}^1(X_{Zar}, \mathcal{O}^*_{X,alg})$ is well defined, independent of the choice of local generators and of the choice of open covering.*

PROOF. First, suppose that different local generators $\{g_i\}$ are taken for the sheaf $\mathcal{F}$ over each U_i. Then for every i there is a regular nowhere zero function s_i such that $g_i = s_i f_i$. Hence on $U_i \cap U_j$ we have $g_i = s_i f_i = s_i t_{ij} f_j = s_i t_{ij} s_j^{-1} g_j$ so that the cocycle computed via the generators $\{g_i\}$ differs from the cocycle computed via the $\{f_i\}$ by the coboundary (s_i/s_j). Hence the cohomology class is the same in $\check{H}^1(\mathcal{U}, \mathcal{O}^*)$, and so of course in $\check{H}^1(X_{Zar}, \mathcal{O}^*)$.

To check that the cohomology class is independent of the choice of open covering it suffices to check that it is invariant under refinement, since any two open coverings have a common refinement. But if $\mathcal{V} = \{V_k\}$ is a refinement of $\mathcal{U}$, with refining map r (so that $V_k \subset U_{r(k)}$ for every index k), then we may choose as a local generator over each V_k the local generator $f_{r(k)}$ for $\mathcal{F}$ over $U_{r(k)}$. Then the cocycle $H(\mathcal{F})$ computed via the covering $\mathcal{U}$ maps (via the cohomology refining map $H(r)$) to the cocycle computed via the covering $\mathcal{V}$. Hence they are equal in the limit group $\check{H}^1(X_{Zar}, \mathcal{O}^*)$. □

We can now close the loop on this circle of ideas.

PROPOSITION 3.7. *Let X be an algebraic curve. Then the map $H_I : \text{Inv}(X) \to \check{H}^1(X_{Zar}, \mathcal{O}^*)$ is an isomorphism of groups. Moreover the composition*

$$\text{Pic}(X) \xrightarrow{\mathcal{O}[-]} \text{Inv}(X) \xrightarrow{H_I} \check{H}^1(X_{Zar}, \mathcal{O}^*)$$

is the isomorphism Δ induced by the connecting homomorphism as in Proposition 3.3.

PROOF. It is enough to check the last statement, that the composition is Δ, since we already know that both Δ and $\mathcal{O}[-]$ are isomorphisms of groups. For this we must show that for a divisor D, $H_I(\mathcal{O}[D]) = \Delta(D)$.

Fix D; we compute $\Delta(D)$ as indicated in the discussion after Proposition 3.3. For this we found a Zariski open covering $\{U_1, U_2\}$ for X such that the support of D was contained entirely in U_1, and $D = \text{div}(f)$ on U_1; we had $D = 0 = \text{div}(1)$ on U_2. Then a cocycle representing $\Delta(D)$ was (t_{ij}), where $t_{12} = 1/f$. (This determines all the t_{ij} by the cocycle conditions.)

The invertible sheaf $\mathcal{O}[D]$ is trivialized over each U_i; we have as local generators $f_1 = 1/f$ on U_1 and $f_2 = 1$ on U_2. On the intersection we have

$$t_{12} = t_{12} \cdot 1 = t_{12} f_2 = f_1 = 1/f,$$

determining the cocycle representing $H_I(\mathcal{O}[D])$. This is the same as that for $\Delta(D)$, proving the proposition. $\square$

Line Bundles Modulo Isomorphism. Let L be a line bundle on X. In the previous section we have described several constructions which assign invertible sheaves and divisor classes to L. We want to weave this all together in the same spirit as was done for invertible sheaves.

Let $\text{LB}(X)$ denote the set of isomorphism classes of line bundles on X. In the last section we saw that the sheaf $\mathcal{O}\{L\}$ of regular sections of L is invertible; this gives a function

$$\mathcal{O}\{-\} : \text{LB}(X) \to \text{Inv}(X).$$

In addition, if s is any rational section of L, then its divisor $\text{div}(s)$ is defined, and by Proposition 2.23, the linear equivalence class of $\text{div}(s)$ depends only on L, not on s; hence we obtain a "divisor class" map, which we will denote by $[\text{div}]$, to the Picard group:

$$[\text{div}] : \text{LB}(X) \to \text{Pic}(X).$$

Finally the transition functions for a line bundle atlas for L gives a 1-cocycle for the sheaf $\mathcal{O}^*_{X,alg}$, and we want to show that this gives a function

$$H_L : \text{LB}(X) \to \check{H}^1(X_{Zar}, \mathcal{O}^*_{X,alg}).$$

To be explicit, if $\pi : L \to X$ is the line bundle map, and $\{\phi_i : \pi^{-1}(U_i) \to \mathbb{C} \times U_i\}$ is a line bundle atlas for L, then the compatibility of the line bundle charts implies that for every i,j, the map

$$\phi_i \circ \phi_j^{-1} : \mathbb{C} \times (U_i \cap U_j) \to \mathbb{C} \times (U_i \cap U_j)$$

has the form

$$(v,p) \mapsto (t_{ij}(p) \cdot v, p)$$

for some regular nowhere zero function $t_{ij} \in \mathcal{O}(U_i \cap U_j)$. If we use z_j for the coordinate in the "fiber" $\mathbb{C}$ for the jth line bundle chart, this equation is expressed by

$$z_i = t_{ij} z_j.$$

In any case, the collection (t_{ij}) satisfies the cocycle conditions and gives a 1-cocycle for the sheaf $\mathcal{O}^*$. If we denote by $\mathcal{U}$ the open covering of X by the supports $\{U_i\}$ of the line bundle atlas, the cohomology class of this 1-cocycle lies in $\check{H}^1(\mathcal{U}, \mathcal{O}^*)$, and hence induces a class in $\check{H}^1(X_{Zar}, \mathcal{O}^*)$. We denote this class by $H_L(L)$.

LEMMA 3.8. *For a line bundle L on an algebraic curve X, the cohomology class $H_L(L)$ in $\check{H}^1(X_{Zar}, \mathcal{O}^*_{X,alg})$ is well defined, independent of the line bundle atlas used to define it.*

PROOF. The proof is quite similar to that of Lemma 3.6; we will only sketch it.

Firstly, one shows that the class depends only on the open covering $\{U_i\}$, not on the particular line bundle charts. This is because if one uses, instead of the line bundle charts ϕ_i, alternate line bundle charts $\phi'_i : \pi^{-1}(U_i) \to \mathbb{C} \times U_i$ (with the same support sets U_i), then the compatibility of ϕ_i with ϕ'_i for each i gives a regular nowhere zero function s_i on U_i for each i, and the 1-cocycle defined using the ϕ'_i charts will differ from that defined using the ϕ_i charts by the coboundary (s_i/s_j). Hence we obtain the same class in $\check{H}^1(\mathcal{U}, \mathcal{O}^*)$, so certainly in $\check{H}^1(X_{Zar}, \mathcal{O}^*)$.

Secondly one remarks that since any two open coverings have a common refinement, it suffices to show that we get the same cohomology class using two line bundle atlases where one open covering is finer than the other.

Finally one checks that if $\mathcal{V} = \{V_k\}$ is a refinement of $\mathcal{U}$, with refining map r (so that $V_k \subset U_{r(k)}$ for every index k), then we may use the same line bundle charts (suitably restricted) for each for the $\{V_k\}$ covering, i.e., define the line bundle chart with support V_k to be the restriction of the line bundle chart with support $U_{r(k)}$. Then the cocycle $H_L(L)$ computed via the covering $\mathcal{U}$ maps (via the cohomology refining map $H(r)$) to the cocycle computed via the covering $\mathcal{V}$. Hence they are equal in the limit group $\check{H}^1(X_{Zar}, \mathcal{O}^*)$. □

Our next task is to show that the cohomology class determines the line bundle (up to isomorphism). In fact:

3. AVATARS OF THE PICARD GROUP

LEMMA 3.9. *The mapping*

$$H_L : \mathrm{LB}(X) \to \check{H}^1(X_{Zar}, \mathcal{O}^*)$$

is a bijection.

PROOF. We showed in the last section that given a collection of transition functions (t_{ij}) satisfying the cocycle conditions, there was a line bundle with those transition functions; moreover the line bundle was unique up to isomorphism. The cocycle conditions are exactly the conditions necessary for the collection (t_{ij}) to be a 1-cocycle for the sheaf $\mathcal{O}^*$; hence the construction of the line bundle from the transition functions exactly gives that the map H_L is onto.

We must show that H_L is 1-1. Suppose that two line bundles $\pi_1 : L_1 \to X$ and $\pi_2 : L_2 \to X$ map to the same cohomology class under H_L. This means that there are coverings $\{U_i^{(1)}\}$ and $\{U_j^{(2)}\}$ supporting atlases $\{\phi_i^{(1)}\}$ and $\{\phi_j^{(2)}\}$ for L_1 and L_2 respectively, such that the transition functions for the $\phi_i^{(1)}$'s and the $\phi_j^{(2)}$'s give cocycles which are equal in cohomology. By passing to a common refinement we may assume that we have a single open covering $\mathcal{U} = \{U_i\}$, so that the two cocycles both live in $\check{H}^1(\mathcal{U}, \mathcal{O}^*)$. Since the cohomology classes of the cocycles are equal in $\check{H}^1(X_{Zar}, \mathcal{O}^*)$, we may pass to a possibly finer covering (which we will also denote by $\mathcal{U} = \{U_i\}$) and assume that the cocycles differ by a coboundary.

This means that there are regular nowhere zero functions s_i on U_i for each i such that if $(t_{ij}^{(1)})$ is the collection of transition functions for the $\phi_i^{(1)}$'s, and $(t_{ij}^{(2)})$ is the collection of transition functions for the $\phi_i^{(2)}$'s, then $t_{ij}^{(1)} s_i/s_j = t_{ij}^{(2)}$ for every i and j.

For each i, define the line bundle automorphism $S_i : \mathbb{C} \times U_i \to \mathbb{C} \times U_i$ by $S_i(z_i, p) = (s_i z_i, p)$. Consider the line bundle chart $\phi_i^{(1a)} : \pi^{-1}(U_i) \to \mathbb{C} \times U_i$ defined by setting

$$\phi_i^{(1a)} = S_i \circ \phi_i^{(1)}.$$

Note that these alternate line bundle charts are compatible with the line bundle charts $\{\phi_i^{(1)}\}$, and so give an alternate line bundle atlas for L_1. Moreover by construction the transition functions for L_1 with respect to this alternate set of line bundle charts are exactly $t_{ij}^{(1a)} = t_{ij}^{(1)} s_i/s_j = t_{ij}^{(2)}$.

What we have shown then is that there are line bundle atlases for both L_1 and L_2 with exactly the same open covering for the supports of the line bundle charts, and exactly the same transition functions. We conclude that $L_1 \cong L_2$, by the uniqueness statement of Proposition 2.9. □

The lemma above has a surprising corollary, when we combine it with Corollary 3.4:

COROLLARY 3.10. *Let L be a line bundle on an algebraic curve X. Then there is a line bundle atlas for L consisting of only two line bundle charts.*

Using the bijection H_L, we can put an abelian group structure on $\mathrm{LB}(X)$ by transferring the group operation from $\check{H}^1(X_{Zar}, \mathcal{O}^*)$. Tautologically we then have that H_L is a group isomorphism.

Since all of these groups are abelian groups, the map sending an element to its inverse is a group isomorphism. In particular, we may define an alternate group isomorphism

$$H'_L : \mathrm{LB}(X) \to \check{H}^1(X_{Zar}, \mathcal{O}^*)$$

by setting $H'_L(L) = 1/H_L(L)$. We only do this because of the following analogue of Proposition 3.7.

PROPOSITION 3.11. *For an algebraic curve X, the maps $H'_L : \mathrm{LB}(X) \to \check{H}^1(X_{Zar}, \mathcal{O}^*)$ and $[\mathrm{div}] : \mathrm{LB}(X) \to \mathrm{Pic}(X)$ are isomorphisms of groups. Moreover the composition*

$$\mathrm{LB}(X) \xrightarrow{[\mathrm{div}]} \mathrm{Pic}(X) \xrightarrow{\Delta} \check{H}^1(X_{Zar}, \mathcal{O}^*)$$

is the isomorphism H'_L obtained by composing the isomorphism H_L with inversion. In other words, if L is a line bundle on X, then $\Delta([\mathrm{div}](L)) = 1/H_L(L)$.

PROOF. As noted above, the fact that H'_L is an isomorphism is by definition. If we show the last statement, then the function $[\mathrm{div}]$ will be equal to the composition $\Delta^{-1} \circ H_L$, and will therefore also be an isomorphism of groups.

It remains then to show that $1/H_L = \Delta \circ [\mathrm{div}]$. Using Corollary 3.10, we may find a line bundle atlas for L consisting of two line bundle charts ϕ_i having supports U_i for $i = 1, 2$. Let $t_{12} = f$ be the transition function; this determines all the t_{ij} by the cocycle conditions. Moreover the cohomology class of (t_{ij}) is exactly the class of $H_L(L)$.

Define a rational section for L by setting $s_1 = f$ and $s_2 = 1$; this satisfies $s_i = t_{ij} s_j$ for all possible pairs i and j (the only one that really counts is $i = 1$ and $j = 2$), and hence induces a global rational section s by Lemma 2.21. The divisor of this section is the divisor D with

$$D|_{U_1} = \mathrm{div}(f)|_{U_1} \quad \text{and} \quad D|_{U_2} = \mathrm{div}(1)|_{U_2} = 0.$$

So this divisor is the divisor of meromorphic functions $f_1 = f$ over U_1 and $f_2 = 1$ over U_2.

The cocycle $(t'_{ij}) = \Delta(D)$ for this divisor is determined by the function

$$t'_{12} = f_2/f_1 = 1/f.$$

Since we have that $t'_{12} = 1/t_{12}$, this cocycle for $\Delta(D)$ is the inverse of the cocycle (t_{ij}) for L. Hence $\Delta \circ [\mathrm{div}] = 1/H_L = H'_L$ as stated. $\square$

Finally we want to get the function $\mathcal{O}\{-\}$ into the act.

PROPOSITION 3.12. *For an algebraic curve X, the map $\mathcal{O}\{-\} : \mathrm{LB}(X) \to \mathrm{Inv}(X)$ is an isomorphism of groups. Moreover the composition*

$$\mathrm{LB}(X) \xrightarrow{\mathcal{O}\{-\}} \mathrm{Inv}(X) \xrightarrow{H_I} \check{H}^1(X_{Zar}, \mathcal{O}^*)$$

is the isomorphism H'_L obtained by composing the isomorphism H_L with inversion. In other words, if L is a line bundle on X, then $H_I(\mathcal{O}\{L\}) = 1/H_L(L)$.

PROOF. Let L be a line bundle on L; we may again assume that L has an atlas consisting of two charts ϕ_1 and ϕ_2 over open sets U_1 and U_2 covering X. If z_i is the fiber coordinate with respect to ϕ_i, and we have

$$z_i = t_{ij} z_j$$

for every i and j, then the collection (t_{ij}) is a 1-cocycle representing the cohomology class $H_L(L)$.

The invertible sheaf $\mathcal{O}\{L\}$ is also trivialized over the two open sets U_i, and local generators for $\mathcal{O}\{L\}(U_i)$ are given by the sections s_i defined by setting $z_i = 1$ identically. We obtain the cocycle representing $H_I(\mathcal{O}\{L\})$ by writing the local generator over U_2 as a multiple of the local generator over U_1. When $z_2 = 1$ (which defines s_2), we have that $z_1 = t_{12}$; this is t_{12} times the section s_1 (which is defined by $z_1 = 1$). Hence

$$s_2 = t_{12} s_1$$

so that the cocycle (t'_{ij}) representing $H_I(\mathcal{O}\{L\})$ has $t'_{21} = t_{12}$. This is the inverse of the cocycle for $H_L(L)$; hence $H_I(\mathcal{O}\{L\}) = 1/H_L(L)$ as claimed.

This then implies that $\mathcal{O}\{-\}$ is an isomorphism of groups, since both H_I and H_L are. □

COROLLARY 3.13. *Let X be an algebraic curve. Then the composition of group isomorphisms*

$$\mathrm{LB}(X) \xrightarrow{[\mathrm{div}]} \mathrm{Pic}(X) \xrightarrow{\mathcal{O}[-]} \mathrm{Inv}(X)$$

is the isomorphism $\mathcal{O}\{-\} : \mathrm{LB}(X) \to \mathrm{Inv}(X)$.

PROOF. We have

$$\begin{aligned}\mathcal{O}\{-\} &= H_I^{-1} \circ H'_L && \text{by Proposition 3.12} \\ &= \mathcal{O}[-] \circ \Delta^{-1} \circ H'_L && \text{by Proposition 3.7} \\ &= \mathcal{O}[-] \circ [\mathrm{div}] && \text{by Proposition 3.11,}\end{aligned}$$

which is the statement to be proved. □

Thus all four of these groups are connected by isomorphisms in a "commuting tetrahedron":

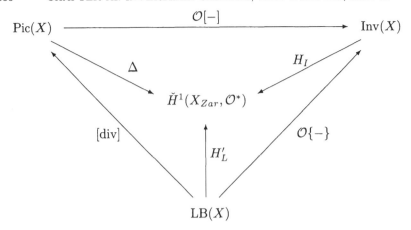

We can use this to identify the invertible sheaf of regular tangent vector fields on X.

COROLLARY 3.14. *Let X be an algebraic curve. Then the sheaf of regular tangent vector fields on X is isomorphic to the tangent sheaf Θ.*

PROOF. Let $\mathbb{T}$ and $\mathbb{K}$ denote the tangent and canonical bundles on X, respectively. Denote by $\mathcal{V}$ the sheaf of regular vector fields. We have seen that $\mathcal{V} = \mathcal{O}\{\mathbb{T}\}$ (Lemma 2.19), that $\Omega^1 = \mathcal{O}\{\mathbb{K}\}$ (Lemma 2.17), and by definition Θ is the inverse sheaf to Ω^1. Moreover by the constructions of the two bundles (via transition functions), they are inverse in the group law of $\mathrm{LB}(X)$; we used inverse cocycles to define them. Hence as classes in $\mathrm{Inv}(X)$, we have

$$\begin{aligned}
\mathcal{V} &= \mathcal{O}\{\mathbb{T}\} \\
&= \mathcal{O}\{\mathbb{K}^{-1}\} \\
&= (\mathcal{O}\{\mathbb{K}\})^{-1} \\
&= (\Omega^1)^{-1} \\
&= \Theta.
\end{aligned}$$

□

This is why we called the inverse sheaf to Ω^1 the tangent sheaf to start with!

The Jacobian. Finally the last avatar of the Picard group for an algebraic curve X is the Jacobian $\mathrm{Jac}(X)$, or more precisely, $\mathrm{Jac}(X) \times \mathbb{Z}$. Here we just remind the reader of Abel's Theorem: the Abel-Jacobi map $A : \mathrm{Pic}^0(X) \to \mathrm{Jac}(X)$ is an isomorphism of groups, where $\mathrm{Pic}^0(X)$ is the group of divisors of degree 0 modulo linear equivalence.

The short exact sequence

$$0 \to \mathrm{Pic}^0(X) \to \mathrm{Pic}(X) \xrightarrow{\deg} \mathbb{Z} \to 0$$

splits, although not canonically. A splitting is afforded by choosing a specific divisor of degree one, and this is commonly taken to be a single point p_0 on X. Then the splitting map takes a divisor D of degree d to the divisor $D - d \cdot p_0$ of degree 0. Composing this with the Abel-Jacobi map A to $\operatorname{Jac}(X)$ and adding the degree information back in via a second coordinate gives a group isomorphism

$$\tilde{A} : \operatorname{Pic}(X) \to \operatorname{Jac}(X) \times \mathbb{Z},$$

sending the class of a divisor D to the pair $(A(D - d \cdot p_0), \deg(D))$. Therefore we have five different groups, all isomorphic to one another:
- the Picard group of divisors modulo linear equivalence: $\operatorname{Pic}(X)$;
- the isomorphism classes of invertible sheaves: $\operatorname{Inv}(X)$;
- the isomorphism classes of line bundles: $\operatorname{LB}(X)$;
- the cohomology group $\check{H}^1(X_{Zar}, \mathcal{O}^*_{X,alg})$;
- the Jacobian (extended by $\mathbb{Z}$): $\operatorname{Jac}(X) \times \mathbb{Z}$.

Only the Jacobian suffers from having a noncanonical isomorphism; for the other four the maps in the commuting tetrahedron are all natural.

Problems XI.3
A. Check that the presheaf $\mathcal{D}iv_{X,alg}$ defined above is in fact a sheaf in the Zariski topology for X.
B. Show that the presheaf $\mathcal{O}^*$ defined above is a sheaf on X.
C. Let $D = n \cdot \infty$ be a divisor on $\mathbb{P}^1$ supported at the single point ∞. Explicitly compute a cocycle representing the class of $\Delta(D)$ in $\check{H}^1(\mathbb{P}^1_{Zar}, \mathcal{O}^*_{X,alg})$.
D. Suppose an explicit cocycle $(f_{ij}) \in \check{H}^1(\mathcal{U}, \mathcal{O}^*_{X,alg})$ is given, where $\mathcal{U}$ is a given open covering of X. How would you compute a divisor D such that $\Delta(D)$ is represented by that cocycle?
E. Show that for the standard open covering $\mathcal{U}$ of $\mathbb{P}^1$, the group $\check{H}^1(\mathcal{U}, \mathcal{O}^*_{X,alg})$ is infinite cyclic.
F. Show that if s is a rational section of the canonical bundle on an algebraic curve X, then $\operatorname{div}(s)$ is a canonical divisor on X.

4. $\check{H}^1$ as a Classifying Space

In this section we want to give an explanation for the occurrence of the cohomology group $\check{H}^1(X_{Zar}, \mathcal{O}^*)$ among the avatars of the Picard group. Taking a slightly broader viewpoint, one realizes that the construction of $\check{H}^1$ for various sheaves will give groups which classify a diverse set of objects.

Why $\check{H}^1(\mathcal{O}^*)$ Classifies Invertible Sheaves and Line Bundles. Both line bundles and invertible sheaves are examples of *locally trivial* objects. A line bundle is locally isomorphic (as a line bundle) to the trivial bundle $\mathbb{C} \times X$; an invertible sheaf is locally isomorphic (as a sheaf of $\mathcal{O}$-modules) to the trivial invertible sheaf of regular functions $\mathcal{O}_X$. What makes a line bundle or an invertible sheaf nontrivial is not its local structure, but its global structure.

It is therefore not a surprise that Čech cohomology, which is designed precisely to organize information which is locally simple but globally complicated, can be used to attack problems of classification for line bundles and for invertible sheaves.

Taking up the case of line bundles first, we see that line bundle charts exactly give the local triviality of a line bundle. The compatibility condition for two line bundle charts exactly says that on the intersection, the composition of the inverse of one line bundle chart with the other line bundle chart is an *automorphism* of the trivial line bundle chart.

What is an automorphism of a trivial line bundle? Take the trivial line bundle $T_U = \mathbb{C} \times U$. A line bundle homomorphism $\alpha : T_U \to T_U$ must have the form

$$\alpha(z, p) = (f(p) \cdot z, p)$$

for some regular function f on U. For α to be an automorphism, one must have that the regular function f be nowhere zero on U, so that the inverse line bundle homomorphism

$$\alpha^{-1}(w, p) = (w/f(p), p)$$

exists. Moreover we see immediately that there is a 1-1 correspondence between the group of line bundle automorphisms of T_U and the group $\mathcal{O}^*(U)$ of nowhere zero regular functions on U; this correspondence is a group isomorphism. Therefore:

LEMMA 4.1. *For an algebraic curve X, the sheaf of automorphisms of the trivial line bundle is isomorphic to the sheaf $\mathcal{O}^*$.*

The case of invertible sheaves is similar: a local generator exactly gives an isomorphism with $\mathcal{O}$, locally. Two different local generators f and g give different isomorphisms with $\mathcal{O}$; the composition of one with the inverse of the other gives an automorphism of $\mathcal{O}$ on the intersection. Specifically, this automorphism is exactly given by multiplication by the nowhere zero regular function h such that $f = h \cdot g$.

In any case, an automorphism of a ring (as a module over itself) is always given by multiplication by a unit in the ring. Since the units in the rings $\mathcal{O}(U)$ form precisely the group $\mathcal{O}^*(U)$, we have that:

LEMMA 4.2. *For an algebraic curve X, the sheaf of automorphisms of the trivial invertible sheaf $\mathcal{O}$ is isomorphic to the sheaf $\mathcal{O}^*$.*

Now it is because of Lemmas 4.1 and 4.2 that $\check{H}^1(X_{Zar}, \mathcal{O}^*)$ classifies both line bundles and invertible sheaves: the sheaf $\mathcal{O}^*$ is the sheaf of automorphisms of the trivial object (to which these objects are locally isomorphic) in both cases. Specifically, cohomology classes get into the act via the transition functions in the case of line bundles (the map H_L constructed in the previous section) and via the ratios of the local generators in the case of invertible sheaves (the map H_I).

4. Ȟ¹ AS A CLASSIFYING SPACE

A collection of trivializations for an invertible sheaf, or a line bundle atlas for a line bundle, gives naturally a 1-cocycle for the sheaf $\mathcal{O}^*$ for a particular covering $\mathcal{U}$ of X. Changing the trivializations of the sheaf over the open sets of the covering, or changing the line bundle atlas (but retaining the same open sets as supports) changes the 1-cocycle by a coboundary. Passing to a finer covering in the cohomology is equivalent to using a finer set of trivializations for the invertible sheaf, or a finer line bundle atlas. Therefore the limit group $\check{H}^1(X_{Zar}, \mathcal{O}^*)$ classifies isomorphism classes of both invertible sheaves and of line bundles.

The reader should check the proofs of Lemmas 3.6 and 3.8 to be reminded how these arguments went in each case.

Locally Trivial Structures. One can apply the principles used in the construction of the bijections H_I and H_L to essentially arbitrary locally trivial objects. One only requires a proper definition of the "trivial" object T; then a locally trivial object is defined to be an object S such that there is an open covering $\mathcal{U} = \{U_i\}$ of X with $S|_{U_i} \cong T|_{U_i}$ for every i.

Of course one sees immediately that one needs to know not only what the object is, but what its sheaf of automorphisms is; otherwise one cannot decide whether $S|_U \cong T|_U$. This is part of the "proper definition".

In the case of invertible sheaves, the trivial object is the sheaf $\mathcal{O}$; in the case of line bundles, the trivial object is the product bundle $\mathbb{C} \times X$. In both cases the sheaf of automorphisms of the trivial object is isomorphic to $\mathcal{O}^*$.

We will give several examples below of locally trivial structures. As an initial example, consider simply two disjoint copies of a Riemann surface X, both mapping via the identity to X:

$$X \coprod X \to X.$$

We take as the local automorphisms only the identity mapping on the disjoint union, together with the map which switches the two copies of X.

With this definition, a locally trivial object is a curve Y together with a map $\pi : Y \to X$ such that there is an open covering $\mathcal{U} = \{U_i\}$ of X and isomorphisms $\phi_i : \pi^{-1}(U_i) \to U_i \coprod U_i$ for every i (making the obvious diagram commute), such that for every pair i, j the composition $\phi_j \circ \phi_i^{-1}$ is either the identity map or the switching map on $(U_i \cap U_j) \coprod (U_i \cap U_j)$.

It is common to call these objects *locally trivial double coverings of X*.

We see that locally there are only two automorphisms for the trivial object: the identity and the switching map. Therefore the sheaf of automorphisms of the trivial object is the constant sheaf $\mathbb{Z}/2\mathbb{Z}$.

We give this example explicitly for a Riemann surface, using the classical topology. We will see below that with the Zariski topology the only locally trivial object is the trivial object.

A General Principle Regarding $\check{H}^1$. Suppose we are in the situation described above, with a proper definition of a trivial object T together with its sheaf of automorphisms $\underline{\mathrm{Aut}}(T)$. We have the following.

PROPOSITION 4.3. *The set of isomorphism classes of locally trivial objects is in 1-1 correspondence with $\check{H}^1(X, \underline{\mathrm{Aut}}(T))$.*

PROOF. The proof is completely formal, simply mimicking what was done for invertible sheaves and line bundles in the previous sections.

Let S be an object locally isomorphic to T. One obtains a cohomology class $H(S)$ as follows: choose an open covering $\mathcal{U} = \{U_i\}$, and trivializations $\phi_i : S|_{U_i} \to T|_{U_i}$ for each i. The compositions $a_{ij} = \phi_i \circ \phi_j^{-1}$ then are automorphisms of $T_{U_i \cap U_j}$, and form a 1-cocycle for the sheaf $\underline{\mathrm{Aut}}(T)$. The class of this 1-cocycle in $\check{H}^1(X, \underline{\mathrm{Aut}}(T))$ is denoted by $H(S)$.

The proof that $H(S)$ is well defined, independent of the choice of open covering and of trivializations, goes exactly like the proofs of Lemmas 3.6 and 3.8. The proof that every 1-cocycle is of the form $H(S)$ for some locally trivial object S involves a glueing procedure identical to the construction of line bundles from transition functions; see the proof of Proposition 2.9. If two different objects S_1 and S_2 have $H(S_1) = H(S_2)$, then one concludes that S_1 and S_2 are isomorphic, using the same argument as was given in Lemma 3.9. □

To continue with the example of locally trivial double coverings of X given above, where the trivial object is $X \coprod X$, we see that the locally trivial double coverings of X are classified by the group $\check{H}^1(X, \mathbb{Z}/2\mathbb{Z})$.

Cyclic Unbranched Coverings. We can easily generalize the double covering situation to coverings of higher degree, maintaining the cyclicity. Specifically, define the trivial object T over X to be N disjoint copies of X, each mapping via the identity to X:

$$T = \pi : X \coprod \cdots \coprod X \to X.$$

We now allow only those automorphisms which act cyclically. Specifically, there are exactly N such automorphisms locally, one for each integer k modulo N: the kth automorphism sends a point in the ith X to itself in the $(i+k)^{th}$ X.

A locally trivial object in this setting will be called a *locally trivial cyclic covering of X of degree N*.

We have as a sheaf of local automorphisms the constant sheaf $\underline{\mathbb{Z}/N\mathbb{Z}}$. Hence:

PROPOSITION 4.4. *The isomorphism classes of locally trivial cyclic coverings of X of degree N are in 1-1 correspondence with the cohomology group $\check{H}^1(X, \underline{\mathbb{Z}/N\mathbb{Z}})$.*

Let us compute this group for a Riemann surface X (using the classical topology). We have a short exact sequence of constant sheaves

$$0 \to \underline{N\mathbb{Z}} \to \underline{\mathbb{Z}} \to \underline{\mathbb{Z}/N\mathbb{Z}} \to 0$$

(where the map on the right is induced by sending an integer to its class modulo N). The kernel sheaf $\underline{N\mathbb{Z}}$ is isomorphic to $\underline{\mathbb{Z}}$; under this isomorphism the map on the left is given by multiplication by N. Hence the long exact exact sequence in cohomology is

$$\begin{aligned} 0 &\to & \mathbb{Z} &\to & \mathbb{Z} &\to & \mathbb{Z}/N\mathbb{Z} &\to \\ &\to & \check{H}^1(X,\underline{\mathbb{Z}}) &\to & \check{H}^1(X,\underline{\mathbb{Z}}) &\to & \check{H}^1(X,\underline{\mathbb{Z}/N\mathbb{Z}}) &\to \\ &\to & \check{H}^2(X,\underline{\mathbb{Z}}) &\to & \check{H}^2(X,\underline{\mathbb{Z}}) &\to & \check{H}^2(X,\underline{\mathbb{Z}/N\mathbb{Z}}) &\to \ldots \end{aligned}$$

The sequence is exact at the $\check{H}^0$ level; moreover the groups $\check{H}^i(X,\underline{\mathbb{Z}})$ are free abelian groups (of rank $2g$ when $i = 1$ and of rank one when $i = 2$). Hence the multiplication-by-N maps are all injective, and we see that

$$\check{H}^1(X,\underline{\mathbb{Z}/N\mathbb{Z}}) \cong (\mathbb{Z}/N\mathbb{Z})^{2g}.$$

Thus in particular there are N^{2g} nonisomorphic cyclic coverings of X of degree N.

Note that if we use the Zariski topology instead of the classical topology, we get that $\check{H}^1(X,\underline{\mathbb{Z}/N\mathbb{Z}}) = 0$ by Proposition 2.1 of Chapter X. Intuitively, since all of the open sets of the Zariski topology are so large, if a covering S splits into a disjoint union over a Zariski open set, it must split globally, and be the trivial object.

Let us remark that one can take locally trivial coverings and remove the cyclic hypothesis; one may allow any permutation of the sheets of the covering. This leads to automorphisms which are sections of the constant sheaf $\underline{S_N}$ of all permutations, and so these will be classified by $\check{H}^1(X, \underline{S_N})$.

The problem with this is that since S_N is not abelian for $N \geq 3$, we have here a sheaf of nonabelian groups. For such things the Čech cohomology produces only a set, not a group. However the general principle is still valid.

Extensions of Invertible Sheaves. As a next example, let us fix two invertible sheaves $\mathcal{F}$ and $\mathcal{G}$, and take as the trivial object their sheaf direct sum $\mathcal{F} \oplus \mathcal{G}$. This is simply the sheaf whose sections over an open set U is the direct sum $\mathcal{F}(U) \oplus \mathcal{G}(U)$. We fix as part of the data the inclusion map of $\mathcal{F}$ into $\mathcal{F} \oplus \mathcal{G}$ and the projection map of $\mathcal{F} \oplus \mathcal{G}$ onto $\mathcal{G}$.

Fixing these two maps means that we want the local automorphisms to preserve these maps. We therefore define a local automorphism (over an open set U of X) to be a sheaf automorphism Φ from $(\mathcal{F} \oplus \mathcal{G})|_U$ to itself, making the diagram

$$\begin{array}{ccccc} \mathcal{F}(V) & \to & (\mathcal{F} \oplus \mathcal{G})(V) & \to & \mathcal{G}(V) \\ \| & & \downarrow \Phi_V & & \| \\ \mathcal{F}(V) & \to & (\mathcal{F} \oplus \mathcal{G})(V) & \to & \mathcal{G}(V) \end{array}$$

commute for all $V \subset U$.

Objects which are locally trivial with these definitions of triviality are called *extensions of $\mathcal{G}$ by $\mathcal{F}$*.

Let's compute the sheaf of local automorphisms. Suppose that $\mathcal{F}$ and $\mathcal{G}$ are both trivialized over an open set U, with local generators f and g. Then locally $\mathcal{F} \oplus \mathcal{G}$ is isomorphic to $\mathcal{O} \oplus \mathcal{O}$ as an $\mathcal{O}$-module. The group of automorphisms as an $\mathcal{O}$-module (ignoring the commuting diagram condition) is therefore just the group of nonsingular 2-by-2 matrices with entries in $\mathcal{O}(U)$: we denote this group by $\mathrm{GL}_2(\mathcal{O}(U))$. A matrix $\begin{pmatrix} a & b \\ c & d \end{pmatrix}$ acts on a pair $(\alpha f, \beta g)$ by sending it to the pair $((a\alpha + b\beta)f, (c\alpha + d\beta)g)$.

Now we can add in the condition that the diagram commutes. In order that the square on the right commute (i.e., the automorphism commutes with the projection to $\mathcal{G}$) we must have that $c\alpha + d\beta = \beta$ for all α and β; this requires $c = 0$ and $d = 1$. In order that the square on the left commute we must have that $a = 1$. Therefore the automorphism must locally have the form

$$\begin{pmatrix} 1 & b \\ 0 & 1 \end{pmatrix}$$

for some regular function $b \in \mathcal{O}(U)$.

We claim that this function b is actually a section of an invertible sheaf. To see this, let $\mathcal{U} = \{U_i\}$ be an open covering on which both $\mathcal{F}$ and $\mathcal{G}$ trivialize; let f_i and g_i be local generators respectively. Suppose that transition functions for $\mathcal{F}$ are t_{ij} and for $\mathcal{G}$ are s_{ij}, so that on $U_i \cap U_j$ we have

$$f_i = t_{ij} f_j \quad \text{and} \quad g_i = s_{ij} g_j$$

for every i and j.

Suppose now that Φ is a global automorphism of the trivial direct sum object above. Then on each U_i there is a regular function b_i such that Φ_{U_i} has the form

$$(\alpha f_i, \beta g_i) \mapsto ((\alpha + b_i \beta) f_i, \beta g_i),$$

or, equivalently given our restrictions, that

$$(0, g_i) \mapsto (b_i f_i, g_i).$$

On the intersection with U_j, we also have that

$$(0, g_j) \mapsto (b_j f_j, g_j).$$

On the other hand, using the change of generators via the transition functions, we see that

$$\begin{aligned} (0, g_i) &= (0, s_{ij} g_j) \\ &\mapsto (s_{ij} b_j f_j, s_{ij} g_j) \\ &= (b_j s_{ij} f_i / t_{ij}, g_i), \end{aligned}$$

which must be equal to $(b_i f_i, g_i)$. Hence we see that we have the compatibility condition

$$b_i = \frac{s_{ij}}{t_{ij}} b_j.$$

Let g'_i be a local generator for the inverse sheaf $\mathcal{G}^{-1}$, which is dual to g_i; note that $g'_i = s_{ij}^{-1} g_j$ for every i and j. Set $h_i = b_i f_i \otimes g'_i$, which is a local section of $\mathcal{F} \otimes_\mathcal{O} \mathcal{G}^{-1}$. On the intersection $U_i \cap U_j$, we have

$$h_i = b_i f_i \otimes g'_i = (\frac{s_{ij}}{t_{ij}} b_j)(t_{ij} f_j) \otimes (s_{ij}^{-1} g'_j) = b_j f_j \otimes g'_j = h_j,$$

so that by the sheaf axiom the local sections h_i patch together to give a global section of the sheaf $\mathcal{F} \otimes_\mathcal{O} \mathcal{G}^{-1}$.

This same computation holds not only for global sections, but also for sections over any open set. Hence we have shown:

LEMMA 4.5. *The sheaf of local automorphisms of the trivial extension of $\mathcal{G}$ by $\mathcal{F}$ is isomorphic to the sheaf $\mathcal{F} \otimes_\mathcal{O} \mathcal{G}^{-1}$.*

What are the locally trivial objects here? The answer should not be unexpected, and we leave it to the reader:

PROPOSITION 4.6. *An extension of $\mathcal{G}$ by $\mathcal{F}$ is a short exact sequence of sheaves of $\mathcal{O}$-modules*

$$0 \to \mathcal{F} \to \mathcal{H} \to \mathcal{G} \to 0.$$

Two such extensions (with middle sheaves $\mathcal{H}_1$ and $\mathcal{H}_2$) are isomorphic if and only if there is an isomorphism $h : \mathcal{H}_1 \to \mathcal{H}_2$ of sheaves of $\mathcal{O}$-modules making the diagram

$$\begin{array}{ccccc} \mathcal{F} & \to & \mathcal{H}_1 & \to & \mathcal{G} \\ \parallel & & \downarrow h & & \parallel \\ \mathcal{F} & \to & \mathcal{H}_2 & \to & \mathcal{G} \end{array}$$

commute.

Such sequences come up frequently in the study of algebraic curves on algebraic surfaces.

In any case, our general principle immediately gives the following:

PROPOSITION 4.7. *If $\mathcal{F}$ and $\mathcal{G}$ are invertible sheaves on an algebraic curve X, the set of isomorphism classes of extensions of $\mathcal{G}$ by $\mathcal{F}$ is in 1-1 correspondence with the cohomology group $\check{H}^1(X, \mathcal{F} \otimes \mathcal{G}^{-1})$.*

First-Order Deformations. How about using these ideas to classify Riemann surfaces themselves? After all, a Riemann surface is some kind of locally trivial object, where the local triviality is expressed exactly by the existence of complex charts. (For this example we will work in the analytic category, with the classical topology.) This seems like a promising train of thought, but to put the problem into the framework of the classifying theory above we need to define a "trivial" object, and a sheaf of local automorphisms which are allowed.

Right away one sees that the trivial object, being one of the objects to be classified after all, should be a Riemann surface; the first difficulty comes in deciding which one. There is no natural choice here, and we get bogged down right at the start.

There *are* constructions of a space M_g which classifies all Riemann surfaces of genus g, but they do not proceed along these lines.

If we insist on pursuing this, we must fall back on choosing a Riemann surface X_0, and try to describe those Riemann surfaces which are "near" to X_0. These should be other Riemann surfaces which occur in a family of Riemann surfaces depending on some parameter t, which when $t = 0$ gives our chosen X_0, and for $t \neq 0$ gives "nearby" Riemann surfaces (at least when t is "near" 0).

A good example to keep in mind would be a family of genus one curves, defined by an equation of the form

$$y^2 = x^3 - a(t)x + b(t),$$

where $a(t) = 1 + a_1(t) + \ldots$ and $b(t) = b_1 t + b_2 t^2 + \ldots$ are holomorphic functions of t with $a(0) = 1$ and $b(0) = 0$. For this family the chosen curve is X_0 defined by $y^2 = x^3 - x$; as t varies we obtain other curves X_t of genus one.

To define a family of curves depending on a parameter properly requires the idea of a complex 2-manifold. We do not want to develop all of the theory of complex 2-manifolds for this example, but you can easily imagine the definition: it is just a second countable Hausdorff space $\mathcal{X}$, with chart functions $\phi : U \to V$ for all U in some open covering of $\mathcal{X}$, which are homeomorphisms onto open sets in $\mathbb{C}^2$, and which are compatible in the sense that the compositions $\phi \circ \psi^{-1}$ for two such charts ϕ and ψ are biholomorphic on the domain (which is an open subset of $\mathbb{C}^2$).

Now a family of Riemann surfaces depending on $t \in V \subset \mathbb{C}$ is a complex 2-manifold $\mathcal{X}$ together with an onto map $\pi : \mathcal{X} \to V$ whose fibers are all Riemann surfaces; if we want the dependence on t to be holomorphic, we require that π be a holomorphic function.

We can now make a provisional definition: a *deformation* of a Riemann surface X_0 over an open neighborhood V of 0 in $\mathbb{C}$ is a complex 2-manifold $\mathcal{X}$ together

4. $\check{H}^1$ AS A CLASSIFYING SPACE

with a commuting diagram

$$\begin{array}{ccc} X_0 & \subset & \mathcal{X} \\ \downarrow & & \downarrow \pi \\ 0 & \in & V \end{array},$$

where π is a holomorphic map, all of whose fibers $X_t = \pi^{-1}(t)$ for $t \in V$ are Riemann surfaces.

The *trivial deformation* over V is simply the product $\mathcal{X} = X_0 \times V$, with the mapping π being the second projection. With this trivial object, what we would be classifying would be *locally trivial deformations* of X_0 over V.

The local triviality of a deformation $\pi : \mathcal{X} \to V$ comes from requiring that for every point $p \in X_0$, there is a neighborhood U of p in X_0 and a neighborhood $\mathcal{U}$ of $p \in \pi^{-1}(0) \subset \mathcal{X}$ in $\mathcal{X}$ such that $\mathcal{U} \cong U \times V$; moreover the isomorphism should make the diagram

$$\begin{array}{ccc} U \times V & \cong & \mathcal{U} \\ \downarrow & & \downarrow \pi|_\mathcal{U} \\ V & = & V \end{array}$$

commute.

What are the local automorphisms of the trivial object? These will be isomorphisms $\Phi : U \times V \to U \times V$ for open sets $U \subset X$ preserving the commuting diagram above; in other words, we require that the diagram

$$\begin{array}{ccccc} U & \hookrightarrow & U \times V & \stackrel{\Phi}{\to} & U \times V \\ \downarrow & & \downarrow \pi & & \downarrow \pi \\ V & = & V & = & V \end{array}$$

commute, and that the map from U to the second $U \times V$ also be the inclusion. Such a function Φ must be of the form

$$\Phi(z, t) = (\theta(z, t), t)$$

for some holomorphic function θ of two variables, if the right side of the diagram is to commute; moreover we must also have

$$\theta(z, 0) = z$$

in order that the map preserve the inclusion of U into $U \times V$.

There are two problems with this which now appear. Firstly, this group of local automorphisms is highly nonabelian, and so we will not have a sheaf of abelian groups. Secondly, the sheaf depends on V in a serious way: if V is larger, then there may well be fewer automorphisms (due essentially to the fact that the relevant holomorphic function θ may not converge for larger t values).

With these issues in mind let us retreat and only consider the *linearization* of the local automorphism Φ. In other words, if we concentrate on just the first-order terms of a Taylor series for θ, can we arrive at a more tractable set of local automorphisms?

To this end expand the function θ as a power series in t to obtain

$$\theta(z,t) = z + \sum_{n=1}^{\infty} \theta^{(n)}(z) t^n$$

because $\theta(z,0) = z$.

Let us define the *first-order part* of the local automorphism Φ to be the holomorphic function $\theta^{(1)}(z)$, which is the coefficient of the linear term of the first coordinate θ of Φ. Two local automorphisms are *equal to first order* if they have the same first-order part. We define the *sheaf of first-order automorphisms* of the trivial deformation $X_0 \times V \to V$ to be the sheaf of local automorphisms up to first order.

Evidently this sheaf is invertible, since a local section Φ of this sheaf is completely determined by the coefficient $\theta^{(1)}(z)$, which can be any holomorphic function a priori. We are now in business! Let us determine the transition functions for this invertible sheaf.

Choose a complex atlas $\phi_i : U_i \to \mathbb{C}$ on X_0, with local coordinates z_i and transition functions T_{ij}, so that $z_i = T_{ij}(z_j)$. Suppose that we have a first-order automorphism Φ for $X_0 \times V$, which then restricts to first-order automorphisms $\Phi_i : U_i \times V \to U_i \times V$ which can be written (up to first order) as

$$\Phi_i(z_i, t) \approx (z_i + \theta_i^{(1)}(z_i)t, t),$$

where we write $\approx$ for equality up through the linear terms.

In order that the automorphisms Φ_i patch together to give Φ, we must have that $\Phi_i(p,t) = \Phi_j(p,t)$ when $p \in U_i \cap U_j$. If p has coordinate z_j in the U_j chart, then (p,t) is mapped via Φ_j to

$$(z_j + \theta_j^{(1)}(z_j)t, t).$$

In this case in the U_i chart the coordinate of p is $z_i = T_{ij}(z_j)$; it is then mapped via Φ_i to

$$(z_i + \theta_i^{(1)}(z_i)t, t).$$

These must be the same to first order; hence we must have

$$T_{ij}(z_j + \theta_j^{(1)}(z_j)t) = z_i + \theta_i^{(1)}(z_i)t$$

to first order. But using Taylor's theorem we have that

$$\begin{aligned} T_{ij}(z_j + \theta_j^{(1)}(z_j)t) &\approx T_{ij}(z_j) + T_{ij}'(z_j)\theta_j^{(1)}(z_j)t \\ &= z_i + T_{ij}'(z_j)\theta_j^{(1)}(z_j)t. \end{aligned}$$

Hence the condition that the Φ_i's patch together to first order is that

$$\theta_i^{(1)}(z_i) = T_{ij}'(z_j)\theta_j^{(1)}(z_j),$$

which exactly means that the local functions $\theta_i^{(1)}$ patch together to give a section of the bundle whose transition functions are the derivatives T_{ij}'.

4. Ȟ¹ AS A CLASSIFYING SPACE

This is exactly the tangent bundle (see Example 2.11)! Hence by Lemma 2.19 and Corollary 3.14, the sheaf of first-order automorphisms of the trivial deformation of X_0 (over any base V) is the tangent sheaf Θ.

The locally trivial objects with this notion of automorphism are clearly the locally trivial deformations of X_0 up to first order. (These are usually called *first-order deformations* of X_0.) Hence applying our general principle we have:

PROPOSITION 4.8. *First-order deformations of a Riemann surface are classified by* $\check{H}^1(X, \Theta)$.

The dimension of this cohomology group can be computed for any curve X:

LEMMA 4.9. *Let X be an algebraic curve of genus g. Then*

$$\dim \check{H}^1(X, \Theta) = \begin{cases} 0 & \text{if } g = 0; \\ 1 & \text{if } g = 1; \text{ and} \\ 3g - 3 & \text{if } g \geq 2. \end{cases}$$

PROOF. Recalling that Θ is the inverse sheaf to $\Omega^1 \cong \mathcal{O}(K)$ for a canonical divisor K, we see by Serre Duality that

$$\dim \check{H}^1(X, \Theta) = \dim \check{H}^0(X, \mathcal{O}(2K)).$$

If $g = 0$, $2K$ has degree -4, which is negative; hence this space is 0. If $g = 1$, then K is a principal divisor, as is $2K$; hence $\mathcal{O}(2K) \cong \mathcal{O}$ and the space has dimension one. If $g \geq 2$, then $2K$ has degree $4g-4$, which is larger than $2g-1$; hence we may use Riemann-Roch to calculate the dimension to be $(4g-4)+1-g = 3g-3$. □

COROLLARY 4.10. *Every first-order deformation of the projective line $\mathbb{P}^1$ is trivial.*

This is perhaps not surprising; after all, all curves of genus zero are isomorphic, so it is not possible to construct families of curves of genus zero which actually vary up to isomorphism.

How should we interpret these other numbers? Suppose that we had succeeded in our original goal to construct a space M_g which classified all Riemann surfaces of genus g. Such a space is called a *moduli space* for curves of genus g. Let us be optimistic and assume that the moduli space M_g is a complex manifold. If we have a locally trivial deformation $\pi : \mathcal{X} \to V$ of X_0 over V, we could then define a function $F : V \to M_g$ sending a point t to the point in M_g which classified the Riemann surface $X_t = \pi^{-1}(t)$. We could be even more optimistic and assume that F is a holomorphic map. We thus obtain a complex "arc" on M_g, namely the image $F(V)$.

Conversely, if we had a holomorphic map $F : V \to M_g$, we could conceivably build a family $\pi : \mathcal{X} \to V$ of curves, by setting $\pi^{-1}(t)$ to be the curve X_t which the point $F(t)$ classified.

We are somewhat out on a limb here, with lots of assumptions; but let us continue to speculate. If we believe the correspondence outlined above, that deformations of X_0 over V correspond to maps $F : V \to M_g$, then given such a deformation, its first-order part will naturally correspond to the tangent vector to M_g in the direction of the curve $F(V)$ on M_g. Hence we arrive at the following heuristic:

First-order deformations of X_0 correspond to tangent vectors to the moduli space M_g at the point corresponding to X_0.

Now if the moduli space M_g is really a complex manifold, then the dimension of its tangent space at a point would be equal to its dimension as a manifold. We are therefore led to the following guess: the moduli space, if it exists, has dimension equal to $\dim \check{H}^1(X,\Theta)$. This line of argument is supported by the count of $3g - 3$ parameters for Riemann surfaces made in Section 2 of Chapter VII for Riemann surfaces of genus $g \geq 2$.

The program of properly constructing the space M_g and verifying all the properties required to make this heuristic argument rigorous has been carried out. The famous number $3g - 3$, which is the number of parameters necessary to describe curves of genus g (for $g \geq 2$), was first computed by Riemann.

Problems XI.4

A. Show that $\mathbb{P}^1$ has no locally trivial cyclic coverings of any degree $N \geq 2$.

B. Let X be the Riemann surface of genus one defined by the equation $y^2 = x^3 - x$. Find all three nontrivial locally trivial cyclic coverings of degree 2.

C. Let $\mathcal{A}$ and $\mathcal{B}$ be two sheaves of abelian groups on X. Show that the direct sum sheaf defined by

$$(\mathcal{A} \oplus \mathcal{B})(U) = \mathcal{A}(U) \oplus \mathcal{B}(U)$$

is a sheaf on X. Show that if $\mathcal{A}$ and $\mathcal{B}$ are sheaves of $\mathcal{O}$-modules, then so is $\mathcal{A} \oplus \mathcal{B}$.

D. Let X be $\mathbb{P}^1$, let $p = [1:0]$, and let $\mathcal{O}[n] = \mathcal{O}[n \cdot p]$. Show that if $n \leq m+1$, the only extension of $\mathcal{O}[n]$ by $\mathcal{O}[m]$ is the trivial one.

E. With the notation as above, write down a nontrivial extension of $\mathcal{O}[0]$ by $\mathcal{O}[-2]$.

F. Let X be the curve of genus one defined by $y^2 = x^3 - x$. Write down a nontrivial extension of $\mathcal{O}$ by itself.

G. Prove Proposition 4.6.

H. Define $\check{H}^1(X,\mathcal{S})$ for a sheaf of nonabelian groups $\mathcal{S}$. Why is this not a group in general?

I. Define vector bundles of rank n on an algebraic curve X. (These should be locally isomorphic to $\mathbb{C}^n \times X$, with automorphisms being linear transformations in the fibers.) Show that the set of isomorphism classes of vector bundles of rank n on X is is 1-1 correspondence with $\check{H}^1(X, \underline{\mathrm{GL}}_n(\mathcal{O}))$, where $\underline{\mathrm{GL}}_n(\mathcal{O})$ is the sheaf whose sections over an open set U is the group $\mathrm{GL}_n(\mathcal{O}(U))$.

Further Reading

The natural extensions of invertible sheaves are the locally free sheaves, and then the *coherent* sheaves, in the algebraic category; see [**Serre55**] and also [**Hartshorne77**].

For line bundles, and more generally vector bundles, [**G-H78**] has a good section, as well as [**Kodaira86**]; see also [**Husemoller94**].

For deformation theory, [**Kodaira86**] has the analytic point of view, while [**Artin76**] and [**Sernesi86**] are good introductions in the algebraic category (which really requires and exploits the theory of schemes). See also [**E-H92**].

The construction of the moduli space M_g for curves of genus g as a projective variety (or, more precisely, as a Zariski open subset of a projective variety) is due to Mumford, see [**Mumford65**]. The reader interested further in moduli questions will want to investigate [**D-M69**], [**Mumford77**], and [**Newstead78**]; an introduction to these papers is found in [**Morrison89**].

References

[Ahlfors66]   L.V. Ahlfors: *Complex Analysis.* McGraw-Hill (1966)
[AS60]   L.V. Ahlfors and L. Sario: *Riemann Surfaces.* Princeton University Press (1960)
[ACGH85]   E. Arbarello, M. Cornalba, P.A. Griffiths, and J. Harris: *Geometry of Algebraic Curves.* Grundlehren der mathematishen Wissenschaften 267, Springer-Verlag, New York (1985)
[Armstrong83]   M.A. Armstrong: *Basic Topology.* Undergraduate Texts in Mathematics, Springer-Verlag, New York (1983)
[Artin76]   M. Artin: *Deformations of Singularities.* TATA Lecture Notes, Vol. 54 (1976)
[Artin91]   M. Artin: *Algebra.* Prentice-Hall (1991)
[AM69]   M. Atiyah and I.G. Macdonald: *Commutative Algebra.* Addison-Wesley Publishing Company (1969)
[Beardon84]   A. F. Beardon: *A Primer on Riemann Surfaces.* London Mathematical Society Lecture Note Series, Vol. 78, Cambridge University Press (1984)
[Bers58]   L. Bers: *Riemann Surfaces.* Courant Institute of Mathematical Sciences, NYU, New York (1958)
[Boas87]   R.P. Boas: *Invitation to Complex Analysis.* Random House (1987)
[B-T82]   R. Bott and L. W. Tu: *Differential Forms in Algebraic Topology.* Graduate Texts in Mathematics, No. 82, Springer-Verlag (1982)
[Brieskorn86]   E. Brieskorn: *Plane Algebraic Curves.* Birkhäuser Verlag (1986)
[Buser92]   P. Buser: *Geometry and Spectra of Compact Riemann Surfaces.* Progress in Mathematics, Vol. 106, Birkhäuser Verlag (1992)
[Castelnuovo1889]   G. Castelnuovo: "Ricerche di geometria sulle curve algebriche". Atti R. Accad. Sci. Torino, Vol. 24 (1889), 196-223.
[Chevalley51]   C. Chevalley: *Introduction to the Theory of Algebraic Functions of One Variable.* Mathematical Surveys VI, American Mathematical Society (1951)
[Clemens80]   C. H. Clemens: *A Scrapbook of Complex Curve Theory.* The University Series in Mathematics, Plenum Press (1980)

[Conway78]   J.B. Conway: *Functions of One Complex Variable*, 2nd edition. Graduate Texts in Mathematics, No. 11, Springer-Verlag (1978)
[Coolidge31] J. L. Coolidge: *A Treatise on Algebraic Plane Curves*. Dover Publications (1959)
[Cornalba89] M. Cornalba: "The Theorems of Riemann-Roch and Abel". In: *Lectures on Riemann Surfaces*, Edited by M. Cornalba, X. Gomez-Mont, and A. Verjovsky; Proceedings of the College on Riemann Surfaces, ICTP, Trieste, Italy, 1987; World Scientific (1989), 302-349.
[C-L-O92]   D. Cox, J. Little, and D. O'Shea: *Ideals, Varieties, and Algorithms*. Undergraduate Texts in Mathematics, Springer-Verlag (1992)
[D-M69]   P. Deligne and D. Mumford: "The Irreducibility of the Space of Curves of Given Genus". Publ. Math. IHES, Vol. 36 (1969), 75-100.
[Deuring73] M. Deuring: *Lectures on the Theory of Algebraic Functions of One Variable*, Lecture Notes in Mathematics, Vol. 314, Springer-Verlag (1973)
[E-H92]   D. Eisenbud and J. Harris: *Schemes: The Language of Modern Algebraic Geometry*. Wadsworth and Brooks-Cole (1992)
[FK80]   H.M. Farkas and I. Kra: *Riemann Surfaces*, 2nd edition. Graduate Texts in Mathematics, No. 71, Springer-Verlag (1992)
[Forster81] O. Forster: *Lectures on Riemann Surfaces*. Graduate Texts in Mathematics, No. 81, Springer-Verlag (1981)
[Fulton69] W. Fulton: *Algebraic Curves*. W. A. Benjamin, New York (1969)
[Godement58] R. Godement: *Topologie algébrique et Théorie des Faisceaux*. Hermann, Paris (1958)
[Gomez-Mont89] X. Gomez-Mont: "Meromorphic Functions and Cohomology on a Riemann Surface". In: *Lectures on Riemann Surfaces*, Edited by M. Cornalba, X. Gomez-Mont, and A. Verjovsky; Proceedings of the College on Riemann Surfaces, ICTP, Trieste, Italy, 1987. World Scientific (1989), 245-301
[Griffiths89] P. Griffiths: *Introduction to Algebraic Curves*. Translations of Mathematical Monographs, vol. 76, American Mathematical Society (1989)
[GA74]   P. Griffiths and J. Adams: *Topics in Algebraic and Analytic Geometry*. Mathematical Notes, Princeton University Press, Princeton, NJ (1974)
[G-H78]   P. Griffiths and J. Harris: *Principles of Algebraic Geometry*. Wiley Interscience (1978)
[Grothendieck] A. Grothendieck (with J. Dieudonné): *Eléménts de Géometrie Algébrique*. Publ. Math. IHES, Vols. 4, 8, 11, 17, 20, 24, 28, 32 (1960-67)

REFERENCES

[G-N76]   J. Guenot and R. Narasimhan: *Introduction à la Théorie des Surfaces de Riemann.* Monographies de L'Enseignement Mathematique No. 23 (1976)

[Gunning66]   R. C. Gunning: *Lectures on Riemann Surfaces.* Princeton University Press (1966)

[Gunning72]   R. C. Gunning: *Lectures on Riemann Surfaces: Jacobi Varieties.* Princeton University Press (1972)

[Gunning76]   R. C. Gunning: *Riemann Surfaces and Generalized Theta Functions.* Ergebnisse der Mathematik und Ihrer Grenzgebiete, New Series, Vol. 91, Springer-Verlag, Berlin (1976)

[Harris80]   J. Harris: "The genus of space curves". Math. Annalen, Vol. 249 (1980), 35-68.

[Harris82]   J. Harris: *Curves in Projective Space.* Les Presses de L'Université de Montréal (1982)

[Harris92]   J. Harris: *Algebraic Geometry - A First Course.* Graduate Texts in Mathematics, No. 133, Springer-Verlag (1992)

[Hartshorne76]   R. Hartshorne: "On the De Rham cohomology of algebraic varieties", Publ. Math. IHES, No. 45 (1976), 5-99

[Hartshorne77]   R. Hartshorne: *Algebraic Geometry.* Graduate Texts in Mathematics, No. 52, Springer-Verlag (1977)

[Hirzebruch66]   F. Hirzebruch: *Topological Methods in Algebraic Geometry.* Grundlehren der mathematishen Wissenschaften 131, Springer-Verlag (1986)

[H-P47]   W. V. D. Hodge and D. Pedoe: *Methods of Algebraic Geometry.* Cambridge University Press (1947)

[Hulek86]   K. Hulek: *Projective Geometry of Elliptic Curves.* Astérisque Vol. 137, Société Mathématique de France (1986)

[Hungerford74]   T. W. Hungerford: *Algebra.* Graduate Texts in Mathematics, No. 73, Springer-Verlag (1974)

[Husemoller87]   D. Husemoller: *Elliptic Curves.* Graduate Texts in Mathematics, No. 111, Springer-Verlag (1987)

[Husemoller94]   D. Husemoller: *Fibre Bundles, 3rd Edition.* Graduate Texts in Mathematics, No. 20, Springer-Verlag (1994)

[Iitaka82]   S. Iitaka: *Algebraic Geometry - An Introduction to Birational Geometry of Algebraic Varieties.* Graduate Texts in Mathematics, No. 76, Springer-Verlag (1982)

[JS87]   G.A. Jones and D. Singerman: *Complex Functions - An algebraic and geometric viewpoint.* Cambridge University Press, Cambridge (1987)

[Kendig77]   K. Kendig: *Elementary Algebraic Geometry.* Graduate Texts in Mathematics, No. 44, Springer-Verlag (1977)

[Kirwan92]   F. Kirwan: *Complex Algebraic Curves.* London Math. Soc. Student Texts No. 23, Cambridge University Press (1992)

REFERENCES

[Klein1894]   F. Klein: *Riemannschen Flächen*, Vol. I and II. Göttingen (1894)

[Kodaira86]   K. Kodaira: *Complex Manifolds and Deformation of Complex Structures.* Grundlehren der mathematishen Wissenschaften 283, Springer-Verlag, New York (1986)

[K-M71]   K. Kodaira and J. Morrow: *Complex Manifolds.* Holt, Rhinehart and Winston (1971)

[Lang82]   S. Lang: *Introduction to Algebraic and Abelian Functions*, 2nd edition. Graduate Texts in Mathematics, No. 89, Springer-Verlag (1982)

[Lang84]   S. Lang: *Algebra*, 2nd edition. Addison-Wesley Publishing Company (1984)

[Lang85]   S. Lang: *Complex Analysis*, 2nd edition. Graduate Texts in Mathematics, No. 103, Springer-Verlag (1985)

[Lang87]   S. Lang: *Elliptic Functions*, 2nd edition. Graduate Texts in Mathematics, No. 112, Springer-Verlag (1987)

[Massey67]   W. Massey: *Algebraic Topology: An Introduction.* Harcourt, Brace, and World, Inc., New York (1967)

[Massey91]   W. Massey: *A Basic Course in Algebraic Topology.* Graduate Texts in Mathematics, No. 127, Springer-Verlag (1991)

[Moishezon67]   B. G. Moishezon: "On n-dimensional compact varieties with n algebraically independent meromorphic functions". American Mathematical Society Translations, Vol. 63 (1967), 51-177.

[Morrison89]   I. Morrison: "Constructing the moduli space of stable curves". In: *Lectures on Riemann Surfaces*, Edited by M. Cornalba, X. Gomez-Mont, and A. Verjovsky; Proceedings of the College on Riemann Surfaces, ICTP, Trieste, Italy, 1987; World Scientific (1989), 201-244.

[Mumford65]   D. Mumford: *Geometric Invariant Theory.* Ergebnisse der Mathematik und Ihrer Grenzgebiete, New Series, Vol. 34, Springer-Verlag, Berlin (1965)

[Mumford76]   D. Mumford: *Algebraic Geometry I – Complex Projective Varieties.* Grundlehren der mathematishen Wissenschaften 221, Springer-Verlag, New York (1976)

[Mumford77]   D. Mumford: "Stability of Projective Varieties". L'Ens. Math., Vol. 23 (1977), 39-100.

[Mumford78]   D. Mumford: *Curves and Their Jacobians.* The University of Michigan Press (1974)

[Mumford83]   D. Mumford: *Tata Lectures on Theta I.* Progress in Mathematics Vol. 28, Birkhäuser Boston (1983)

[Munkres75]   J. R. Munkres: *Topology - A First Course.* Prentice-Hall, New Jersey (1975)

[Munkres84]   J. R. Munkres: *Elements of Algebraic Topology.* Benjamin-Cummings, Menlo Park, CA (1984)

REFERENCES

[Munkres91]   J. R. Munkres: *Analysis on Manifolds*. Addison-Wesley, Reading, MA (1991)

[Namba84]   M. Namba: *Geometry of Projective Algebraic Curves*. Monographs and textbooks in pure and applied mathematics, Vol. 88, M. Dekker (1984)

[Narasimhan92]   R. Narasimhan: *Compact Riemann Surfaces*. Lectures in Mathematics, ETH Zürich, Birkhäuser Verlag (1992)

[Newstead78]   P. E. Newstead: *Introduction to Moduli Problems and Orbit Spaces*. Tata Institute of Fundamental Research Lecture Notes, Springer-Verlag (1978)

[O-O81]   G. Orzech and M. Orzech: *Plane Algebraic Curves: An Introduction via Valuations*. Pure and Applied Mathematics, Vol. 61, M. Dekker (1981)

[Pfluger57]   A. Pfluger: *Theorie der Riemannschen Flächen*. Springer-Verlag (1957)

[R-F74]   H. E. Rauch and H. M. Farkas: *Theta-Functions with Applications to Riemann Surfaces*. Williams and Wilkins, Baltimore (1974)

[Reid88]   M. Reid: *Undergraduate Algebraic Geometry*. London Math. Soc. Student Texts No. 12, Cambridge University Press (1988)

[Reyssat89]   E. Reyssat: *Quelques Aspects des Surfaces de Riemann*. Progress in Mathematics, Vol. 77, Birkhäuser Press (1989)

[Riemann1892]   B. Riemann: *Gesammelte Mathematische Werke*, 2nd edition. Teubner (1892; Dover reprint 1953)

[Samuel69]   P. Samuel: *Lectures on Old and New Results on Algebraic Curves*. Lectures on Mathematics and Physics, Mathematics Vol. 36, Tata Institute of Fundamental Research (1969)

[Seidenberg68]   A. Seidenberg: *Elements of the Theory of Algebraic Curves*. Addison-Wesley (1968)

[S-K59]   J. G. Semple and G. T. Kneebone: *Algebraic Curves*. Oxford Clarendon Press (1959)

[S-N70]   L. Sario and M. Nakai: *Classification Theory of Riemann Surfaces*. Grundlehren der mathematishen Wissenschaften 164, Springer-Verlag (1970)

[S-R49]   J. G. Semple and L. Roth: *Introduction to Algebraic Geometry*. Oxford University Press (1949)

[Sernesi86]   E. Sernesi: *Topics on Families of Projective Schemes*. Queen's Papers in Pure and Applied Mathematics, No. 73 (1986)

[Serre55]   J. P. Serre: "Faisceaux algébriques cohérents", Ann. of Math. **61** (1955) 197-278.

[Serre56]   J. P. Serre: "Geometrie Algebrique et Geometrie Algebrique", Ann. Inst. Fourier, Vol. 6. (1956), 1-42

[Serre59]   J. P. Serre: *Groupes Algébriques et Corps de Classes*. Hermann

(1959); English Translation: *Algebraic Groups and Class Fields*. Graduate Texts in Mathematics, No. 117, Springer-Verlag (1988)

[Serre73] J. P. Serre: *A Course in Arithmetic*. Graduate Texts in Mathematics, No. 7, Springer-Verlag (1973)

[Shafarevich77] I. R. Shafarevich: *Basic Algebraic Geometry*. Springer-Verlag, Berlin (1977)

[Shokurov94] V. V. Shokurov: "Riemann Surfaces and Algebraic Curves". In: *Algebraic Geometry I*, edited by I. R. Shafarevich. Encyclopedia of Mathematical Sciences, Vol. 23, Springer-Verlag (1994)

[Sieradski92] A. Sieradski: *An Introduction to Topology and Homotopy*. PWS-Kent Publishing Company (1992)

[Silverman86] J. H. Silverman: *The Arithmetic of Elliptic Curves*. Graduate Texts in Mathematics, No. 106, Springer-Verlag (1986)

[Smith89] R. Smith: "The Jacobian Variety of a Riemann Surface and its Theta Geometry". In: *Lectures on Riemann Surfaces*, Edited by M. Cornalba, X. Gomez-Mont, and A. Verjovsky; Proceedings of the College on Riemann Surfaces, ICTP, Trieste, Italy, 1987. World Scientific (1989), 350-427

[Springer57] G. Springer: *Riemann Surfaces*. Chelsea Publishing Company (1957)

[tomDieck87] T. tom Dieck: *Transformation Groups*, De Gruyter Studies in Mathematics, No. 8. Walter De Gruyter and Co., Berlin (1987)

[Walker50] R. J. Walker: *Algebraic Curves*. Dover Publications (1950)

[Warner71] F.W. Warner: *Foundations of Differentiable Manifolds and Lie Groups*. Scott-Foresman (1971)

[Weil38] A. Weil: "Sur les fonctions algébrique à corps de constantes finis". Comptes Rendus, Vol. 210, (1940), 129-133.

[Wells73] R. O. Wells: *Differential Analysis on Complex Manifolds*. Prentice-Hall (1973)

[Weyl55] H. Weyl: *The Concept of a Riemann Surface*, 3rd edition. Addison-Wesley Publishing Company, (1955)

[Yang91] K. Yang: *Complex Algebraic Geometry: An Introduction to Curves and Surfaces*. Monographs and Textbooks in Pure and Applied Mathematics, vol. 149, M. Dekker (1991)

[Z-S60] O. Zariski and P. Samuel: *Commutative Algebra*. Van Nostrand (1960)

Index of Notation

$\mathbb{C}$	the field of complex numbers
$\mathbb{C}^*$	the group of nonzero complex numbers
$\mathbb{R}$	the field of real numbers
$\mathbb{Q}$	the field of rational numbers
$\mathbb{Z}$	the ring of integers
S^2	the unit 2-sphere in $\mathbb{R}^3$
$\mathbb{C}_\infty$	The Riemann Sphere
$\mathbb{P}^1$	the projective line
$[x:y]$	homogeneous coordinates in $\mathbb{P}^1$
$\mathbb{P}^2$	the projective plane
$[x:y:z]$	homogeneous coordinates in $\mathbb{P}^2$
$\mathbb{P}^n$	projective n-space
$[x_0:\cdots:x_n]$	homogeneous coordinates in $\mathbb{P}^n$
$\mathcal{O}_X(W)$	the ring of holomorphic functions on open set $W \subseteq X$
$\mathcal{M}_X(W)$	the ring of meromorphic functions on open set $W \subseteq X$
$\mathrm{ord}_p(f)$	the order of a meromorphic function f at a point p
$\theta(z)$	theta function
$\mathrm{mult}_p(F)$	the multiplicity of a holomorphic map F at a point p
$d_y(F)$	the sum of the multiplicities of F at $F^{-1}(y)$
$\deg(F)$	the degree of a holomorphic map F between compact Riemann surfaces
$e(S)$	the Euler number of a compact manifold S
$X \coprod Y$	the disjoint union of X and Y
$X \coprod Y/\phi$	glueing together X and Y via ϕ
$\mathrm{Aut}_0(X)$	the group of automorphisms of a torus X fixing 0
$G:X$	a group G acting on a set X
$G \cdot p$	the orbit of a point p under the action of G
G_p	the stabilizer of a point p under the action of G
X/G	the quotient of X under the action of G
$\mathrm{Aut}(X)$	the group of automorphisms of a Riemann surface X

$\mathrm{ord}_p(\omega)$	the order of a meromorphic 1-form ω at a point p
$\mathcal{E}(U)$	C^∞ functions $f: U \to \mathbb{C}$
$\mathcal{E}^{(1)}(U)$	C^∞ 1-forms defined on U
$\mathcal{E}^{(1,0)}(U)$	C^∞ 1-forms of type $(1,0)$ defined on U
$\mathcal{E}^{(0,1)}(U)$	C^∞ 1-forms of type $(0,1)$ defined on U
$\mathcal{E}^{(2)}(U)$	C^∞ 2-forms defined on U
$\Omega^1(U)$	holomorphic 1-forms on U
$\mathcal{M}(U)$	meromorphic functions f defined on U
$\mathcal{M}^{(1)}(U)$	meromorphic 1-forms defined on U
$\int_\gamma \omega$	the integral of a 1-form ω along a path γ
$\mathrm{Res}_p(\omega)$	the residue of a meromorphic 1-form ω at a point p
$\iint_D \eta$	the (double) integral of a 2-form over a triangulable set D
$\pi_1(X, p)$	the first homotopy group of X with base point p
$H_1(X)$	the first homology group of X
$\mathrm{CLCH}(X)$	the group of closed chains on X
$\mathrm{BCH}(X)$	the subgroup of boundary chains on X
$\mathrm{Div}(X)$	the group of divisors on X
$\deg(D)$	the degree of a divisor D
$\mathrm{Div}_0(X)$	the group of divisors of degree 0 on X
$\mathrm{div}(f)$	the divisor of a meromorphic function f
$\mathrm{div}_0(f)$	the divisor of zeroes of a meromorphic function f
$\mathrm{div}_\infty(f)$	the divisor of poles of a meromorphic function f
$\mathrm{div}(\omega)$	the divisor of a meromorphic 1-form ω
$\mathrm{div}_0(\omega)$	the divisor of zeroes of a meromorphic 1-form ω
$\mathrm{div}_\infty(\omega)$	the divisor of poles of a meromorphic 1-form ω
$\mathrm{PDiv}(X)$	the subgroup of principal divisors on X
$\mathrm{KDiv}(X)$	the set of canonical divisors on X
$F^*(f)$	the pullback of a function f
$F^*(\omega)$	the pullback of a form ω
$F^*(D)$	the pullback of a divisor D
R_F	the ramification divisor of a holomorphic map F
B_F	the branch divisor of a holomorphic map F
$\mathrm{div}(G)$	the intersection divisor of a homogeneous polynomial G
$D_1 \sim D_2$	linear equivalence of two divisors D_1 and D_2
$\deg(X)$	the degree of a projective curve X
A	the Abel-Jacobi mapping for a complex torus X
$L(D)$	the space of meromorphic functions with poles bounded by D
$|D|$	the complete linear system of a divisor D
$L^{(1)}(D)$	the space of meromorphic 1-forms with poles bounded by D
ϕ_f	the holomorphic map to $\mathbb{P}^n$, given by $f = (f_0, \ldots, f_n)$
$|\phi|$	the linear system of a holomorphic map to $\mathbb{P}^n$
g_d^r	a linear system of dimension r and degree d

INDEX OF NOTATION

ϕ_D	the holomorphic map to $\mathbb{P}^n$, given by a basis for $L(D)$
$\phi_D^*(H)$	the hyperplane divisor of a holomorphic map to $\mathbb{P}^n$
$\mathbb{P}V$	the projectivization of a vector space V
$(\mathbb{P}V)^*, \mathbb{P}(V^*)$	the dual projective space
$\mathcal{T}(X)$	the group of Laurent tail divisors on X
$\mathcal{T}[D](X)$	the group of Laurent tail divisors whose top terms are bounded by $-D$
$t_{D_2}^{D_1}$	the truncation map on groups of Laurent tail divisors
μ_f^D	the multiplication isomorphism of groups of Laurent tail divisors
α_D	the Laurent tail divisor map
$H^1(D)$	the cokernel of α_D
$H^1(D_1/D_2)$	the kernel of the natural map $H^1(D_1) \to H^1(D_2)$ when $D_1 \leq D_2$
Res_ω	the linear functional on $H^1(D)$ if $\omega \in L^{(1)}(-D)$
Res	the Serre Duality isomorphism from $L^{(1)}(-D)$ to $H^1(D)^*$
$\text{span}(D)$	the span of a divisor on a smooth projective curve
$\mathbb{G}(1,n)$	the Grassmann variety of lines in $\mathbb{P}^n$
$\mathbb{G}(k,n)$	the Grassmann variety of k-planes in $\mathbb{P}^n$
$\text{Symm}^k(V)$	homogeneous polynomial expressions of vectors in V
$G_p(D)$	set of gap numbers
$W_z(g_1,..,g_\ell)$	the Wronskian determinant
$L^{(n)}(D)$	space of n-fold differentials with poles bounded by D
$W(Q)$	the Wronskian n-fold differential of the linear system Q
$w_p(Q)$	inflectionary weight
$\int_{[c]}$	the integration map from $\Omega^1(X)$ to $\mathbb{C}$.
Λ	the subgroup of periods inside $\Omega^1(X)^*$
$\text{Jac}(X)$	the Jacobian of X, $= \Omega^1(X)^*/\Lambda$
A	the Abel-Jacobi map from X or $\text{Div}(X)$ to $\text{Jac}(X)$
A_0	the Abel-Jacobi map from $\text{Div}_0(X)$ to $\text{Jac}(X)$
$\text{Tr}(h)$	the trace of a meromorphic function h
$\text{Tr}(\omega)$	the trace of a meromorphic 1-form ω
$F^*\gamma$	the pullback of a path γ
$A_i(\sigma)$	an a-period of a 1-form σ
$B_i(\sigma)$	a b-period of a 1-form σ
$\mathbf{A}$	the matrix of a-periods
$\mathbf{B}$	the matrix of b-periods
$\text{Pic}(X)$	the Picard group of divisors on X mod linear equivalence
$\text{Pic}^0(X)$	the Picard group of degree 0 divisors mod linear equivalence
$\mathcal{F}(U)$	the sections of a sheaf or presheaf $\mathcal{F}$ over an open set U
ρ_V^U	the restriction map in a sheaf or presheaf

INDEX OF NOTATION

$\mathcal{F}(X)$	the global sections of a sheaf $\mathcal{F}$ on X
G^X	the presheaf of all functions from X to a group G
$\mathcal{C}_X^\infty$	the sheaf of $\mathcal{C}^\infty$ functions on X
$\mathcal{O}_X$	the sheaf of holomorphic functions on X
$\mathcal{O}_X^*$	the sheaf of nowhere zero holomorphic functions on X
$\mathcal{M}_X$	the sheaf of meromorphic functions on X
$\mathcal{M}_X^*$	the sheaf of nonzero meromorphic functions on X
$\mathcal{H}_X$	the sheaf of harmonic functions on X
$\mathcal{O}_X[D]$	the sheaf of meromorphic functions on X with poles bounded by D
G_X	for a group G, the constant presheaf $G_X(U) = G$ for every U
$\mathcal{E}_X^1$	the sheaf of $\mathcal{C}^\infty$ 1-forms on X
$\mathcal{E}_X^{1,0}$	the sheaf of $\mathcal{C}^\infty$ $(1,0)$-forms on X
$\mathcal{E}_X^{0,1}$	the sheaf of $\mathcal{C}^\infty$ $(0,1)$-forms on X
Ω_X^1	the sheaf of holomorphic 1-forms on X
$\overline{\Omega}_X^1$	the sheaf of complex conjugates of holomorphic 1-forms on X
$\mathcal{M}_X^{(1)}$	the sheaf of meromorphic 1-forms on X
$\Omega_X^1[D]$	the sheaf of meromorphic 1-forms on X with poles bounded by D
$\mathcal{E}_X^2$	the sheaf of $\mathcal{C}^\infty$ 2-forms on X
$\underline{G}$	the sheaf of locally constant G-valued functions
$\underline{\mathbb{Z}}$	the sheaf of locally constant $\mathbb{Z}$-valued functions
$\underline{\mathbb{R}}$	the sheaf of locally constant $\mathbb{R}$-valued functions
$\underline{\mathbb{C}}$	the sheaf of locally constant $\mathbb{C}$-valued functions
$\mathbb{C}_p$	the skyscraper sheaf $\mathbb{C}$ supported at a single point p
$\mathcal{D}iv_X$	the sheaf of divisors on X
$\mathcal{T}_X[D]$	the sheaf of Laurent tail divisors on X bounded above by $-D$
$\mathcal{T}_X[D_1/D_2]$	the sheaf of Laurent tail divisors on X with terms between $-D_2$ and $-D_1$
$U_{i_0, i_1, \ldots, i_n}$	the intersection of $U_{i_0}, \ldots, U_{i_n}$
$\check{C}^n(\mathcal{U}, \mathcal{F})$	the Čech cochain group
$\check{Z}^n(\mathcal{U}, \mathcal{F})$	the Čech cocycle group
$\check{B}^n(\mathcal{U}, \mathcal{F})$	the Čech coboundary group
$\check{H}^n(\mathcal{U}, \mathcal{F})$	the Čech cohomology group with respect to a covering $\mathcal{U}$.
ϕ_*	the induced map on cohomology given a map ϕ on sheaves
$\mathcal{V} \prec \mathcal{U}$	the covering $\mathcal{V}$ is finer than $\mathcal{U}$
$\tilde{r}$	the induced map on cochains for a refining map r
$H(r)$	the induced map on cohomology for a refining map r
$H_\mathcal{V}^\mathcal{U}$	the map on cohomology when $\mathcal{V}$ is finer that $\mathcal{U}$.
$\varinjlim_{a \in A} G_a$	the direct limit of a direct system of groups

INDEX OF NOTATION

$\check{H}^n(X, \mathcal{F})$	the nth cohomology group of $\mathcal{F}$ on X
Δ	the connecting homomorphism in the long exact sequence
$H^k_d(X)$	the kth DeRham cohomology group
$H^{p,q}_{\bar\partial}(X)$	the (p,q) Dolbeault cohomology group
$\mathcal{O}_{X,alg}$	the sheaf of regular functions on an algebraic curve X
$\mathcal{O}_{X,alg}[D]$	the sheaf of rational functions on X with poles bounded by D
$\mathcal{M}_{X,alg}$	the (locally constant) sheaf of rational functions
$\Omega^1_{X,alg}$	the sheaf of regular 1-forms on an algebraic curve X
$\Omega^1_{X,alg}[D]$	the sheaf of rational 1-forms on X with poles bounded by D
$\mathcal{M}^{(1)}_{X,alg}$	the (locally constant) sheaf of rational 1-forms
X_{Zar}	an algebraic curve X with the Zariski topology
$\check{H}^n(X_{Zar}, \mathcal{F})$	the Čech cohomology group for a sheaf in the Zariski topology
$\mathcal{T}_{X,alg}[D]$	the sheaf of global Laurent tails
$\alpha_{D,alg}$	the map sending a rational function to its Laurent tail divisor
$F \otimes_R G$	the tensor product of two modules F and G over a ring R
$\mathcal{F} \otimes_\mathcal{O} \mathcal{G}$	the tensor product of two invertible sheaves
$\mathcal{F}^{-1}$	the inverse, or dual, of an invertible sheaf
Θ_X	the tangent sheaf on an algebraic curve X
$\mathrm{Inv}(X)$	the group of isomorphism classes of invertible sheaves on X
$[\mathcal{F}]$	the isomorphism class of an invertible sheaf $\mathcal{F}$
$\mathrm{LB}(X)$	the group of isomorphism classes of line bundles on X
L_n	a line bundle on $\mathbb{P}^1$
$\mathbb{T}_X$	the tangent bundle on X
$\mathbb{K}_X$	the canonical bundle on X
$\mathcal{O}\{L\}$	the invertible sheaf of regular sections of a line bundle L
$\mathrm{ord}_p(s)$	the order of a rational section s of a line bundle
$\mathrm{div}(s)$	the divisor of a rational section s of a line bundle
$\mathcal{D}iv_{X,alg}$	the algebraic sheaf of finite divisors on X
Δ	the isomorphism from $\mathrm{Pic}(X)$ to $\check{H}^1(X_{Zar}, \mathcal{O}^*)$
H_I	the isomorphism from $\mathrm{Inv}(X)$ to $\check{H}^1(X_{Zar}, \mathcal{O}^*)$
H_L	the isomorphism from $\mathrm{LB}(X)$ to $\check{H}^1(X_{Zar}, \mathcal{O}^*)$
H'_L	the composition of H_L with inversion

Index of Terminology

1-1 sheaf map, 282

a-period, 257, 259, 262
Abel's Theorem, 247, 250, 263, 266, 356
    for a torus, 140
    for curves of genus one, 265
    for the Riemann Sphere, 140
    proof of necessity, 255
    proof of sufficiency, 257, 260
    the high road, 319, 320
Abel-Jacobi map, 249, 250, 255, 260, 319, 356, 357
    for a curve of genus one, 265
    for a torus, 140
    has canonical map as derivative, 264
    independent of the base point, 250
    is 1-1 on X, 264, 265
    is holomorphic, 265, 267
    is surjective, 263
    on divisors, 250
    on divisors of degree d, 264
    on points, 249
Affine conic, 13
Affine plane curve, 11, 14, 15, 144
    1-forms on, 111
    irreducible, 12
    meromorphic functions on, 35, 36
    node of, 67
    of degree 2, 13
    projections are holomorphic, 22
    ramification of projection, 46
    smooth part, 12
Algebraic curve, 169, 195, 254, 309, 325
    complex torus, 170
    computing the function field, 177
    existence of 1-forms, 200, 202, 203
    function field has tr. deg. one, 174
    hyperelliptic, 170
    is projective, 196
    of genus 0, 197
    of genus 1, 197
    of genus 2, 198
    of genus 3, 206
    of genus 4, 207
    of genus 5, 209
    Riemann Sphere, 170
    smooth projective curve, 170
    trigonal, 209
Algebraic set, 95
Analytic genus, 192

Arithmetic genus, 192
Atlas, 3, 341
    C^∞, 5
    n-dimensional complex, 6
    equivalence of, 4
    line bundle, 333, 336, 339, 352–355
    on $\mathbb{P}^n$, 16
    on X/G, 78
    on a graph, 10
    on a smooth affine plane curve, 11
    on a torus, 9
Automorphism, 40
    168, 84
    first-order, 366
    group is finite, 82, 243
    linear, 97
    number of fixed points, 210
    of $\mathbb{P}^1$, 43
    of $\mathcal{O}_X$, 358
    of a torus, 64, 65
    of a trivial extension, 363
    of the trivial line bundle, 358

b-period, 257, 259
    normalized matrix, 262
Base point
    canonical system has none, 200
    gap number criterion, 234
    of a linear system, 157
    of fundamental group, 84
    removing from a linear system, 160
Base-point-free linear system, 157
Bezout's Theorem, 143
Biholomorphism, 40
Bitangent hyperplane, 220
Boundary chain, 123, 126
    group, 126, 247
Boundary divisor of a chain, 133
Branch divisor, 134
Branch point, 45
    and monodromy, 88, 91
    behaviour of the trace, 251
    of a quotient map, 80
    of hyperelliptic projection, 61
Branched covering, 49, 90
    number of parameters for, 213

Canonical bundle, 337
    sheaf of sections, 340
Canonical class, 139

Canonical divisor, 131
    H^1 has dimension one, 191
    from a section of the canonical bundle, 357
    has degree $2g - 2$, 133, 139, 191
    on a torus, 137
    pullback formula, 135
Canonical map, 203
    for a curve of genus 3, 206
    for a curve of genus 4, 207
    for a curve of genus 5, 209
    for a hyperelliptic curve, 204
    for a nonhyperelliptic curve, 203
    is the derivative of Abel-Jacobi, 264
Castelnuovo curves, 230
Castelnuovo's bound, 229
Čech coboundary, 292
Čech cochain, 291
Čech cocycle, 292
Čech cohomology, 296
    for $\mathcal{C}^\infty$ sheaves, 302
    for constant sheaves, 304
    for skyscraper sheaves, 303
    long exact sequence, 298, 299
    with respect to a cover, 292
    Zariski topology, 313
Center of projection, 98
Chain, 120
Chart
    centered at a point, 1
    complex, 1
    complex n-dimensional, 6
    defining a Riemann surface, 8
    domain, 1
    hole, 66
    line bundle, 331
    on $\mathbb{P}^1$, 8
    on $\mathbb{P}^n$, 16
    on a covering, 89
    on a quotient, 78
    on a smooth affine plane curve, 11
    on a torus, 9
    on the Riemann Sphere, 3
    real, 5
$\mathcal{C}^\infty$ function, 27
$\mathcal{C}^\infty$ real manifold, 5
$\mathcal{C}^\infty$ structure, 5
Closed 1-form, 114
Closed chain, 126
Closed path, 118
Coboundary operator, 291
Cocycle conditions, 335
Collinear, 96
Collinearity point, 266
Compact Riemann surface, 6

Compatibility
    of complex charts, 2, 6
    of line bundle charts, 332
    of real charts, 5
Complete linear system, 147
    as a fiber of Abel-Jacobi, 264
    is a projective space, 147
Complex manifold, 6
Complex plane, 4
Complex structure, 4
    n-dimensional, 6
Complex torus, 9
    g-dimensional, 263
    Abel's Theorem, 140
    automorphisms, 64
    canonical divisor is principal, 137
    elliptic normal curve, 165
    every curve of genus one is, 198, 265
    function field, 177
    has divisible group law, 12
    is an algebraic curve, 170
    isomorphic to $\mathbb{C}/(\mathbb{Z} + \mathbb{Z}\tau)$, 44
    meromorphic functions on, 25, 35, 42, 50
Concatenation of paths, 119
Conic, 57, 58
    affine, 13
    is isomorphic to $\mathbb{P}^1$, 58
    is planar, 216
Connecting homomorphism, 297
Constant presheaf, 271
Constant sheaf, 273
    Čech cohomology, 304, 313
Continuous action, 75
Covering map, 48
Covering space, 84
    cyclic unbranched, 360
Cubic, 57
    group law, 267
Cusp, 72
Cyclic covering, 73
    unbranched, 360

De Rham cohomology, 305, 320
Deformation, 364
    first-order, 367
    locally trivial, 365
    trivial, 365
Degree
    invariant under linear equivalence, 138
    minimal, 216
    of a canonical divisor is $2g - 2$, 133, 139
    of a covering map, 86
    of a divisor, 129

of a holomorphic map, 48
of a homogeneous polynomial, 14
of a line bundle, 345
of a principal divisor is zero, 130
of a projective curve, 142
of a smooth projective plane curve, 16
of the image of a map to $\mathbb{P}^n$, 164
Differential
    n-fold, 236
    form, *see* form
    of a function, 113
Dimension of a linear system, 147
Dimension of a subspace, 96
Dimension theorem, 211
Direct limit, 295
Direct system, 295
Divisor, 129
    boundary divisor of a chain, 133
    branch, 134
    canonical, 131
    complete linear system of, 147
    degree, 129
    fixed, 161
    general, 210
    hyperplane, 136, 142, 159
    intersection, 136
    inverse image, 133
    invertible sheaf of, 271, 325
    linear equivalence, 138
    map, 279, 319, 346
    of a meromorphic 1-form, 131
    of a meromorphic n-fold differential, 238
    of a meromorphic function, 130
    of a rational section of a line bundle, 343
    of Laurent tails, 179
    of poles, 131, 132
    of zeroes, 131, 132
    partial ordering, 136
    principal, 130
    pullback, 134
    ramification, 134
    skyscraper sheaf, 274
    space of forms with poles bounded by, 148
    space of functions with poles bounded by, 146
    span, 208
    special, 198
    very ample, 163, 195
Dolbeault cohomology, 306
Dolbeault's Lemma, 117, 284

Effective action, 75

Elliptic function, 25, 54
Elliptic normal curve, 165
    achieves the Castelnuovo bound, 229
Embedding, 163
Endpoint, 118
Equivalence of atlases, 5, 6
Equivalence of line bundle atlases, 333
Essential singularity, 23
Euler number, 50, 70
Euler's Formula, 14
Exact 1-form, 113
Exact sequence, 181
Exact sequence of sheaves, 286
Exponential map, 281
Extensions of invertible sheaves, 362

Fermat curve, 54, 69, 93
Fiber coordinate, 332
First homology group, 125, 126, 247
Fixed part of a linear system, 161
Flex point, 219, 233, 241, 266
    nine on a cubic, 266
Flexed tangent line, 266
Form, *see* one-form, two-form
Free linear system, 157
Function field, 171, 177, 309
    has tr. deg. one, 174
    separating points and tangents, 169
Fundamental group, 84, 125
    abelianization is homology, 125
    of $\mathbb{P}^1$ minus n points, 91
    of a curve of genus g, 94
    of a punctured disc, 86
    of a torus, 86

GAGA Theorem, 316
Gap number, 233
General divisor, 210
General Position Lemma, 225
Genus
    analytic, 192
    arithmetic, 192
    of a plane curve, 70
    topological, 6, 192
Global sections, 269
Glueing
    line bundles, 335
    locally trivial objects, 360
    Riemann surfaces, 59
g^r_d, 157, 234
Graph of a function, 10
Grassmann variety, 212
Group action, 75
Group law on a cubic, 267

Harmonic function, 27, 114

Hausdorff, 4
Hessian, 232
Higher-order cusp, 72
Higher-order tacnode, 72
Hodge filtration, 317
Hole chart, 66
Holomorphic
    action, 75
    embedding in $\mathbb{P}^n$, 36
    function, 21
        is constant on a compact surface, 29
        of several variables, 6
        on an affine plane curve, 22
        preserves orientation, 5
        sheaf, 270
    map, 38
        1-1 criterion, 161
        between tori, 63
        branch divisor, 134
        branch point, 45
        defined by a linear system, 160
        defined by monodromy, 91
        degree, 48
        degree of the image, 164
        embedding, 163
        embedding criterion, 163
        has discrete preimages, 41
        hyperplane divisor, 159
        inverse image divisor, 133
        is an isomorphism if 1-1, 40
        is an isomorphism if degree one, 48
        linear system of, 156
        local normal form, 44
        monodromy representation, 87
        multiplicity, 45
        ramification divisor, 134
        ramification point, 45
        the canonical map, 203
        to $\mathbb{P}^1$, 166
        to $\mathbb{P}^n$, 153
        to the Riemann Sphere, 41
        Veronese map, 165
        without coordinates, 166
    one-form, 105, 106
        on a hyperelliptic curve, 193
        on a projective plane curve, 193
        on a torus, 193
        period of, 248
        sheaf, 271
    tangent vector field, 341
Homogeneous coordinates, 13, 94
Homogeneous ideal, 96
Homogeneous polynomial, 14, 31
Homomorphism
    of line bundles, 334
    of sheaves, 278
Homotopic loops, 84
Homotopy of paths, 124
Hurwitz formula, 52, 135, 244
Hurwitz's theorem on automorphisms, 82
Hyperelliptic involution, 61
Hyperelliptic surface, 61, 92, 167
    Abel-Jacobi fibers, 264
    and equality in Clifford, 202
    automorphisms, 243
    function field, 178
    homology generators, 249
    is an algebraic curve, 170
    meromorphic function on, 62
    one-form on, 193
    one-forms on, 112
    the canonical map, 204
    Weierstrass points and gap numbers, 245
Hyperplane, 96
    bitangent, 220
    containing a divisor, 208
    divisor, 136
        form the linear system, 159
        monodromy of, 222
        of a holomorphic map, 159
    flexed, 233
    general, 221
    monodromy representation, 222
    tangent, 217, 233
    transverse, 218, 219
Hypersurface, 16, 95, 226
    containing the canonical curve, 204

Identity Theorem, 29, 40
Implicit Function Theorem, 10
Imposing conditions on hypersurfaces, 226
Imposing independent conditions, 226
Incidence space, 212
Index of speciality, 198
Infinitely near triple point, 72
Inflection point, 219, 234
    counting, 241
    Wronskian criterion, 235
Inflectionary basis, 234
Inflectionary weight, 240
Initial point, 118
Integral
    around a closed chain, 248
    depends holomorphically on the endpoints, 198
    of a 1-form along a chain, 121
    of a 1-form along a path, 119
    of a 2-form, 122
    of a trace, 255

Intersection divisor, 136
Inverse image divisor, 133
Inverse of an invertible sheaf, 328
Invertible sheaf, 324
    cocycle for, 350
    extensions, 362
    group, 330
    inverse, 328
    is locally trivial, 358
    local generator, 325
    of a divisor, 325, 348
    of sections of a line bundle, 338, 355
    of tangent vector fields, 341, 356, 367
    tensor product, 327
    trivialization, 324
Irreducible polynomial, 11
Isomorphism
    between $L(D)$'s, 148
    between $L^{(1)}(D)$ and $L(D+K)$, 149
    between $L^{(1)}(D)$'s, 148
    of covering spaces, 85
    of line bundles, 334
    of Riemann surfaces, 40
    of sheaves, 287
Isotropy subgroup, 75

Jacobian, 248, 264, 319, 356

Kernel of a sheaf map, 282
Kernel of an action, 75

Lattice, 9, 43
    hexagonal, 64
    homothetic to $\mathbb{Z} + \mathbb{Z}\tau$, 44
    in $\mathbb{C}^g$, 263
    square, 63
Laurent series, 25
Laurent Series Approximation, 173
Laurent tail, 171
Laurent tail 1-form divisors, 201
Laurent tail divisor, 179
Line, 8, 18, 57, 96, 241
    flexed, 266
    secant, 99
    tangent, 100, 241, 266
    transverse, 266
Line bundle, 333
    atlas, 333
    automorphism, 358
    canonical bundle, 337, 339
    chart, 331
    compatibility of charts, 332
    defined by transition functions, 336
    fiber coordinate, 332
    glueing, 335
    group, 351
    homomorphism, 334
    rational section, 342
    regular section, 337
    structure, 333
    support of a chart, 331
    tangent bundle, 337, 341
    tautological, 334
    transition function, 332
    trivial, 333
Linear equivalence, 138
Linear isomorphism of projective spaces, 97
Linear subspace of $\mathbb{P}^n$, 95
Linear system, 147
    base-point-free, free, 157
    canonical, 200
    complete, 147
    defining a holomorphic map, 160
    dimension, 147
    fixed part/divisor, 161
    gap number, 233
    moving part, 161
    of a holomorphic map, 156
Linearly dependent subset of $\mathbb{P}^n$, 96
Linearly independent subset of $\mathbb{P}^n$, 96
Local complete intersection curve, 18, 37, 205
Local generator for an invertible sheaf, 325
Local Normal Form, 44
Locally constant sheaf, 273
Locally finite covering, 299
Long exact sequence, 298, 299
Loop, 84

Maximum Modulus Theorem, 29
Meromorphic
    n-fold differential, 236
    function, 24
        divisor of, 130
        Laurent tail divisor of, 179
        multiplicity one, 169
        on $\mathbb{P}^1$, 24, 32
        on a cyclic covering of the line, 74
        on a hyperelliptic curve, 62
        on a projective curve, 25, 36, 37
        on a torus, 25, 35, 50, 150
        on an affine plane curve, 36
        on the Riemann Sphere, 24, 30, 149
        order, 26
        separating points, 169
        separating tangents, 169
    sheaf, 271
    sum of the orders is zero, 31, 33, 42, 49, 124

trace, 251
  with poles bounded by D, 146
  with prescribed Laurent tails, 173
  with prescribed orders, 173
one-form, 106, 107
  divisor of, 131
  on a cyclic covering of the line, 112
  on a hyperelliptic curve, 112
  on an affine plane curve, 111
  on an projective plane curve, 112
  on an projective plane curve with nodes, 112
  on the Riemann Sphere, 111
  order, 107
  product, 237
  pullback order formula, 115
  ratio of two is a function, 131
  residue, 121
  sheaf, 271
  sum of the residues is zero, 123
  trace, 252
  with poles bounded by D, 148
  with prescribed Laurent tails, 200
Minimum of a set of divisors, 136
Mittag-Leffler Problem, 180
Moduli space for curves, 215, 367
Monodromy representation
  defining a covering space, 89
  defining a holomorphic map, 91
  of a covering map, 86
  of a holomorphic map, 87
  of hyperplane divisors, 222
  surjectivity, 223
Monomial singularity, 71
Moving part of a linear system, 161
Multiplicity of a holomorphic map, 45
  and the order, 47
  derivative formula, 45
Multiplicity of a hyperplane meeting a projective curve, 219, 241
Multiplicity one function, 169

Net, 147
Node of a plane curve, 67, 102, 144
  meromorphic one-form, 112
  resolving, 69
Nonsingular polynomial, 11
Nonspecial divisors, 210
Nullstellensatz, 35

One-form
  $\mathcal{C}^\infty$, 109
  ∂-closed, 114
  $\bar{\partial}$-closed, 114
  closed, 114
  defined with an atlas, 106, 107, 110
  differential of, 114
  exact, 113
  holomorphic, 105, 106
  integral along a chain, 121
  integral along a path, 119
  meromorphic, 106, 107
  of type $(0,1)$, 110
  of type $(1,0)$, 110
  pullback, 115
  rational, 309
    with bounded poles, 310
  regular, 310
  residue, 121
  wedge product, 113
Onto sheaf map, 282
Open Mapping theorem, 40
Orbit, 75
Order
  of a meromorphic 1-form, 107
  of a meromorphic n-fold differential, 237
  of a meromorphic function, 26
  of a rational section of a line bundle, 343
Ordinary cusp, 72
Ordinary plane curve singularity, 72
Ordinary triple point, 72
Orientable, 6

Paracompact, 299
Partition of a path, 119
Partition of unity, 302
Path, 84, 118
Path-connected, 5
Path-lifting property, 85
Pencil, 147
Period, 248, 257
  mapping, 126
  matrix, 259
Picard group, 263, 345
  and the Jacobian, 356
  as the group of invertible sheaves, 348
  as the group of line bundles, 354
  isomorphic to $\check{H}^1(X, \mathcal{O}^*)$, 347
Plücker's Formula, 70
Plücker's formula, 144
k-plane, 96
Plane curve
  affine, see Affine plane curve
  projective, see Projective plane curve
Plugging a hole, 66
Poincaré's Lemma, 117, 284, 306
Pole, 23
  of a meromorphic 1-form, 107

INDEX OF TERMINOLOGY

of a meromorphic n-fold differential, 238
of a meromorphic function, 26
of a meromorphic function is isolated, 29
Presheaf, 269
    constant, 271
Principal divisor, 130
Principal part, 181
Projection, 98
Projective n-space, 16, 94
    dual, 166
    holomorphic embedding in, 36
    holomorphic map to, 153
Projective algebraic set, 95
Projective curve, 36
Projective line, 8
    automorphism, 43
    has no deformations, 367
    holomorphic map to, defined by monodromy, 91
    isomorphic to the Riemann Sphere, 40
    meromorphic function, 24, 32
Projective plane, 13
Projective plane curve, 14
Projectively normal, 231
Projectivization of a vector space, 94, 147, 166
Properly discontinuous action, 83
Pullback of a 1-form, 115
Pullback of a divisor, 134
Pullback of a path, 254

Ramification divisor, 134
Ramification point, 45
Rational function field, *see* Function field
Rational normal curve, 165, 216
    achieves the Castelnuovo bound, 229
    has no inflection points, 245
    is the canonical image of a hyperelliptic curve, 204
Rational section of a line bundle, 342
    divisor of, 343
    order of, 343
Refinement of a triangulation, 51
    common refinements exist, 51
    elementary, 51
Refinement of an open covering, 293
Refining map, 293
Regular
    function, 309
    one-form, 310
    section of a line bundle, 337
    tangent vector field, 341, 356
Removable singularity, 23

Reparametrization of a path, 118
Residue, 121
    of a trace, 253
Residue map, 188
Residue Theorem, 123, 186, 200
    algebraic proof, 253
Resolving a node, 69
Restriction maps, 116, 269
Reversal of a path, 118
Riemann bilinear relations, 262
Riemann Sphere, 4
    as a rational normal curve, 165
    computation of $L(D)$, 149
    finite groups acting on, 80
    function field, 177
    has trivial Jacobian, 248
    holomorphic maps to, 41
    is an algebraic curve, 170
    is the only curve of genus zero, 197
    isomorphic to $\mathbb{P}^1$, 40
    linear equivalence on, 140
    meromorphic function on, 24, 30
    meromorphic one-form, 111
Riemann surface, 4
Riemann-Roch problem, 186
    motivating Čech cohomology, 290
Riemann-Roch Theorem, 186, 192
    Geometric Form, 208
    implies algebraicity, 195

Scroll, 209
Secant line, 99
Second countable, 4
Section
    global, 269
    of a line bundle, 337
    of a presheaf, 269
    of the canonical bundle, 339
    of the tangent bundle, 341
    rational, of a line bundle, 342
Separates points, 169
Separates tangents, 169
Serre Duality, 188
Sheaf, 272
    algebraic, 309
    axiom, 272
    constant, 273
    direct sum, 277
    homomorphism, 278
    invertible, 324
    isomorphism, 287
    locally constant, 273
    map, 278
        1-1, 282
        induces a cochain map, 291

induces a cohomology map, 293, 296
kernel, 282
onto, 282
short exact sequence, 284
of $\mathcal{C}^\infty$ functions, 270
of $\mathcal{O}$-modules, 323
of forms, 271
of harmonic functions, 271
of holomorphic functions, 270
of meromorphic functions, 271
of meromorphic functions with poles bounded by D, 271
of nonzero holomorphic functions, 271
of not identically zero meromorphic functions, 271
of rational 1-forms, 310
of rational functions, 310
of regular functions, 309
of regular one-forms, 310
restriction to an open subset, 276
skyscraper, 273
stalk, 277
totally discontinuous, 273
Short exact sequence, 181
Short exact sequence of sheaves, 284
Simple plane curve singularities, 72
Simple tangency, 219
Simply connected, 84
Singularity, 23
Skyscraper sheaf, 273
Small loop around a point, 88
Small path enclosing a point, 118
Smooth complete intersection curve, 17
Smooth projective curve, 36
Smooth projective plane curve, 16
Span of a divisor, 208
Special divisor, 198
Stabilizer, 75
Stalk, 277
Standard identified polygon, 247
Stoke's Theorem, 123
Subchart, 1
Support of a divisor, 129
Support of a function, 302
Support of a line bundle chart, 331

Tacnode, 72
Tangent bundle, 337, 341, 367
Tangent hyperplane, 217
Tangent line, 100, 241, 266
Tangent sheaf, 329
Tangent vector field, 341
Tautological line bundle, 334
Tensor product of invertible sheaves, 327
Theta-function, 34, 50

Totally discontinous sheaf, 273
Trace
of a 1-form, 253
of a function, 251
Transform
of a 2-form, 111
of a $\mathcal{C}^\infty$ 1-form, 109
of a holomorphic 1-form, 105
of a meromorphic 1-form, 106
of a meromorphic n-fold differential, 236
Transition function between charts, 2
Transition function for line bundle charts, 332
cocycle conditions, 335
defining a line bundle, 336
for the canonical bundle, 337
for the tangent bundle, 337
Transverse hyperplane, 218, 219
Transverse line, 266
Triangulation, 50
Trigonal algebraic curve, 209
Triple point, 72
Trivialization of an invertible sheaf, 324
Twisted cubic curve, 17, 100, 102, 145, 165, 216
Two-form
$\mathcal{C}^\infty$, 110
defined with an atlas, 111
integral, 122
on a Riemann surface, 111
pullback, 115
sheaf, 271
transform of, 111

Universal cover, 85
of a compact Riemann surface with $g \geq 2$, 83
of a torus, 62

Veronese map, 165, 204
Very ample divisor, 163, 195

Web, 147
Wedge product, 113
Weierstrass points, 242

Zariski topology, 95, 311
Zero
of a meromorphic 1-form, 107
of a meromorphic n-fold differential, 238
of a meromorphic function, 26
of a meromorphic function is isolated, 29
Zero Mean Theorem, 318